Lecture Notes in Computer Science   14191

Founding Editors

Gerhard Goos
Juris Hartmanis

The series Lecture Notes in Computer Science (LNCS), including its subseries Lecture Notes in Artificial Intelligence (LNAI) and Lecture Notes in Bioinformatics (LNBI), has established itself as a medium for the publication of new developments in computer science and information technology research, teaching, and education.

LNCS enjoys close cooperation with the computer science R & D community, the series counts many renowned academics among its volume editors and paper authors, and collaborates with prestigious societies. Its mission is to serve this international community by providing an invaluable service, mainly focused on the publication of conference and workshop proceedings and postproceedings. LNCS commenced publication in 1973.

Gernot A. Fink · Rajiv Jain · Koichi Kise ·
Richard Zanibbi
Editors

Document Analysis
and Recognition –
ICDAR 2023

17th International Conference
San José, CA, USA, August 21–26, 2023
Proceedings, Part V

Editors
Gernot A. Fink
TU Dortmund University
Dortmund, Germany

Rajiv Jain
Adobe
College Park, MN, USA

Koichi Kise
Osaka Metropolitan University
Osaka, Japan

Richard Zanibbi
Rochester Institute of Technology
Rochester, NY, USA

ISSN 0302-9743   ISSN 1611-3349 (electronic)
Lecture Notes in Computer Science
ISBN 978-3-031-41733-7   ISBN 978-3-031-41734-4 (eBook)
https://doi.org/10.1007/978-3-031-41734-4

This Springer imprint is published by the registered company Springer Nature Switzerland AG
The registered company address is: Gewerbestrasse 11, 6330 Cham, Switzerland

Foreword

We are delighted to welcome you to the proceedings of ICDAR 2023, the 17th IAPR International Conference on Document Analysis and Recognition, which was held in San Jose, in the heart of Silicon Valley in the United States. With the worst of the pandemic behind us, we hoped that ICDAR 2023 would be a fully in-person event. However, challenges such as difficulties in obtaining visas also necessitated the partial use of hybrid technologies for ICDAR 2023. The oral papers being presented remotely were synchronous to ensure that conference attendees interacted live with the presenters and the limited hybridization still resulted in an enjoyable conference with fruitful interactions.

ICDAR 2023 was the 17th edition of a longstanding conference series sponsored by the International Association of Pattern Recognition (IAPR). It is the premier international event for scientists and practitioners in document analysis and recognition. This field continues to play an important role in transitioning to digital documents. The IAPR-TC 10/11 technical committees endorse the conference. The very first ICDAR was held in St Malo, France in 1991, followed by Tsukuba, Japan (1993), Montreal, Canada (1995), Ulm, Germany (1997), Bangalore, India (1999), Seattle, USA (2001), Edinburgh, UK (2003), Seoul, South Korea (2005), Curitiba, Brazil (2007), Barcelona, Spain (2009), Beijing, China (2011), Washington, DC, USA (2013), Nancy, France (2015), Kyoto, Japan (2017), Sydney, Australia (2019) and Lausanne, Switzerland (2021).

Keeping with its tradition from past years, ICDAR 2023 featured a three-day main conference, including several competitions to challenge the field and a post-conference slate of workshops, tutorials, and a doctoral consortium. The conference was held at the San Jose Marriott on August 21–23, 2023, and the post-conference tracks at the Adobe World Headquarters in San Jose on August 24–26, 2023.

We thank our executive co-chairs, Venu Govindaraju and Tong Sun, for their support and valuable advice in organizing the conference. We are particularly grateful to Tong for her efforts in facilitating the organization of the post-conference in Adobe Headquarters and for Adobe's generous sponsorship.

The highlights of the conference include keynote talks by the recipient of the IAPR/ICDAR Outstanding Achievements Award, and distinguished speakers Marti Hearst, UC Berkeley School of Information; Vlad Morariu, Adobe Research; and Seiichi Uchida, Kyushu University, Japan.

A total of 316 papers were submitted to the main conference (plus 33 papers to the ICDAR-IJDAR journal track), with 53 papers accepted for oral presentation (plus 13 IJDAR track papers) and 101 for poster presentation. We would like to express our deepest gratitude to our Program Committee Chairs, featuring three distinguished researchers from academia, Gernot A. Fink, Koichi Kise, and Richard Zanibbi, and one from industry, Rajiv Jain, who did a phenomenal job in overseeing a comprehensive reviewing process and who worked tirelessly to put together a very thoughtful and interesting technical program for the main conference. We are also very grateful to the members of the Program Committee for their high-quality peer reviews. Thank you to our competition chairs, Kenny Davila, Chris Tensmeyer, and Dimosthenis Karatzas, for overseeing the competitions.

The post-conference featured 8 excellent workshops, four value-filled tutorials, and the doctoral consortium. We would like to thank Mickael Coustaty and Alicia Fornes, the workshop chairs, Elisa Barney-Smith and Laurence Likforman-Sulem, the tutorial chairs, and Jean-Christophe Burie and Andreas Fischer, the doctoral consortium chairs, for their efforts in putting together a wonderful post-conference program.

We would like to thank and acknowledge the hard work put in by our Publication Chairs, Anurag Bhardwaj and Utkarsh Porwal, who worked diligently to compile the camera-ready versions of all the papers and organize the conference proceedings with Springer. Many thanks are also due to our sponsorship, awards, industry, and publicity chairs for their support of the conference.

The organization of this conference was only possible with the tireless behind-the-scenes contributions of our webmaster and tech wizard, Edward Sobczak, and our secretariat, ably managed by Carol Doermann. We convey our heartfelt appreciation for their efforts.

Finally, we would like to thank for their support our many financial sponsors and the conference attendees and authors, for helping make this conference a success. We sincerely hope those who attended had an enjoyable conference, a wonderful stay in San Jose, and fruitful academic exchanges with colleagues.

August 2023

David Doermann
Srirangaraj (Ranga) Setlur

Preface

Welcome to the proceedings of the 17th International Conference on Document Analysis and Recognition (ICDAR) 2023. ICDAR is the premier international event for scientists and practitioners involved in document analysis and recognition.

This year, we received 316 conference paper submissions with authors from 42 different countries. In order to create a high-quality scientific program for the conference, we recruited 211 regular and 38 senior program committee (PC) members. Regular PC members provided a total of 913 reviews for the submitted papers (an average of 2.89 per paper). Senior PC members who oversaw the review phase for typically 8 submissions took care of consolidating reviews and suggested paper decisions in their meta-reviews. Based on the information provided in both the reviews and the prepared meta-reviews we PC Chairs then selected 154 submissions (48.7%) for inclusion into the scientific program of ICDAR 2023. From the accepted papers, 53 were selected for oral presentation, and 101 for poster presentation.

In addition to the papers submitted directly to ICDAR 2023, we continued the tradition of teaming up with the International Journal of Document Analysis and Recognition (IJDAR) and organized a special journal track. The journal track submissions underwent the same rigorous review process as regular IJDAR submissions. The ICDAR PC Chairs served as Guest Editors and oversaw the review process. From the 33 manuscripts submitted to the journal track, 13 were accepted and were published in a Special Issue of IJDAR entitled "Advanced Topics of Document Analysis and Recognition." In addition, all papers accepted in the journal track were included as oral presentations in the conference program.

A very prominent topic represented in both the submissions from the journal track as well as in the direct submissions to ICDAR 2023 was handwriting recognition. Therefore, we organized a Special Track on Frontiers in Handwriting Recognition. This also served to keep alive the tradition of the International Conference on Frontiers in Handwriting Recognition (ICFHR) that the TC-11 community decided to no longer organize as an independent conference during ICFHR 2022 held in Hyderabad, India. The handwriting track included oral sessions covering handwriting recognition for historical documents, synthesis of handwritten documents, as well as a subsection of one of the poster sessions. Additional presentation tracks at ICDAR 2023 featured Graphics Recognition, Natural Language Processing for Documents (D-NLP), Applications (including for medical, legal, and business documents), additional Document Analysis and Recognition topics (DAR), and a session highlighting featured competitions that were run for ICDAR 2023 (Competitions). Two poster presentation sessions were held at ICDAR 2023.

As ICDAR 2023 was held with in-person attendance, all papers were presented by their authors during the conference. Exceptions were only made for authors who could not attend the conference for unavoidable reasons. Such oral presentations were then provided by synchronous video presentations. Posters of authors that could not attend were presented by recorded teaser videos, in addition to the physical posters.

Three keynote talks were given by Marti Hearst (UC Berkeley), Vlad Morariu (Adobe Research), and Seichi Uchida (Kyushu University). We thank them for the valuable insights and inspiration that their talks provided for participants.

Finally, we would like to thank everyone who contributed to the preparation of the scientific program of ICDAR 2023, namely the authors of the scientific papers submitted to the journal track and directly to the conference, reviewers for journal-track papers, and both our regular and senior PC members. We also thank Ed Sobczak for helping with the conference web pages, and the ICDAR 2023 Publications Chairs Anurag Bharadwaj and Utkarsh Porwal, who oversaw the creation of this proceedings.

August 2023

Gernot A. Fink
Rajiv Jain
Koichi Kise
Richard Zanibbi

Organization

General Chairs

David Doermann   University at Buffalo, The State University of New York, USA

Srirangaraj Setlur   University at Buffalo, The State University of New York, USA

Executive Co-chairs

Venu Govindaraju   University at Buffalo, The State University of New York, USA

Tong Sun   Adobe Research, USA

PC Chairs

Gernot A. Fink   Technische Universität Dortmund, Germany (Europe)

Rajiv Jain   Adobe Research, USA (Industry)

Koichi Kise   Osaka Metropolitan University, Japan (Asia)

Richard Zanibbi   Rochester Institute of Technology, USA (Americas)

Workshop Chairs

Mickael Coustaty   La Rochelle University, France

Alicia Fornes   Universitat Autònoma de Barcelona, Spain

Tutorial Chairs

Elisa Barney-Smith   Luleå University of Technology, Sweden

Laurence Likforman-Sulem   Télécom ParisTech, France

Competitions Chairs

Kenny Davila   Universidad Tecnológica Centroamericana,
                                UNITEC, Honduras
Dimosthenis Karatzas   Universitat Autònoma de Barcelona, Spain
Chris Tensmeyer   Adobe Research, USA

Doctoral Consortium Chairs

Andreas Fischer   University of Applied Sciences and Arts Western
                                Switzerland
Veronica Romero   University of Valencia, Spain

Publications Chairs

Anurag Bharadwaj   Northeastern University, USA
Utkarsh Porwal   Walmart, USA

Posters/Demo Chair

Palaiahnakote Shivakumara   University of Malaya, Malaysia

Awards Chair

Santanu Chaudhury   IIT Jodhpur, India

Sponsorship Chairs

Wael Abd-Almageed   Information Sciences Institute USC, USA
Cheng-Lin Liu   Chinese Academy of Sciences, China
Masaki Nakagawa   Tokyo University of Agriculture and Technology,
                                Japan

Industry Chairs

Andreas Dengel   DFKI, Germany
Véronique Eglin   Institut National des Sciences Appliquées (INSA)
                                de Lyon, France
Nandakishore Kambhatla   Adobe Research, India

Publicity Chairs

Sukalpa Chanda	Østfold University College, Norway
Simone Marinai	University of Florence, Italy
Safwan Wshah	University of Vermont, USA

Technical Chair

Edward Sobczak	University at Buffalo, The State University of New York, USA

Conference Secretariat

University at Buffalo, The State University of New York, USA

Program Committee

Senior Program Committee Members

Srirangaraj Setlur	Apostolos Antonacopoulos
Richard Zanibbi	Lianwen Jin
Koichi Kise	Nicholas Howe
Gernot Fink	Marc-Peter Schambach
David Doermann	Marcal Rossinyol
Rajiv Jain	Wataru Ohyama
Rolf Ingold	Nicole Vincent
Andreas Fischer	Faisal Shafait
Marcus Liwicki	Simone Marinai
Seiichi Uchida	Bertrand Couasnon
Daniel Lopresti	Masaki Nakagawa
Josep Llados	Anurag Bhardwaj
Elisa Barney Smith	Dimosthenis Karatzas
Umapada Pal	Masakazu Iwamura
Alicia Fornes	Tong Sun
Jean-Marc Ogier	Laurence Likforman-Sulem
C. V. Jawahar	Michael Blumenstein
Xiang Bai	Cheng-Lin Liu
Liangrui Peng	Luiz Oliveira
Jean-Christophe Burie	Robert Sabourin
Andreas Dengel	R. Manmatha
Robert Sablatnig	Angelo Marcelli
Basilis Gatos	Utkarsh Porwal

Program Committee Members

Harold Mouchere
Foteini Simistira Liwicki
Vernonique Eglin
Aurelie Lemaitre
Qiu-Feng Wang
Jorge Calvo-Zaragoza
Yuchen Zheng
Guangwei Zhang
Xu-Cheng Yin
Kengo Terasawa
Yasuhisa Fujii
Yu Zhou
Irina Rabaev
Anna Zhu
Soo-Hyung Kim
Liangcai Gao
Anders Hast
Minghui Liao
Guoqiang Zhong
Carlos Mello
Thierry Paquet
Mingkun Yang
Laurent Heutte
Antoine Doucet
Jean Hennebert
Cristina Carmona-Duarte
Fei Yin
Yue Lu
Maroua Mehri
Ryohei Tanaka
Adel M. M. Alimi
Heng Zhang
Gurpreet Lehal
Ergina Kavallieratou
Petra Gomez-Kramer
Anh Le Duc
Frederic Rayar
Muhammad Imran Malik
Vincent Christlein
Khurram Khurshid
Bart Lamiroy
Ernest Valveny
Antonio Parziale

Jean-Yves Ramel
Haikal El Abed
Alireza Alaei
Xiaoqing Lu
Sheng He
Abdel Belaid
Joan Puigcerver
Zhouhui Lian
Francesco Fontanella
Daniel Stoekl Ben Ezra
Byron Bezerra
Szilard Vajda
Irfan Ahmad
Imran Siddiqi
Nina S. T. Hirata
Momina Moetesum
Vassilis Katsouros
Fadoua Drira
Ekta Vats
Ruben Tolosana
Steven Simske
Christophe Rigaud
Claudio De Stefano
Henry A. Rowley
Pramod Kompalli
Siyang Qin
Alejandro Toselli
Slim Kanoun
Rafael Lins
Shinichiro Omachi
Kenny Davila
Qiang Huo
Da-Han Wang
Hung Tuan Nguyen
Ujjwal Bhattacharya
Jin Chen
Cuong Tuan Nguyen
Ruben Vera-Rodriguez
Yousri Kessentini
Salvatore Tabbone
Suresh Sundaram
Tonghua Su
Sukalpa Chanda

Mickael Coustaty
Donato Impedovo
Alceu Britto
Bidyut B. Chaudhuri
Swapan Kr. Parui
Eduardo Vellasques
Sounak Dey
Sheraz Ahmed
Julian Fierrez
Ioannis Pratikakis
Mehdi Hamdani
Florence Cloppet
Amina Serir
Mauricio Villegas
Joan Andreu Sanchez
Eric Anquetil
Majid Ziaratban
Baihua Xiao
Christopher Kermorvant
K. C. Santosh
Tomo Miyazaki
Florian Kleber
Carlos David Martinez Hinarejos
Muhammad Muzzamil Luqman
Badarinath T.
Christopher Tensmeyer
Musab Al-Ghadi
Ehtesham Hassan
Journet Nicholas
Romain Giot
Jonathan Fabrizio
Sriganesh Madhvanath
Volkmar Frinken
Akio Fujiyoshi
Srikar Appalaraju
Oriol Ramos-Terrades
Christian Viard-Gaudin
Chawki Djeddi
Nibal Nayef
Nam Ik Cho
Nicolas Sidere
Mohamed Cheriet
Mark Clement
Shivakumara Palaiahnakote
Shangxuan Tian

Ravi Kiran Sarvadevabhatla
Gaurav Harit
Iuliia Tkachenko
Christian Clausner
Vernonica Romero
Mathias Seuret
Vincent Poulain D'Andecy
Joseph Chazalon
Kaspar Riesen
Lambert Schomaker
Mounim El Yacoubi
Berrin Yanikoglu
Lluis Gomez
Brian Kenji Iwana
Ehsanollah Kabir
Najoua Essoukri Ben Amara
Volker Sorge
Clemens Neudecker
Praveen Krishnan
Abhisek Dey
Xiao Tu
Mohammad Tanvir Parvez
Sukhdeep Singh
Munish Kumar
Qi Zeng
Puneet Mathur
Clement Chatelain
Jihad El-Sana
Ayush Kumar Shah
Peter Staar
Stephen Rawls
David Etter
Ying Sheng
Jiuxiang Gu
Thomas Breuel
Antonio Jimeno
Karim Kalti
Enrique Vidal
Kazem Taghva
Evangelos Milios
Kaizhu Huang
Pierre Heroux
Guoxin Wang
Sandeep Tata
Youssouf Chherawala

Reeve Ingle
Aashi Jain
Carlos M. Travieso-Gonzales
Lesly Miculicich
Curtis Wigington
Andrea Gemelli
Martin Schall
Yanming Zhang
Dezhi Peng
Chongyu Liu
Huy Quang Ung
Marco Peer
Nam Tuan Ly
Jobin K. V.
Rina Buoy
Xiao-Hui Li
Maham Jahangir
Muhammad Naseer Bajwa

Oliver Tueselmann
Yang Xue
Kai Brandenbusch
Ajoy Mondal
Daichi Haraguchi
Junaid Younas
Ruddy Theodose
Rohit Saluja
Beat Wolf
Jean-Luc Bloechle
Anna Scius-Bertrand
Claudiu Musat
Linda Studer
Andrii Maksai
Oussama Zayene
Lars Voegtlin
Michael Jungo

Program Committee Subreviewers

Li Mingfeng
Houcemeddine Filali
Kai Hu
Yejing Xie
Tushar Karayil
Xu Chen
Benjamin Deguerre
Andrey Guzhov
Estanislau Lima
Hossein Naftchi
Giorgos Sfikas
Chandranath Adak
Yakn Li
Solenn Tual
Kai Labusch
Ahmed Cheikh Rouhou
Lingxiao Fei
Yunxue Shao
Yi Sun
Stephane Bres
Mohamed Mhiri
Zhengmi Tang
Fuxiang Yang
Saifullah Saifullah

Paolo Giglio
Wang Jiawei
Maksym Taranukhin
Menghan Wang
Nancy Girdhar
Xudong Xie
Ray Ding
Mélodie Boillet
Nabeel Khalid
Yan Shu
Moises Diaz
Biyi Fang
Adolfo Santoro
Glen Pouliquen
Ahmed Hamdi
Florian Kordon
Yan Zhang
Gerasimos Matidis
Khadiravana Belagavi
Xingbiao Zhao
Xiaotong Ji
Yan Zheng
M. Balakrishnan
Florian Kowarsch

Mohamed Ali Souibgui
Xuewen Wang
Djedjiga Belhadj
Omar Krichen
Agostino Accardo
Erika Griechisch
Vincenzo Gattulli
Thibault Lelore
Zacarias Curi
Xiaomeng Yang
Mariano Maisonnave
Xiaobo Jin
Corina Masanti
Panagiotis Kaddas
Karl Löwenmark
Jiahao Lv
Narayanan C. Krishnan
Simon Corbillé
Benjamin Fankhauser
Tiziana D'Alessandro
Francisco J. Castellanos
Souhail Bakkali
Caio Dias
Giuseppe De Gregorio
Hugo Romat
Alessandra Scotto di Freca
Christophe Gisler
Nicole Dalia Cilia
Aurélie Joseph
Gangyan Zeng
Elmokhtar Mohamed Moussa
Zhong Zhuoyao
Oluwatosin Adewumi
Sima Rezaei
Anuj Rai
Aristides Milios
Shreeganesh Ramanan
Wenbo Hu

Arthur Flor de Sousa Neto
Rayson Laroca
Sourour Ammar
Gianfranco Semeraro
Andre Hochuli
Saddok Kebairi
Shoma Iwai
Cleber Zanchettin
Ansgar Bernardi
Vivek Venugopal
Abderrhamne Rahiche
Wenwen Yu
Abhishek Baghel
Mathias Fuchs
Yael Iseli
Xiaowei Zhou
Yuan Panli
Minghui Xia
Zening Lin
Konstantinos Palaiologos
Loann Giovannangeli
Yuanyuan Ren
Shubhang Desai
Yann Soullard
Ling Fu
Juan Antonio Ramirez-Orta
Chixiang Ma
Truong Thanh-Nghia
Nathalie Girard
Kalyan Ram Ayyalasomayajula
Talles Viana
Francesco Castro
Anthony Gillioz
Huawen Shen
Sanket Biswas
Haisong Ding
Solène Tarride

Contents – Part V

Posters: Text and Document Recognition

Transductive Learning for Near-Duplicate Image Detection in Scanned Photo Collections

Francesc Net[1], Marc Folia[2], Pep Casals[2], and Lluis Gómez[1(✉)]

[1] Computer Vision Center, Universitat Autònoma de Barcelona, Bellaterra, Catalunya, Spain
{fnet,lgomez}@cvc.uab.cat
[2] Nubilum, Gran Via de les Corts Catalanes 575, 1r 1a, 08011 Barcelona, Spain
{marc.folia,pep.casals}@nubilum.es

Abstract. This paper presents a comparative study of near-duplicate image detection techniques in a real-world use case scenario, where a document management company is commissioned to manually annotate a collection of scanned photographs. Detecting duplicate and near-duplicate photographs can reduce the time spent on manual annotation by archivists. This real use case differs from laboratory settings as the deployment dataset is available in advance, allowing the use of transductive learning. We propose a transductive learning approach that leverages state-of-the-art deep learning architectures such as convolutional neural networks (CNNs) and Vision Transformers (ViTs). Our approach involves pre-training a deep neural network on a large dataset and then fine-tuning the network on the unlabeled target collection with self-supervised learning. The results show that the proposed approach outperforms the baseline methods in the task of near-duplicate image detection in the UKBench and an in-house private dataset.

Keywords: Image deduplication · Near-duplicate images detection · Transductive Learning · Photographic Archives · Deep Learning

1 Introduction

Historical archives are the storehouses of humankind's cultural memory, and the documents stored in them form a cultural testimony asset of incalculable value. From all the documents stored and managed in Historical Archives, photographs represent an ever-growing volume. This increase in volume has led to a critical need for efficient and effective methods for managing the data contained in these archives, particularly in terms of annotating them with useful metadata for indexing.

When a new document collection of scanned photographs arrives in an archive, archivists will typically need to perform certain manual tasks of classification, annotation, and indexing. Detecting duplicate and near-duplicate photographs can alleviate the burden on archivists by reducing the amount of time

G. A. Fink et al. (Eds.): ICDAR 2023, LNCS 14191, pp. 3–17, 2023.
https://doi.org/10.1007/978-3-031-41734-4_1

spent annotating the same image multiple times. This is where computer vision techniques can play a critical role in automating the process and improving the efficiency of archival work.

As[1] illustrated in Fig. 1, duplicate and near-duplicate images can appear in a given photo collection for different reasons and the number of them, while obviously varying for each collection, can represent a considerable percentage.

This paper presents a comparative study of various near-duplicate image detection techniques in the context of a real use case scenario. In particular, the study has been carried out in a document management company that has been commissioned to manually annotate a collection of scanned photographs from scratch. Furthermore, this real-world use case differs slightly from laboratory settings that are typically used to evaluate near-duplicate image detection techniques in the literature. Specifically, here the data set in which near-duplicate detection techniques are deployed is available in advance, which allows the use of transductive learning.

Transductive learning [2] is a type of learning that takes into account the information from the test data to improve the performance of the model. In contrast, in traditional inductive learning, the model is trained on a separate training set and then applied to a test set to evaluate its generalization capabilities.

In this paper, we consider fine-tuning a pre-trained model on the target collection with self-supervision as a form of transductive learning. The idea is that even though we do not have annotations on the test data set, we can use self-supervision techniques to learn representations that are useful for solving the near-duplicate detection task. In this way, as the model is adapted to the specific characteristics of the test data during the training process, this can result in improved performance compared to using a model previously trained on a different dataset.

Deep learning [1] is a helpful and effective technique for a range of applications, but it also has drawbacks and limits when it comes to detecting near-duplicate images. One issue is that deep learning models must be trained using a significant quantity of labeled data. The model has to be trained on a broad and representative set of photos, including near-duplicate and non-duplicate images, in order to recognize near-duplicate images successfully. This procedure can take a lot of time and resources, especially if the data set is not yet accessible or if the photos are challenging to label appropriately.

Another issue is the possibility of overfitting, which occurs when a model performs well with training data but badly with fresh, untested data. In the case of recognizing near-duplicate photos, this might be especially troublesome, as the model may become extremely specialized in identifying the particular images on which it has been trained and may not generalize well to new images.

Fortunately, in real applications like the one explained above, being able to access the collection of images gives us an advantage that we can exploit to mitigate these problems. Our focus in this article is to evaluate the different state-of-the-art vision architectures for near-duplicate image detection in these real use case scenarios. Our experiments show that the use of self-supervision

---

[1] https://digitaltmuseum.se/.

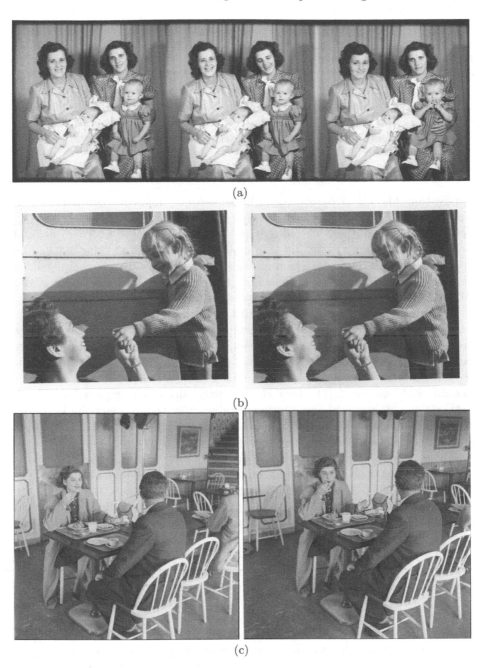

Fig. 1. Duplicate and near-duplicate images can appear in a given photo collection for different reasons: (a) shots from the same studio session, (b) scanned images from two paper originals with slight variations due to the paper wear over time, (c) different shots in a sequence of the same scene, etc. The images shown in this figure are reproduced from the DigitaltMuseum with Creative Common licenses (CC BY-NC-ND and CC Public Domain). The respective authors are: (a) Carl Johansson, (b) Anna Riwkin-Brick, and (c) Sune Sundahl.

techniques on the retrieval data set effectively improves their results. Although in our experiments we have used photographic collections, our study is also relevant for the detection of near duplicates in document images. In fact, some of the samples in the datasets we use contain textual information (scene/handwritten).

The rest of the paper is organized as follows: the most recent state-of-the-art in near-duplicate detection is found in Sect. 2. In Sect. 3, a list of the models employed is provided. In Sect. 4, the experiments and the results are shown. Finally, Sect. 5 contains the conclusions, discussions, and future work.

2   Related Work

Applications like image search, copyright defense, and duplicate picture detection can all benefit from locating and classifying photos that are similar to or nearly identical to one another. The idea is to spot near-duplicate photographs and group them together so that just one image is maintained and the rest can be deleted or flagged for closer examination. Typically, the work is completed by comparing image properties, such as histograms, textures, or CNN features, and utilizing similarity metrics to gauge how similar the images are to one another.

The evolution of near-duplicate image detection systems over the last two decades has been marked by a shift towards the use of deep learning techniques and an increasing emphasis on the development of more efficient and accurate methods for image feature extraction and matching. Before the use of deep learning, the primary approach was to use CBIR (Content-Based Image Retrieval) techniques. Chum et al. [11], for instance, makes use of an upgraded min-Hash methodology for retrieval and a visual vocabulary of vector quantized local feature descriptors (SIFT) to perform near-duplicate image detection. Similarly, Dong et al. [12] use SIFT features with entropy-based filtering.

Thyagharajan et al. [3] provides a survey of various techniques and methods for detecting near-duplicate images, including content-based, representation-based, and hashing-based methods and all of their variations. They discuss the issues and challenges that arise when dealing with near-duplicate detection, including the trade-off between precision and recall, approach scalability, and the impact of data distribution.

Liong et al. [4] and Haomiao et al. [5] use a deep neural network to learn compact binary codes for images, by seeking multiple hierarchical non-linear transformations. The methods are evaluated in medium-scale datasets (e.g. CIFAR-10). Similarly, Dayan et al. [6] and Fang et al. [7] map images into binary codes while preserving multi-level semantic similarity in medium-scale multi-label image retrieval. Both methods are evaluated in the MIRFLICKR-25K dataset.

Global and local CNN features can be used as in [8] to detect near-duplicates. In order to filter out the majority of irrelevant photos, the method uses a coarse-to-fine feature matching scheme, where global CNN features are first created via sum-pooling on convolutional feature maps and then matched between a given query image and database images. On the other hand, Lia and Lambreti et al. [9] discuss the performance of deep learning-based descriptors for unsupervised near-duplicate image detection. The method tests different descriptors on a range of datasets and benchmarks and compares their specificity using Receiver Operating Curve (ROC) analysis. The results suggest that fine-tuning deep convolutional networks are a better choice than using off-the-shelf features for this task.

The applications of near-duplicate image detection can be very wide. He *et al.* [13], for instance, performs vehicle re-identification to distinguish different cars with nearly identical appearances by creating a custom CNN.

More recent publications [10] use spatial transformers in comparison with convolutional neural networks (CNN) models. The method, which is designed to make better use of the correlation information between image pairs, utilizes a comparing CNN framework that is equipped with a cross-stream to fully learn the correlation information between image pairs. Additionally, to address the issue of local deformations caused by cropping, translation, scaling, and non-rigid transformations, the authors incorporate a spatial transformer module into the comparing CNN architecture. The method is tested on three popular benchmark datasets, California, Mir-Flickr Near Duplicate, and TNO Multiband Image Data Collection, and the results show that the proposed method achieves superior performance compared to other state-of-the-art methods on both single-modality and cross-modality tasks.

3 Methods

The overview of our near-duplicate detection system is shown in Fig. 2. The idea is to predict for each image of the dataset an embedding and then compare through them in order to extract the near-duplicates on the images. In this section, several models to generate embeddings will be discussed.

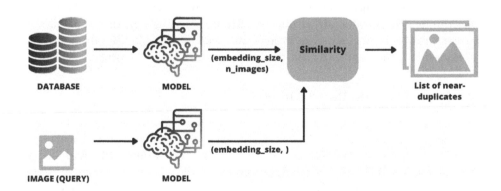

Fig. 2. An overview of the near-duplicates detection system.

3.1 Hashing Methods

Traditional hand-crafted methods for detecting near-duplicate images are based on hashing algorithms that extract compact representations of the image and compare them to determine their similarity. Hashing algorithms create fixed-length representations, or hashes, of digital objects like images. For tasks like data deduplication, where the goal is to find and get rid of duplicate copies of data to save space and reduce redundancy, hash algorithms are frequently used.

The simple Average Hash algorithm reduces the size of the image to 8 × 8 square, then reduces its color information by converting it to grayscale, and finally constructs a 64-bit hash where each bit is simply set based on whether the color value of a given pixel is above or below the mean. Average Hash is very fast to compute and robust against scale and aspect ratio changes, brightness/contrast modifications, or color changes. More robust algorithms extend this simple idea using other block-based statistics or the discrete cosine transform (DCT) to further reduce the frequencies of the image [14]. In this paper, we use two different hashing algorithms' implementations[2] as baselines: Perceptual Hashing and Block Mean Hashing.

The ability of hashing algorithms to be quick and efficient is one of their benefits. Hash algorithms are suitable for tasks like data deduplication where speed is important because they are built to generate a hash quickly and with minimal computational resources. Another benefit is that, in comparison to deep learning, hash algorithms can be fairly straightforward and simple to implement. They may therefore be a good choice for applications where simplicity and ease of use are crucial factors.

However, there are some drawbacks as well. They might not always be as accurate as alternative learning-based methods, particularly for images that are very similar but not identical. Hashing algorithms may struggle to distinguish minute differences between similar images because they are built to find exact duplicates. Hashing algorithms' sensitivity to changes in the input data is another drawback. For instance, even if an image is only slightly altered or its size is reduced, the resulting hash may differ greatly from the original hash.

3.2 Deep Learning-Based Methods

The current state of the art approach to compute image similarity involves using deep neural networks. Typical approaches include using off-the-shelf representations from pre-trained models [15], transfer learning [16], and image retrieval models [17]. In this paper we consider the use of two popular computer vision architectures: ResNet convolutional neural networks (CNN) [18], and Vision Transformers (ViTs) [19]. For both cases we consider supervised and self-supervised training strategies.

---

[2] https://docs.opencv.org/3.4/d4/d93/group_img_hash.html.

Supervised Learning. In order to train a model to predict a certain output given a collection of inputs, supervised learning approaches need labeled data. Training a model to identify whether a picture is a near duplicate or not is one way to find near-duplicate photos using supervised learning. The features of the photos may then be retrieved and fed into a deep learning model, such as a convolutional neural network (CNN), using this method. After training, the model may be used to categorize the photos based on their attributes.

To train our near-duplicate image detection models with supervised learning we consider the following two loss functions:

- Categorical Cross-Entropy loss: in this case we consider each sub-set of annotated near-duplicate images as a class, and compute $-\sum_{c=1}^{M} y_{i,c} \log(p_{i,c})$, where M is the number of classes, y is a binary indicator (0 or 1) for the correct class label being c, and p is the predicted probability for class c.
- Triplet loss: in this case the model is trained with triplets of images $\{x^a, x^p, x^n\}$ to make the distance low between the embeddings of anchor image $f(x^a)$ and positive image $f(x^p)$ (near-duplicate), while the distance between the embeddings of anchor image $f(x^a)$ and negative image $f(x^n)$ (non near-duplicate) is kept large: $\sum_{i}^{N} [\|f(x_i^a) - f(x_i^p)\|_2^2 - \|f(x_i^a) - f(x_i^n)\|_2^2 + \alpha]$, where α is the margin parameter.

Self-supervised Learning. Thanks to self-supervised learning, models can be trained into large sets of unlabelled data, using the data itself as supervision. The objective is to learn useful visual features from the data without any manual annotations. This is especially useful in scenarios such as the real use case that we study in this article, since normally the new collections that arrive at a document archive manager to be cataloged lack annotations. What distinguishes our use case from self-supervised methods per se is that here we have the data set on which we want to apply our duplicate detection model, and therefore self-supervised learning techniques are a tool of ideal transductive learning.

Two popular self-supervised models will be compared in this paper for our use case scenario: SimCLR (Simple Contrastive Learning) [20] and Masked Autoencoders (MAE) [21].

In SimCLR, a neural network is trained to determine whether a pair of pictures is a positive pair (i.e., they are different views of the same image) or a negative pair (i.e., they are different images). A contrastive loss function is used to train the model, which motivates it to create identical representations for positive pairings and dissimilar representations for negative pairs.

As illustrated in Fig. 3, given an image x the first step is to obtain two views of the image (x_1 and x_2) by applying data augmentation techniques. Then, both views are processed by the visual feature extraction model (e.g. a CNN + MLP) to obtain their embedding representations (z_1 and z_2). The loss function for a positive pair of examples (i, j) is defined as:

$$l_{i,j} = -\log \frac{\exp sim(z_i, z_j)/\tau}{\sum_{k=1}^{2N} \mathbb{1}_{k \neq 1} \exp sim(z_i, z_k)/\tau} \tag{1}$$

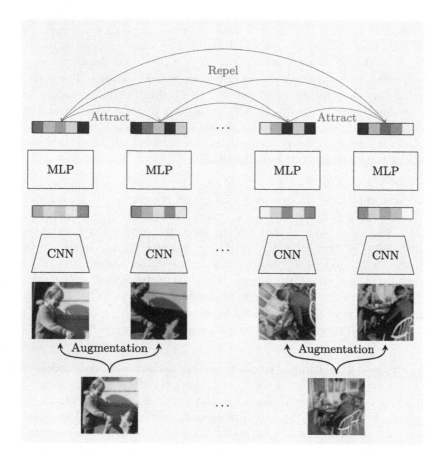

Fig. 3. An illustration of the SimCLR training process.

where N is the number of images in a given mini-batch, $sim(z_i, z_j)$ is the dot product between ℓ_2 normalized z_i and z_j, τ is a temperature parameter, and $\mathbb{1}_{k \neq 1} \in \{0, 1\}$ is an indicator function evaluating to 1 iff $k \neq i$.

Masked Autoencoders is another state of the art self-supervised learning technique in which an autoencoder is trained to rebuild an input picture using a portion of its original pixels. A division into non-overlapping patches is created, and then a subset of these patches is randomly sampled and masked so that the model has no access to them. The model is trained to reconstruct the masked image from the unmasked pixels as shown in Fig. 4, and the difference between the input picture and the reconstructed image is used as a supervisory signal.

The encoder is a ViT that is applied only to visible (not masked) patches and the decoder has to predict all the tokens (visible and masked). The model is trained with a loss function that computes the mean squared error (MSE) between the reconstructed and original image in the pixel space.

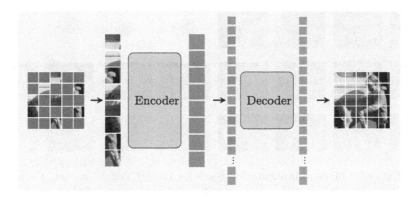

Fig. 4. An illustration of the Masked Autoencoders' workflow.

4   Experiments

4.1   Datasets

Two datasets with similar dimensions have been used in our experiments, one is a well-known public benchmark dataset for near-duplicate image detection and the other is a confidential dataset that represents our real use case scenario:

· **UKBench** [23]: contains 10.200 images, including indoor and outdoor scenes, objects, and people, and is designed to test the ability of retrieval systems to find visually similar images. For each image, 3 near-duplicates are provided (i.e. different points of view, camera rotations, etc.) as shown in Fig. 5. The dataset is divided into 60/20/20 for train, validation, and test sets. We would like to note that, as seen in Figs. 5 and 7, some of the images in this data set contain textual information, either in the form of scene text or handwritten text.

· **In-house private dataset**: consists of 41.039 images of portraits and artworks. Several views of the same people or similar artworks are provided. In addition, 1014 manual annotations of near-duplicates are given. These annotations contain a query image and one near-duplicate for this specific query. It is important to notice here that the annotations for this dataset are incomplete, we only have one near-duplicate annotation for each of those 1014 queries but other (many) near-duplicate images might exist for each of them. Images from this data set cannot be displayed here due to copyright issues. To get an idea of the type of images in this collection we refer the reader to Fig. 1 (a) where some examples of public images that are similar to the private data set are shown.

4.2   Implementation Details

In all our experiments we make use of two different base architectures: a ResNet50 CNN and a ViT-L/16 Transformer. In both cases we use models pre-trained on the ImageNet [24] dataset and fine-tuned on the target datasets. In both the supervised and self-supervised learning settings the models are trained

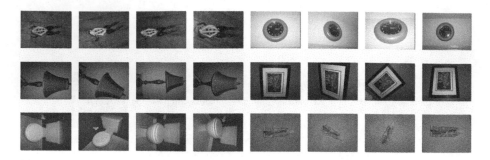

Fig. 5. Examples of near-duplicate images from the UKBench dataset. Each image query has 3 near-duplicates that correspond to different points of view, camera rotations, illumination changes, etc.

under the same hyperparameters: learning rate of $3 \cdot 10^{-4}$ and Adam optimizer, decaying the learning rate of each parameter group by 0.1 every 8 epochs.

For data augmentation in the SimCLR model, we apply a set of 5 transformations following the original SimCLR setup: random horizontal flip, crop-and-resize, color distortion, random grayscale, and Gaussian blur. The masking ratio for the MAE model is set to 75%.

4.3   Metrics

In order to evaluate the performance of our near-duplicate images detection system we make use of the following metrics:

· **Precision@K**: Precision@K is defined as the percentage of near-duplicate items retrieved among the top-K retrieved items.

· **Mean Average Precision (mAP)**: mAP summarizes the overall performance of an image retrieval system of near-duplicates. It represents the mean of the precision values for each query.

In our experiments, we use mAP@4 for the UKBench dataset, since we know that all queries have exactly four near-duplicate images in the retrieval set. There are actually three near duplicates of each query image, but the query image itself is also on the retrieval set, which means there are four relevant images per query. On the other hand, in the private dataset, we evaluate our models using Precision@K for different values of K. This is due to the fact that the dataset is only partially annotated and therefore it is much more difficult to interpret the performance of the models from a single mAP metric value.

4.4   Results

Table 1 shows the results for different models on the UKBench dataset when using inductive training, i.e. when they are trained/fine-tuned only with the training data.

Table 1. Overview of the results for the UKBench dataset.

Model	Pre-training		Fine-tuning		Metrics
	dataset	loss	dataset	loss	mAP@4
pHash	–	–	–	–	0.263
BlockMeanHash	–	–	–	–	0.321
ResNet50	–	–	UKBench	C. Entropy	0.765
ResNet50	ImageNet	C. Entropy	UKBench	C. Entropy	**0.943**
ResNet50	ImageNet	C. Entropy	UKBench	Triplet	0.931
ViT-L-16	ImageNet	C. Entropy	UKBench	C. Entropy	0.898
MAE (ViT-L-16)	–	–	UKBench	MSE	0.723
MAE (ViT-L-16)	ImageNet	C. Entropy	UKBench	MSE	0.810
SimCLR(ResNet50)	ImageNet	C. Entropy	UKBench	Contrastive	0.727
SimCLR(ViT-L-16)	ImageNet	C. Entropy	UKBench	Contrastive	0.806

For the supervised learning setting we evaluate a ResNet50 model trained from scratch, two ResNet50 models pre-trained on ImageNet and fine-tuned on the UKBench dataset (with cross-entropy and triplet losses respectively), and a ViT-L/16 also pre-trained on ImageNet and fine-tuned on the UKBench dataset with cross-entropy. For the self-supervised learning setting, we evaluate two MAEs with ViT-L/16 (one pre-trained on ImageNet and the other trained from scratch), and two models trained with SimCLR: SimCLR(ResNet50) and SimCLR(ViT-L-16). We also show the results of hand-crafted hashing algorithms for comparison.

We appreciate that hashing techniques yield poor outcomes and the best results are obtained with supervised learning. The ResNet50 is working better than the Visual Transformer, which can be explained by the fact that the training dataset is quite small. Self-supervised models exhibit promising outcomes that lag behind the best results of supervised models. At this point, it is clear that supervised learning is the best choice when training data is available for the target domain. Moreover, pre-training on ImageNet is always beneficial for all models.

In the next experiment, we compare the performance of the self-supervised MAE model in the inductive and transductive scenarios for the UKBench dataset. Somehow, what we want is to check what would happen if we didn't have training data annotated for UKBench and therefore we couldn't use supervised learning. This is in fact the scenario we are faced with in our private dataset, for which we have no annotated training data.

As shown in Table 2 when transductive learning is used in self-supervised MAE models, outcomes are improved significantly. This is because transductive learning allows the model to learn representations directly from the retrieval set, rather than relying solely on generalization to new images. By learning from the retrieval set, the model is able to identify similarities between pairs of images more effectively, resulting in improved performance. Note that the inductive learning results are the same as in Table 1 for MAE (ViT-L-16).

Table 2. Overview of the results when changing the type of learning used for the UKBench dataset

Model	Partition in which the model is trained	**mAP@4**
MAE (ViT-L-16)	Train (inductive)	0.810
	Train and test	0.824
	Test (transductive)	**0.843**

Finally, in Table 3 we show the performance comparison of different models in our internal private dataset using transductive learning. As discussed before in this dataset there are not enough annotations to train supervised models. The few available annotations are only used for evaluation purposes, and precision is computed for the top-K retrieved items. The pre-training and fine-tuning data in Table 3 are identical to those in Table 1, but omitted here to prevent the table from being too large.

Both Table 1 and 3 exhibit some correlations when comparing self-supervised models. Models trained with MAEs perform better than the ones trained with SimCLR. ViT-L/16 outperforms ResNet50 in this setting. Overall, results in Table 3 are lower than in the UKBench dataset because our in-house private dataset is only partially annotated, many near-duplicates for the same image are present but only one is actually labeled.

Table 3. Overview of the results for the in-house private dataset.

Model	Metrics			
	Precision@1	Precision@5	Precision@10	Precision@50
pHash	0.004	0.007	0.011	0.024
BlockMeanHash	0.015	0.027	0.045	0.091
SimCLR(ResNet50)	0.041	0.113	0.165	0.279
SimCLR(ViT-L-16)	0.052	0.142	0.198	0.341
MAE (ViT-L-16)	0.062	**0.156**	**0.218**	0.337

In Figs. 6 and 7 we present qualitative results of different methods for two given queries of the UKBench dataset. We appreciate that qualitative and quantitative results are consistent: hashing is the method that yields subpar outcomes while deep learning models trained with supervised or self-supervised learning strategies exhibit comparable results. In Fig. 7 we show an example where the query image contains handwritten text, demonstrating the potential use of the studied techniques in document collections as well as photo collections.

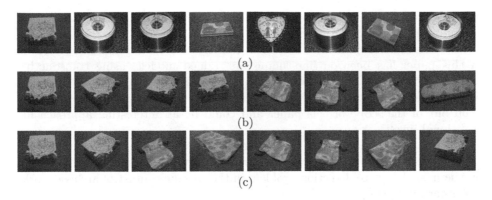

Fig. 6. Examples of qualitative results on the UKBench dataset for a given query image. We show the top-8 retrieved images ordered by closest distance from left to right for: (a) Block mean hashing, (b) Supervised learning (ResNet50), and (c) Self-supervised learning (MAE). In all cases, the left-most image is both the query and the first retrieved image.

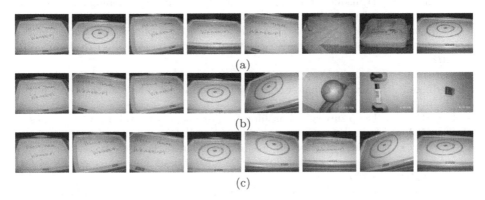

Fig. 7. Examples of qualitative results on the UKBench dataset for a given query image. We show the top-8 retrieved images ordered by closest distance from left to right for: (a) Block mean hashing, (b) Supervised learning (ResNet50), and (c) Self-supervised learning (MAE). In all cases, the left-most image is both the query and the first retrieved image. Note that in this example the query contains handwritten text and, as can be appreciated, the MAE model has no problem finding its near duplicates.

5   Conclusions

The present study provides a comprehensive evaluation of various models for embedding images in order to detect near-duplicates. The results demonstrate that training a self-supervised model is a suitable approach when labels are missing, as it enables the learning of the entire dataset's characteristics without the need for manual labeling. The proposed transductive learning approach, which involves fine-tuning a deep neural network on the target retrieval set with self-supervision, shows promising results in terms of retrieval metrics, outperforming the inductive learning baselines.

While synthetic data can be a useful tool for pre-training models, the focus in this paper has been on fine-tuning pre-trained models, using transductive learning on real data from the retrieval collection. While the use of GANs or Stable Diffusion can increase the size and diversity of the pre-training set, it has been out of the scope of this paper and could be an interesting approach for additional investigation.

Future research could also investigate the impact of using ViT foundational models pre-trained on larger datasets, fine-tuned using self-supervised learning, on the near-duplicate detection problem in the specific context of archival work. An other interesting

References

1. Goodfellow, I., Bengio, Y., Courville, A.: Deep learning. MIT press (2016)
2. Joachims, T.: Transductive learning via spectral graph partitioning. In: Proceedings of the 20th International Conference on Machine Learning (ICML-03) (2003)
3. Thyagharajan, K.K., Kalaiarasi, G.: A review on near-duplicate detection of images using computer vision techniques. Archives Comput. Methods Eng. **28**(3), 897–916 (2021)
4. Erin Liong, V., et al.: Deep hashing for compact binary codes learning. In: Proceedings of the IEEE Conference on Computer Vision and Pattern Recognition (2015)
5. Liu, H., et al.: Deep supervised hashing for fast image retrieval. In: Proceedings of the IEEE Conference On Computer Vision And Pattern Recognition (2016)
6. Wu, D., et al.: Deep supervised hashing for multi-label and large-scale image retrieval. In: Proceedings of the 2017 ACM on International Conference on Multimedia Retrieval (2017)
7. Zhao, F., et al.: Deep semantic ranking based hashing for multi-label image retrieval. In: Proceedings of the IEEE Conference on Computer Vision and Pattern Recognition (2015)
8. Zhou, Z., et al.: Near-duplicate image detection system using coarse-to-fine matching scheme based on global and local CNN features. Mathematics **8**(4), 644 (2020)D
9. Morra, L., Lamberti, F.: Benchmarking unsupervised near-duplicate image detection. Expert Syst. Appl. **135**, 313–326 (2019)
10. Zhang, Y., et al.: Single-and cross-modality near duplicate image pairs detection via spatial transformer comparing CNN. Sensors **21**(1), 255 (2021)
11. Chum, O., Philbin, J., Zisserman, A.: Near duplicate image detection: Min-hash and TF-IDF weighting. In: Bmvc, vol. 810, pp.81-815 (2008)
12. Dong, W., et al.: High-confidence near-duplicate image detection. In: Proceedings of the 2nd ACM International Conference on Multimedia Retrieval (2012)
13. He, B., et al.: Part-regularized near-duplicate vehicle re-identification. In: Proceedings of the IEEE/CVF Conference on Computer Vision and Pattern Recognition (2019)
14. Zauner, C.: Implementation and Benchmarking of Perceptual Image Hash Functions (2010)

15. Sharif Razavian, A., et al.: CNN features off-the-shelf: an astounding baseline for recognition. In: Proceedings of the IEEE Conference on Computer Vision and Pattern Recognition Workshops (2014)
16. D Yosinski, J., et al.: How transferable are features in deep neural networks? In: Advances in Neural Information Processing Systems, vol. 27 (2014)
17. Dubey, S.R.: A decade survey of content based image retrieval using deep learning. IEEE Trans. Circuits Syst. Video Technol. **32**(5), 2687–2704 (2021)
18. He, K., et al.: Deep residual learning for image recognition. In: Proceedings of the IEEE Conference On Computer Vision And Pattern Recognition (2016)
19. Dosovitskiy, A., et al.: An image is worth 16x16 words: Transformers for image recognition at scale. arXiv preprint arXiv:2010.11929 (2020)
20. Chen, T., et al.: A simple framework for contrastive learning of visual representations. In: International Conference on Machine Learning. PMLR (2020)
21. He, K., et al.: Masked autoencoders are scalable vision learners. In: Proceedings of the IEEE/CVF Conference on Computer Vision and Pattern Recognition (2022)
22. Radford, A., et al.: Learning transferable visual models from natural language supervision. In: International Conference on Machine Learning, PMLR (2021)
23. Nister, D., Stewenius, H.: Scalable recognition with a vocabulary tree. In: 2006 IEEE Computer Society Conference on Computer vision and pattern recognition, Vol. 2. IEEE (2006)
24. Deng, J. et al.: Imagenet: A large-scale image database. In: 2009 IEEE Conference on Computer Vision and Pattern Recognition, pp. 248–255 (2009)

Incremental Learning and Ambiguity Rejection for Document Classification

Tri-Cong Pham[1,2]([☒]) [ID], Mickaël Coustaty[1] [ID], Aurélie Joseph[2] [ID],
Vincent Poulain d'Andecy[2], Muriel Visani[1] [ID], and Nicolas Sidere[1] [ID]

[1] La Rochelle Université, L3i, Avenue Michel Crépeau, 17042 La Rochelle, France
{cong.pham,mickael.coustaty,muriel.visani,nicolas.sidere}@univ-lr.fr
[2] Yooz, 1 Rue Fleming, 17000 La Rochelle, France
{aurelie.joseph,vincent.poulaindandecy}@getyooz.com

Abstract. Along with the innovation and development of society, millions of documents are generated daily, and new types of documents related to new activities and services appear regularly. In the workflow for processing these documents, the first step is to classify the received documents to assign them to the relevant departments or staff. Therefore, two major problems arise: 1) the document classification algorithm is required to not only learn new categories efficiently but also have low error rate on them; 2) document classification needs to have the ability to detect when new classes appear. To address the first problem, we propose a class incremental learning method combined with ambiguity rejection to reduce the error rate of document classification. In this method, the ambiguity rejection module uses the classifier's probabilities and statistical analysis to optimize per-class thresholds for filtering out (rejecting) uncertain results, thereby increasing the accuracy of non-filtered documents. Our method is thoroughly evaluated on a public business document dataset and on our private administrative document dataset including 36 categories of more than 23 thousand images. On our private dataset, the method increases the classification accuracy of 0.82%, from 98.28% to 99.10%, while the rejection rate is only 3.42%. Additionally, with incremental learning scenario, it can achieve error and rejection rates of 1% and 8.98%, respectively. This result has great significance, can also be applied to similar document datasets, and especially has great potential for real-world applications.

Keywords: Ambiguity Rejection · Document Classification · Class Incremental Learning · Administrative Document · Imbalanced Dataset

1 Introduction

Document-formatted files are amongst the most widely used and shared, for any kind of human activity. Therefore, there are companies specializing in handling such files, especially administrative documents and documents related to banking, finance and insurance. In order to process the documents, the first

G. A. Fink et al. (Eds.): ICDAR 2023, LNCS 14191, pp. 18–35, 2023.
https://doi.org/10.1007/978-3-031-41734-4_2

step is usually document classification, *i.e.* assigning a given document to its category, and any error at this stage might lead to processing errors at later stages. But, in real-life applications, several challenges arise for document classification. Amongst those challenges, one can cite the large quantity of document categories, and the fact that new categories appear regularly, and that existing categories evolve with time (new template, different layout, new logo, ...). Therefore, document classification needs not only high accuracy (or low error rate) but also the ability to be re-trained on documents from new categories in an effective way. Figure 1 shows some examples of documents along with their categories, from our private dataset.

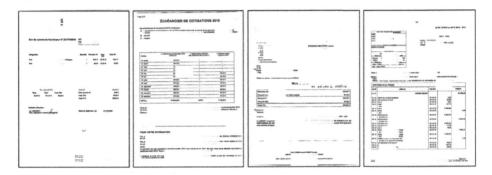

Fig. 1. Samples of different categories in our private administrative document dataset. From left to right: Order, Account Due, Invoice, and Bank Statements.

By leveraging big data, deep learning networks have been achieving a high level of performance in various scientific domains, including computer vision [1] and NLP [2–5]. Such algorithms are also widely applied in document understanding systems such as [6,8,10], in particular for document classification [5–14]. When applied in a static learning (STL) framework, deep learning-based methods can achieve an accuracy up to 96% by using several sources of information including textual, layout, and visual information, such as [6–8,10–12,14]. But, in real-life applications, datasets are constantly evolving. Thus, we need classifiers which are able to learn new document categories in an effective way, *i.e.* class incremental learning models. However, most class incremental learning methods have been proposed in the general field of computer vision such as iCARL [21]. There are only few studies about incremental learning for document images.

In addition, the algorithms, although achieving high accuracy, are mostly built on public datasets (like RVL-CDIP [15]) with a small number of categories (16 for RVL-CDIP) and need a large training set to converge. When applied on datasets of specific domains, other languages, or with many more categories with few samples, the accuracy decreases. Meanwhile, industrial systems require to reach an error rate as low as possible (often maximum 1%), to avoid manual checking.

Our main contributions in this paper are the following:

1. we propose a unified document classification framework using ambiguity rejection, specifically designed so as to be effective even in the presence of imbalanced datasets. This allows to reduce the error rate of document classification, both when using static learning and class incremental learning. Our proposed method uses either only text, or all text, layout and image information. The ambiguity rejection module rejects uncertain results based on per-class threshold of the predicted probabilities. This method is evaluated on both a public document dataset and our private administrative document dataset;

2. we demonstrate the effectiveness of the proposed framework for document classification, not only with static learning, but also with class-incremental learning. It achieves high performances in terms of accuracy, mean precision, mean F1 (respectively 98.28%, 98.27%, and 98.26% with static learning and 96.12%, 96.59%, and 96.24% with incremental learning on the private dataset);

3. we demonstrate the effectiveness of the proposed ambiguity rejection module: it reduces the error rate to only 1%, with a high coverage rate. The framework achieves a coverage rate at select accuracy of 99% (COV@SA99) of 96.58% and 91.02% with static learning and incremental learning, respectively.

2 Related Work

2.1 Document Classification

Document classification is a problem that has been attracting many researchers and has significant achievements applied in practice. Recently, several studies proposed solutions using only one modality (textual, layout or visual) from a document [2–5,13]. For the textual approaches, several BERT-Based models were used [5,13] on the OCR output of the document images. Meanwhile, Deep CNN approaches [15–17] used an image classifier for document image classification. With the development of deep learning and transformers-based architectures, some multimodal approaches were proposed by combining at least two of these modalities [11,12,14] in end-to-end trainable deep architectures [6,8,10]. The multimodal deep learning approaches [11,12,14] used some fusion methods to leverage the visual and the textual content into a joint embedding space to perform entity recognition on documents. The most recent studies use end-to-end trainable multimodal deep architectures such as LayoutLMv2, LayoutLMv3 [7, 8], SelfDoc [10], DocFormer [6]. These architectures are based on a combination of textual, layout, and visual information into a single deep architecture and are pre-trained on very large document datasets such as IIT-CDIP [15], DocVQA [19] and PubLayNet [20].

In order to compare the performances of different document classifiers, researchers often use the publicly available business document dataset named RVL-CDIP [15]. This dataset includes 400,000 gray-scale images of documents from 16 categories, of which the validation and test datasets have 40,000 images each, the rest being the training dataset. DocFormer, with 96.17%, performed the best accuracy on RVL-CDIP, followed by LayoutLMv3 with a 95.44% accuracy. SelfDoc proposed by Li et al. [10] is the next with an accuracy of 93.81%. All of these three methods leverage self-supervised learning from millions of unlabeled document images from public datasets to pre-train the models, then fine-tune them for the downstream tasks such as document classification.

However, with a much smaller administrative document dataset (about 23 000 document images) spread into 47 categories, Mahamoud et al. [12] proposed a multimodal (image + text) approach which achieves an accuracy of 98.2%, while its counterpart based on textual information only achieves a 97.5% accuracy, which represents a loss in accuracy of only 0.7%, showing that in some cases the textual contents of the documents bear most of the information that is required for classification.

While most of the methods cited in this section deal with a static learning framework, most real-life applications require incremental learning, as explained in Sect. 1. Next section thus presents some incremental learning methods.

2.2 Incremental Learning

Incremental learning stems from the desire that real-world AI systems need to utilize resources to learn new knowledge when new data arrives, without forgetting old knowledge. In the presence of new data, instead of using the entire dataset to re-train the model (like in static learning frameworks), incremental learning only uses a part of the data or knowledge previously learned, combined with all the new data, to retrain the network. This reduces memory and computing consumptions. Put simply, there are three main incremental learning scenarios [23] including: Task-incremental learning, Domain-incremental learning, and Class-incremental learning. In 2017, Rebuffi et al. [21] proposed iCaRL, an approach for class incremental learning that allows learning continuously from an evolving data stream in which new classes occur. The authors introduced a new training strategy to mitigate catastrophic forgetting by knowledge distillation [22] and a replay-based approach. In order to keep old knowledge, replay-based approaches store exemplar samples in memory, which are replayed to re-train the network for new knowledge while alleviating forgetting. One major insight in iCARL's experimental results is that, using exemplar images, the network can effectively learn new knowledge while still effectively mitigating catastrophic forgetting of old knowledge. Several successive studies also demonstrated the effectiveness of replay-based approaches, e.g. studies from Hou et al. [24], Wu et al. [25], Liu et al. [26], and Mittal et al. [27]. To the best of our knowledge, there is only little research effort that is dedicated to incremental document classification, especially using both textual and visual information from scanned document.

2.3 Ambiguity Rejection

In real-life applications, classification models usually cannot achieve an accuracy of 100%, and there are always errors in results. Thus, it is interesting to know which outputs are the most certain (with highest confidence), and conversely, which results are the most uncertain (with lowest confidence). Indeed, then the latter can be filtered out (rejected) so as to increase the reliability of the results.

Ambiguity rejection is a problem that has been studied since the 1970s and the seminal work of Chow [28]. In this study, a single threshold is used so as to reject the samples with lowest confidence. It is easy to see that the threshold value will affect strongly the error and rejection rates. In practice, we seek for models with small rejection rates and low error rates, which in this context raises the problem of choosing the most appropriate threshold for a given dataset. For binary classification problems, the threshold can be easily determined based on receiver operating characteristic (ROC) curve analysis [29,30]. However, with multi-class classification problems, per-class threshold estimation is more complicated [18,31–33]. Even though only little research effort is dedicated to ambiguity rejection for document classification, one should mention the rejection method proposed by Shu et al. [18], based on Gaussian fitting from predicted probabilities of Sigmoid function.

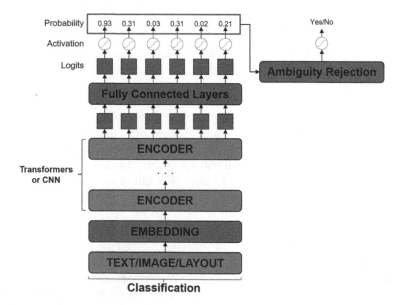

Fig. 2. Overview of the proposed unified document classification framework using ambiguity rejection. The classification module is based on BERT-based transformers, deep convolutional neural networks, or multimodal end-to-end trainable deep architectures and the ambiguity rejection module is based on threshold-based score functions. The activation functions can be Softmax or Sigmoid.

3   Proposed Method

3.1   Unified Document Classification Framework Using Ambiguity Rejection, and Class Incremental Learning

In this study, in order to handle ambiguity rejection for document categories described by few samples, and with a high imbalance between classes, we propose a unified document classification framework using ambiguity rejection. As shown in Fig. 2, our proposed architecture includes two main components: a document classifier, and an ambiguity rejector. In the framework, classifiers are trained in both static learning and class incremental learning scenarios, the latter being illustrated in Fig. 3.

In our framework, in order to build the classification modules, we decided to use different modalities including the text extracted through OCR and then analyzed using BERT-Based transformers [2] or FlauBERT [3], document images analyzed using deep convolutional neural networks (CNN) [15], or multimodal end-to-end trainable deep architectures such as LayoutLMv2 [8]. Depending on the architecture used, the input information can be textual information, visual information, layout information, or a combination of the three. Our classifiers can be trained either as multi-class classification models by using Softmax activation function or as multi-class classifiers with 1-vs-rest by using Sigmoid activation function at the last layer (see Fig. 2). In this system, the ambiguity rejector uses the output probabilities of the classifiers as an input to determine whether a result should be rejected or not. In this study, we use two per-class confidence threshold based methods for the ambiguity rejection module including Gaussian fitting and Binomial Cumlative Distribution Function based methods.

In this study, we use the Replay-based method for Class Incremental Learning (CIL). This method has been used in many studies and proved its effectiveness in solving various problems relying on incremental learning. There are three important concepts in Replay-based method of CIL which are base classes (CIL-BC), new classes (CIL-NC), and memory size (CIL-MZ). Typically, with CIL-BC $= 3$, CIL-NC $= 3$, and CIL-MZ $= 200$, the three initial classes and next three new classes for each incremental iterations are selected in the descending order of the number of samples per class. At the first training step of each class, all samples of the class are used to train model. From the next step on, only a subset of maximum CIL-MZ (here 200) randomly chosen samples from the category are used for training.

3.2   Ambiguity Rejector

The outputs of a multi-label classification model can be considered as a probabilities set $P = \{p_1, p_2, .., p_n\}$ where each p_i denotes the classifier's predicted probabilities of the ith category, and the final classified result is defined as a function $f(d) = argmax(p_i)$, with $i \in \{1, 2, .., n\}$, where n is the total number of categories. When integrated with rejection function by per-class confidence threshold method, the function $f(d)$ is computed as in Eq. (1).

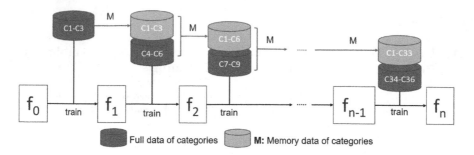

Fig. 3. Replay method of class incremental learning with CIL-BC = 3, CIL-NC = 3, and CIL-MZ = 200: the three initial categories and next three new categories for each incremental iterations are selected in the descending order of the number of samples per category. At the first training step of each category, all samples of the category are used to train model.

$$f(d) = \begin{cases} reject & \text{if } p_i \leq \lambda_i, \forall i \in \{1, 2, .., n\} \\ argmax(p_i), i \in \{1, 2, .., n\} & \text{otherwise} \end{cases} \quad (1)$$

where $\lambda = \{\lambda_1, \lambda_2, .., \lambda_n\}$ denotes confidence thresholds, of which λ_i is the threshold of ith category (c_i).

Binomial Cumulative Distribution Function Based Method (B-CDF). Motlagh et al. [32] proposed in 2022 the B-CDF method to optimize the λ value of per-class thresholds based on a Binomial distribution. The most important question to address is how to decide whether a given possible threshold is acceptable or not? (Q1). With a given possible threshold of a given category, we have the total number of rejected samples m and the total number of wrongly rejected samples k, whose predicted label is equal to the true label. We want to make sure that there are no more than k failures in the m samples. If the probability of having more than k failures in m is greater than a given significance level δ, we discard it. That mean, if we call the probability that there are less than or equal to k failures in m samples $ProbWrong(k, m)$, then question Q1 can be transformed into: if $ProbWrong(k, m) > 1 - \delta$, not be accepted otherwise accepted. If the probability that a sample is wrongly rejected is p (any rejected sample may randomly be wrong so $p = 0.5$), then the corresponding Binomial Cumlative Distribution Function (Formula 2) can be used to assess the probability of having at most k failures.

$$binom.cdf(k; m, p) = \sum_{i=0}^{k} \binom{m}{i} p^i (1 - p)^{m-i} \quad (2)$$

Algorithm 1 optimizes the threshold λ_i of the ith category to maximize the selected accuracy. For a given δ, we have λ set of thresholds for all categories when applying the algorithm for all categories. The higher the δ, the smaller the SA. The parameter δ is chosen depending on the data and requirements of the problem. This is the first study use this approach for ambiguity rejection of document classification.

Gaussian fitting method is used in DOC [18] to estimate λ_i for each c_i. The predicted probability p_i of c_i follows one half of the Gaussian distribution with mean of 1.0. The main idea of this method is, if p_i is too far away from mean, then c_i is rejected. So, to do the λ_i estimation, we calculate the standard deviation std_i of the class c_i based on the dataset on the train set by Gaussian fitting, then determine $\lambda_i = max(MIN, 1.0 - SCALE * std_i)$, where MIN = 0.5, and SCALE is an adjustify parameter. The higher the SCALE, the higher the SA. In DOC [18], SCALE is 3.0, however, for our study, we adjust the SCALE to ensure select accuracy at 99% or 90% then calculate coverage rate to compare with other methods.

Algorithm 1. Ambiguity Rejection using B-CDF

Find the best threshold λ_i of ith category to get the highest select accuracy and coverage

1: **function** CLASS_THRESHOLD_ESTIMATION($Y_{true}, Y_{pred}, P, \delta$)
2: $\quad D^{incorrect} \leftarrow Y_{true} \neq Y_{pred}$ $\qquad\qquad\qquad$ ▷ incorrectly predicted documents
3: $\quad P_{max} \leftarrow max(P)$ $\qquad\qquad$ ▷ list max predicted probabilities of documents
4: $\quad \lambda_{possible} \leftarrow unique(P_{max}[D^{incorrect}_{index}])$ $\qquad\qquad$ ▷ list possible thresholds
5: $\quad \lambda_{best} \leftarrow 0$ $\qquad\qquad\qquad\qquad\qquad\qquad$ ▷ init best threshold
6: $\quad COV_{best} \leftarrow -1$ $\qquad\qquad\qquad\qquad\qquad$ ▷ init best coverage
7: $\quad SA_{best} \leftarrow accuracy(Y_{true}, Y_{pred})$ $\qquad\qquad$ ▷ init best select accuracy
8: $\quad$ **for** $\lambda \in \lambda_{possible}$ **do**
9: $\qquad Select \leftarrow P_{max} > \lambda$
10: $\qquad Reject \leftarrow P_{max} \leq \lambda$
11: $\qquad Reject_{NG} \leftarrow Y_{true}[Reject_{index}] = Y_{pred}[Reject_{index}]$ $\qquad$ ▷ wrong rejection
12: $\qquad m \leftarrow count(Reject)$
13: $\qquad k \leftarrow count(Reject_{NG})$
14: $\qquad prob \leftarrow binom.cdf(k, m, p)$ $\qquad\qquad\qquad\qquad$ ▷ with p = 0.5
15: $\qquad$ **if** $prob \leq 1 - \delta$ **then** $\qquad\qquad\qquad\qquad$ ▷ accepted
16: $\qquad\quad SA \leftarrow accuracy(Y_{true}[Select_{index}], Y_{pred}[Select_{index}])$ ▷ select accuracy
17: $\qquad\quad COV \leftarrow count(Select)/count(Y_{true})$ $\qquad\qquad$ ▷ coverage rate
18: $\qquad\quad$ **if** $(SA > SA_{best})$ or $(SA = SA_{best}$ and $COV > COV_{best})$ **then**
19: $\qquad\qquad \lambda_{best} \leftarrow \lambda$
20: $\qquad\qquad SA_{best} \leftarrow SA$
21: $\qquad\qquad COV_{best} \leftarrow COV$
22: $\qquad\quad$ **end if**
23: $\qquad$ **end if**
24: $\quad$ **end for**
25: $\quad$ **return** λ_{best}
26: **end function**

---

4 Experimental Results

4.1 Materials

In this study, we use two datasets to evaluate the effectiveness of our ambiguity rejection framework. The first one is a private corpus of French administrative documents. This dataset includes 23,528 document images (15,885 for training and 1,766 for validation) of 36 categories The fixed test dataset consists of 5,877 document images. This dataset is naturally imbalanced where the biggest category has 4,540 documents (1,135 documents in test set), and the smallest category has only 25 documents (6 documents in test set). The top 18 categories have more than or close to 200 documents per category in the train set, whereas the other categories have 160 or less documents.

The second dataset is the well-known RVL-CDIP [15], large public dataset of 400,000 document images split in 16 categories. This is a balanced dataset of 25,000 documents per category. It is split into train, validation, and test sets with respectively 320,000 documents in the training step (40,000 per class) and 40,000 for the validation and test set (2,500 per class). To simulate a small and imbalanced dataset with similar distribution of our private dataset, we collected the data of the first 90,000 records in the train set, and filter the images with at least 10 words in the text. Then, we divided 16 categories into 4 groups including: 1) Letter, Form, Email, Specification; 2) Handwritten, Report, Publication, News; 3) Advertisement, Budget, Invoice, Presentation; and 4) File, Questionnaire, Resume, Memo; with turn data rates of 100%, 50%, 10% and 5%, respectively. With the validation and test datasets, we collected all the samples with at least 10 words in the text. As a result, the train set has 36,966 document images, the biggest category has 5,572 images, the smallest category has 285 images. The validation and test sets have 8,319 and 8,389 images, respectively.

4.2 Evaluation Metrics

In order to evaluate our proposed class incremental learning with ambiguity rejection, we use Accuracy; Weighted Precision (Precision); Weighted F1 (F1); and $\Delta_{\text{Performance}}$ for classification performance and forgetting. Besides, in order to evaluate the ambiguity rejection, we use Coverage (COV), Select Accuracy (SA), and Reject Accuracy (RA). The error rate (ERR) is calculated from SA by $ERR = 1.0 - SA$. A real-world application always needs to have a low ERR and acceptable COV because it will take human effort to process misclassified documents, while uncovered or rejected documents can be transferred to other efficient systems or algorithms to process. COV@SA99 and COV@SA90 stand for the coverage rate at select accuracy of 99% and 90%, respectively. Mathematically, the measures can be expressed as follows:

$$\Delta_{\text{Performance}} = \text{Performance}_{\text{previous step}} - \text{Performance}_{\text{current step}} \quad (3)$$

$$\text{COV} = \frac{\text{The number of all samples } \in D_{select}}{\text{The number of all samples}} \quad (4)$$

$$SA = \frac{\text{The number of correctly classified samples} \in D_{select}}{\text{The number of all} D_{select}} \tag{5}$$

$$RA = \frac{\text{The number of incorrectly classified samples} \in D_{reject}}{\text{The number of all} D_{reject}} \tag{6}$$

if $\Delta_{\text{Performance}} > 0$ then the model is forgetting, else the model is learning better than before without forgetting. We have three forgetting measures consists of Δ_{Accuracy}, $\Delta_{\text{Precision}}$, and Δ_{F1} for Accuracy, Precision, and F1, respectively.

4.3 Experimental Setting

For textual models (namely Text-Modality), we use FlauBERT$_{BASE}$ [3] and BERT [2] for French and English datasets, respectively. LayoutLMv2$_{BASE}$ [8] (namely Multi-Modality), a multi-modal transformer model using the document text, layout, and visual information, is used to compare with textual information based methods. In order to optimize for our administrative document in French, we customize two modules of LayoutLMv2 including: 1) OCR engine is in French language; 2) Tokenizer uses FlauBERT's French vocabulary. Besides, Text-Modality is combined with two activation functions, Softmax and Sigmoid, while Multi-Modality is only combined with Softmax activation function. Text-Modality is trained by 20 epochs, while Multi-Modality is trained by 50 epochs for both static learning and incremental learning and the model of the final epoch is used to evaluate the effectiveness of proposed algorithm. For class-incremental learning, we use CIL-BC $= 3$, CIL-NC $= 3$, and CIL-MZ $= 200$ for our private administrative document dataset and use CIL-BC $= 4$, CIL-NC $= 4$, and CIL-MZ $= 400$ for RVL-CDIP small-imbalanced dataset.

4.4 Performance Evaluation and Discussions

In this paper, three classifiers are integrated with two ambiguity rejection methods. The models are evaluated with both static learning and class incremental learning scenarios. We evaluate the effectiveness of document classification task and then evaluate the effect of ambiguity rejection on our private dataset and RVL-CDIP small-imbalanced dataset.

Document Classification Performances. Table 1 shows the performance of three classifiers with STL and CIL on two test datasets. Overall, STL has higher performance than CIL with all classifiers on both datasets. This is understandable because of the forgetting problem [21] in incremental learning. In addition, Text-Modality achieves higher performance than Multi-Modality on two datasets. Last but not least, it has better performance on our private administrative document dataset than that of RVL-CDIP small-imbalanced dataset.

Table 1. Document classification performance on administrative document dataset of 36 categories and RVL-CDIP small-imbalanced dataset of 16 categories. The performance is calculated based by same test documents on the last model of the final step of static learning and incremental learning.

#	Method	Static Learning			Incremental Learning		
		Accuracy	Precision	F1	Accuracy	Precision	F1
Administrative document dataset							
1	Text-Modality Softmax	98.15	98.12	98.11	**96.12**	**96.59**	**96.24**
2	Text-Modality Sigmoid	**98.28**	**98.27**	**98.26**	95.78	96.39	95.95
3	Multi-Modality Softmax	91.22	91.49	91.19	81.86	88.10	82.91
RVL-CDIP small-imbalanced dataset							
1	Text-Modality Softmax	80.16	**82.67**	80.26	**79.76**	81.75	**80.45**
2	Text-Modality Sigmoid	**80.26**	82.62	**80.43**	78.07	**81.83**	79.46
3	Multi-Modality Softmax	77.01	81.13	76.71	72.76	76.83	73.08

Static learning vs class-incremental learning because of forgetting problem [21] in incremental learning, CIL is less effective than STL, however the gap is different depending on classifiers. While the exemplars are effective in helping Text-Based models prevent catastrophic forgetting, CIL only reduces about 1–2% compared to STL on both datasets, such as Text-Modality-Softmax, it has accuracy of STL vs CIL on the private dataset and RVL-CDIP small-imbalanced dataset at 98.15% vs 96.12% and 80.16% vs 79.76%, respectively. However, Multi-Modality has a very big gap in terms of accuracy between STL and CIL, which are 9.36% (91.22% vs 81.86%) and 4.25% (77.01% vs 72.76%) on private dataset and RVL-CDIP small-imbalanced dataset, respectively.

Text-Modality vs Multi-Modality the Text-Based models achieve higher performance than Multi-Modality in both scenarios of STL and CIL on both datasets. On private document dataset, Text-Based models of Softmax and Sigmoid activation functions have stable and excellent accuracy, Precision, and F1 above 98% with STL, and reduces a bit to about 96% with CIL. In which, Sigmoid activation function is slightly better for STL, but worse for CIL than Softmax activation function, however the gap is not significant. The Text-Modality-Softmax combination is more stable than Text-Modality-Sigmoid on both scenarios of STL and CIL. Meanwhile, Multi-Modality has lower accuracy, Precision, and F1 on both scenarios than Text-Modality does. Especially, Multi-Modality has problem of catastrophic forgetting. There is big gap in terms of accuracy between STL and CIL (9.36% and 4.25% on private dataset and RVL-CDIP small-imbalanced dataset, respectively). On the RVL-CDIP, we can see the same pattern, only with lower performance than on private dataset. This result demonstrates that Multi-Modality requires a larger amount of documents to train.

Administrative dataset vs RVL-CDIP small-imbalanced dataset there is a big difference of performance between private and public datasets, it is better on private

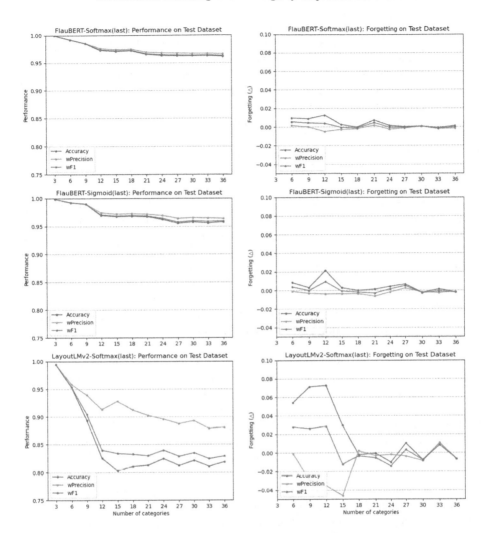

Fig. 4. Performance and Forgetting of our class-incremental learning with replay method on administrative document dataset of 36 categories. The x-axis shows the number of categories model learned. The y-axis represents the efficiency (the left column), and Forgetting of old categories (the right column).

dataset than RVL-CDIP dataset with both Text-Modality & Multi-Modality and both STL & CIL scenarios. The reason is that the private data is related to the administrative documents, which includes information that is clear, distinguishable, and within a narrow timeframe. Meanwhile, the public dataset is collected over a wider period of time, such as advertising documents from 197x to the present, and covers a larger information region. Therefore, these documents have bias in terms of information and visual style, so a lot of data is needed to train our models to get accurate performance.

Incremental learning and forgetting analysis CIL only uses a small number of examplars of old categories when re-training the model on samples of new categories, so the performance of old categories will be reduced in the new model when new categories are added. Figure 4 illustrates the performance and forgetting of the models at each training iteration of CIL on our private administrative document dataset. We can see that, Text-Based models achieve stable and excellent performance of accuracy, Precision, and F1 above 96%. And the $\Delta_{Performance}$ of these metrics are very small, at less than 1%. This shows that text information is useful for classification task of administrative document dataset. This also explains why the final Text-Based model of 36 categories of CIL reduces only 2% of accuracy compared to STL.

In contrast, with Multi-Modality, the efficiency of the model decreases markedly after each training iteration, when increasing the number of categories from 3 to 15. The biggest $\Delta_{Accuracy}$ is when the number of categories reach 12, then $\Delta_{Accuracy}$ is above 8%. This leads to the accuracy of the models, with the number of classes from 12 onwards, reaching less than 85%. The 12th, 15th, and 16th categories of our dataset have 360, 230, and 203 documents in training set, respectively. This experimental result shows that the Multi-Modality model needs a lot of data for each category to train the network. The less documents the category has in training set, the less efficient the model achieves.

Ambiguity rejection performances Table 2 is a performance comparison of ambiguity rejection task of three classifiers combined with two per-class confidence threshold methods and basic method (without threshold). The parameters of the confidence threshold methods are optimized so that the SA reaches at least 99% on our private administrative document dataset. On RVL-CDIP small-imbalanced dataset, the SA level is 90% because the accuracies of classifiers are only about 80%. For methods that cannot reach these SA value, the parameters with the highest SA will be taken. Overall, Text-Based models achieve higher performance than Multi-Modality with both STL and CIL. Besides, at the same SA of around 99%, Text-Based classifiers achieve stable and excellent COV above 90% on private dataset with both learning methods. Meanwhile, Multi-Modality cannot achieve SA at 99%, and has a very low COV only reaches 83.15% at the highest SA of 97.99%. This is easy to explain because the accuracy of classification task of Multi-Modality is much lower than that of Text-Modality. The same insight can be seen on RVL-CDIP small-imbalanced dataset at SA of 90%.

On private administrative document dataset, Table 2 shows that, with STL method, the combination of Text-Modality-Sigmoid classifier and B-CDF rejection method achieves the highest COV of 96.58% at the SA of 99.10%. Compared with the basic method, this method helps increase 0.82% of accuracy (99.10% vs 98.28%), while losing 3.42% of documents. With CIL, the best combination is Text-Modality-Softmax classifier and B-CDF rejection method. This combination has low ERR of 1.05% at COV 91.02%, and improves 2.83% of SA (98.95% vs 96.12%). The combination also achieves high performance on static learning with COV and SA at 94.54% and 98.99%, respectively.

Table 2. Ambiguity rejection performance on administrative document dataset of 36 categories and on RVL-CDIP small-imbalanced dataset of 16 categories.

#	Method	Static Learning				Incremental Learning			
		COV↑	SA↑	RA↑	ERR↓	COV↑	SA↑	RA↑	ERR↓
Administrative document dataset @SA99									
1	Text-Modality Softmax$_{Basic}$	100	98.15	–	01.85	100	96.12	–	03.88
2	Text-Modality Softmax$_{Gaussian}$	62.06	99.56	04.17	00.44	96.04	97.96	48.50	02.04
3	Text-Modality Softmax$_{B\text{-}CDF}$	94.54	98.99	16.51	01.01	**91.02**	**98.95**	32.58	**01.05**
4	Text-Modality Sigmoid$_{Basic}$	100	98.28	–	01.72	100	95.78	–	04.22
5	Text-Modality Sigmoid$_{Gaussian}$	66.31	99.00	03.13	01.00	98.50	96.77	69.32	03.23
6	Text-Modality Sigmoid$_{B\text{-}CDF}$	**96.58**	**99.10**	24.88	**00.90**	92.09	98.28	33.33	01.72
7	Multi-Modality Softmax$_{Basic}$	100	91.22	–	08.78	100	81.86	–	18.14
8	Multi-Modality Softmax$_{Gaussian}$	98.18	92.58	82.24	07.42	80.91	95.18	74.60	04.82
9	Multi-Modality Softmax$_{B\text{-}CDF}$	83.15	97.99	42.22	02.01	75.96	97.38	67.16	02.62
RVL-CDIP small-imbalanced dataset @SA90									
1	Text-Modality Softmax$_{Basic}$	100	80.16	–	19.84	100	79.76	–	20.24
2	Text-Modality Softmax$_{Gaussian}$	94.83	81.62	46.54	18.38	92.63	83.54	67.80	16.46
3	Text-Modality Softmax$_{B\text{-}CDF}$	80.57	90.12	61.10	09.88	**78.66**	**90.03**	58.10	**09.97**
4	Text-Modality Sigmoid$_{Basic}$	100	80.26	–	19.74	100	78.07	–	21.93
5	Text-Modality Sigmoid$_{Gaussian}$	96.77	81.67	61.99	18.33	97.57	79.41	75.98	20.59
6	Text-Modality Sigmoid$_{B\text{-}CDF}$	**80.68**	**90.13**	60.95	**09.87**	76.28	90.01	60.35	09.99
7	Multi-Modality Softmax$_{Basic}$	100	77.01	–	22.99	100	72.76	–	27.24
8	Multi-Modality Softmax$_{Gaussian}$	97.71	78.39	82.29	21.61	03.60	99.34	28.23	00.66
9	Multi-Modality Softmax$_{B\text{-}CDF}$	74.50	90.54	62.55	09.46	62.87	90.29	56.92	09.71

On RVL-CDIP small-imbalanced dataset, B-CDF rejection method also is the best method, and achieves much more efficient than Gaussian and Basic methods. The method can achieve SA at 90% while Gaussian method can not when combines with all Text-Modality and Multi-Modality classifiers. On the dataset, the best models of STL and CIL scenarios are Text-Modality classifier combining with B-CDF rejection method with COV@SA90 at 80.68% and 78.66%, respectively. In other words, our ambiguity rejection method improves accuracy by 9.87% and 10.27%, while losing 19.32% and 21.34% of documents to be processed. This trade-off is still very effective when we have to manually process whenever we have misclassified documents.

Discussion: This study thoroughly evaluates multiple modalities and trains models on two datasets. Its results indicate that the proposed B-CDF ambiguity rejection is the best and possesses the highest SA (or the lowest ERR) compared to other rejection method with both STL & CIL scenarios on both datasets. B-CDF is effective for rejecting ambiguity of many types of classifiers, from low accuracy classifiers such as classifiers on public dataset to high accuracy classifiers such as classifiers on private dataset. Besides, like most classification methods, our method achieves better results for STL than for CIL. However, exemplars of Replay-based method are effective in helping Text-Modality mitigate catastrophic forgetting. As a result, compared with STL, CIL of Text-Modality achieves excellent and stable performance with a small gap of accuracy by using small number of exemplars in memory on private dataset. Besides, Sigmoid displays the best results with STL, although it is not much different from Softmax, because it uses Binary Cross Entropy loss function to train model as 1-vs-rest strategy. This method compresses the knowledge spaces and tightens the decision boundaries [18], thus, improves effectiveness of ambiguity rejection. Lastly, Multi-Modality achieves lower performance in both scenarios than Text-Modality does, and has the problem of catastrophic forgetting where there is big gap in terms of performance between STL and CIL on both datasets.

5   Conclusion

This paper proposes a unified document classification framework using ambiguity rejection, based on classifiers trained in both STL and CIL scenarios to reduce the error rate and be able to learn new categories. The method is evaluated on an imbalanced simulation of the public dataset RVL-CDIP containing documents from 16 categories, and our private (highly imbalanced) administrative document dataset containing 36 document categories. Our experimental results demonstrate that, in the presence of a document dataset with many categories with only few samples to train, textual information is useful to train models. The model trained only with the text modality achieves stable and excellent in classification task with both STL and CIL scenarios.

In order to further reduce the error rate by filtering out the documents for which the classifier's performance is too low, our proposed B-CDF ambiguity rejection learns per-class probability thresholds and achieves the best performance on both datasets, both in static and incremental learning contexts. In particular, on our private administrative document dataset and with static learning, the combination of B-CDF and Text-Modality-Sigmoid reduces error rate to only 0.9% with a high coverage of 96.8%. With incremental learning, Text-Modality-Softmax can archives error rate at 1% with coverage at 91.02% (which is a reduction of only 3.52% when compared with static learning). These results have great potential for real-world applications when considering the context of imbalanced datasets with many categories but very few documents inside some categories to train classifiers.

Acknowledgement. This work was supported by the French government in the framework of the France Relance program and by the YOOZ company under the grant number ANR-21-PRRD-0010-01. We also would like to thank Guénael Manic, Mohamed Saadi, Jonathan Ouellet and Jérôme Lacour from YOOZ for their support.

References

1. Krizhevsky, A., Sutskever, I., Hinton, G.E.: ImageNet classification with deep convolutional neural networks. In: Neural Information Processing Systems, vol. 25, pp. 1097–1105 (2012). https://doi.org/10.1145/3065386
2. Devlin, J., Chang, M.W., Lee, K., Toutanova, K.: BERT: pre-training of deep bidirectional transformers for language understanding. arXiv arXiv:cs.CL/1810.04805 (2019)
3. Le, H., et al.: FlauBERT: unsupervised language model pre-training for French. In: The 12th Language Resources and Evaluation Conference. LREC 2020, pp. 2479–2490 (2020)
4. Martin, L., et al.: CamemBERT: a tasty French language model. In: The 58th Annual Meeting of the Association for Computational Linguistics. ACL 2020, pp. 7203–7219 (2020)
5. d'Andecy, V.P., Joseph, A., Ogier, J.M.: InDUS: incremental document understanding system focus on document classification. In: Proceedings of 13th IAPR International Workshop on Document Analysis Systems, DAS 2018, pp. 239–244 (2018). https://doi.org/10.1109/DAS.2018.53
6. Appalaraju, S., Jasani, B., Kota, B.U., Xie, Y., Manmatha, R.: DocFormer: end-to-end transformer for document understanding. In: 2021 IEEE/CVF International Conference on Computer Vision (ICCV), ICCV 2021, pp. 973–983 (2021). https://doi.org/10.1109/ICCV48922.2021.00103
7. Huang, Y., Lv, T., Cui, L., Lu, Y., Wei, F.: LayoutLMv3: pre-training for document AI with unified text and image masking. arXiv arXiv:abs/2204.08387 (2022)
8. Xu, Y., et al.: LayoutLMv2: multi-modal pre-training for visually-rich document understanding. arXiv arXiv:abs/2012.14740 (2021)
9. Xu, Y., Li, M., Cui, L., Huang, S., Wei, F., Zhou, M.: LayoutLM: pre-training of text and layout for document image understanding. In: Proceedings of the 26th ACM SIGKDD International Conference on Knowledge Discovery & Data Mining, pp. 1192–1200 (2020). https://doi.org/10.1145/3394486.3403172
10. Li, P., et al.: SelfDoc: self-supervised document representation learning. In: IEEE/CVF Conference on Computer Vision and Pattern Recognition (CVPR), pp. 5648–5656 (2021). https://doi.org/10.1109/CVPR46437.2021.00560
11. Bakkali, S., Ming, Z., Coustaty, M., Rusiñol, M.: EAML: ensemble self-attention-based mutual learning network for document image classification. Int. J. Doc. Anal. Recogn. (IJDAR) **24**, 251–268 (2021). https://doi.org/10.1007/s10032-021-00378-0
12. Mahamoud, I.S., Voerman, J., Coustaty, M., Joseph, A., d'Andecy, V.P., Ogier, J.M.: Multimodal attention-based learning for imbalanced corporate documents classification. In: Lladós, J., Lopresti, D., Uchida, S. (eds.) Document Analysis and Recognition - ICDAR 2021. ICDAR 2021. LNCS, vol. 12823, pp. 223–237. Springer, Cham (2021). https://doi.org/10.1007/978-3-030-86334-0_15
13. Voerman, J., Mahamoud, I.S., Joseph, A., Coustaty, M., d'Andecy, V.P., Ogier, J.M.: Toward an incremental classification process of document stream using a cascade of systems. In: Barney Smith, E.H., Pal, U. (eds.) Document Analysis and Recognition - ICDAR 2021 Workshops. ICDAR 2021. LNCS, vol. 12917, pp. 240–254. Springer, Cham (2021). https://doi.org/10.1007/978-3-030-86159-9_16
14. Bakkali, S., Ming, Z., Coustaty, M., Rusiñol, M.: Visual and textual deep feature fusion for document image classification. In: IEEE/CVF Conference on Computer Vision and Pattern Recognition Workshops (CVPRW 2020), pp. 2394–2403 (2020). https://doi.org/10.1109/CVPRW50498.2020.00289

15. Harley, A., Ufkes, A., Derpanis, K.: Evaluation of deep convolutional nets for document image classification and retrieval. In: 13th International Conference on Document Analysis and Recognition (ICDAR), ICDAR 2015, pp. 991–995 (2015). https://doi.org/10.1109/ICDAR.2015.7333910

16. Afzal, M.Z., Kölsch, A., Ahmed, S., Liwicki, M.: Cutting the error by half: investigation of very deep CNN and advanced training strategies for document image classification. In: 2017 14th IAPR International Conference on Document Analysis and Recognition (ICDAR), vol. 1, pp. 883–888 (2017). https://doi.org/10.1109/ICDAR.2017.149

17. Das, A., Roy, S., Bhattacharya, U., Parui, S.K.: Document image classification with intra-domain transfer learning and stacked generalization of deep convolutional neural networks. In: 24th International Conference on Pattern Recognition (ICPR), pp. 3180–3185. IEEE (2018). https://doi.org/10.1109/ICPR.2018.8545630

18. Shu, L., Xu, H., Liu, B.: DOC: deep open classification of text documents. In: Proceedings of the 2017 Conference on Empirical Methods in Natural Language Processing, EMNLP 2017, pp. 2911–2916 (2017). https://doi.org/10.18653/v1/D17-1314

19. Mathew, M., Karatzas, D., Jawahar, C.V.: DocVQA: a dataset for VQA on document images. In: WACV (2021)

20. Zhong, X., Tang, J., Yepes, A.J.: PubLayNet: largest dataset ever for document layout analysis. In: International Conference on Document Analysis and Recognition (ICDAR), ICDAR 2019 (2019). https://doi.org/10.1109/ICDAR.2019.00166

21. Rebuffi, S.A., Kolesnikov, A., Sperl, G., Lampert, C.H.: iCaRL: incremental classifier and representation learning. In: IEEE/CVF Conference on Computer Vision and Pattern Recognition (CVPR), pp. 5533–5542 (2017). https://doi.org/10.1109/CVPR.2017.587

22. Hinton, G., Vinyals, O., Dean, J.: Distilling the knowledge in a neural network. In: NIPS Deep Learning and Representation Learning Workshop (2015)

23. Van de Ven, G.M., Tuytelaars, T., Tolias, A.S.: Three types of incremental learning. Nat. Mach. Intell. 4, 1185–1197 (2022). https://doi.org/10.1038/s42256-022-00568-3

24. Hou, S., Pan, X., Change, C., Wang, Z., Lin, D.: Learning a unified classifier incrementally via rebalancing. In: IEEE Conference on Computer Vision and Pattern Recognition (2019)

25. Wu, Y., et al.: Large scale incremental learning. In: IEEE Conference on Computer Vision and Pattern Recognition (2019)

26. Liu, Y., Su, Y., Liu, A.A., Schiele, B., Sun, Q.: Mnemonics training: multi-class incremental learning without forgetting. In: IEEE Conference on Computer Vision and Pattern Recognition (2020)

27. Mittal, S., Galesso, S., Brox, T.: Essentials for class incremental learning. In: IEEE Conference on Computer Vision and Pattern Recognition (2021)

28. Chow, C.K.: On optimum recognition error and reject tradeoff. IEEE Trans. Inf. Theory 16, 41–46 (1970)

29. Pietraszek, T.: Optimizing abstaining classifiers using ROC analysis. In: ICML (2005)

30. Pham, T.C., Luong, C.M., Hoang, V.D., Doucet, A.: AI outperformed every dermatologist in dermoscopic melanoma diagnosis, using an optimized deep-CNN architecture with custom mini-batch logic and loss function. Sci. Rep. 11, 17485 (2021). https://doi.org/10.1038/s41598-021-96707-8

31. Fumera, G., Roli, F., Giacinto, G.: Multiple reject thresholds for improving classification reliability. In: Ferri, F.J., Iñesta, J.M., Amin, A., Pudil, P. (eds.) SSPR /SPR 2000. LNCS, vol. 1876, pp. 863–871. Springer, Heidelberg (2000). https:// doi.org/10.1007/3-540-44522-6_89
32. Motlagh, N.K., Davis, J., Anderson, T., Gwinnup, J.: Learning when to say "I Don't Know". arXiv arXiv:2209.04944 (2022)
33. Franc, V., Prusa, D., Voracek, V.: Optimal strategies for reject option classifiers. J. Mach. Learn. Res. **24**, 1–49 (2023)

EEBO-Verse: Sifting for Poetry in Large Early Modern Corpora Using Visual Features

Danlu Chen(✉)[iD], Nan Jiang[iD], and Taylor Berg-Kirkpatrick[iD]

UC San Diego, La Jolla, CA 92092, USA
danlu@ucsd.edu

Abstract. One branch of important digital humanities research focuses on the study of poetry and verse, leveraging large corpora to reveal patterns and trends. However, this work is limited by currently available poetry corpora, which are restricted to few languages and consist mainly of works by well-known classic poets. In this paper, we develop a new large-scale poetry collection, EEBO-verse (Code and dataset is available on https://github.com/taineleau/ebbo-verse), by automatically identifying the poems in a large Early Modern books collection — English Early-modern Printed Books Online (EEBO). Instead of training text-based classifiers to sub-select the 3.5% of EEBO that actually consists of poetry, we develop an image-based classifier that can operate directly on page scans, removing the need to perform OCR – which, in this domain, is often unreliable. We leverage large visual document encoders (DiT and BEiT), which are pretrained on general domain document images, by fine-tuning them on an in-domain annotated subset of EEBO. In experiments, we find that an appropriately trained image-only classifier performs as well or better than text-based poetry classifiers on human transcribed text, and far surpasses the performance of text-based classifiers on OCR output.

Keywords: historical document classification · dataset · poetry · pretraining

1 Introduction

Ever sinceCode and dataset is available on https://github.com/taineleau/ebbo-verse. the publication of Aristotle's Poetics, the study of poetry has been a popular research topic in the field of literature and humanities [2]. More recently, poetry corpora have been used in digital humanities research to analyze various aspects of poetry ranging from poetic meter [1] to influence and authorship (e.g. comprehending the relative contributions of Shakespeare and Fletcher in Henry VIII [23]). This type of research is sometimes referred to as distant reading, a type of corpus-driven analysis that uses computational methods to study literature and poetry. Distant readings are restricted, however, by the availability of corpora. In the domain of poetry, the few corpora that are available represent

G. A. Fink et al. (Eds.): ICDAR 2023, LNCS 14191, pp. 36–52, 2023.
https://doi.org/10.1007/978-3-031-41734-4_3

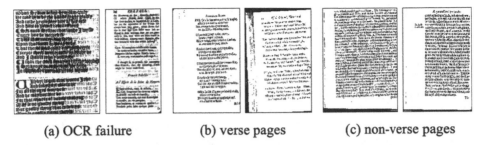

(a) OCR failure   (b) verse pages   (c) non-verse pages

Fig. 1. Sample pages from the EEBO data. (a) Blurry and over-inked examples on which modern OCR systems fail to detect the text correctly. (b) Typical example pages labelled as verse. (c) Typical examples of non-verse.

only a few languages and are dominated by well-known classic poets (details in Table 1). To further facilitate the digital study of poetry, we aim to build machine learning tools that make poetry identification in large collections more feasible, and to use these tools to create new, large-scale, poetry corpora.

Early-modern Printed English Books Online (EEBO) is an ideal collection for this purpose. While consisting mostly of prose, it still contains a substantial amount of poetry distributed throughout. It consists of scans of books printed from 1475 through to 1700, and contains 148,000 volumes. 56,000 of the 148,000 volumes in EEBO have been transcribed into text and annotated with genre labels by qualified experts (including a label for poetry). The remaining 92,000 volumes consists of more than 4 million scanned pages that have not been transcribed. Based on the annotated portion, we estimate that 3.5% of the page scans in EEBO are poetry. Further, while dominated by English, EEBO also contains Latin, Ancient Greek, and Old French. This makes it an ideal resource for creating a large and diverse poetry corpus. However, this approach depends on being able to accurately and automatically identify which pages are poetry in the unannotated portion.

Based on the annotated portion of the EEBO data, we could train a text-based binary classifier to identify poetry in the unannotated portion of EEBO and in other large-scale unannotated collections. This would require access to the text itself, either through human transcription or through OCR. However, as previous work on digitizing EEBO [4] has shown, even state-of-the-art OCR systems yield unreliable outputs on Early Modern documents: many of the scans are from microfilm and are therefore blurred and damaged (Fig. 1a). A further challenge with this text-based approach is generalizing to collections in other languages – text-based classifiers are unlikely to readily generalize to languages they were not trained on.

Thus, in this paper, we propose an image-based approach for detecting poems in unannotated pages. Specifically, given a scanned page image we would like to detect if there are any poetic lines on the page without ever observing the transcribed text or any additional meta-data. Since poetry is often formatted differently than prose, and often uses distinct fonts, visual features may be quite successful. To study whether image-based poetry identification is indeed feasible, we leverage powerful document image encoders from past work: DiT [15] and BEiT v2 [22]. These models were pretrained via self-supervision on general domain (non-historical) images. We take them, and fine-tune them on in-domain labeled data from EEBO. Our experiments indicate that the resulting system learns to identify poetry directly using image features and that it performs as well or better than text-based models without having access to the transcribed text. As shown in Fig. 1b-c, we found that our proposed model learns to leverage, for example, text alignment and density to identify poetry

We summarize the main contributions of this paper as follows:

1. We build **EEBO-Verse**, a challenging large-scale document classification dataset of poetry, containing $143,816$ annotated pages of poems, which is at least 10 times larger than existing publicly available poetry datasets (details in Table 2). Even not restricted to the domain of poetry, very few datasets [19] in the historical document image domain have size comparable to that of EEBO-Verse. We also release our model's prediction for the unannotated part in EEBO, which can be used for distant reading in digital humanities.
2. We conduct an empirical analysis of text-based, image-based and multimodal models for poetry detection and establish that image-only detection approaches can be as performant as their traditional text-only counterparts. This is true even when text-only methods use with human text transcriptions. OCR-based approaches substantially underperform.
3. Our results show that the models can recognize poems without textual information due to the rhythmic text's varied sentence lengths and spacing between paragraphs. Our technique could be used on other corpora to create additional poetry corpora or discover poetry in existing large collections where meta-data is absent. Our best system is able to achieve 59.19% precision at 76.24% recall on EEBO.

The remaining sections of the paper are organized as follows: we first review and compare related work on poetry detection and historical document analysis in Sect. 2; then we describe the EEBO dataset in detail with statistics and data visualization. In Sect. 4, we introduce different machine learning methods to detect verse. Finally, we present our empirical results and error analysis in Sect. 5.

2   Related Work

We first summarize existing poetry datasets and methods on poetry detection. Secondly, as we frame the poetry detection problem as a document classification task, we briefly review two major document classification methods used in prior work – text-based and image-based.

Table 1. Comparison between existing early modern poetry collections and the EEBO-Verse dataset. The statistics for EEBO-Verse (unannotated) are estimated from the poem v.s. non-poem rate in EEBO-Verse. "Poetry collection only" does not count pages appear out of a collection, e.g., on book preface, while "all annotated" counts all pages with label in EEBO. The "unannotated" part counts pages in EEBO without annotation.

	Poems	Books	Authors	Words	Period	Format
Women Writers Online (WWO) [17]	400				1400–1850	text only
English Broadside Ballad Archive (EBBA) [8]	400				1700–1800	text only
Representative Poetry Online (RPO)	4,700		723		1477–1799	text only
Verse Miscellanies Online	400				1700–1800	text only
EEBO-Verse (poetry collection only)		5,716	1,489	17.6M	1475–1700	paired image pages and text
EEBO-Verse (all annotated)		8,395	2,925	60M		
EEBO–Verse (unannotated)		~20k	~20k	~120M	1475–1700	image pages only

Poetry Datasets. Lamb [12] provides a detailed review on poetry datasets, whose summary is shown in Table 1. The most similar work is by Jacobs [11], who manually cleaned a collection of more than 100 poetry text extracted from the Gutenberg project, a library of over 60,000 free digitized books. Another dataset, Verse miscellanies online[1] [21], use the EEBO-TCP data to make a critical edition of seven printed poetry collections from sixteenth- and seventeenth-century England. Unfortunately, none of these are available in machine-readable format and they only contain positive examples (i.e., poetry only). [14] collected classical Chinese poems and the accompany paintings. The only available poetry dataset built for machine learning is [9], which contains only $2,779$ annotated pages.

ML for Poetry Detection. Poetry detection is usually considered as a classification task. [9] used digital books' metadata, such as book margin, and handcrafted features, such as the number of punctuation and capitalized words, to detect poems from text. Without these features, a Char-LSTM performs poorly on poetry detection. However, for image-only EEBO classification, we cannot use these kinds of handcrafted features because we do not have access to the text. [16,26] attempted to detect poems with image snippets in historical newspapers by using a two layer MLPs and obtained good performance. However, their data is balanced – half of the snippets contain poems, which is unrealistic in most real world cases: for example, only 3.5% of the pages in EEBO contains verse. Further, they heavily pre-processed the newspaper images by han-designing a binarization threshold and then segmenting them into smaller snippets, making

---

[1] http://versemiscellaniesonline.bodleian.ox.ac.uk/.

each sample in the dataset relative clean. In our method, rather, we do not apply any handcrafted features or heavy pre-processing steps on our data. While the imaged-based model we propose likely learns similar features, our model operates directly on raw page images and learns features automatically through its pretrained neural encoder.

Historical Document Analysis. Other related work, while not poetry-specific, has developed processing techniques for historical document images from the same or similar domains, focusing on tasks like Name Entity Recognition [7] and document layout analysis and classification [20]. Other examples of analyses for early modern printed books are font group detection [25] and content analysis of early modern printed Indian documents [5]. In our work, we focus on early modern English printed books.

Pre-training for Document Analysis. Another line of work in NLP/CV has demonstrated that pretraining foundation models on either labelled or unlabelled data, followed by subsequent fine-tuning on a target task, can yield pronounced gains in various downstream tasks. In the domain of document classification, for instance, ResNeXt [27] is the start-of-the-art standalone image classification model pretrained with supervision on ImageNet. The most related to our work is a sub-field in machine learning called visually-rich document understanding, which usually uses both textual and visual feature for document analysis. Different from the existing visually-rich document datasets such as FUNSD², which are fine-grained annotation of scanned single pages, in contrast, the dataset we develop, EEBO-verse, consists of transcriptions of thousands of full books with block-level annotations. Moreover, the EEBO-Verse data usually contains tags that span across multiple pages.

Recently, LayoutLM [10,28] was developed with a focus on detecting meaningful units in document images like dates and figures in medical records or research papers. This model achieved state-of-the-art performance on FUNSD. However, it assumes a good OCR system is available for data prepossessing to obtain the 2D coordinates of the text. As stated earlier, recognizing text for historical documents is much harder than that for modern documents because there are usually lacuna and damage on historical document images. BEiT [3], v2 [22] and Document image Transformer (DiT) [15] represent another line of work which seeks to encode document images without text transcriptions. These methods use self-supervised pretraining objectives to create a general-purpose document image encoder. We leverage both DiT and BEiT in our experiments, fine-tuning these models for the purpose of image-based poetry classification.

3   Dataset Creation

As mentioned, we use the EEBO data and its human annotation to build our poetry dataset. We describe the procedure of dataset creation in Sect. 3.1 and show basic statistics of the dataset in Sect. 3.2.

---

² https://guillaumejaume.github.io/FUNSD/.

3.1 Building EEBO-Verse from EEBO-TCP

Fig. 2. An example of the EEBO-TCP annotation. **Right:** the scanned page. **Left:** The TCP texts are encoded in TEI, an XML-like language. We leverage EEBO-TCP as supervision for poetry classification.

The[3] EEBO-TCP project[4] presented a collection of manually transcribed texts for the Early English Books Online (EEBO) collections, which preserves a large portion of the printed books from 1475 to 1700. There are about $148,000$ volumes in total, and $61,053$ (about 37.8%) of them of them are part of EEBO-TCP, and are therefore transcribed and annotated by trained human. Figure 2 is an example of a pair of scanned pages and annotation from EEBO-TCP. In addition to text transcriptions, EEBO-TCP also provides labels as XML tags for blocks of text specifying a category, e.g. preface and elegy. We automatically paired the transcription and the scanned books using Python and the XML parser package `BeautifulSoup`[5]. The pairing procedure is as followed:

1. We keep component <div> with **type** and page break indicator <pb>, and discard all other XML annotations.
2. Of the labels types in EEBO-TCP, we manually map 22 to the 'verse' category in our dataset[6], and all the rest of the text as 'prose'.
3. Finally, we split the text into pages and pair it to the scanned pages according to the `book_id`.

Our final extraction consists of $52,456$ annotated volumes made up of of $4,223,808$ pages. We name the resulting benchmarking dataset EEBO-Verse.

---

[3] Text Encoding Initiative (TEI): https://tei-c.org/.

[4] https://textcreationpartnership.org/tcp-texts/eebo-tcp-early-english-books-online/.

[5] https://www.crummy.com/software/BeautifulSoup/bs4/doc/.

[6] All 22 types are: `poems`, `poem`, `song`, `songs`, `pastoral`, `authors_poem`, `biographical_poem`, `verse_response`, `acrostic_dedicatory_poem`, `verse_prologue`, `introductory_poem`, `canto`, `sonnets`, `sonnet`, `epithalamium`, `Psalm`, `elegy`, `encomium`, `hymn`, `ode` and `ballad`

Table 2. Statistics of the EEBO-Verse dataset, produced by combining EEBO and EEBO-TCP and mapping labels to verse vs. no-verse coarse categories. EEBO-Verse includes images for all pages, and human labels and text transcriptions for the annotated portion.

volumes			pages (annotated)		pages (unannotated)
all	poems (collection)	poems (sporadic)	all	poems	
52,456	5,716	2,679	4,223,808	143,816	~8,800,000

3.2   EEBO-Verse Statistics

One might expect that the poetry detection problem can be solved by classifying only the title of a book – i.e. that poetry lives in volumes that consist only of collections of poetry. However, poems can appear not only in a collection, but also in the preface, middle of a chapter, or the epilogue of a book (see Fig. 3). We visualize the distribution of poetry in books in EEBO (see Fig. 4a). In Fig 4b, two peaks in the histogram indicate that poems are either presenting everywhere (possibly a collection of poems) in a book or on just a few pages (citing in the dedication or chapters). Even for a poetry collection book, not 100% of the pages contain poetry. In Fig 4b, we also show the relative position of the poems presenting in a book, which indicates that most of the poems appear in the first 20% of a book. Fig 3b also shows a histogram of which genres poems appear in most frequently.

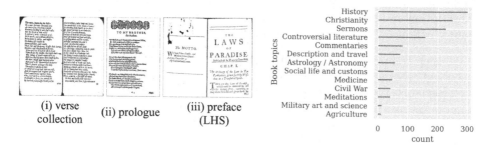

Fig. 3. (a): Examples of poems appearing in the EEBO dataset. (b): Topic keyword frequency count from books that sporadically contain poems.

4   Methods

We tackle the poetry detection problem as a binary classification problem. Given either a scanned page or its transcription, we would like to classify whether the page is 'verse' or 'non-verse'. For our dataset and task, we say that a page is verse if it contains at least one line of verse. Otherwise, we categorize it as non-verse. Thus, we use a page as the minimal predictable unit.

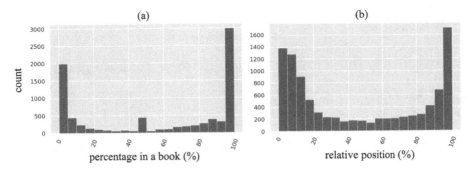

Fig. 4. Two histogram shows how poetry distributed in EEBO-Verse dataset. (a) Book-level histogram showing fraction of pages within a book that contain poetry, across books in EEBO-Verse. (b) Page-level histogram showing relative position of pages containing poetry within their encompassing book, across pages in EEBO-Verse.

4.1 Self-supervised Pre-trained Models for Image Classification

As mentioned earlier, we would like to build a model only using visual features to predict poetry. BEiT v1 [3] and v2 [22] are self-supervised pre-trained image encoders that achieve the state-of-the-art performance on image classification tasks on ImageNet. Inspired by the text-based BERT's masked Language modeling (MLM) objective, BEiT divides an image into patches and uses a masked image modeling (MIM) objective to train the model. DiT [15] uses the same architecture as BEiT v1 but is pretrained on a modern document dataset RVL-CDIP [13], whereas BEiT is trained on ImageNet-21k.

All of these pretraining models have similar architectures and use similar training procedures, as shown in Fig. 5. There are three steps during the training procedure, which are described as follows. Since these pre-trained models are just encoders that must be fine-tuned to perform downstream tasks, the third step is task-specific. Here, we describe the appropriate fine-tuning step for poetry classification. We describe the training details of the first two steps for completeness. The resulting pretrained models were produced in prior work. We take these pre-existing models and fine-tune them on our EEBO-Verse datasets, before conducting error analysis.

1. **Training the codebook.** A codebook is a many-to-one mapping from a continuous vector space to a discrete set of indices. Discretizing continuous feature representations in this manner can facilitate certain downstream operations in computational models. For document encoders, the continuous input is an image patch consisting of $4 \times 4 = 16$ image patches, whose size is 56×56. Each patch of an image is passed through a CNN to create a feature vector. A dense feature vector for each patch is then mapped to a discrete value $d \in [1, 9182]$. In BEiTv1 and DiT, Discrete VAE from DALL.E [24] with a reconstruction loss is used to train the codebook. For BEiT v2, a teacher model CLIP [22] is used to guide the training for the codebook and thus make the codebook more compact.

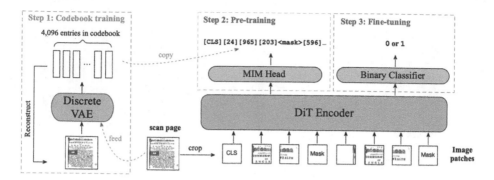

Fig. 5. An overview of the training procedure for the image pre-trained models. Step 1 consists of first learning the codebook of images patches. Step 2 consists of pre-training the image encoder using a masked-reconstruction objective that uses a cross-entropy loss on each masked codebook entry. Step 3 uses a binary classification loss to fine-tune the model with the EEBO-verse data.

2. **Pre-training on unlabelled images.** The codebook from step 1 is used to look up the discrete values for image patches (grey dash arrows in the overview). The encoder is a vanilla Transformer architecture. During the pre-training, they use the image patches as input and the discrete values as output, where a cross-entropy classification loss is used to distinguish the correct masked codebook index from the set of all codebook indices as depicted in Fig. 5.

3. **Fine-tuning on labelled images.** We use the same encoder and weights in step 2 and change the decoder into a binary classifier. We obtain the weights from the released models[7], and continue fine-tuning both the binary classifier and the encoder using our labelled EBBO-Verse data. We train by back-propagating into entire encoder, not just the classification head. However, the codebook and initial CNN encoders remain fixed.

4.2 Baselines

In this paper, we are interested in two questions related to sifting out poetry from the large scale books collections: (i) Do pretraining methods help improve performance compared to models trained from scratch? (ii) Can an image-based model perform as well as a text-based model?

In order to help answer our first question, we conduct experiments with text-only baselines such as Char-LSTMs and 2 layer MLPs. To fairly compare between image-based and text-based models, we also include BERT [6], a state-of-the-art text classification model using self-supervised pretraining techniques, as a textual baseline. Further, we also includes ResNeXt, the state-of-the-art image classification model using supervised pretraining techniques. Specifically,

---

[7] The pre-trained model is obtained from https://github.com/microsoft/unilm/tree/master/dit.

the various systems and baselines we will conduct experiments with consist of the following:

1. **Char-LSTMs**[8]. According to [9], we include a character-level LSTM (Char-LSTM) for poetry classification. We use one layer LSTM with hidden size 128, followed by a linear layer for binary classification.
2. **BERT**[9]. BERT is a state-of-the-art pretrained text encoder model that uses a multi-layer transformer architecture. BERT is pre-trained using plain text corpus with both next sentence prediction and masked language modeling (MLM) objectives.
3. **Two-layer MLPs.** Following [26], we build a two layer multi-layer perceptrons (MLPs), whose layer has 20 neurons, for image classification.
4. **ResNeXt**[10]. ResNeXt is a multi-layer convolutional networks (CNNs) using skip connections and aggregated transformation. We use ResNeXt-101 and initialize the model with weights trained on dataset ImageNet-21k.

All of the models described above are text or document image encoders, and we attach a one-layer binary dense layer for classification and use a binary cross entropy loss to fine-tune on the corresponding labeled EEBO-verse dataset, where the gradient is always back-propagated into the encoders.

5   Experiments and Analysis

5.1   Data Preparation

We use all the transcribed $4,223,808$ pages as our data, with a split ratio 90/5/5 for training/validation/testing. Note that we split the data volume-wise, i.e. pages in the same volume appear in the same split. Table 3 shows detailed statistics about the EEBO-verse dataset. The label distribution is extremely imbalanced, with only 3.5% of the labels being poetry. We also create a subset consisting of only $10,000$ positive examples from the training dataset to produce a small-scale training benchmark EEBO-verse-10k for development.

Table 3. An overview of the training, validation and test set split. EEBO-verse-10k is a sub-sample from the training set and contains only positive examples.

Num. of Pages	All	Training	Validation	Test
EEBO-verse	$4,223,808$	$3,721,403$	$207,186$	$206,262$
- Positive (verse)	$143,816$	$126,730$	$200,933$	$200,181$
- Negative (non-verse)	$4,079,992$	$3,594,673$	$6,253$	$6,081$
EEBO-verse-10k	$10,000$	$10,000$	–	–
- Positive (verse)	$10,000$	$10,000$	–	–

---

[8] Implementation is based on https://huggingface.co/keras-io/char-lstm-seq2seq, whose decoder is replaced with a classification head.
[9] https://huggingface.co/bert-base-uncased.
[10] https://pytorch.org/hub/pytorch_vision_resnext/.

In experiments, we would like to also evaluate how much the text-only model's performance degrades when using OCR output. This is a more realistic setting because many large historical document corpora do not have ground-truth transcriptions readily available. Further, again to match real-world constraints, we would like to run a realistic OCR engine – specifically, one that is freely available and relatively fast. Thus, we used Tesseract[11] to obtain the detected **OCR text**. It took about 20 d to OCR all the 4 million annotated pages using an AMD 3960X (24-Core) with parallelism. And it is also worth noting that Tessearct is only able to detect about 65.6% of the text by comparing the number of characters of annotated transcription.

For either **human transcription** or **OCR transcription**, we use BPE (Byte Pair Encoder) to process the text, which uses subwords as the unit to tokenize the text. We only keep up to 512 tokens per each page for the BERT model. In Char-LSTM, we keep the first 512 words instead of subwords. For the **grey scale scanned pages**, we use on-the-fly augmentation. We randomly resize and crop the high resolution images down to size 224 × 224. The prepossessing and data augmentation are the same for 20-layer MLP, ResNeXt, BEiT v2 and DiT.

5.2 Experimental Setup

We use 8x GTX 2080 Ti (11GB) for training and testing. All experiments are implemented using PyTorch. Details can be found in our codebase[12]. The training configuration can be found in Table 4. We train for a certain number of steps instead of epochs to make the result comparable as the negative ratio affects the number of training examples. We follow BEiT and DiT's training recipe, except the cosine learning scheduler is set to use 50, 000 warm-up iterations.

For all the models, we fine-tune the hyper-parameters and the binary decision threshold on the validation set and use the same threshold to report F_1 score on test set. We use early stopping based on validation F_1 score for all models to mitigate overfitting.

Table 4. Training configurations for all models.

	Pre-training dataset	Pre-training supervision	LR	Step	Effective batchsize	LR scheduler	Optimizer
Char-LSTM	N/A	N/A	1^{-3}	20k	256	N/A	ADAM
BERT	common crawl	self-supervised	2^{-5}	60k	128	N/A	ADAMW
2 Layer MLPs	N/A	N/A	1^{-3}	20k	512	stepLR	SGD w/ momentum
ResNeXt-101	ImageNet-21k	supervised	1^{-3}	50k	512	stepLR	SGD w/ momentum
BEiT v2	ImageNet-21k	self-supervised	5^{-4}	150k	256	cosine	ADAMW
DiT	RVL-CDIP	self-supervised	5^{-4}	150k	256	cosine	ADAMW

---

[11] We use pytesseract (v0.3.10), a Python wrapper for Tesseract, whose underlying engine is Tesseract v5.3.0. https://pypi.org/project/pytesseract/0.3.10/.
[12] https://github.com/taineleau/ebbo-verse.

Data Imbalance. The ratio of positive examples in the EEBO-verse dataset is 3.5%. To mitigate the data imbalance issue of the EEBO-Verse data, we employed both loss re-weighting and negative sampling during the training. The negative sampling rate that matches the frequency of poetry in EEBO is $\frac{1}{3.5\%} = 28.36$. We explored negative sampling ratio 1, 4, and 16. The results showed that a negative sampling ratio of 4 is optimal for BERT, while a ratio of 1 is the best for DiT and BEiT v2.

In addition to different negative sampling ratios, we also explored the effects of label smoothing and loss-re-weighting. Specifically, we experiment on a negative sampling ratio of 4 using both label smoothing and loss re-weighting when fine tuning on the self-pretrained image models. *Label*

Table 5. Ablation study of label smoothing and loss re-weighting.

	F_1	P	R
Default	65.83	58.57	75.15
+ smooth $\epsilon = 0.1$	**66.18**	60.17	73.52
+ reweighing	65.22	58.04	74.44

smoothing [18] uses ϵ and $1 - \epsilon$ as the target for training instead of 0 and 1. We explored $\epsilon = 0.1, 0.2, 0.4$, and found $\epsilon = 0.1$ is the best. *Loss re-weighing* is to multiply the loss with a class-related scaling factor of $\frac{N_{total\_example}}{N_{example\_of\_class}}$. As shown in Table 5, we found that the loss reweighing not only performs worse than label smoothing alone, but also worse than the default settings.

5.3 Results

Table 6. F_1 score on the EEBO-Verse test set. The best F_1 for image-based models is in **bold** and the overall best model is in ***italic bold***. Human or OCR indicates human or OCR transcription respectively.

neg: pos ratio	1:1			4:1			16:1		
	F_1	P	R	F_1	P	R	F_1	P	R
Char-LSTMs (human)	37.45	27.29	59.66	37.55	29.12	52.82	33.32	23.74	55.84
BERT (human)	66.91	56.29	82.47	***67.24***	56.60	82.81	66.93	56.22	82.68
BERT (OCR)	48.20	34.68	79.01	57.04	48.18	69.91	59.75	58.77	60.77
2 Layer MLPs	34.70	24.95	56.94	33.28	23.04	56.80	30.32	21.91	49.23
ResNeXt	55.93	45.66	72.16	57.57	50.19	67.50	57.58	49.15	69.50
BEiT v2	64.81	54.96	78.95	64.79	55.00	78.81	64.60	54.82	78.62
DiT	**66.64**	59.19	76.24	66.36	57.76	77.98	65.81	57.71	76.54

Table 6 shows the main results for poetry detection on the test set with different negative sampling ratios. We summarize observations from the experimental results as the follows:

1. *Pre-training always helps.* Despite being trained on general (mostly modern) data, all the pre-trained models achieve substantial improvements over the two models that were trained from scratch.

2. *Relevant data can enhance performance.* While past work has found that BEiT v2 is state-of-the-art on modern document image classification tasks, we find that DiT outperforms BEiT v2 on our data because it is pre-trained on large scale unlabelled modern document data while BEiT v2 is trained on general image data.

3. *Image-based models can achieve comparable performance with text-based models.* While past work has found that BEiT is state-of-the-art on modern document image classification tasks, we find that DiT achieve 66.64% on F_1, which is very close to the best result 67.29%s achieved by BERT on EEBO-verse. Generally, the textual models seem to be more sensitive to positive data and thus have higher recall but slightly worse precision on detecting poetry.

4. *Self-supervised pre-training outperforms supervised pre-training.* ResNeXt is also pre-trained supervisedly on ImageNet-21k, but the performance is about 10% worse than models with self-supervised pre-training.

5. *Poor transcription quality from OCR degrades classification performance.* Comparing the result from BERT using the human and OCR transcription, there is 10 % F_1 score drop if using OCR. It is not surprising to see such a huge performance gap between human and OCR transcription because OCR quality on historical documents is often poor, as mentioned earlier. Another important observation is that the OCR text-based model is the only model whose performance rises as the ratio of negative sampling increases.

Table 7. The performance boost (diff.) based on OCR transcription for our dataset and a modern document dataset (RVL-CDIP). The best F_1 scores for each model are copied from Table 6, and the reference accuracy for RVL-CDIP is copied from [15].

	EEBO-verse		RVL-CDIP		Speed
	F_1	Diff.	Acc.	Diff.	
BERT (OCR)	59.75		89.92		30.2 s/it
BERT (human)	67.24	+12.53%	—N/A—		1.8 s/it
DiT	66.64	+11.53%	92.69	+3.08%	3.1 s/it

Table 8. Performance of textual and visual model on validation set for all data and non-English portion only.

		F_1	P	R
BERT	all	52.62	42.16	69.97
	non-EN	47.19	32.64	85.16
DiT	all	65.54	61.40	70.27
	non-EN	81.04	82.67	79.49

5.4   Analysis

Historical v.s. Modern Document Dataset. As shown in Table 7, we run the same models on our EEBO-verse and compared it to the performance on RVL-CDIP, a modern document classification datasets. Since our data comes with human annotated transcription, we are able to see the low quality of OCR text decrease the performance of the BERT model. The performance gap between BERT (OCR) and DiT on modern dataset is only 3.08%, but is 11.53% for our historical dataset, which further support our claim that OCR system on historical documents is very poor and popular multi-modal approaches may not work well on historical data.

Out-of-Distribution Languages. There are 4,493 pages in the validation set that are in non-English languages such as Latin and French. As the visual features of English and non-English verse are likely similar, we expect the image-only model to exhibit similar performance. However, for text-based model, we expect the opposite since it relies on text-level features which will be different across languages. Table 8 shows that the F_1 score of the BERT model drops greatly on out-of-distribution languages, but the visual model (DiT) actually performs better on the non-English subset.

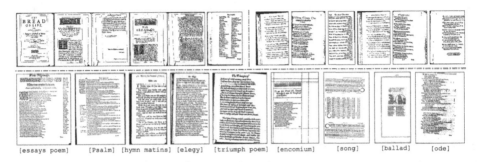

[essays poem]   [Psalm] [hymn matins] [elegy]   [triumph poem] [encomium]   [song]   [ballad]   [ode]

Fig. 6. Pages are sampled from EEBO-verse's validation set. **Upper row**: Visualization of true positive and true negative examples. **upper left**: DiT's prediction confidence of containing poetry $\leq 0.01\%$; **upper right**: DiT's confidence $\geq 95\%$. **Lower row**: Visualization of the false negative examples for DiT.

Data and Computation Efficiency. We sample 1k and 10k positive examples from the training set and train BERT and DiT with a negative sampling ratio of 1. As shown in Table 9, we can see that when training in a low resource setting, where we only have 2k (1k positive) training examples, the performance of DiT and BERT are similar. But when the number of training dataset increased to 20k (10k positive), DiT's performance is greatly boosted, while BERT requires almost the full data to achieve its best performance. This experiment shows image-based features are more data efficient than text-based models for this task. Moreover, from Table 7, we can see the speed of the BERT-based approach is significantly slower than DiT because it also includes OCR run time.

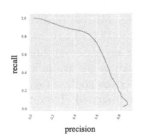

Fig. 7. PRcurve for DiT.

Table 9. F_1 score on the EEBO-Verse validation set using 1k and 10k positive training examples with negative sampling ratio 1.

Training split	1k			10k			Full (143k)		
	F_1	P	R	F_1	P	R	F_1	P	R
BERT	53.96	41.57	76.89	55.70	46.03	80.62	66.91	41.89	83.10
DiT	54.96	47.21	65.76	63.64	54.45	76.57	66.64	59.19	76.24

Qualitative Analysis. In Fig 6, we visualize examples of pages that our best visual model (DiT) finds most and least likely to be poetry. Consistent with our expectations, pages with high probability are often loosely spaced and display varying line lengths. In contrast, pages with low probabilities are significantly more dense and lack the characteristics of poetry lines.

Precision and Recall Trade-Off. The trade-off between precision and recall is plotted in Fig. 7, where the best F_1 can be achieved at 66.64% with a precision of 59.19% and recall 76.24%. It is possible to set the threshold at a recall of 88% and a precision of 40% in order to construct a poetry dataset. The annotation effort and cost can be greatly reduced as humanities researcher can first identify poetry pages and then correct the OCR transcription. That is being said, we can retrieve 88% of the verse from the unannotated EEBO, but only transcribed $\frac{3.5\%}{40\%} = 8.75\%$ of the pages instead of all pages, potentially leading to a savings of over 90% of the budget.

False Negative Analysis. We can take a closer look at the fine-grained categories of the false negative from the models in Fig. 8. The visual model DiT fails most on religious text such as Palsm, hymn and encomium, while the text model BERT fails most on song because it usually comes with side notes and the music scores. Figure 6 shows examples of false negatives for different sub-categories.

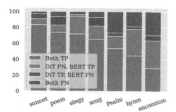

Table 10. True negative and false positive statistics of DiT and BERT.

	BERT TN	BERT FP	total
DiT TN	192,389 (95.87%)	4,112 (2.05%)	196,501 (97.92%)
DiT FP	776 (0.39%)	3,396 (1.69%)	4,172 (2.08%)
total	193,165 (96.26%)	7,508 (3.74%)	200,673 (100.0%)

Fig. 8. False negative analysis of different categories for DiT and BERT.

False Positive Analysis. From Table 10 we can see the false positive rate is 2.08% for DiT and 3.74%. We sample 50 pages of the false positive from both models respectively, and found that 49 of DiT and (46 of BERT) are indeed verses (True Positive). The error mostly comes from the inconsistent human annotation of EEBO-TCP. For example, these false negative pages are labelled generally as text, letter or book. We also find annotation errors such as tags for two adjacent pages are flipped and page number is wrongly assigned. Therefore, the machine learning model we trained is potentially a good tool to discover annotation error and label inconsistency.

6   Conclusion

In this paper, we did an large empirical analysis focused on the problem of classifying poetry in historical corpora. We leveraged EEBO and EEBO-TCP, which are very valuable corpora for further document analysis research. We showed that image-only poetry classification can be highly successful, pointing the way for related work in sifting through large corpora for other types of texts without having to rely on costly human transcriptions or noisy OCR. We release labels for EEBO that define EEBO-verse, a large-scale historical poetry corpus that may support future research in digital humanities.

References

1. Agirrezabal, M., Alegria, I., Hulden, M.: Machine learning for metrical analysis of english poetry. In: Proceedings of COLING 2016, the 26th International Conference on Computational Linguistics: Technical Papers, pp. 772–781 (2016)
2. Arduini, B., Magni, I., Todorovic, J.: Interpretation and Visual Poetics in Medieval and Early Modern Texts: Essays in Honor of H. Wayne Storey, Brill (2021)
3. Bao, H., Dong, L., Piao, S., Wei, F.: Beit: Bert pre-training of image transformers. arXiv preprint arXiv:2106.08254 (2021)
4. Christy, M., Gupta, A., Grumbach, E., Mandell, L., Furuta, R., Gutierrez-Osuna, R.: Mass digitization of early modern texts with optical character recognition. J. Comput. Cultural Heritage (JOCCH) 11(1), 1–25 (2017)
5. Clausner, C., Antonacopoulos, A., Derrick, T., Pletschacher, S.: Icdar 2017 competition on recognition of early indian printed documents - reid2017. In: 2017 14th IAPR International Conference on Document Analysis and Recognition (ICDAR). vol. 01, pp. 1411–1416 (2017). https://doi.org/10.1109/ICDAR.2017.230
6. Devlin, J., Chang, M.W., Lee, K., Toutanova, K.: Bert: Pre-training of deep bidirectional transformers for language understanding. arXiv preprint arXiv:1810.04805 (2018)
7. Ehrmann, M., Hamdi, A., Pontes, E.L., Romanello, M., Doucet, A.: Named entity recognition and classification on historical documents: A survey. CoRR abs/2109.11406 (2021). http://arxiv.org/abs/2109.11406
8. Finley, P.: English broadside ballad archive. Ref. Rev. 32(1), 25–26 (2018)
9. Foley IV, J.J.: Poetry: Identification, entity recognition, and retrieval (2019)
10. Huang, Y., Lv, T., Cui, L., Lu, Y., Wei, F.: Layoutlmv3: Pre-training for document ai with unified text and image masking (2022). https://doi.org/10.48550/ARXIV.2204.08387, http://arxiv.org/abs/2204.08387
11. Jacobs, A.M.: The gutenberg english poetry corpus: exemplary quantitative narrative analyses. Front. Digital Humanities 5, 5 (2018)
12. Lamb, J.P.: Digital resources for early modern studies. SEL Studies in English Literature 1500–1900 58(2), 445–472 (2018)
13. Lewis, D., Agam, G., Argamon, S., Frieder, O., Grossman, D., Heard, J.: Building a test collection for complex document information processing. In: Proceedings of the 29th Annual International ACM Sigir Conference On Research and Development in Information Retrieval, pp. 665–666 (2006)
14. Li, D., et al.: Paint4poem: A dataset for artistic visualization of classical Chinese poems. arXiv preprint arXiv:2109.11682 (2021)

15. Li, J., Xu, Y., Lv, T., Cui, L., Zhang, C., Wei, F.: Dit: Self-supervised pre-training for document image transformer. arXiv preprint arXiv:2203.02378 (2022)
16. Lorang, E.M., Soh, L.K., Datla, M.V., Kulwicki, S.: Developing an image-based classifier for detecting poetic content in historic newspaper collections (2015)
17. MCCARTHY, E.A.: Women writers online. database. Renaissance and Reformation / Renaissance et Réforme **42**(4), 163–165 (2019). www.jstor.org/stable/26894251
18. Müller, R., Kornblith, S., Hinton, G.E.: When does label smoothing help? In: Advances in Neural Information Processing Systems, vol. 32 (2019)
19. Nikolaidou, K., Seuret, M., Mokayed, H., Liwicki, M.: A survey of historical document image datasets (2022). arxiv.org/abs/2203.08504
20. Nikolaidou, K., Seuret, M., Mokayed, H., Liwicki, M.: A survey of historical document image datasets (2022). arxiv.org/abs/2203.08504
21. O'Callaghan, M.: Verse miscellanies online: a digital edition of seven printed poetry collections from sixteenth-and seventeenth-century england. J. Early Modern Cultural Stud. **13**(4), 148–150 (2013)
22. Peng, Z., Dong, L., Bao, H., Ye, Q., Wei, F.: Beit v2: Masked image modeling with vector-quantized visual tokenizers. arXiv preprint arXiv:2208.06366 (2022)
23. Plecháč, P.: Relative contributions of shakespeare and fletcher in henry viii: An analysis based on most frequent words and most frequent rhythmic patterns. Digital Scholarship Humanities **36**(2), 430–438 (2021)
24. Ramesh, A., et al.: Zero-shot text-to-image generation. In: International Conference on Machine Learning, pp. 8821–8831. PMLR (2021)
25. Seuret, M., Limbach, S., Weichselbaumer, N., Maier, A., Christlein, V.: Dataset of pages from early printed books with multiple font groups. In: Proceedings of the 5th International Workshop on Historical Document Imaging and Processing. p. 1–6. HIP '19, Association for Computing Machinery, New York, NY, USA (2019). https://doi.org/10.1145/3352631.3352640
26. Soh, L.K., Lorang, E., Liu, Y.: Aida: intelligent image analysis to automatically detect poems in digital archives of historic newspapers. In: Proceedings of the AAAI Conference on Artificial Intelligence, vol. 32 (2018)
27. Xie, S., Girshick, R., Dollár, P., Tu, Z., He, K.: Aggregated residual transformations for deep neural networks. arXiv preprint arXiv:1611.05431 (2016)
28. Xu, Y., Li, M., Cui, L., Huang, S., Wei, F., Zhou, M.: Layoutlm: Pre-training of text and layout for document image understanding. In: Proceedings of the 26th ACM SIGKDD International Conference on Knowledge Discovery & Data Mining, pp. 1192–1200 (2020)

A Graphical Approach to Document Layout Analysis

Jilin Wang[1]([✉]), Michael Krumdick[1]([✉]), Baojia Tong[2], Hamima Halim[1],
Maxim Sokolov[1], Vadym Barda[1], Delphine Vendryes[3], and Chris Tanner[1]

[1] Kensho Technologies, Cambridge, MA 02138, USA
{jilin.wang,michael.krumdick,hamima.halim,maxim.sokolov,
vadym,chris.tanner}@kensho.com
[2] Meta, Cambridge, MA 02140, USA
[3] Google, Los Angeles, CA 90291, USA

Abstract. Document layout analysis (DLA) is the task of detecting the distinct, semantic content within a document and correctly classifying these items into an appropriate category (e.g., text, title, figure). DLA pipelines enable users to convert documents into structured machine-readable formats that can then be used for many useful downstream tasks. Most existing state-of-the-art (SOTA) DLA models represent documents as images, discarding the rich metadata available in electronically generated PDFs. Directly leveraging this metadata, we represent each PDF page as a structured graph and frame the DLA problem as a graph segmentation and classification problem. We introduce the Graph-based Layout Analysis Model (GLAM), a lightweight graph neural network competitive with SOTA models on two challenging DLA datasets - while being an order of magnitude smaller than existing models. In particular, the 4-million parameter GLAM model outperforms the leading 140M+ parameter computer vision-based model on 5 of the 11 classes on the DocLayNet dataset. A simple ensemble of these two models achieves a new state-of-the-art on DocLayNet, increasing mAP from 76.8 to 80.8. Overall, GLAM is over 5 times more efficient than SOTA models, making GLAM a favorable engineering choice for DLA tasks.

1 Introduction

Much of the world's information is stored in PDF (Portable Document Format) documents; according to data collected by the Common Crawl[1], PDF documents are the most-popular media type on the web other than HTML.

A standard PDF file contains a complete specification of the rendering of the document. All objects (e.g., texts, lines, figures) in the PDF file include their associated visual information, such as font information and location on the page. However, a PDF file does not always contain information about the *relationships*

J. Wang and M. Krumdick—The authors contributed equally to this work.

[1] https://commoncrawl.github.io/cc-crawl-statistics/plots/mimetypes.

© The Author(s), under exclusive license to Springer Nature Switzerland AG 2023
G. A. Fink et al. (Eds.): ICDAR 2023, LNCS 14191, pp. 53–69, 2023.
https://doi.org/10.1007/978-3-031-41734-4_4

between those objects, which is critical for understanding the structure of the document. For example, all text is stored at the individual character level. Sentences from the characters and consequently paragraphs or titles formed by the sentences are not included in the metadata. Therefore, despite PDF files containing explicit, structured representations of their content, correctly structuring this content into human-interpretable categories remains a challenging problem and is the crux of Document Layout Analysis (DLA).

Existing approaches to PDF extraction typically consist of two main components—a PDF parser and DLA model—as depicted in Fig. 1 [6,32]:

1. **PDF parser** (e.g., *PDF Miner*): extracts the document's underlying raw data, often doing so by producing fine-grained, pertinent contiguous blocks of information referred to as "objects" (e.g., a line of text). For a standard (non-scanned) PDF document, the PDF parser is able to generate information about each "object": its location on the page, the raw content contained within it, and some metadata like font and color about how the object is displayed.

2. **DLA model:** aims to group together objects that concern semantically similar content (e.g., a table, figure, paragraph of text), essentially turning an unordered collection of data into a coherent, structured document. The output is a collection of bounding boxes, each representing one of the semantic objects (e.g., a paragraph).

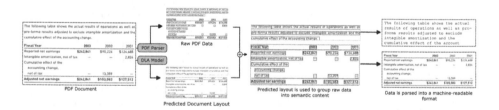

Fig. 1. Visualization of the data extraction pipeline. We use *pdfminer.six* as the PDF parser in our pipeline.

DLA models frequently take a computer vision-based approach [15,21,30]. These approaches rely on treating documents as images and ignoring much of the rich, underlying metadata.

Generally, PDF extraction can be evaluated by the quality of the DLA model's segmentation predictions. The standard approach to evaluating the segmentation performance is to use the mean average precision (mAP) metric to assess the overlap between the predicted and ground-truth bounding boxes. Although this is standard practice, mAP is sensitive to the bounding boxes' exact pixel values. Therefore, bounding box-level metrics are only proxies for measuring segmentation performance, and ultimately a user wishes to measure the performance of clustering objects to semantic segments. In this work, we approach the problem by modeling the document not as a collection of pixels but as a structured graphical representation of the output from a PDF parser.

Specifically, we frame the DLA task as a joint graph segmentation and node classification problem. Using this formulation, we develop and evaluate a graph neural network (GNN) based model – the Graph-based Layout Analysis Model (GLAM). Our model is able to achieve performance that is comparable in mAP to the current SOTA methods while being lightweight, easier to train, and faster at inference time.

Our primary contributions are:

1. Propose a novel formulation of DLA, whereby we create a structured graphical representation of the parsed PDF data and treat it as a joint node classification problem and graph segmentation problem.
2. Develop a novel Graph-based Layout Analysis Model (GLAM) for this problem formulation and evaluate it on two challenging DLA tasks: PubLayNet and DocLayNet.
3. Demonstrate that by ensembling our model with an object detection model, we are able to set a new state-of-the-art (SOTA) on DocLayNet, increasing the mAP on this task from 76.8 to 80.8.
4. Validate that GLAM is compact in size and fast at speed with a comparison against SOTA models, making it efficient and applicable in industrial use cases.

2   Related Work

2.1   Document Layout Analysis

As a field, document layout analysis (DLA) is mainly aimed toward enterprise use cases [3]. Businesses often need to process massive streams of documents to complete complex downstream tasks, such as information retrieval, table extraction, key-value extraction, etc. Early literature in this space focused on parsing-based solutions that would attempt to understand the document layout from bounding boxes and textual data parsed from born-digital files, leveraging features from the native file type [6].

2.2   Object Detection-Based Methods in VRDU

Recent work has focused on Visually Rich Document Understanding (VRDU), which models documents as images. Image-based object detection formulations of this problem have been particularly favorable and popular because of their generality; effectively any type of human-readable document can be converted into an image. Further, using a traditional, simple convolutional neural network

(CNN) for feature extraction works well in generating representations of documents. Early iterations of *LayoutLM* and others [11,31] all adopt this method.

Optionally, textual information can be captured by an optical character recognition (OCR) engine, avoiding the need to integrate text parsing engines for multiple document media types [8]. This OCR step, whose accuracy is critical for any model that uses textual information [20], could be fine-tuned on specific formats but will also introduce an upper bound of OCR accuracy.

Recently, there has been an emphasis on using large pre-trained document understanding models that leverage vast, unlabeled corpora, such as the IIT-CDIP test collection [19]. The Document Image Transformer (DiT) [21] model adopts a BERT-like pre-training step to train a visual backbone, which is further adapted to downstream tasks, including DLA. There have been additional efforts in creating pre-trained multi-modal models that leverage both the underlying document data and the visual representation. The LayoutLM series of models [16,30] have been particularly successful. LayoutLMv3 [15] is pre-trained with Masked Language Modeling (MLM), Masked Image Modeling (MIM), and Word-Patch Alignment (WPA) training objectives. The LayoutLMv3 is integrated as a feature backbone in a Cascade R-CNN detector [5] with FPN [23] for downstream DLA tasks.

2.3   Graph Neural Networks in Document Understanding

Graphs and graphical representations have been used extensively in document understanding (DU) [4], especially toward the problem of table understanding. [17] extracted saliency maps using a CNN and then used a Conditional Random Field (CRF) model to reason about the relationships of the identified salient regions for table and chart detection. Holecek et al. [14] used the parsed data from a PDF to create a graphical representation of a document and then a GNN for table detection and extraction. Qasim et al. [26] uses a computer-vision feature augmented graph and a GNN for segmentation in order to perform table structure recognition.

Graphs have been used in DLA as well. Zhang et al. [33] use data from a PDF parser and a module based on graph neural networks to model the relationship of potential candidate layout outputs as part of their larger, multi-modal fusion architecture. A key component of any graph-based method is the formulation of the graphs. While the edges have diverse interpretations, the nodes will typically represent some logically contiguous region of the document. In image-based methods, this can be obtained by computer vision, as in [22], or by a more modular OCR task as leveraged by [24]. In contrast, Wei et al. [29] and this work parse the ground truth text boxes if focusing on digital-generated documents.

Typically, two nodes are connected by an edge if they satisfy certain spatial criteria (e.g., based on their vertical or horizontal locations). Riba et al. [27] notes that the resulting graph structures are similar to visibility graphs, which represent nodes as geometric blocks of space that are connected only if they have some vertical or horizontal overlap with no other nodes in between.

Most similar to our work are [10,22]. Li et al. [22] also frames DLA as a graph segmentation and node classification problem but uses computer vision in order to build its "primitive" nodes rather than using PDF parsing. Gemelli et al. [10] uses a GNN on a representation of the document similar to this work. However, they focus only on node classification for table extraction. Our model is able to perform the entire DLA task in an end-to-end fashion by modeling both node-level classifications and graph segmentation.

3   Methodology

3.1   Graph Generation

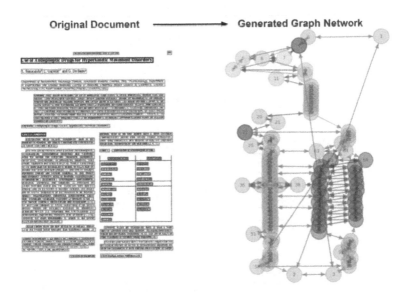

Fig. 2. Example document graph from PubLayNet with associated node and edge labels. Left: text boxes parsed from the PDF, whereby categories are labeled in different colors (blue for *title*, orange for *text*, and magenta for *table*). Right: corresponding graph generated. Each node represents a text box, and the arrows between nodes represent the edges. Illustratively, green edges indicate a link (positive), and grey edges represent there is no link (negative). (Color figure online)

We use a PDF parser to extract all text boxes, each of which also includes 79 associated features, such as bounding box location, text length, the fraction of numerical characters, font type, font size, and other simple heuristics. Using this parsed metadata, we construct a vector representation for each individual box, which constitutes a node in our graph.

Bidirectional edges are established between nodes by finding each node's nearest neighbor in each possible direction (up, down, left, and right). Additional edges are added to approximate the potential reading order of the underlying text boxes in the document (i.e. the heuristic order parsed from PDF documents – up to down, left to right). Each edge also has its own vector representation, which includes the edge direction and the distance between the two nodes that it connects. A visualization of the resulting graph can be seen in Fig. 2.

3.2   Graph-Based Layout Analysis Model

There are two key parts to the Graph-based Layout Analysis Model (GLAM): a graph network and a computer vision feature extractor.

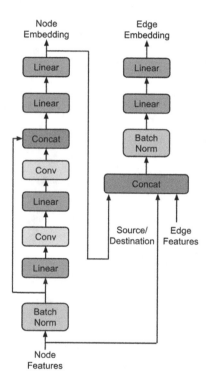

Fig. 3. Visualization of the GNN architecture used in GLAM.

Graph Network. Our graph network model consists of a graph convolutional network (GCN) [18] and two classification heads – one for node classification and one for edge classification. The model structure is shown in Fig. 3. We apply batch normalization to the node features before inputting them into a series of linear and topology adaptive graph (TAG) convolutional [9] layers. We set the TAG's first hidden layer to have 512 dimensions when training on PubLayNet, and we increase this to 1024 dimensions for the more complex DocLayNet dataset. As the GCN network gets deeper, we scale down the number of parameters in both linear and TAG convolutional layers by a factor of 2. We concatenate this output with the original normalized features and then apply a series of linear layers to create our node embeddings. A classification head then projects this into a distribution over the possible segment classes (e.g., title, figure, text). This model is very lightweight, totaling just over one million parameters.

To create the edge embeddings, we extract the relevant node embeddings and concatenate them with the original edge features. We batch-normalize this representation and then apply another two linear layers to generate the final edge embeddings.

The model is trained with a weighted, joint loss function Eq. (1) from the two classification tasks.

$$L = L_{node} + \alpha L_{edge} \tag{1}$$

where L_{node} and L_{edge} are a cross-entropy loss for node and edge classification, and the edge loss scale $\alpha = 4$ is applied to improve edge classification results for better segmentation.

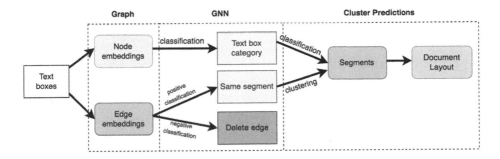

Fig. 4. Visualization of the model pipeline.

At inference time, segment clustering follows the steps shown in Fig. 4. For a given input graph, the model will classify each node into one of the given segment categories and predict whether each edge is positive or negative. The negative edges are removed from the graph. The segments are then defined by the remaining connected components in the graph. Each segment is classified by the majority vote of the classification of all of its component nodes. This is finally converted back into a COCO-style annotation by using the minimum spanning box of the nodes within the segment as the bounding box, the selected class label as the label, and the mean probability of that label over all of the nodes as the score.

Computer Vision Feature Extraction. One of the main limitations of PDF-parser-based data pre-processing is that it is unable to capture all of the visual information encoded in the PDF. All of the text boxes are only represented by their metadata, which does not account for many of the semantic visual cues within a layout. These cues include background colors, lines, dots, and other non-textual document elements. To include this visual information, we extend our node-level features with a set of visual features. We render the PDF as an image and extract a visual feature map using ResNet-18 [12]. Specifically, to extract the image features corresponding to each input node's bounding box, we use Region of Interest (ROI) extraction with average pooling [13].

We concatenate these image features with our metadata-based feature vectors, then apply a simple three-layer attention-based encoder model [28] to these features to give them global context before concatenating them with the features associated with the nodes in our graph. This maintains the overall graphical structure while adding the ability to leverage additional visual cues. The network is also relatively small, totaling only slightly over three million parameters. This feature extraction process does not create any additional nodes. Thus, while the model may be able to include some information from image-rich regions of the document—such as figures and pictures—the model is still limited in its ability to account for them.

4   Experiments

4.1   Datasets

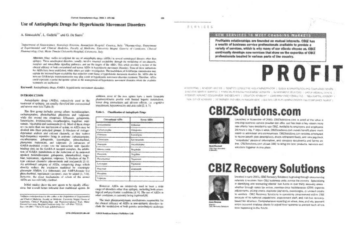

Fig. 5. Example document pages from PubLayNet (left) and DocLayNet (right) datasets

PubLayNet [34] is a dataset consisting of 358,353 biomedical research articles in PDF format (approximately 335k for training, 11k for development, and 11k for testing). Each document is annotated with bounding boxes and labels for every distinct semantic segment of content (i.e., text, title, list, figure, and table). These bounding boxes are automatically generated by matching the PDF format and the XML format of the articles. In practice, this leads to bounding boxes that are often not perfectly aligned with the underlying document. [22] finds that this process severely degrades the performance of any method that generates boxes using data directly from the underlying document. For an equal comparison across baselines, we maintain these boxes without adjustment and report our raw mAP (mean average precision) score (Fig. 5).

In this work, we primarily focused on using the DocLayNet [25] dataset, due to it having a diverse set of unique layouts. While PubLayNet only contains consistently formatted journal articles, DocLayNet spans six distinct genres (i.e., financial reports, manuals, scientific articles, laws & regulations, patents, and government tenders). It provides layout annotation bounding boxes for 11 different categories (caption, footnote, formula, list-item, page footer, page header, picture, section header, table, text, and title) on 80,863 unique document pages. The documents are divided into a training set (69,375 documents), a development set (6,489 documents), and a testing set (4,999 documents).

Unlike PubLayNet, DocLayNet is a human-annotated dataset; the ground truth bounding boxes are generated by human annotators hand-drawing boxes on top of the documents. These human-drawn boxes then are adjusted to match the underlying extracted text boxes, removing any additional whitespace or annotator errors in the process. This ensures a perfect correlation between the bounding boxes defined by the extracted data and the ground truth labels. For classes that are not well aligned with the underlying text boxes (e.g., picture and table), the boxes are left unmodified. The current state-of-the-art DLA model for this dataset is an off-the-shelf YOLO v5x6 [2] model.

4.2 Data Pre-processing

In conventional VRDU-based methods, PDFs are first converted into images before inference. However, many ground truth features—not only the raw text but text and artifact formatting—can be lost in the image conversion step if the original file information is discarded. Thus, we parse text boxes directly from the PDF documents, which allows us to acquire precise text content, position, and font information.

When using the DocLayNet dataset, we directly use the provided parsed text boxes, which include the bounding box location, the text within the box, and information about its font. These boxes are only provided for textual content, as opposed to figures, tables, etc.

When using the PubLayNet dataset, text boxes are parsed directly from original PDF pages using an in-house PDF parser based on *pdfminer.six* [1]. To better match DocLayNet, we ignore image-based objects in the document entirely. We apply a series of data cleaning steps (e.g. ignore random symbols in some documents by removing very small bounding boxes smaller than 10 pixels in width and height along with extra white space) on the parsed boxes. To reduce the number of unnecessary edge classifications between adjacent nodes, we use simple distance-based heuristics to merge adjacent boxes. This simplifies the graph structure and reduces the overall memory requirements during training and inference.

Both DocLayNet and PubLayNet contain ground truth data in the COCO format. In order to convert this format to labels on our graph, we take the bounding box for each annotation and find all of the text boxes in the document that have some overlap. We compute the minimum spanning box of this set of text boxes and compare it with the original ground truth. For DocLayNet, if the IoU of the ground truth box and the box formed by all the overlapping text boxes is less than 0.95, we remove text boxes in order of their total area outside the annotation box and select the configuration that achieves the highest IoU.

Once we have associated every annotation with its set of text boxes and their corresponding nodes on the graph, we label each node on the graph with the class of the annotation. Every edge that connects two nodes within the same annotation is labeled a positive edge, while every edge connecting nodes from different annotations is labeled as a negative edge. An example of the labeled graph is presented in Fig. 2.

Table 1. Model performance mAP@IoU[0.5:0.95] on the DocLayNet test set, as compared to metrics reported in [25].

	MRCNN R50	MRCNN R101	FRCNN R101	YOLO v5x6	GLAM	GLAM + YOLO v5x6
caption	68.4	71.5	70.1	**77.7**	74.3	77.7
footnote	70.9	71.8	73.7	**77.2**	72.0	77.2
formula	60.1	63.4	63.5	66.2	**66.6**	66.6
list item	81.2	80.8	81.0	**86.2**	76.7	86.2
page footer	61.6	59.3	58.9	61.1	**86.2**	86.2
page header	71.9	70.0	72.0	67.9	**78.0**	78.0
picture	71.7	72.7	72.0	**77.1**	5.7	77.1
section header	67.6	69.3	68.4	74.6	**79.8**	79.8
table	82.2	82.9	82.2	**86.3**	56.3	86.3
text	84.6	85.8	85.4	**88.1**	74.3	88.1
title	76.6	80.4	79.9	82.7	**85.1**	85.1
overall	72.4	73.5	73.4	**76.8**	68.6	**80.8**

Table 2. Model performance mAP@IoU[0.5:0.95] on the PubLayNet val set.

	LayoutLM v3 [15]	MBC [22]	VSR [33]	YOLO v5x6	GLAM
text	94.5	88.8	**96.7**	91.7	87.8
title	90.6	52.8	**93.1**	79.1	80.0
list item	**95.5**	88.1	94.7	86.9	86.2
table	**97.9**	97.7	97.4	96.5	86.8
figure	**97.0**	84.0	96.4	94.7	20.6
overall	**95.1**	82.3	**95.7**	89.8	72.2

4.3 Results

DocLayNet. As shown in Table 1, GLAM achieves competitive performance while being over 10× smaller than the smallest comparable model—and over 35× smaller than the current state-of-the-art. Our model outperforms all of the object detection models on five of the 11 total classes, yielding an overall mAP of 68.6. The SOTA model achieves a 76.8.

Nine of the 11 classes have annotations that involve text box snapping; the tables and pictures classes do not. When considering only these nine classes, our model achieves an mAP score of 77.0, compared to 75.7 for the next-best model. As expected, our model does not perform as well for tables, and it is particularly low-performing for pictures. When only considering the parsed text of tables (and thus disregarding visual markers such as table borders), our model yields a low mAP for the table class. In fact, GLAM performs well in extracting the compact regions of text from table contents (Fig. 6). If we consider all 10 text-based classes (including tables but excluding pictures), GLAM achieves an mAP score of 74.9, compared to 76.8 from the SOTA model. By creating a simple ensemble with GLAM and YOLO v5x6, we are able to increase the state of the art from 76.8 to 80.8.

PubLayNet. The results on PubLayNet are shown in Table 2. On this dataset, our model is significantly handicapped by the misalignment between the text boxes and the ground truth boxes as shown in Fig. 7. Since it cannot improve the alignment of the original text boxes, this enacts a hard upper bound on performance. The model can only achieve an IoU as high as the overlap between the ground-truth label and the underlying PDF data. Although our method may correctly identify and label all of the underlying nodes, it may still not achieve an IoU high enough to register as a correct classification at most of the thresholds.

We can get a more accurate picture of performance by looking at the lower IoU thresholds. At a threshold of 0.5, GLAM achieves an mAP higher than 90 for all of the text classes. For the "title" class, the gap between mAP at this threshold and the final value averaged over the full threshold range is over 17%. As a comparison, this gap on DocLayNet is less than 7% for every class. Even

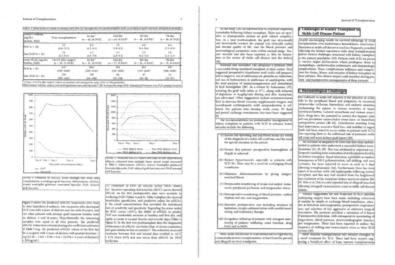

Fig. 6. Examples of GLAM predictions on PubLayNet documents. Bounding boxes in orange, blue, green, magenta, and red colors represent predicted segments in text, title, list, table, and figure categories, respectively. Due to the inability to parse images from PDF, the GLAM model underperforms in figure categories. (Color figure online)

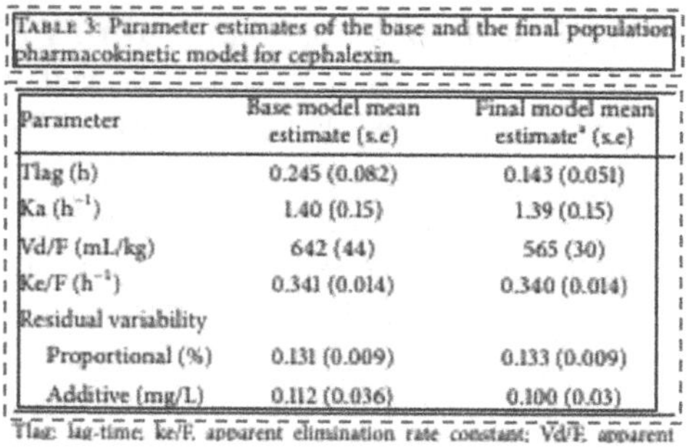

Parameter	Base model mean estimate (s.e)	Final model mean estimate* (s.e)
TABLE 3: Parameter estimates of the base and the final population pharmacokinetic model for cephalexin.		
Tlag (h)	0.245 (0.082)	0.143 (0.051)
Ka (h^{-1})	1.40 (0.15)	1.39 (0.15)
Vd/F (mL/kg)	642 (44)	565 (30)
Ke/F (h^{-1})	0.341 (0.014)	0.340 (0.014)
Residual variability		
Proportional (%)	0.131 (0.009)	0.133 (0.009)
Additive (mg/L)	0.112 (0.036)	0.100 (0.03)

Tlag: lag-time; Ke/F: apparent elimination rate constant; Vd/F: apparent

Fig. 7. Examples of bounding box misalignment between GLAM prediction and annotation in PubLayNet. Red bounding boxes are from annotation, and blue bounding boxes are from GLAM prediction. GLAM predictions have tight bounding boxes around text-based segments. (Color figure online)

with this misalignment, the GLAM model still is able to achieve 88% of the performance of the SOTA model on the text-based categories.

Analysis. In general, our model performs strongest on small, text-box-based classes, such as titles and headers. We hypothesize that this is mostly due to its ability to perfectly match its bounding box to the ground truth label. Since GLAM has access to the exact text boxes in question, it can output pixel-perfect

Table 3. CV Feature extraction Ablation, mAP@IoU[0.5:0.95] on DocLayNet test set.

	GLAM	GLAM-CV	Δ
caption	74.3	60.6	13.7
footnote	72.0	61.1	10.9
formula	66.6	43.8	22.8
list item	76.7	61.5	15.2
page footer	86.2	87.3	−1.1
page header	78.0	78.1	−0.1
picture	5.7	4.3	1.4
section header	79.8	70.1	9.7
table	56.3	50.2	6.1
text	74.3	68.4	5.9
title	85.1	72.9	12.2
overall	68.6	59.9	8.79

Table 4. CV Feature extraction Ablation, mAP@IoU[0.5:0.95] on PubLayNet val set.

	GLAM	GLAM - CV	Δ
text	87.8	85.9	1.9
title	80.0	79.8	0.2
list item	86.2	84.8	1.4
table	86.8	86.1	0.7
figure	20.6	18.3	2.3
overall	72.2	71.0	1.2

annotations by simply segmenting the relevant boxes. Object detection models on the other hand are known to struggle with smaller objects [7].

Conversely, our model performs worst on classes that are not well captured by the underlying text boxes in the document. This is a fundamental limitation of our structural representation of the document. This discussion is continued in Sect. 5.

4.4   CV Feature Ablation

We include an ablation study of the computer vision feature extractor in Tables 3 and 4. For these models, we only train the graph network component of GLAM, without the CV features. The GNN component represents only one-fourth of the total parameters of the GLAM. Removing the vision feature extraction leads to a consistent decrease in performance across both tasks, albeit only slightly on PubLayNet. This is likely the result of two factors: (1) our model nearing

Table 5. Comparison of model size and efficiency. All numbers are bench-marked on a G4 AWS instance, with NVIDIA Tesla T4 GPU.

	Model Size (M)	Inference Time (ms)		Pages/sec	
		GPU	CPU	GPU	CPU
LayoutLMv3	133	687	6100	1.5	0.2
YOLO v5	140.7	56	179	17.8	5.6
GLAM	4	10	16	98.0	63.7
GLAM - CV	1	4	9	243.9	109.9

the performance limit imposed by the misaligned boxes in PubLayNet; and (2) the visual complexity of the DocLayNet data. DocLayNet layouts contain many visual cues that are not captured by the underlying text boxes (e.g., bullet points, lines, and colored backgrounds). Results on the PubLayNet dataset (Table 4) indicate that the GNN component in GLAM already performs well on the visual-simple type of documents like research articles.

4.5 Model Size and Efficiency

Although large, pre-trained models often lead to high performance when adapted to downstream tasks, they tend to be relatively large and slow in training and inference. Both the state-of-the-art object detection models on PubLayNet and DocLayNet have over 130 million parameters. To highlight the efficiency of our graph-based model, in Table 5 we include additional benchmarks regarding inference speed and throughput. We baseline our model against the two highest-performing, open-source models on each task.

In general, GLAM is much more efficient than the baseline models. The full GLAM model, which achieves competitive performance to the baseline models, is able to process a document page in 10 ms on a single GPU[2], compared to 56 ms and 687 ms for YOLO v5x6 and LayoutLMv3, respectively. Removing the CV feature extraction drops the inference time to 4 ms, allowing the model to process over 243 document pages per second on a single GPU. Such efficient DLA models significantly reduce the engineering complexity in real-world use cases and enable the possibility of on-device model deployments.

5   Limitations

The main limitation of GLAM is that it is highly dependent on the parsing quality of the document. The benefit of using off-the-shelf object detection models is that their performance is much less sensitive to the structure of the underlying document. That is, they can account for the misaligned ground-truth boxes, such as the ones in PubLayNet. Importantly, they can also properly work with

---

[2] We use an NVIDIA Tesla T4 as GPU in our experiments.

scanned documents and other documents with little or no structure – alleviating the need to address pre-processing decisions. For the same reason, GLAM is unable to capture many regions of vision-rich segments (e.g., figures and pictures) in PDF documents. GLAM also does not take into account any of the available document semantic information. To address this limitation, we could use an additional text feature extractor (text embedding) for the text boxes.

6 Conclusion

In this work, we introduce GLAM, a lightweight graph-network-based method for performing document layout analysis on PDF documents. We build upon a novel, structured representation of the document created from directly parsing the PDF data. We frame the document layout analysis task as performing node and edge classification on document graphs. With these insights, GLAM performs comparably to competitive baseline models while being orders of magnitude smaller. The model works particularly well for text-based layout elements such as headers and titles. By ensembling this model with an off-the-shelf object detection model, we are able to set state-of-the-art performance in DocLayNet. We demonstrate that GLAM is a highly-efficient and effective approach to extracting and classifying text from PDF documents, which makes it competitive, especially in real-world applications with limited computational resources.

References

1. pdfminer.six (2022). http://github.com/pdfminer/pdfminer.six
2. YOLOv5 SOTA realtime instance segmentation (2022). http://github.com/ultralytics/yolov5
3. Binmakhashen, G.M., Mahmoud, S.A.: Document layout analysis: a comprehensive survey. ACM Comput. Surv. (CSUR) **52**(6), 1–36 (2019)
4. Bunke, H., Riesen, K.: Recent advances in graph-based pattern recognition with applications in document analysis. Pattern Recogn. **44**(5), 1057–1067 (2011). ISSN 0031-3203. https://doi.org/10.1016/j.patcog.2010.11.015.
5. Cai, Z., Vasconcelos, N.: Cascade R-CNN: delving into high quality object detection. In: Proceedings of the IEEE Conference on Computer Vision and Pattern Recognition, pp. 6154–6162 (2018)
6. Chao, H., Fan, J.: Layout and content extraction for PDF documents. In: Marinai, S., Dengel, A.R. (eds.) DAS 2004. LNCS, vol. 3163, pp. 213–224. Springer, Heidelberg (2004). https://doi.org/10.1007/978-3-540-28640-0_20
7. Chen, G., et al.: A survey of the four pillars for small object detection: multiscale representation, contextual information, super-resolution, and region proposal. IEEE Trans. Syst. Man Cybern. Syst., 1–18 (2020). https://doi.org/10.1109/TSMC.2020.3005231
8. Cui, L., Xu, Y., Lv, T., Wei, F.: Document AI: benchmarks, models and applications. arXiv preprint arXiv:2111.08609 (2021)
9. Du, J., Zhang, S., Wu, G., Moura, J.M.F., Kar, S.: Topology adaptive graph convolutional networks. arXiv preprint arXiv:1710.10370 (2017)

10. Gemelli, A., Vivoli, E., Marinai, S.: Graph neural networks and representation embedding for table extraction in pdf documents. In: 2022 26th International Conference on Pattern Recognition (ICPR), pp. 1719–1726 (2022). https://doi.org/10.1109/ICPR56361.2022.9956590

11. Gu, J., et al.: Unified pretraining framework for document understanding. arXiv e-prints, pages arXiv-2204 (2022)

12. He, K., Zhang, X., Ren, S., Sun, J.: Deep residual learning for image recognition. In: 2016 IEEE Conference on Computer Vision and Pattern Recognition (CVPR), pp. 770–778 (2016). https://doi.org/10.1109/CVPR.2016.90

13. He, K., Gkioxari, G., Dollár, P., Girshick, R.: Mask R-CNN. In: 2017 IEEE International Conference on Computer Vision (ICCV), pp. 2980–2988 (2017). https://doi.org/10.1109/ICCV.2017.322

14. Holecek, M., Hoskovec, A., Baudiš, P., Klinger, P.: Table understanding in structured documents. In: 2019 International Conference on Document Analysis and Recognition Workshops (ICDARW), vol. 5, pp. 158–164 (2019). https://doi.org/10.1109/ICDARW.2019.40098

15. Huang, Y., Lv, T., Cui, L., Lu, Y., Wei, F.: LayoutLMv3: pre-training for document AI with unified text and image masking. arXiv preprint arXiv:2204.08387 (2022)

16. Huang, Y., Lv, T., Cui, L., Lu, Y., Wei, F.: LayoutLMv3: pre-training for document AI with unified text and image masking. In: ACM Multimedia 2022, October 2022. www.microsoft.com/en-us/research/publication/layoutlmv3-pre-training-for-document-ai-with-unified-text-and-image-masking/

17. Kavasidis, I., et al.: A saliency-based convolutional neural network for table and chart detection in digitized documents. In: Ricci, E., et al. (eds.) Image Analysis and Processing - ICIAP 2019, pp. 292–302. Springer, Cham (2019). https://doi.org/10.1007/978-3-030-30645-8_27. ISBN 978-3-030-30645-8

18. Kipf, T.N., Welling, M.: Semi-supervised classification with graph convolutional networks. In: 5th International Conference on Learning Representations, ICLR 2017, Toulon, France, 24–26 April 2017, Conference Track Proceedings. OpenReview.net (2017). http://openreview.net/forum?id=SJU4ayYgl

19. Lewis, D., Agam, G., Argamon, S., Frieder, O., Grossman, D., Heard, J.: Building a test collection for complex document information processing. In: Proceedings of the 29th Annual International ACM SIGIR Conference on Research and Development in Information Retrieval, SIGIR 2006, pp. 665–666. Association for Computing Machinery, New York (2006). ISBN 1595933697. https://doi.org/10.1145/1148170.1148307

20. Li, C., et al.: StructuralLM: structural pre-training for form understanding. In: Proceedings of the 59th Annual Meeting of the Association for Computational Linguistics and the 11th International Joint Conference on Natural Language Processing (Volume 1: Long Papers), pp. 6309–6318, August 2021. https://doi.org/10.18653/v1/2021.acl-long.493

21. Li, J., Xu, Y., Lv, T., Cui, L., Zhang, C., Wei, F.: DIT: self-supervised pre-training for document image transformer. arXiv preprint arXiv:2203.02378 (2022)

22. Li, X.-Hu., Yin, F., Liu, C.-L.: Page segmentation using convolutional neural network and graphical model. In: Bai, X., Karatzas, D., Lopresti, D. (eds.) Document Analysis Systems, pp. 231–245. Springer, Cham (2020). ISBN 978-3-030-57058-3. https://doi.org/10.1007/978-3-030-57058-3_17

23. Lin, T.-Y., Dollár, P., Girshick, R., He, K., Hariharan, B., Belongie, S.: Feature pyramid networks for object detection. In: Proceedings of the IEEE Conference on Computer Vision and Pattern Recognition, pp. 2117–2125 (2017)

24. Liu, X., Gao, F., Zhang, Q., Zhao, H.: Graph convolution for multimodal information extraction from visually rich documents. In: NAACL (2019)
25. Pfitzmann, B., Auer, C., Dolfi, M., Nassar, A.S., Staar, P.W.J.: DocLayNet: a large human-annotated dataset for document-layout analysis. arXiv preprint arXiv:2206.01062 (2022)
26. Qasim, S., Mahmood, H., Shafait, F.: Rethinking table recognition using graph neural networks. In: 2019 International Conference on Document Analysis and Recognition (ICDAR), pp. 142–147. IEEE Computer Society, Los Alamitos, CA, USA, September 2019. https://doi.org/10.1109/ICDAR.2019.00031
27. Riba, P., Dutta, A., Goldmann, L., Fornés, A., Ramos, O., Lladós, J.: Table detection in invoice documents by graph neural networks. In: 2019 International Conference on Document Analysis and Recognition (ICDAR), pp. 122–127 (2019). https://doi.org/10.1109/ICDAR.2019.00028
28. Vaswani, A., et al.: Attention is all you need. In: Guyon, I., et al. (eds.) Advances in Neural Information Processing Systems, vol. 30. Curran Associates Inc. (2017). http://proceedings.neurips.cc/paper/2017/file/3f5ee243547dee91fbd053c1c4a845aa-Paper.pdf
29. Wei, M., He, Y., Zhang, Q.: Robust layout-aware IE for visually rich documents with pre-trained language models (2020)
30. Xu, Y., et al.: LayoutLMv2: multi-modal pre-training for visually-rich document understanding. In: Proceedings of the 59th Annual Meeting of the Association for Computational Linguistics and the 11th International Joint Conference on Natural Language Processing (Volume 1: Long Papers), pp. 2579–2591, August 2021. Association for Computational Linguistics. https://doi.org/10.18653/v1/2021.acl-long.201
31. Xu, Y., Li, M., Cui, L., Huang, S., Wei, F., Zhou, M.: Layoutlm: Pre-training of text and layout for document image understanding. In: Proceedings of the 26th ACM SIGKDD International Conference on Knowledge Discovery & Data Mining, pp. 1192–1200 (2020)
32. Yang, H., et al.: Pipelines for procedural information extraction from scientific literature: towards recipes using machine learning and data science. In: 2019 International Conference on Document Analysis and Recognition Workshops (ICDARW), vol. 2, pp. 41–46 (2019). https://doi.org/10.1109/ICDARW.2019.10037
33. Zhang, P., et al.: VSR: a unified framework for document layout analysis combining vision, semantics and relations. In: Lladós, J., Lopresti, D., Uchida, S. (eds.) Document Analysis and Recognition - ICDAR 2021, pp. 115–130. Springer, Cham (2021). https://doi.org/10.1007/978-3-030-86549-8_8. ISBN 978-3-030-86549-8
34. Zhong, X., Tang, J., Yepes, A.J.: PublayNet: largest dataset ever for document layout analysis. In: 2019 International Conference on Document Analysis and Recognition (ICDAR), pp. 1015–1022. IEEE (2019)

Gaussian Kernels Based Network for Multiple License Plate Number Detection in Day-Night Images

Soumi Das[1], Palaiahnakote Shivakumara[2(✉)], Umapada Pal[3],
and Raghavendra Ramachandra[4]

[1] Technology Innovation Hub (TIH), Indian Statistical Institute, Kolkata, India
[2] Faculty of Computer Science and Information Technology, University of Malaya, Kula Lumpur, Malaysia
shiva@um.edu.my
[3] Computer Vison and Pattern Recognition Unit, Indian Statistical Institute, Kolkata, India
umapada@isical.ac.in
[4] Norwegain University of Science and Technology, Trondheim, Norway
raghavendra.ramachandra@ntnu.no

Abstract. Detecting multiple license plate numbers is crucial for vehicle tracking and re-identification. The reliable detection of multiple license plate numbers requires addressing the challenges like image defocusing and varying environmental conditions like illumination, sunlight, shadows, weather conditions etc. This paper aims to develop a new approach for multiple license plate number detection of different vehicles in day and night scenarios. The proposed work segments the vehicle region containing license plate numbers based on a multi-column convolutional neural network and iterative clustering to reduce the background challenges and the presence of multiple vehicles. To address challenges of font contrast variations and text-like objects in the background, the proposed work introduces the Gaussian kernels that represent a text pixel distribution to integrate with a proposed deep learning model for detection, Experimental results on benchmark datasets of day and night license plate number show that the proposed model is effective and outperforms the existing methods.

Keywords: Text detection · Vehicle detection · Text segmentation · Deep learning · Gaussian kernels · Multiple license plate number detection

1 Introduction

Automatic license plate detection and recognition have received lots of attention because of real-time applications such as automatic toll fee collection day and night, traffic monitoring day and night, identifying illegal car parking day and night and vehicle re-identification day and night [1, 2]. In the applications mentioned above, except for toll fee collection, the camera captures multiple vehicles in a single frame. Therefore, the critical challenge of images captured with multiple vehicles in day and night conditions include

G. A. Fink et al. (Eds.): ICDAR 2023, LNCS 14191, pp. 70–87, 2023.
https://doi.org/10.1007/978-3-031-41734-4_5

defocusing, use of headlights and street light effects, loss of visibility due to spatially scattered vehicles, and tiny number plates due to distance variations from the camera [1, 2]. Several existing methods have indicated robustness to noise, low contrast, poor quality, small font size, variations of vehicle speed, and uneven illumination developed in the past for addressing several challenges of license plate number detection [3]. Most methods focus on images containing a single vehicle like a car, truck, etc. but not images with multiple vehicles. However, a few approaches focus on multiple vehicles in a single image, but the scope of the existing techniques is limited to day images [2, 3].

In the same way, a few methods consider images captured during day and night for license plate number detection, but the scope is limited to images with a single vehicle [1]. To better illustrate the challenges of license plate detection from multiple vehicles captured in day and night conditions, we have presented a qualitative result in Fig. 1(a) and (b) comparing the proposed approach with the existing approach (based on Yolo and FasterRCNN) [4]. It is worth noting that the performance of the existing method is [4] degraded in detecting the license plate numbers. In contrast, the proposed method indicates reliable performance in detecting multiple license plates. The degraded performance of the existing techniques can be attributed to the need for more robustness to handle image degradations due to changes in the day and night conditions with multiple vehicles in dense traffic scenarios. Similar to license plate number detection, there exist several methods for scene text detection in the literature [5–7]. Although scene text and license plate number detection appear the same, the methods may not work well for license plate number detection because the license plate number does not provide semantic information while scene text provides semantic information. Therefore, scene text detection methods are not suitable for license plate number detection. Hence, it can be concluded that the existing models are not suitable for detecting multiple license plate numbers captured day and night.

In this work, we aim to develop a new model that can work well irrespective of the number of vehicles, type of vehicles, and day-night. Motivated by the work [3], where to improve license plate detection performance for multiple vehicles, the model segments vehicles in the image, we explore the same idea by proposing a new Multi-Column Convolutional Neural Network (MCNN) in this work. This step helps us reduce the background complexity and the effect of objects like license plate numbers by segmenting the region of interest. For the segmented region of interest, based on the observation that the distribution of text pixels represents Gaussian distribution, we define different-sized Gaussian kernels for tackling the challenges of font, contrast, and small font variations and the proposed model integrates different Gaussian kernels with deep learning models for successful detection. To achieve this, we propose a Dilated Convolution Neural Network (DCNN) known as CSRNet. Overall, modifications to MCNN and CSRNet, such as deciding our own novel kernels to MCNN and Gaussian kernels to CSRNet, are the main contributions compared to the existing methods.

Thus, the following are the key contributions. (i) This is the first work for detecting multiple license plate numbers in day and night images. (ii) Exploring a multi-input neural network for segmenting vehicle regions containing license plate numbers is new compared to the state-of-the-art methods. (iii) Use Gaussian kernels to detect license plate numbers from the segmented region of interest.

The rest of the work is organized as follows. A review of different methods for license plate detection is presented in Sect. 2. In Sect. 3, the model for the region of interest and detection is illustrated. Section 4 presents experimental analysis and validation for evaluating the proposed and existing methods. Section 5 summarizes the contribution of the proposed work.

(a) Existing model [4] for multiple vehicles of night images

(b) Existing model [4] for multiple vehicles of day images

(c) Proposed model for multiple vehicles of night images

(d) Proposed model for multiple vehicles of day images.

Fig. 1. Illustrating complexity of multiple license plate number of detection in night and day images

2 Related Work

A wide variety of techniques have been developed for license plate number detection in the literature. Recently, methods based on deep learning models are popular and can address complex issues. Therefore, we review the recent deep-learning approaches devised for license plate number detection. Lin et al. [8] explored CNN for developing an efficient license plate number recognition system. The main weakness of the method is that it follows the conventional approach to solving the problem, which needs to be more robust to handle complex scenes. Kumari et al. [9] proposed a traditional idea for license plate number detection, which follows feature extraction, binarization, and then traditional classifiers, such as KNN and simple CNN. Srilekha et al. [10] developed the model by combining the features and CNN model for license plate number detection. The primary goal of this work is to detect helmet and non-helmet riders but not license plate number detection. If the method detects helmet riders and non-helmet riders using an optical character recognizer. Since the performance of the above methods depends on binarization, the method cannot be robust to handle defocused images and complex background images. Charan et al. [11] used Yolo-v4 and tesseract OCR for license plate number detection. The aim of the work is to detect vehicle traffic violations rather than license plate detection. However, the tesseract OCR does not work well for degraded text and complex background images. Xu et al. [12] aim to develop a method for handling large datasets because most methods use small datasets to achieve the results; hence, the techniques need more scalability. To address this challenge, the authors proposed an end-to-end deep neural network.

Mokayed et al. [13] pointed out that the methods developed in the past focus only on images captured by standard cameras but not images captured by drone cameras. Therefore, the approach explores the combination of discrete cosine transform and phase congruency model for license plate detection in drone images. Usama et al. [4] discussed different models for license plate detection using Yolo architecture and fasterRCNN. Since the main application of this work is toll collection, the model focuses on a particular dataset and type. Tote et al. [14] used a tensor flow developed by Google for license plate number detection. The method follows a conventional approach for detection, such as preprocessing and number plate localization. However, since the main target of the work is the Indian license plate number, the model is confined to a particular type of vehicle. Gong et al. [15] noted from the past methods that the recent deep learning-based methods are suitable for images with English license plate numbers but not Chinese license plate numbers. Therefore, the approach explores ResNet, pyramid network, and LSTM to achieve the best results for Chinese license plate number detection. This technique is mainly used to detect the type of license plate number and the reported results indicate the limited generalizability of the technique. Gizatullin et al. [16] focused on car license plate detection using an image-weight model. This approach explores multiple scale wave-let transforms and the gray image is considered as a weight image to improve the performance. However, the model works well for a particular type of vehicle but not different vehicles.

It is noted from the above literature study that most methods consider a day image for license plate detection but not night images. Zhang et al. [17] used a fusion method to address the challenge of night images to enhance low-contrast details in the input images.

The approach fuses the output of spatial and frequency do-main-based models to exploit the spatial and frequency information. However, the method works well for night images rather than day-night images. To overcome this problem, Chowdhury et al. [18] used the combination of U-Net and CNN to enhance low-contrast pixels. However, the method's performance could be better because it is not robust to noise and degradation. Chowdhury et al. [1] developed a model for license plate detection in day and night images based on augmentation. The augmentation is applied by expanding the window iteratively to detect dominant pixels irrespective of day and night images. The approach explores the gradient vector flow concept to detect candidate pixels and the recognition approach to improve license plate detection. The combination has been used for enhancing the fine edge detail in the images and then the deep learning model has been used for detection. Chowdhury et al. [2] observed that none of the methods consider images with multiple vehicles for license plate detection. Therefore, the approach explores an adaptive progressive scale expansion-based graph attention network. Kumar et al. [3] also address the problem of multiple license plate detection based on Yolo-v5 architecture and the combination of wavelet transform and phase congruency model. The Yolo-v5 has been used for vehicle detection and the wavelet transform-phase congruency model is used for license plate detection. The above methods [2, 3] are confined only to day images but not night images. Therefore, these techniques may not be effective for images of multiple vehicles captured in day and night scenarios.

In summary, one can infer from the above review that a few methods addressed the challenges of multiple vehicle license plate number detection but ignored day and night images. At the same time, a few techniques focused on the challenges of day and night images but ignored images with multiple vehicles in a single image. Therefore, detecting multiple license plate number in day and night images is still an open challenge and hence this work focus on developing a new approach to address the same effectively.

3 Proposed Model

As discussed in the previous section, detecting multiple license plate numbers in a day and night images is a complex problem. Inspired by the work [3], where vehicle regions are segmented to reduce the complexity of license plate number detection in video, we explored the same idea for segmenting vehicle regions containing license plate numbers as the region of interest. Motivated by the model [19], where saliency has been used for detecting faces in the crowd, we explore the model for obtaining a density map for the multiple vehicles in a single image. For generating density maps, the proposed work uses multi-input convolutional neural networks and fuses feature maps given by each neural network. For the density map, the proposed work employs K-means clustering with $K = 2$ iteratively for classifying each vehicle region as a region of interest. In this work, the cluster that has the maximum mean is considered a region of interest. The motivation for using the K-means clustering is that the density map provides high pixel values for the license plate area and vehicle area while low values for the non-license plate and vehicle area.

This step outputs the vehicle region containing the license plate number of each vehicle in the image by localizing the license plate region. However, the region of interest contains license plate numbers and other background information, which may result in an increased number of false positives. Therefore, to detect license plate numbers by fixing tight bounding boxes without affecting false positives from the segmented region of interest, the proposed work defines Gaussian kernels of different sizes, which represent the pixel distribution of license plate numbers. Integrating Gaussian kernels and deep learning models helps us segment the exact license plate region area from its background. For fixing tight bounding box for the license plate number, we follow the instruction in [2], which works based on polygon boundary fixing. The framework of the proposed method can be seen in Fig. 2.

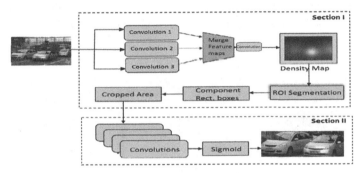

Fig. 2. The proposed framework

3.1 Region of Interest Segmentation

For the input of day and night images, the proposed work uses different-sized kernels to extract features with the help of a Multi-column Convolutional Neural Network (MCNN) as shown in Fig. 3. The reason for using different-sized kernels is to capture the license plate number of different font sizes. When the single image contains multiple vehicles, it is evident that the size of the license plate number varies from one vehicle to another. To cope with this challenge, different-sized kernels are proposed in this work.

The output of each kernel is fused to obtain the final density map for segmenting vehicle region as a region of interest containing license plate numbers. Where Θ is the learning parameter of the model. The number of training set images denoted by n. For every input image $X = \{X_1, X_2, ..., X_n\}, f(X_i, \Theta)$ is the predicted density map and f_i is the ground truth density map. For generating the ground truth density map, each image X_i is annotated with two points (x_{min}, y_{min}) and (x_{max}, y_{max}) which indicate the bounding box of the vehicle region. The center point of each bounding boxes calculated as $c_i = (\frac{(x_{max}+x_{min})}{2}, \frac{(y_{max}+y_{min})}{2})$. If every image contains m number of bounding boxes $C = \{c_1, c_2, ..., c_m\}$, the ground truth density map of image X_i is formulated as

$$\forall_c \in X_i, f_j = \sum_{j=0}^{m} \mathcal{N}^{gt}(c_j; \mu, \sigma^2)$$

(1)

where $\mathcal{N}^{gt}$ is a gaussian kernel with mean μ and standard deviation σ. The aim of our model is to learn to minimize the difference between the predicted density map f (X_i) and the ground truth density map f_i by using the loss function L(Binary Cross Entropy Loss).

Since the density maps generated by the above process provide high values for the license plate area and low values for the non-license plate area, the proposed work employs K-means clustering K = 2 to segment vehicle regions containing license plate numbers. Out of two clusters, the cluster with a high mean value is considered a cluster of vehicle region. To reduce the influence of the background of a region of the license plate number, the proposed work performs the K-means clustering operation iteratively by considering the output of each iteration as input to the next iteration. The iterative process stops when the mean current and the next cluster converge.

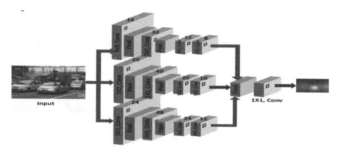

Fig. 3. The proposed architecture for region interest segmentation

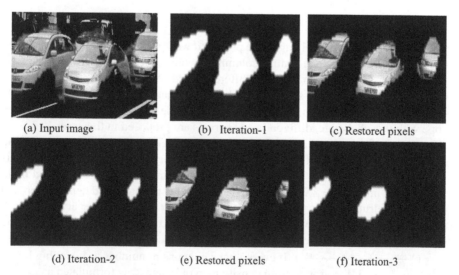

(a) Input image   (b) Iteration-1   (c) Restored pixels

(d) Iteration-2   (e) Restored pixels   (f) Iteration-3

Fig. 4. The results iterative k-means clustering for segmenting region of interest.

The steps of iterative k-means clustering for segmenting Region of Interest (ROI) are illustrated in Fig. 4(a)–(f), where for the input image in (a), the ROI are segmented

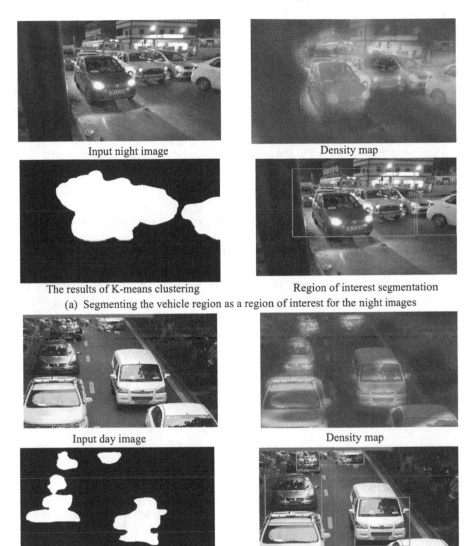

Input night image

Density map

The results of K-means clustering

Region of interest segmentation

(a) Segmenting the vehicle region as a region of interest for the night images

Input day image

Density map

The results of K-means clustering

Segmenting vehicle regions

(b) Segmenting the vehicle region as a region of interest for the day images

Fig. 5. Multi column convolutional neural network for region of interest segmentation.

from density map of (a) as shown in (b). This is the result of iteration-1. The pixels in the input images are restored for the corresponding pixels in (b) as shown in Fig. 4(c). The proposed method applies k-means clustering on the density map of (c) to segment a region of interest as shown in (d), where one can see the background (other than text) has been reduced compared to the results in (b). This is the result of iteration-2. Similarly, for the resorted pixels corresponding to pixels in (e), region of interest is obtained as shown in (f), where background pixels are still reduced compared to (d). This is the result of iteration-3.

The effect of region of interest segmentation for the night and day images can be seen in Fig. 5(a) and Fig. 5(b), respectively, where one can see the density maps highlight vehicle region containing license plate numbers, K-means clustering segments each vehicle region containing license plate number as one component and then bounding boxes are fixed for each component given by clustering. It is observed from Fig. 5 that the proposed step works well for both night and day images containing multiple vehicles. The segmented regions are fed to the subsequent license plate number detection.

3.2 License Plate Number Detection

The previous step reduces the complexity of the background and the presence of multiple vehicles in a single image. However, variation in license plate size remains a challenge for detection. To overcome this problem, we define Gaussian kernels of different sizes to detect license plate numbers. To achieve this, we propose a Dilated Convolution Neural Network (DCNN) [20, 21], which is known as CSRNET. This model comprises two major components, namely, a VGG16 convolution neural network (CNN) for feature extraction, which is the front end and a dilated CNN for the back end, which uses dilated kernels to deliver larger reception fields to replace pooling operations. The architecture is shown in Fig. 6. The first 13 layers of VGG16 include 10 Gaussian kernel-based convolution layers with three max pooling layers to obtain feature maps of size $1/8^{th}$ of its original size. The back end dilated convolution layer extracts the deeper information and outputs a high-resolution image. The role of each component in obtaining the feature maps is represented in Eq. (2).

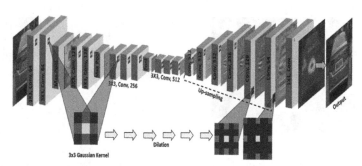

Fig. 6. Dilated convolutional neural network architecture for license plate number detec-tion (CSRNet).

$$Y(i,j) = \sum_{i=0}^{H} \sum_{j=0}^{W} X(i + (r \times \text{m}), j + (r \times \text{n})) \times k(n, m) \qquad (2)$$

where, X (i, j) is the input of the convolution layer and Y (i, j) is the output of the layer. The kernel k (m, n) is defined by height, H, width, W and dilation rate, r. The kernel is magnified to k + (k − 1)× (r − 1). In this work, a 3 × 3 kernel is used with dilation rate r = 1, 2, 3, which generates three different sized kernels of 3 × 3, 5 × 5 and 7 × 7, respectively. If this study considers other way as 16 for 5 × 5, 20 for 7 × 7 and 24 for 9 × 9, computational complexity increases. Furthermore, a Sigmoid layer followed by a 1 × 1 convolution layer output heat-map and hence this step outputs an image with the resolution which is same as the input image.

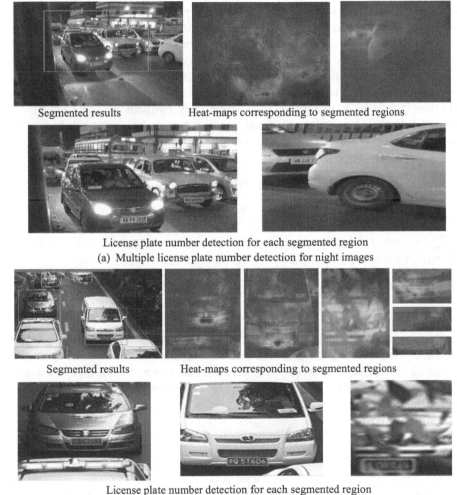

Fig. 7. The proposed model for multiple license plate number detection for the night and day images.

The outcome of the above step is illustrated in Fig. 7(a) and Fig. 7(b), where for the segmented results, heat-maps sharpen the license plate area for accurate license plate number detection for the night and day images, respectively. It is noted from the results in Fig. 7, the proposed model detects license plate numbers irrespective of day, night images with different foreground and background complexities.

4 Experimental Results

To evaluate the proposed method for multiple license plate number detection in day and night images, we construct our own dataset, which includes images of different vehicles affected by illumination, headlights, streetlights, and perspective distortions. Our dataset consists of 391, 109, and 200 day-night and multiple vehicles images, respectively. In total, 500 images are considered for experimentation. To test the generic property of the proposed method, we consider a standard dataset called CCPD [12], which includes 200,000 images with blur, high quality, low quality, day and night time, and one more standard dataset called CRPD [15], which contain 26 thou-sand images with single and multiple vehicles from different time within a day. Furthermore, one more dataset, namely LP-Night [1], provides only night images for single and multiple vehicles. The characteristics of each dataset, including our dataset, are included in Table 1.

Table 1. Details of different license plate datasets

Dataset	Training samples	Test samples	Day	Night	Single	Multiple	Blur	Illumination	perspective	Orientation
Our dataset	300	200	Yes	Yes	Yes	Yes	Yes	Yes	Yes	Yes
CCPD	193000	97000	Yes	No	Yes	No	Yes	No	No	Yes
CRPD	17300	8700	Yes	Yes	Yes	Yes	No	No	No	No
LP-Night	120	80	No	Yes	Yes	Yes	Yes	Yes	Yes	Yes

We calculate standard measures, namely, Precision, Recall and F-measure for evaluating the performance of the proposed and existing methods. The threshold used for comparing the results of the proposed method with the ground truth is 0.5, which is the standard threshold used for text detection [2]. To show the proposed method is effective and valuable, we compare the results of the proposed method with the results of state-of-the-art models. The existing techniques include both scene text detection methods [5–7] and license plate number detection methods [1, 5, 17–19].

Implementation Details: During the training procedure, the input images are normalized to a standard size of 512×512 with three channels (RGB). The ratio between training and testing samples for all the datasets is 7:3. All the models are implemented by PyTorch. The optimizer of the network is SGD with the learning rate of 0.0001, and momentum 0.95. The network is trained for 500 iterations with batch size 16. We use Processor: Intel i5-3470, Ram: 10 GB, GPU: Single NVIDIA GTX 1070 Ti with 8 GB memory for all the experiments.

4.1 Ablation Study

The proposed approach is based on two key architectures, an adapted Multi-Column Convolutional Neural Network (MCNN) and an adapted Dilated Convolutional Neural Network (CSRNET), to address multiple license plate number detection. To validate the effectiveness of each architecture compared to baseline architectures, we conducted the following experiments using our dataset. (i) Baseline CNN for the region of interest segmentation and baseline of CSRNET for license plate number detection. (ii) Proposed MCNN for segmentation and baseline CSRNET for license plate number detection. (iii) Feeding full input image to the proposed CSNR for license plate detection and (iv) The proposed MCNN for segmentation and proposed CSRNET for detection. The results are reported in Table 2, where it is observed that the proposed method (iv) is the best compared to the results of all other experiments (i)-(iii). This shows that the adapted MCNN and CSRNET are better than the baseline MCNN and CSRNET for multiple license plate detection. It is reasonable because when we compare the results of (ii)-(iii) with the results of (i), the results of (ii) and (iii) are better than (i).

Table 2. Ablation experiments for the key steps to validate the effectiveness for detection.

Experiment	Segmentation model	Detection model	P	R	F
(i)	Single column CNN	CSRNet Baseline	0.62	0.54	0.57
(ii)	Proposed MCNN	CSRNet Baseline	0.76	0.59	0.66
(iii)	No segmentation	Proposed CSRNET	0.83	0.63	0.71
(iv)	Proposed MCNN	Proposed CSRNET	**0.90**	**0.84**	**0.87**

4.2 Experiments on Region of Interest Segmentation

Qualitative results of the region of interest segmentation are shown in Fig. 8, where one can see the step detects vehicle region containing license plate numbers for the images of multiple vehicles and different situations. This shows that the proposed segmentation is capable of handling complex situations. Since the ground is not available for quantitative results, the measures are counted manually, and the results are reported in Table 3. It is noted from Table 3 that for both datasets, the proposed segmentation step achieves promising results. It can also be inferred that the segmentation step works well irrespective of night and day datasets.

4.3 Experiments on Detection

Qualitative results of the proposed detection for different datasets are shown in Fig. 9, where it can be seen that for the sample of all the datasets, the proposed method detects license plate numbers accurately in all the images. This infers that the proposed method is independent of day, night, and the number of vehicles in a single image. The same

| Our License Plate dataset | CCPD dataset |

| CRPD dataset | LP-Night dataset |

Fig. 8. Segmentation result of sample images for different datasets

Table 3. Evaluating the step of region of interest segmentation

Datasets	P	R	F
LP-Night	0.83	0.86	0.84
Our dataset	0.79	0.92	0.85

conclusions can be drawn from the quantitative results reported in Table 4 on all four datasets. Table 4 shows that for our dataset, the proposed method is the best in terms of all three measures compared to the scene text detection method [5] and license plate detection method [4]. The existing methods result in poor results because the techniques are developed for specific datasets and objectives but not day, night, and multiple license plate number detection. On the other hand, the proposed method adapted with MCNN and CSRNET to cope with the challenges of different situations, indicating the best performance.

<table>
<tr><td>Our License Plate dataset</td><td>CCPD dataset</td></tr>
<tr><td>CRPD dataset</td><td>LP-Night dataset</td></tr>
</table>

Fig. 9. Proposed model detection result for the images of different datasets.

For CCPD, CRPD, and LP-Night datasets, the proposed method achieves the best precision and F-measure compared to existing methods. However, the method [5] obtains the highest recall for all three above datasets compared to other methods including the proposed method. This is due to the method being developed for scene text detection, which shares almost the same challenges as license plate number detection. However, the method [5] does not score better precision and F-measure compared to the proposed method. In the same way, since other existing methods were proposed for specific dataset and objectives, the methods do not score better results for the complex dataset of license plate numbers.

To validate efficiency of the proposed method, the number of floating point operations in terms of GigaFLOP and number of parameters of the proposed and existing methods are reported in Table 5. It is noted from Table 5 that the efficiency of proposed method is higher compared to the existing methods in terms of GFLOPs and the Parameters. Thus, one can infer that the proposed model is efficient for multiple license plate number detection in day and night images.

When the images are affected by severe blur, too low contrast, and loss of license plate number information as shown in Fig. 10, the proposed model may not work well. The reason is that the step of vehicle region detection fails to contain the license plate number. This is because the step requires fine details like edge information for successful segmentation. However, blur, too low contrast, and loss of license plate number degrade the vital edge information. This shows that there is scope for improvement further.

Table 4. Performance of the proposed and existing methods on different datasets for license plate number detection

Methods	Our dataset			CCPD dataset				CRPD dataset			LP-Night Dataset			
	P	R	F	Methods	P	R	F	P	R	F	Methods	P	R	F
LPD [4]	0.63	0.78	0.69	Scene [7]	0.67	0.79	0.73	0.82	0.88	0.85	LPD [18]	0.82	0.41	0.55
Scene [5]	0.87	0.83	0.85	Scene [5]	0.75	**0.84**	0.79	0.89	**0.97**	0.92	Scene [5]	0.67	**0.78**	0.72
LPD [18]	0.89	0.56	0.69	Chinese LPD [15]	0.72	0.75	0.73	0.84	0.95	0.89	LPD [1]	0.81	0.72	0.76
Proposed	**0.90**	**0.84**	**0.87**	Proposed	**0.82**	0.79	**0.81**	0.91	0.96	**0.93**	Proposed	**0.87**	0.73	**0.79**

Table 5. Efficiency of the proposed and existing methods on our dataset. M indicates Million.

Model Name	GFLOPs	Parameters (M)
U-Net [18]	30.71	17.27
YOLOv3 [4, 15]	154.9	61.5
YOLOv4 [4, 15]	107.1	61.35
Resnet-50 [7]	234	37.7
RetinaNet [7, 15]	206	38
Proposed	**21.56**	**16.26**

Fig. 10. Limitations of the proposed method

5 Conclusion and Future Work

We have proposed a new model for detecting multiple license plate numbers of different vehicles based on the region of interest segmentation and a deep learning model for detection. To reduce the background complexity, the proposed work segments vehicle region containing license plate numbers as a region of interest. For segmentation, the multi-column convolutional neural network has been introduced with kernels of different sizes, which outputs density maps for the input images. The K-means clustering with K = 2 is employed iteratively on a density map to segment the region of interest. For each segmented region of interest, the proposed work introduces different Gaussian kernels to convolve with the network to detect license plate numbers in the images. The results on our standard datasets of night and day-night images containing multiple vehicles show that the proposed method outperforms the existing methods in terms of accuracy. However, for the images affected by severe blur, too low contrast, and loss of license plate information due to different rotations of vehicles, the method does not perform well. To alleviate this limitation, we plan to propose an end-to-end transformer to cope with the challenges of severe blur and other adverse effects in the near future.

Acknowledgement. This work is partially supported by IDEAS-Technology Innovation Hub grant, Indian Statistical Institute, Kolkata, India.

References

1. Chowdhury, P.N., Shivakumara, P., Pal, U., Lu, T., Blumenstein, M.: A new augmentation-based method for text detection in night and day license plate images. Multimedia Tools Appl. **79**(43–44), 33303–33330 (2020). https://doi.org/10.1007/s11042-020-09681-0

2. Chowdhury, P.N., et al.: Graph attention network for detecting license plates in crowded street scenes. Pattern Recogn. Let. 18–25 (2020)

3. Kumar, A., Shivakumara, P., Pal, U.: RDMMLND: a new robust deep model for multiple license plate number detection in video, In Proceesings of the ICPRAI, pp. 489–501 (2022)

4. Usama, M., Anwar, H., Anwar, A., Anwar, S.: Vehicle and License Plate Recognition with novel dataset for Toll Collection, pp. 1–13 (2022), https://doi.org/10.48550/arXiv.2202.0563

5. Du, Y., et al.: PP-OCRv2: bag of tricks for ultra-lightweight OCR system. arXiv preprint arXiv:2109.03144 (2021)

6. Deng, D., Liu, H., Li, X., Cai, D.: Pixellink: detecting scene text via instance segmentation. In: Proceeding of the AAAI, vol. 32, No. 1 (2018)

7. Deng, L., Gong, Y., Lu, X., Lin, Y., Ma, Z., Xie, M.: STELA: a real-time scene text detector with learned anchor. IEEE Access **7**, 153400–153407 (2019)

8. Lin, C.H., Lin, Y.S., Liu, W.C.: An efficient license plate recognition system using convolution neural networks. In: Proceedings of the ICASI, pp. 224–227 (2018)

9. Kumari, P.L., Tharuni, R., Vasanth, I.V.S.S., Kumar, M.V.: Automatic license plate detection using KNN an convolutional neural networks. In: Proceedings of the ICCMC, pp. 1634–1639 (2202)

10. Srilekha, B., Kiran, K.V.D., Padyala, V.V.P.: Detection of license plate numbers and identification of non-helmet riders using Yolo v2 and OCR method. In: Proceedings of the ICEARS, pp. 1539–1549 (2022)

11. Charan, R.S., Dubey, R.K.: Two-wheeler vehicle traffic violations detection and augmented ticketing for Indian road scenario. IEEE Trans. Intell. Transport. Syst. 22002–22007 (2022)

12. Xu, Z., et al.: Towards end-to-end license plate detection and recognition: a large dataset and baseline. In: Ferrari, V., Hebert, M., Sminchisescu, C., Weiss, Y. (eds.) ECCV 2018. LNCS, vol. 11217, pp. 261–277. Springer, Cham (2018). https://doi.org/10.1007/978-3-030-01261-8_16

13. Mokayed, H., Shivakumara, P., Woon, H.H., Kankanhalli, M., Lu, T., Pal, U.: A new DCT-PCM method for license plate number detection in drone images. Pattern Recogn. Let. 45–53 (2021)

14. Tote, A.S., Pardeshi, S.S., Patange, A.D.: Automatic number plate detection using tensorFlow in Indian scenario: an optical character recognition approach. Mater. Today: Proc. 1073–1078 (2023)

15. Gong, Y., et al.: Unified Chinese license plate detection and recognition with high efficiency. J. Visual Commun. Image Represent. (2022)

16. Gizatullin, Z.M., Lyasheva, M., Shleymovich, M., Lyasheva, S.: Automatic car license plate detection based on the image weight model. In: Proceedings of the EIConRus, pp. 1346–1349 (2022)

17. Zhang, C., Shivakumara, P., Xue, M., Zhu, L., Lu, T., Pal, U.: New fusion based enhancement for text detection in night video footage. In: Proceedings of the PRCM, pp. 46–56 (2018)

18. Chowdhury, P.N., Shivakumara, P., Raghavendra, R., Pal, U., Lu, T., Blumenstein, M.: A new U-net based license plate enhancement model in night and day images. In: Proceedings of the ACPR, pp. 749–763 (2020)

19. Zhang, Y., Zhou, D., Chen, S., Gao, S., Ma, Y.: Single-image crowd counting via multi-column convolutional neural network. In: Proceedings of the CVPR, pp. 589–597 (2016)

20. Li, Y., Zhang, X., Chen, D.: Csrnet: dilated convolutional neural networks for understanding the highly congested scenes. In: Proceedings of the CVPR, pp. 1091–1100 (2018).
21. Xiong, J., Po, L.M., Yu, W.Y., Zhou, C., Xian, P., Ou, W.: CSRNet: Cascaded Selective Resolution Network for real-time semantic segmentation. Expert Syst. Appl. **211**, 118537 (2023)

Ensuring an Error-Free Transcription on a Full Engineering Tags Dataset Through Unsupervised Post-OCR Methods

Mathieu Francois[1,2(✉)] and Véronique Eglin[1(✉)]

[1] Univ. Lyon, INSA Lyon, CNRS, UCBL, LIRIS, UMR5205, 69621 Villeurbanne, France
veronique.eglin@insa-lyon.fr
[2] Orinox, Vaulx-en-Velin, France
francois.mathieu@orinox.com

Abstract. The digital transformation of engineering documents is an ambitious research topic in the industrial world. The representation of component identifiers (tags), which are textual entities without a language model is one of the major challenges. Most of OCR use dictionary-based correction methods so they fail at recognizing hybrid entities composed by numerical and textual characters. This study aims to adapt OCR results on language-free strings with a specific semantics and requiring an efficient post-OCR correction with unsupervised approaches. We propose a two-step methodology to face the questions of post-OCR correction in engineering documents. The first step focuses on the alignment of OCR transcriptions producing a single prediction refined from all OCR predictions. The second step presents a combined incremental clustering & correction approach achieving a continuous correction of tags' transcriptions relatively to their assigned cluster. For both steps, the dataset was produced from a set of 1,600 real technical documents and made available to the research community. When compared to the best state-of-art OCR, the post-OCR approach produced a gain of 9 % of WER.

Keywords: Post-OCR correction · Incremental Clustering · Merging OCR · engineering P&ID diagrams

1 Introduction

In the industrial world, the creation of a digital twin of technical engineering documents is still a subject of interest. Research on this subject has been conducted for decades, but due to the complexity and lack of structure of these documents, the technical challenges in this area are still being studied by researchers around the world. Among these, precise transcription of texts without a language model, referred to as tags, is particularly sought after by industrial actors.

G. A. Fink et al. (Eds.): ICDAR 2023, LNCS 14191, pp. 88–103, 2023.
https://doi.org/10.1007/978-3-031-41734-4_6

Among these, precise transcription of texts without a language model, referred to as tags, is particularly sought after by industrial actors. These actors need an error-free transcription of these texts in order to identify which component the tag represents and to refer to the project documentation for all the technical characteristics.

In this paper, we will attempt to provide innovative solutions to this problem in the form of an end-to-end system. First, by publishing a dataset representing this use case. Then, the proposed pipeline is divided into three main steps. The first one is a novel approach to align multiple OCR predictions. Next, we will propose an initial approach to tag correction based on clustering and correction through the tag structure. Finally, we will propose an evolutionary approach to our Post-OCR correction.

This paper is presented as follows: first, section two will present the related works on the topics of context retrieval on semantic-free texts. Section three will present the dataset submitted to the scientific community. The fourth section will present the different approaches proposed to reduce the number of transcription errors on semantic-free strings. And finally, part five will present the experiments performed with the results step by step. The results of this part are encouraging. Between the raw output of the OCR and the end of our end-to-end pipeline, we obtain a 9.16% gain in WER on the tags.

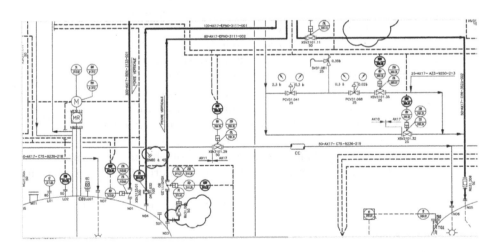

Fig. 1. Sample of a technical engineering drawing.

2 Related Works

Nowadays, OCRs have become very efficient, especially on printed text. OCR prediction is now considered a mature field and related work is now focused on post-OCR correction or more directly on the post OCR processing of even noisy outputs. If we consider the particular nature of the texts of the industrial *P&ID* (Piping and Instrumentation Diagrams), beyond even the many noises that spoil the images of these texts, one must resolve to correct all OCR errors, even sparse, because they strongly compromise the quality of the indexation. In our industrial context, an accuracy close to 100% of correct recognition is explicitly required to assure secured industrial processes. A post-OCR processing is therefore definitely required.

2.1 Post-OCR Processing and Specificity of Short-Text

In recent years, the literature has presented many works focused on the impact of noisy transcriptions on information extraction or retrieval and NLP-based tasks (e.g. question answering, text summarization). In the general case of information extraction, it is essential to have transcriptions that are as faithful as possible to the original data, whether they correspond to long texts (complete semantic sentences) or short words (named entities, tags embedded in graphical documents, information present in tables, etc.). A notably valuable analysis was produced in 2021 by Van Strien et al. [14] confirms that a low quality of transcriptions of short texts had a negative impact on an information extraction/retrieval task. While information retrieval performance can remain satisfactory even with high WER (word error rate) estimates on long texts, it has been shown that the performance drastically decreases on short texts, [15]. They attempt to prove that an error rate of 5% leads to linearly increase indexing (and thus information retrieval & extraction) errors. The main limitations that prevent OCR tools from reaching 100% accuracy mostly concern text background appearence (incrusted in colored backgrounds, presenting sometimes colorful patterns); blurry texts; skewed or non-oriented documents; presence of a large variety of letters (uncommon font types and sozes, rare alphabets, cursivity or handwritten-like aspects...); look-alike characters (OCR tools fail to distinguish between the number "0" and the letter "O" for example), [4]. This is a common criterion usually solved by the use of dictionaries or language models. A less common but highly significant factor can be added to this list. It is related to technical material (text-graphic documents, part lists, bills...) presenting texts written as fragments (partial words, hybrid sequences of letters and numbers...) with no apparent semantics from a linguistic point of view, but with a "domain specific" semantics that only an expert can decipher. Our work is part of this context. More precisely, it can be depicted as *"isolated-word approaches"*, according to the taxonomy of Nguyen in [4]. For this class of approaches, the post-OCR correction techniques rely on observations of single tokens. Given the poor semantic and linguistic context of these tokens, merging OCR outputs techniques or lexicon-based approaches are often proposed [18]. In our situation, the lack of contextual information around words (isolated tags) encourages to privilege competitive methodologies aiming at a selection of most suitable transcriptions. In that context, we can mention Lund et al.'s work in [1]. Authors proposed to use the A* algorithm to align different OCR predictions. Their algorithm can be used for shortest path problems but it can also be applied to text comparison. One of the strong points of this algorithm is that it will determine which solution is the most optimal without having to explore all possible paths. Wemhoener et al. [2] also proposed their own alignment method, based on selecting a pivot from the OCR outputs. The other predictions are then individually aligned with the pivot. This makes it possible to link all predictions and to perform a comprehensive alignment. Broadly speaking, voting strategies or ensemble methods for multiple input selection have become well-established standard options for post-OCR error processing [16]. Nevertheless, as some small texts can be devoid of semantics (such as isolated tags in engineering diagrams), a voting strategy can also be combined with a lexical alignment technique [17]. Since our tag dataset could not be supplied with any lexicon and a voting strategy could not provide a reliable answer, we designed a hybrid technique combining a subsequence-based alignment of OCR outputs and the support of incremental clustering before correction.

2.2 Incremental Similarity Clustering

Since the documents in our study contain texts that are heavily lacking context - and thus do not allow for reliable transcriptions -, we turned to clustering approaches to identify frequent patterns/tags and contribute to their accurate recognition. The incrementality associated with clustering enables us to address real engineering situations where the data must be processed in a continuous flow. At the start of the process, we do not have enough training or initialization data that could guarantee an immediate and stable clustering. Clustering methods are widely used in many areas of science today. However, when data is constantly being added or updated, these methods can be limited. Most algorithms process all data in one pass, and when the information is updated, the clustering is recalculated on the updated data, resulting in a large consumption of resources. To reduce the computational costs and to better consider the evolution of data as they come in, *incremental clustering* approaches have been introduced. Prasad et al. [7] proposed an incremental adaptation of the K-Means algorithm. The principle of their approach is to adapt the value of the seeds one by one. The center of the cluster is updated at each iteration by minimizing the Sum of Squared Error (SSE), overcoming in that way the tedious initialization problem of k-means method. Authors show that this method allows a better composition of clusters with a total SSE estimated over all the clusters to be decreased.

Sun et al. [8] proposed an *Incremental Affinity Propagation* method (*AP*). To evaluate their approach, they performed their experiments on real world time series. Their method uses the traditional *Affinity Propagation algorithm* to process the first data. Once inserted, data are processed through two approaches. The first one is based on an association between *AP* and K-Medoids, exploiting *AP* to define the clusters and K-Medoids to evolve them. The second approach proposed is an association between *AP* and the Nearest Neighbor Assignment. This last approach allows that two similar data not only belong to the same cluster but also have the same status.

Chakraborty in [9] has worked on an incremental *DBSCAN* method. This approach will initially form the clusters by given radius and minimum number of points per cluster. When the data is modified, the clusters are also updated by calculating the minimum mean distance between the existing cluster data and the newly inserted or updated data.

Taking into account the scalability of the incoming data of a digitization chain, and the nature of the clusters (number and quality) obtained respecting the diversity of the tags of technical *P&ID* documents, we opted for a cooperation between $\mathcal{AP}$ (for the initialization step without any a priori number of classes) and an iterative cluster adjustment through incremental K-means (see Sect. 4.3).

3   Tags_PID: A New Public Dataset for *P&ID* Documents

Data from engineering documents are really sensitive. To our knowledge, there was no public annotated tags dataset that was truly representative of what can be found in the industrial world for *P&ID* documents. This section presents *Tags_PID*, a new tags image dataset obtained from real P&ID documents sources and made available to the scientific community[1].

3.1   Context

Tags_PID dataset is composed of tags images. The tags are the identifiers of the engineering technical components which are language model free and incomprehensible if we have neither engineering knowledge nor information about the project. These data come from real industrial projects from different sectors of activity such as: water treatment, gas and pharmaceutical.

The goal of *Tags_PID* is to represent a real industrial use case, which has been imagined as follows: An industrial engineer digitizes the drawings of a project and is particularly interested in tags. He has no indication of the tagging rules employed for this project. Considering the quantity of data to process, he wants a fully automated digitization process. Several days later, he retrieves new documents and wants to extract the tags from them. This engineer needs a transcription as close as possible to the ground truth to correctly identify the technical component.

The tag images and ground truth were originally extracted on searchable PDFs. Thanks to the PDFMiner library, the coordinates and the transcription of each text could be extracted via the documents' metadata. After that, each tag sub-image are converted into an image format and associated with their transcription.

3.2   Content of *Tags_PID*

Tags_PID is composed by data separated in 2 groups. The first one, named *Tags_1*, represents the tags from 30 drawings, i.e. 1570 images. The second one, named *Tags_2*, corresponds to the tags of 7 other drawings from the same project, i.e. 125 images.

---

[1] https://github.com/mathieuF789/dataset_tags.

The images are named under the form: 'eval_X_Y.png' with 'X', an iterator representing the drawing from which the image comes from and 'Y', an iterator representing the text. The data is stored in a *.csv* format with the name of the image, the ground-truth and the different OCR predictions.

Fig. 2. Examples of tag images from *Tags_PID*.

To evaluate the quality of OCR predictions compared to the ground-truth, the CER and WER metrics were used. Even if some tags can be considered as several words when calculating the metrics, we fixed the fact that a tag always represents a single string that can include spaces. Below are the formulas to calculate the CER and WER with S the number of substitutions, D of deletions, I of insertions and N of reference characters. The results of the OCRs on the dataset can be found in Sect. 5.1.

$$CER = (S + D + I)/N \tag{1}$$

$$WER = \begin{cases} 0 & \text{if } CER = 0 \\ 100 & \text{otherwise} \end{cases}$$

The WER is the most significant metric in this use case because it is important that there are no errors in the tag transcriptions. Indeed, the tags allow to refer to the technical documentation and to identify which component is represented on the drawing. If there is any error then the identified component and its characteristics will not be correct.

4   Our Approach

The proposed approach is an end-to-end system and it is divided into several steps. The main objective of this approach is to ensure an error-free transcription on a engineering tags dataset thanks to OCR Merging and Post-OCR correction methods. Additionally, it will provide a continuous evolution of the system each time new data is inserted. First, this allows to correct the new data via the knowledge already acquired previously. Additionally, new tags can allow for the creation, deletion, or modification of existing clusters. This will allow for the correction of tags that were already present in the clusters and considered as correct. An explanatory illustration of the complete system can be found in Fig. 3.

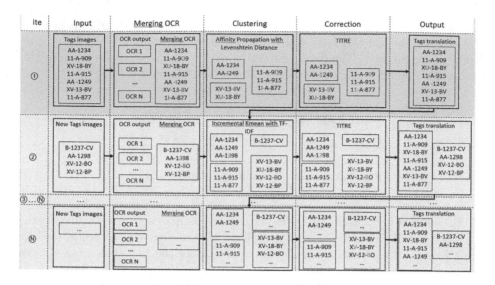

Fig. 3. Diagram of the proposed complete process.

4.1   Selection of the Best OCR Output

Our approach relies on the setting of multiple OCR views on a word input and the challenge is to select the best segments of each OCR output by focusing regions of mismatching. Given three OCR views of the same word i namely $S_1^i = (c_{1,1}^i, c_{2,1}^i, ...c_{n,1}^i)$, $S_2^i = (c_{1,2}^i, c_{2,2}^i, ...c_{n,2}^i)$, and $S_3^i = (c_{1,3}^i, c_{2,3}^i, ...c_{n,3}^i)$, where $c_{p,3}^i$ is the pth characters of the third OCR prediction, our approach consists in two main steps:

(i) alignment of OCR outputs and spotting of differences and common substrings (determination of the common fixed basis of the predictions)

(ii) picking the majority choice for each substrings and concatenation into the final output.

Each sequence S_1^i, S_2^i and S_3^i is then decomposed into the same number of subsequences thanks to the *BaseFix* detection algorithm, thus leading to the division of S_1^i, S_2^i and S_3^i into $S_1^i = (id_{1,1}; s_{id_{1,1}}); (id_{2,1}; s_{id_{2,1}}); ...; (id_{m,1}; s_{id_{m,1}})$, $S_2^i = (id_{1,2}; s_{id_{1,2}}); (id_{2,2}; s_{id_{2,2}}); ...; (id_{m,2}; s_{id_{m,2}})$ and $S_3^i = (id_{1,3}; s_{id_{1,3}}); (id_{2,3}; s_{id_{2,3}})$; ...; $(id_{m,3}; s_{id_{m,3}})$, with for S_1^i m pairs $(id_i; s_id_i)$ where s_id_i is a subsequence and id_i is a subsequence identifier in the word. In real situations, for large sequences composed by a maximum of 25 characters, we have $m <= 9$.

The goal is to segment predictions into consistent aligned substrings (common sub-sequences). We allow here small variations between two substrings from S_1^i, S_2^i and S_3^i transcriptions (addition, deletion, substitution of characters). To figure out the process, an example of sequences alignment considering three transcriptions for each word is presented in Fig. 4. It is splitted into four steps also developed in the Algorithm 1. The first step, *BaseFix Identification*, is the localisation of strictly identical substrings among the three predictions. To be admitted as a fixed base, a substring must be composed of three or more characters and be present in all OCR outputs. These sequences are kept in memory and the remaining characters are isolated into secondary groups during the step *Sub-parts Isolation*. After this step, each group of strings is individually processed to determine the position of one or more potential offsets in the *Sub-parts alignment* phase. If the substrings are not the same size then the system will align them by adding a 'dummy' character. This one will be placed once again according to the similarities of the sequences calculated via a score system. Finally, we concatenate the aligned sub-sequences and the fixed bases previously kept in memory: the strings are now aligned. A voting system is finally applied to build the final string, which will serve as input of the next incremental clustering step.

Algorithm 1. Alignment of several OCR predictions for three OCR outputs

---

Require: Ω : list composed by S_1, S_2, S_3 the 3 OCR predictions
  B = findIdenticalSubSeq(S_1, S_2, S_3) /* baseFix identification */
  if B not $\varnothing$ **then**
    for $B_i \in$ B **do**
      for $S_j \in \Omega$ **do**
        $\Psi \cup \{ S_j \notin B_i \}$ /* sub-parts isolation */
      end for
    end for
  else
    $\Psi = \Omega$
  end if
  for $X_i \in \Psi$ **do**
    X_i = addCarac(X_i) /* sub-parts alignment */
  end for
  $\Omega = \{B \cup \Psi\}$ /* Concatenation */
  return Ω

4.2 Post-OCR Correction

In this section, a first phase of non-dynamic post-OCR correction is presented which is summarized in Algorithm 2. To propose a correction for OCR prediction without a language model, a first clustering of tag predictions had to be done. For this purpose, we used the same clustering algorithm as in our previous work based on Affinity Propagation, [5: anonymous reference]. To form the very first clusters in the first instance, we implement $\mathcal{AP}$ with the Levenshtein distance as metric (other could also been proposed but less adapted to string comparision. This metric proposes a very consistent clustering able to group tags of same composition even if they present punctual transcription errors.

Following this, we proposed a refinement of the clusters quality to improve the relevance of the corrections made to the tag. The first step was to define a regular expression that best represents each cluster. Tags that do not conform

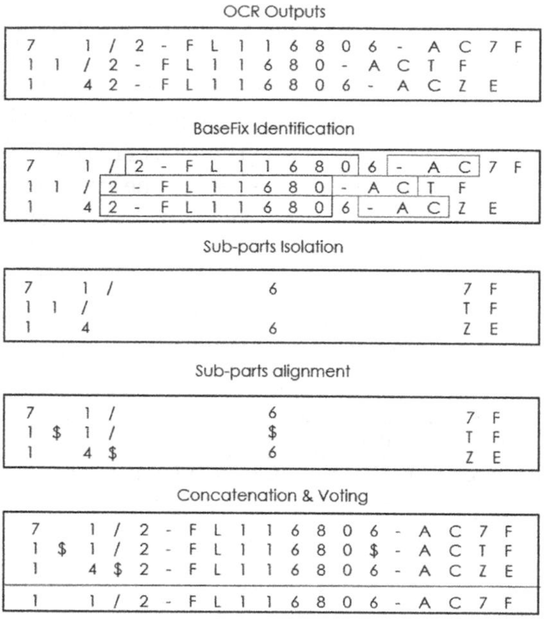

OCR Outputs

```
7    1 / 2 - F L 1 1 6 8 0 6 - A C 7 F
1 1  / 2 - F L 1 1 6 8 0 - A C T F
1    4 2 - F L 1 1 6 8 0 6 - A C Z E
```

BaseFix Identification

```
7    1 / | 2 - F L 1 1 6 8 0 6 | - | A C | 7 F
1 1  / | 2 - F L 1 1 6 8 0 | - | A C | T F
1    4 | 2 - F L 1 1 6 8 0 6 | - | A C | Z E
```

Sub-parts Isolation

```
7    1 / /           6                7 F
1 1  /                                T F
1    4               6                Z E
```

Sub-parts alignment

```
7    1 / /           6                7 F
1 $  1 /             $                T F
1    4 $             6                Z E
```

Concatenation & Voting

```
7    1 / 2 - F L 1 1 6 8 0 6 - A C 7 F
1 $  1 / 2 - F L 1 1 6 8 0 $ - A C T F
1    4 $ 2 - F L 1 1 6 8 0 6 - A C Z E

1    1 / 2 - F L 1 1 6 8 0 6 - A C 7 F
```

Fig. 4. Illustration of the proposed alignment method in four steps.

to the regular expression of the cluster from which they originate are placed in a rejection class. This last group of tags is isolated and two tag cleaning steps are performed in their respective clusters.

First, the process starts with the exploration of tags of each cluster and determines the similarities that we will call: *BaseFix*. In some cases, there is a hesitation if some characters are a *BaseFix* or not due to some tags not conforming to the whole cluster. We then use the *Ocr weighted Levenshtein distance* [10] to quantify the probability that this difference is due to a transcription error made by the OCR. From a threshold, we consider the error is real and we correct the corresponding tag(s). The second step is the identification of clusters where a recurring error in the OCR transcripts has occurred. This creates clusters with all tags having the same error. We study the proximity of all clusters in order to detect this case. If two clusters are very close and their difference, calculated using the *Ocr weighted Levenshtein distance*, is below a defined threshold, then the largest cluster is considered to be the one with the correct transcription. The tags considered false are modified and re-injected in the right cluster.

Following the refinement of the clusters, the tags of the rejection class can be analyzed. Using the alignment method proposed in Sect. 4.1 we compare the tags with the regular expressions. We check for each tag if it can match the properties of a cluster with an addition, a deletion or a replacement of a character. If it is the case, we make the choice to modify this character or not thanks to a score system based on more characteristics such as: is the character part of a BaseFix, the distance between the possibly false tag and the one it is closest in the cluster and finally, the consistency of the character to be added, deleted or replaced.

Algorithm 2. Post-OCR correction for tags

Require: T : list composed of tag transcriptions
Γ = AffinityPropagation(T)
E = $\emptyset$ /* regular expression representing the clusters*/
K = $\emptyset$ /* reject class */
for $C_i \in \Gamma$ **do**
   E = E $\cup$ { regEx(C_i) }
   for $s_j \in C_i$ **do**
     if $S_j \notin$ r **then**
       K = K $\cup$ S_j
       C_i = C_i - { $C_i \cap S_j$}
     end if
   end for
end for
Γ = transformWrongCluster(Γ)
for $k_j \in$ K **do**
   if alignment(k_j, C_i) $\in C_i$ **then**
     if score(k_j, C_i) $\leq$ threshold **then**
       k_j = correc(k_j, C_i)
       C_i = $C_i \cup$ { k_j }
       K = K - K $\cap k_j$
     end if
   end if
end for
T = $\Gamma \cup$ K
return T

4.3   Contextualizing Tags Through Incremental Clustering

In accordance with the use case proposed in Sect. 3.1 we wanted the tag clusters to be able to evolve if new data is injected. This has two main objectives: to continuously correct the tags by refining the delimitations of the clusters and also, to avoid starting the process again from scratch when new data are injected downstream from the first analysis. The incremental clustering track has therefore been studied.

An Incremental K-Means is used for the incremental part with the *term frequency-inverse document frequency* (TF-IDF) chosen as tags proximity measure. TF-IDF allows to evaluate the importance of a term contained in a set of text.

This second metric is complementary to the Levenshtein applied in the first clustering by $\mathcal{AP}$. By changing the metrics concerning the clustering methods, we will show in the experimental parts in what extend it brings a kind of freshness in the construction of clusters, especially it allows to detect mistranscribed tags that were not picked up in the first step.

Once the clusters were built, the same correction method as presented in part 4.2 when correcting the tags from the rejection class was applied. But depending on the constitution of the clusters, the results could be very different. This is the reason why all measurements are systematically taken on 40 batches.

5   Experiments and Discussions

In this section, we present the step-by-step results of the methods presented in Part 4. For each tag images, 4 OCR predictions are performed (EasyOCR [13], Tesseract [11], Paddle_OCR [12] and Paddle_OCR with binarized images). The results of these systems can be found in Table 1 on three datasets, *Tags_PID*, SROIE [3] and CORD [6].

The CER and WER metrics are used in this study to quantify the results. As mentioned before, we considered that a tag is a single string even if it contains spaces. Therefore, we have adjusted the WER calculation so that even the slightest error in the tag will set the WER to 100. This first table is not discussed here, as it will serve as a baseline for comparison with the changes brought about by our proposal.

Table 1. Comparing OCR results with different datasets.

Dataset	Tags_1		SROIE [3]		CORD [6]	
Method	CER	WER	CER	WER	CER	WER
EasyOCR (1)	8.71	53.79	14.06	43.19	17.09	38.74
Tesseract (2)	2.63	18.60	15.34	32.08	26.08	37.54
PaddleOCR (3)	**1.98**	**15.68**	**5.70**	**21.61**	**5.72**	**15.08**
PaddleOCR_b (4)	3.85	29.89	10.44	25.32	13.37	32.32

5.1   Merging OCR Results

In this section, we present and compare the results of the approach proposed in Sect. 4.1 (selection of the best OCR output). The OCR predictions on the different datasets (Table 1) are used to the alignment phase of strings. They allow to form four different combinations consisting of three OCR predictions. For each of the combinations, the following alignment methods were applied: MinDist [1], Pivot [2] and our approach.

For each case, we apply the same voting system after the alignments so that the evaluation is as fair as possible. The voting method is basic (and works as described in Sect. 4.1 by sliding slices of three characters, allowing for some shifts depending on the recognition results). We first sort the aligned strings according to the reliability of the OCR transcription. In other words, the first string will be the one from the OCR with the best results and the third will be from the OCR with the worst results, then, we apply the following condition:

For a given word i, if a character $c^i_{p^j,2}$ (at the p^{jth} position) of the second string is equal to the character $c^i_{p^j,3}$ (at the p^{jth} position, with a possible negative or positive shift of 1) of the third string then the character of the final string will be this one, otherwise it will be the character $c^i_{p^j,1}$ of the first string (more confident result).

The results are presented in Table 2. The combination of OCRs is expressed as a number to lighten the table with : EasyOCR (1), Tesseract (2), PaddleOCR (3) and PaddleOCR_b (4). Our alignment method outperforms the 2 methods seen in the state-of-art. It is obvious to notice that the better the quality of the original OCR transcripts, the better the merging-OCR result.

Table 2. Comparing alignment methods using different OCR combinations.

Dataset	Tags_1						SROIE [3]						CORD [6]					
Method	MinDist		Pivot		Approach		MinDist		Pivot		Approach		MinDist		Pivot		Approach	
OCRs	CER	WER	CER	WER	CER	WER	CER	WER	CER	WER	CER	WER	CER	WER	CER	WER	CER	WER
(1,2,3)	1.70	14.34	1.95	15.78	**1.37**	**13.73**	5.57	19.08	7.92	23.89	**5.15**	**18.69**	5.77	16.31	9.63	22.79	**5.48**	**16.17**
(1,2,4)	2.20	16.21	2.47	19.62	**2.00**	**15.49**	8.66	23.54	9.42	26.02	**8.41**	**23.66**	12.02	26.46	13.00	29.34	**11.33**	**26.16**
(1,3,4)	1.97	15.72	2.18	19.08	**1.63**	**15.44**	5.12	17.53	6.82	20.59	**4.71**	**16.72**	5.32	16.28	8.11	20.88	**4.92**	**15.46**
(2,3,4)	1.72	14.61	1.68	14.94	**1.39**	**13.78**	4.81	15.75	6.21	17.69	**4.37**	**14.55**	5.65	15.16	8.52	20.17	**5.16**	**14.26**

5.2 Post-OCR Result-First Correction

As seen in the state-of-art, post-OCR correction of strings with poor semantics is a topic that is usually not addressed very well, as it concerns very specific situations of decontextualized data. Indeed, it is very hard to find methods to compare with that are not based on supervised learning, a lexical approach, or a language model. We consider here a very unique use-case but nevertheless essential in the engineering world.

Consequently, the overall comparison of our approach with public datasets is tenous to achieve. On datasets like SROIE [3], CORD [6], the data have no context between them, each image comes from a different environment. So, for this part, we only compare our proposition with the basic solution (based on AP Clustering) without incrementality.

The results presented in Table 3 have been realized on the Tags_1 dataset with the Merging OCR output (2,3,4) of Table 2 as input. We see a clear improvement quantified by a 6.5% gain on the WER.

Table 3. Results of post-OCR methods.

Methodes	CER	WER
Best result of 5.1	1.39	13.78
[5]	1.39	13.72
Approach	0.91	7.20

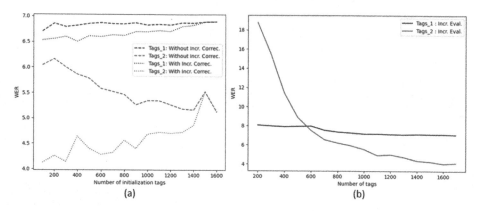

Fig. 5. (a) Comparing incremental correction and final correction. (b) Step-by-step evaluation of data insertion.

5.3   Incremental Clustering Results

To carry out these studies we used the 2 datasets presented in part 3. The blue curves represent the results for the dataset Tags_1. These are the tags treated in the previous parts and we take back the transcript of the tags from the output of the part 5.3. The Tags_2 dataset is represented by the red curve. The transcriptions of the tags of this dataset are done thanks to the OCR Merging seen in part 5.2. The results are available in Table 4.

The first study is the comparison between a correction performed when all the tags are inserted in the clusters and an incremental correction. Figure 4a. aims to measure the WER for a number n of initial tags (varying along x-axis). The x-axis represents the number of tags initially given to the Incremental Clustering algorithm and for each case, 200 tags are added at each new iteration. That is to say that for number of initial tags equal to 500 for example, the Incremental K-means will go through the iterations: 500,700,900...1700. The tags are inserted in a random way. Therefore, the results on a single case can differ completely depending on the order in which the tags are placed, which is why the results presented on the curve are the result of the average of 40 different distributions. The difference between the two methods is that the curves with the dashes were calculated thanks to the correction made when all the tags were sorted, whereas the curves with the dotted lines were calculated thanks to a correction phase that was done at each iteration. For the incremental correction case, the best results are when the number n of considered tags is at its lowest. This seems quite obvious since there are more correction phases. The curve without incremental correction has an inverse trajectory. Depending on the distribution, the initial tags may not be very representative of all the tags. So, when we iterate too much on this basis the clusters are not ideal and the correction is less efficient. The curves logically meet at 1500 because there is only one iteration and therefore only one correction for both cases.

Figure 4b. represents the curve with an incremental correction, however the calculation of the metrics was performed at each iteration. As in the previous study, the results presented here are based on an average of 40 random tag distributions. We can see that both curves are descending and confirm that the more tags are inserted, the more representative the clusters are and therefore the better the correction. We just see a bigger progression for the second dataset, this is due to the fact that these tags were not processed by the part 5.3 and therefore more errors in the transcription were still present.

5.4 End-to-End Results

Table 4 shows the step-by-step results of each method presented above. PaddleOCR, which is supposed to be the best OCR we selected, reacts badly with the second dataset of tags. This is due to presence of many tags that are composed of spaces and these are poorly detected by PaddleOCR.

We have a 9.15% WER gain between the best OCR results and the end of the whole process for the first dataset. However, we can notice that the biggest increase is in the Post-OCR part. And when we inject the second set of tags, we have a gain of 24.68% of the WER between the output of the OCR Merging and the output of the incremental clustering and correction. These promising results allow us to conclude that on a perfectly representative set of real P&ID data, considering the data in batches and adapting the correction according to the evolution of the clusters outlines in an incremental way is much more effective than attempting a post-correction by operating the whole diversity of the dataset in one go. The gradual evolution of the description of the tag clusters is an additional dimension that contributes to the quality of the correction.

Table 4. Overall results of the proposed process on the *Tags_PID* dataset.

Dataset	Tags_1		Tags_2	
Methodes	CER	WER	CER	WER
PaddleOCR	1.98	15.68	25.15	79.03
Merging-OCR	1.39	13.78	3.45	28.8
Post-OCR	0.91	7.20	–	–
Incr. Clustering	0.83	6.53	0.25	4.12

6   Conclusion

In this paper, we propose an end-to-end approach to ensure the quality of the transcription of semantic-free strings from technical engineering P&ID documents.

First, we propose in this study a complete dataset of tags namely *Tags_PID* representing real textual entities what can be found in an industrial context. The dataset has been made available to the scientific community. This type of engineering data is quite sensitive and, to our knowledge, there was no public dataset of text/tag images with these specific characteristics.

After a merging-OCR method that aligns three different OCR predictions of tags and shows improvement compared to state-of-the-art results, our last contribution is a post-OCR correction method based on tag sorting using the Affinity Propagation algorithm with the Levenshtein distance metric. A correction method based on the tag structure and a Levenshtein distance weighted for OCR errors was then applied.

Finally, as we wanted our approach to be scalable, we applied an incremental clustering to improve correction efficiency using TF-IDF metric as distance between OCR predictions. This change of metrics introduced after the initialization step of the clusters by affinity propagation adds a new dimension to the evolution of the clusters boundaries and their generation. This incremental approach has also proven to be particularly effective as a support for post-OCR correction.

For future work, the next step is to study the graphical context around the tags and the interpretation of diagrams, providing additional information on the tags and helping to correct any remaining transcription errors.

References

1. Lund, W.B., Ringger, E.K.: Improving optical character recognition through efficient multiple system alignment. In: Proceedings of the 9th ACM/IEEE-CS Joint Conference on Digital libraries (JCDL 2009), pp 231–240. Association for Computing Machinery (2009)
2. Wemhoener, D., Yalniz, I.Z., Manmatha, R.: Creating an improved version using noisy OCR from multiple editions. In: 2013 12th International Conference on Document Analysis and Recognition, pp. 160–164 (2013)
3. Huang, Z., et al.: ICDAR2019 Competition on Scanned Receipt OCR and Information Extraction. In: 2019 International Conference on Document Analysis and Recognition (ICDAR), pp. 1516–1520 (2019)
4. Nguyen, T.T.H., Jatowt, A., Coustaty, M., Doucet, A.: Survey of Post-OCR Processing Approaches. ACM Comput. Surv. **54**(6), Article 124 (2021)
5. Francois, M., Eglin, V., Biou, M.: Text detection and post-OCR correction in engineering documents. In: Document Analysis Systems: 15th IAPR International Workshop, DAS 2022, La Rochelle, France, May 22–25 (2022)
6. Park, S., et al.: Cord: A consolidated receipt dataset for post-ocr parsing. In: Workshop on Document Intelligence at NeurIPS 2019 (2019)
7. Prasad, R., Sarmah, R., Chakraborty, S.: Incremental k-Means Method (2019)

8. Leilei, S., Chonghui, G.: Incremental affinity propagation clustering based on message passing. IEEE Trans. Knowl. Data Eng. **26**, 2731–2744 (2014)
9. Chakraborty, S.: Analysis and study of incremental DBSCAN clustering algorithm. Int. J. Enterprise Comput. Bus. Syst. **1** (2011)
10. Ocr weighted Levenshtein distance.Joan Capell Garcia. Version 1.0.0. July. 13 (2022). http://github.com/zas97/ocr_weighted_levenshtein
11. Smith, R.: an overview of the tesseract OCR engine. In: Ninth International Conference on Document Analysis and Recognition (ICDAR 2007), pp. 629–633 (2007)
12. Yuning, D.: PP-OCR: A Practical Ultra Lightweight OCR System (2020)
13. EasyOCR. JaidedAI. Nov. 15 (2022). http://github.com/JaidedAI/EasyOCR
14. van Strien, D., Beelen, K., Coll Ardanuy, M., Hosseini, K., McGillivray, B., Colavizza, G.: Assessing the impact of OCR quality on downstream NLP Tasks. In Proceedings of the 12th International (2020)
15. Torresan Bazzo, G., Acauan Lorentz, G., Suarez Vargas, D., Moreira, V.P.: Assessing the impact of OCR errors in information retrieval. In: ECIR 2020: Advances in Information Retrieval, pp 102–109 (2020)
16. Xu, S., Smith, D.: Retrieving and combining repeated passages to improve OCR. In Proceedings of the 2017 ACM/IEEE Joint Conference on Digital Libraries (JCDL 2017) (2017)
17. Gupta, H., Del Corro, L., Broscheit, S., Hoffart, J., Brenner, E.: unsupervised multi-view post-ocr error correction with language models. In: Proceedings of the 2021 Conference on Empirical Methods in Natural Language Processing, pp. 8647–8652 (2021)
18. Rijhwani, S., Rosenblum, D., Neubig, G.: Lexically aware semi-supervised learning for OCR post-correction. Comput. Sci. Trans. Assoc. Comput. Ling. (2021)

Unraveling Confidence: Examining Confidence Scores as Proxy for OCR Quality

Mirjam Cuper(✉) [iD], Corine van Dongen, and Tineke Koster

KB, National Library of the Netherlands, Prins Willem-Alexanderhof 5,
2595 BE The Hague, The Netherlands
mirjam.cuper@kb.nl

Abstract. While performing Optical Character Recognition (OCR), most engines provide confidence scores. These scores give an indication on how certain an engine is that a word or character is correctly determined. The practical application of this score is not yet clear and various studies have discussed the (un)usability of these confidence score as an estimation of OCR quality. Using a dataset of 2000 historical Dutch newspapers we investigated different aspects of the confidence score as provided by ABBYY Finereader, while also looking for a way to use the confidence score as an indication of quality. Such an indication could be used by institutions to determine which part of their collection would benefit from re-OCRing or post-processing. We found that the reliability of the confidence score as a measure of quality is largely dependent on the way the engine has been configured. In addition we show that when there is a high enough correlation between the word confidence and the Word Character Error (order independent) the word confidence can be used to calculate a proxy measure for categorizing digitized texts. However, such a measure must be recalculated for individual OCR engine set ups and producers. For our dataset this proxy measure performs well for the separation of digitized texts into categories of those with a very good and those with a very bad quality with total accuracy of 83%.

Keywords: ocr quality · confidence score · quality indication · digitisation workflow

1 Introduction

Mass digitization is often used by heritage institutions to digitize their textual collections. As there is a continuous growth in methods and techniques to analyze these digitized texts automatically, the need for high quality digitized texts is increasing [11,12,14,19]. However, automatic quality measurement of these texts is still an unsolved issue, and the question of how to automatically determine what parts of collections are candidate for re-OCRing or post-processing is still unanswered [3,11,16].

Some of the frequently used Optical Character Recognition (OCR) software engines that produce these digitized texts provide a measure, the confidence

G. A. Fink et al. (Eds.): ICDAR 2023, LNCS 14191, pp. 104–120, 2023.
https://doi.org/10.1007/978-3-031-41734-4_7

score, that indicates how certain the engine is that the suggested word is the correct word [7]. Numerous studies mention the use of these confidence scores as an indication of OCR quality, however there is no consensus on the applicability. Some studies report a relation between the word confidence and quality [4,5,12, 17], but other studies suggest caution when using them [11,15,19]. Overall, there is no agreement about whether these confidence scores are a reliable indication of quality. To make matters more complex, there is little to no transparency on how the confidence scores are calculated, and this calculation may vary between producers or software versions [5,6,16,19]. This makes it unlikely to find a single solution for all OCR engines.

In order to improve our understanding of the usefulness of the confidence score we conducted several exploratory analyses. From our institutions' perspective the quality of the whole page is very important, as this can be used to determine which pages need to be re-OCRed of post-processed. We therefore try to determine a way to measure if the confidence score can be used as an accurate proxy measure for quality on the page level.

Furthermore, in order to get a better insight into the specific performance of the OCR engine, we analyzed the results of the OCR on word level, comparing OCR results with a Dutch lexicon and to Ground Truth.

2 Related Work

Finding information about the confidence scores of OCR engines is extremely hard, and there appears to be no standard. The documentation on the confidence score and what a confidence score exactly represents differs per company. ABBYY Finereader describes their word confidence as an estimated probability that the chosen word variant is correct. However, they point out that this metric is only useful for in-page word comparisons [2]. The documentation of Kofax stated little information about their word confidence. They mention that 0% means low confidence and 100% means a high confidence. Furthermore, they mention that the confidence measure is not comparable between different engines [10]. For other engines we could not find information about whether they use a confidence score or how the confidence score is calculated.

We were unable to find any documentation on how the different OCR engines actually calculate the confidence scores. Some engines mention that bonuses and/or penalties are also incorporated, but a precise method is not given. It appears that the user is also able to influence the confidence calculation by changing certain settings [2]. Due to the lack of standardization and transparency for the confidence calculation a comparison between engines is not useful in our setting. In fact, as confidence scores are apparently defined by the recognition engine, even similar confidence scores may not translate well between engines and represent a different confidence for each engine.

Various studies have mentioned the use of confidence score as a indication of OCR quality. Some studies report positively about the usability of confidence scores as quality indication. [17] showed that the mean confidence score correlates well with the character error rate (CER), as well as with lexicality. Similar research was done by [5]. While they did not use the confidence scores as it was not clear what exactly was being measured, they did see a promising correlation. [12] performed an extensive survey of post-OCR processing approaches where they saw an important role for the confidence score in detecting where post-OCR improvements could be made. Word confidence has also been used to finetune the selection of bounding boxes by [4] where a multimodel relation was shown between the word confidence and the noise within bounding boxes.

On the other hand, experiments from [15] show that the word confidence score as provided by the OCR engine deviates from the true word recognition. Consequently, they conclude that the word confidence score are a limited estimator. Supporting this, [11] compared the OCR confidence score for each page with the CER and found that the confidence score had at best a slight correlation with the CER and was not useful as a parameter for quality. [19] also found no clear use for the confidence score provided with OCRed texts, not only from a theoretical point, but also its practical implementation as many user interfaces do not support filtering or automatic extraction of the confidence. They also noticed that the method of calculation and documentation for the calculation of the confidence score is quite nontransparent, making it hard to trust in the measure.

[6] mentions that confidence scores are often used as a substitute for accuracy by lack of other, more accurate, metrics. They also present a method to transform the confidence score into a proxy measure. They suggest taking a subset from which the quality is known, and from which the confidence scores are available. Using this subset an algorithm needs to be written to correlate the two scores and calculate a proxy accuracy measure. This proxy measure can then be used on the collection as a quality indication. Regular checks on the algorithm are necessary, to see if it still fits the collection.

3 Methodology

In this study, we analyse the confidence scores using various different viewpoints to get a better insight in the usability of the confidence scores as quality indicators.

First, we examine to what extent the average confidence score on page level can be used by institutions to support decision making regarding which parts of the collections are most suited to be re-OCRed or post-processed. We therefore follow the suggested approach of [6]: to correlate the actual accuracy with the extracted confidence scores and use these scores to create an algorithm that provides a proxy measure.

Furthermore, to get a more detailed view on the performance of the OCR engine and to better understand the word confidence, we zoom in on the word level, using a lexicon lookup and the Ground Truth to find discrepancies between word confidence and the probable 'correctness' of a word.

3.1 Dataset

We used a set of 2000 newspaper pages ranging from 1631 to 1995 [20]. These pages were originally digitized by two different companies, using three different versions of ABBYY Finereader [1]. The full 2000 pages were re-OCRed by a third producer with a newer version of ABBYY Finereader leading to a dataset of 4000 pages digitized by three different producers and four different ABBYY Finereader versions. In addition to the OCRed documents, there is also manually created Ground Truth available for each document.

The data contains newspapers from a broad range of years and contains both scans from microfilms and scans from paper. The years are divided in three time ranges, based on spelling changes in Dutch.

Table 1 shows an overview of the pages divided by producer and ABBYY version. The OCR is stored in Alto XML files. These files provide the confidence score of each detected word. These word confidence scores are used to calculate the average word confidence of per page.

Table 1. Distribution of pages among producer and ABBYY Finereader version.

Producer	ABBYY version	Number of pages
A	8.1	1325
A	9	31
A	10	92
B	10	552
C	12	2000

3.2 Analysis on Page Level

For our analysis on page level, we compared the average word confidence per page with the WER_oi per page. The average word confidence was based on the individual word confidence of all words on a page. The WER_oi was determined for all OCRed document with the use of the Ground Truth and the ocrevalU-Ation tool [8] with the default settings. We choose to use the WER_oi for the comparison instead of the WER, as for this analysis, we are more interested in the correct prediction of each individual word than in the correct order of words.

To determine if the average word confidence is a reliable indication of the quality of a page, we calculate the correlation between the average word confidence and the WER_oi. If these measures correlate it could imply that the word confidence can be used in a similar way as the WER_oi. In order to exclude that certain characteristics have a strong influence on the correlation, we also determined the correlation between the word confidence and WER_oi for each producer, ABBYY Finereader version, year group and whether it was a microfilm scan.

Selection of the appropriate correlation type is done by plotting the average word confidence and the WER_oi in a scatterplot. These plots are then manually inspected to determine which correlation coefficient to use. If a linear relation is detected Pearsons r will be used. If a non-linear monotonic relation is detected, Spearmans rho will be used. If no relation is detected, the page level analysis will be aborted.

Based on the outcome of the correlation, we will either continue with exploring a proxy measure (see Sect. 3.3) or discard the exploration and continue to an analysis on word level (see Sect. 3.4). We consider a correlation with a coefficient of 0.8 or higher as a strong enough correlation.

3.3 Exploring a Proxy Measure

Based on the outcomes of the previous section, we choose a set corresponding to an ABBYY Version with an high enough correlation (>0.8) and a sufficient amount of pages. We choose an engine based selection as this is the most practical selection method to use as institution. If other stratification levels, such as year group, point to a significant disturbance of the correlation, these are removed from the dataset before continuing further analysis.

For this subset, we will explore if a proxy WER_oi measure can be calculated with use of the average word confidence, to classify the quality of a page.

To do this, we will compare three approaches:

1. Naive conversion to convert the average word confidence (wc) into a proxy WER_oi (proxy) measurement with the formula:

$$proxy = (1 - wc) * 100 \tag{1}$$

2. The use of simple linear regression to convert the average word confidence (wc) to a proxy WER_oi measurement with the formula:

$$proxy = a * wc + b \tag{2}$$

3. The use of polynomial regression to convert the average word confidence (wc) to a proxy WER_oi measurement with the formula:

$$proxy = a * wc^2 + b * wc + c \tag{3}$$

Before calculating the proxy WER_oi, we divided our dataset randomly in a train (70%) and test (30%) set. The train set was used to determine the formula for calculating the proxy WER_oi measure, the test set was kept apart for testing the formula.

As the word confidence represents the confidence that a word is correct, and the WER_oi is the percentage of word errors, OCR confidence score were inverted into a measure of 'unconfidence' or confidence that the word is wrong. This is then multiplied by 100 to achieve a WER_oi proxy measure on the same scale as the WER_oi.

For both linear regression and polynomial regression the formula obtained from the regression is used to predict the WER_oi proxy measure based on the average word confidence.

The performance of the three methods was determined by calculating the Mean Square Error and R^2 based on the test set. The method with the best results was chosen for further analysis.

These comparisons utilize a continuous scale. However, for most institutions it is more relevant to have a broader classification. For example, knowing if a page is good enough to be presented online, or if it is desirable to re-digitize it. Therefore, to get an indication of the practical usefulness of our best performing method, a step was added in which the classification performance of the selected method was tested.

We started by categorizing the pages based on the WER_oi. We based our cut-offs on the recommendations of [18]. They recommended an OCR quality of at least above 80%, but preferably over 90% for downstream tasks and analysis. To translate this to WER_oi, the inverse of these quality cut-offs is taken as the WER_oi corresponds to the percentage of faults and their cut-offs are based on the percentage quality. This results in cut-off values of WER_oi equal than or lower than 10 for the desired ('Good') category, between 10 and 20 for the minimal required ('Average') category, or bigger than 20 as low quality category.

We then calculated the WER proxy measure using the correlation formula of the best performing method and categorized all pages into on of the above categories. The categorization performance of the method was determined using a confusion matrix from which the accuracy, precision and recall were calculated.

3.4   Analysis on Word Level

For our analysis on word level, we chose one or multiple potential interesting subsets based on the results of Sect. 3.2. All words of the subset were extracted with their corresponding word confidence score. Then, we pre-processed the words to be able to perform a decent lexicon comparison. We executed the following steps:

- If a word is more than one character long, and it starts or end with a punctuation mark, this punctuation mark is removed;
- If there are any uppercase characters in the word, they are transformed to lower case.

Then, we performed a lexicon comparison for all words. The lexicon we used was a combination of a modern lexicon from OpenTaal [13] and a historical lexicon from the Dutch Language Institute [9]. We labeled every word as either 'found in lexicon' or 'not found in lexicon'. When a word was purely numeric, or contained only numbers and periods or commas, it was labeled as 'found in lexicon'.

A boxplot and a histogram are used to determine how well the two categories correspond to the word confidence.

Any anomalies that are found in the data, such as words with a high confidence that are not found in the lexicon, will be further investigated. Such words are compared to the Ground Truth and investigated to determine if these words have specific characteristics.

4   Results

4.1   Exploratory Analysis on Page Level

To determine if the word confidence score can be used as a proxy for quality, we started by investigating the correlation between the WER_oi and the word confidence. The analyses were stratified by various variables (manufacturer, OCR software version, year group and he presence of a microfilm) in order to investigate strong effects on the confidence score.

As can be observed in Fig. 1, there was a clear, slightly non-linear, monotonic relation between the average word confidence and the WER_oi. Therefore we used Spearmann's rho to determine the strength of the correlation of the total set and the various subsets. Table 2 provides an overview.

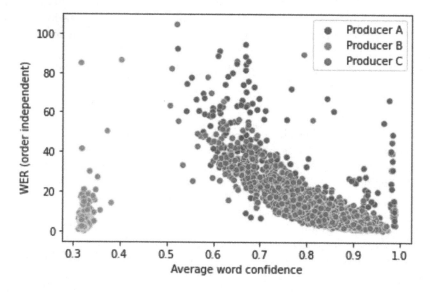

Fig. 1. Correlation between average word confidence and WER_oi per producer

As can be seen in the plot, the confidence score is strongly dependent on the producer and producer B appears to be a discrete subset (Fig. 1). Despite a high correlation of the larger subsets, the correlation of the total set is quite low with a confidence interval of −0.440 to −0.489. This is caused by the extreme low and inverted correlation of the word confidence from producer B to the

WER_oi (0.367). As the correlation of producer B is very different from the other producers it could have a distorting effect when determining differences in version, year or presence of microfilm. Therefore we excluded this producer for the remaining subsets.

Table 2. Distribution of pages per time period. For the sets marked with an *, producer B was excluded

	# pages	Spearmans' rho	p-value	confidence interval
Total	4000	−0.465	≤0.001	[−0.440, −0.489]
Stratified by producer				
Producer A	1448	−0.780	≤0.001	[−0.759, −0.799]
Producer B	552	0.367	≤0.001	[0.437, 0.292]
Producer C	2000	−0.854	≤0.001	[−0.841, −0.865]
Stratified by ABBYY Finereader version				
ABBYY 8	1325	−0.774	≤0.001	[−0.752, −0.795]
ABBYY 9	31	−0.934	≤0.001	[−0.866, −0.968]
ABBYY 10*	92	−0.876	≤0.001	[−0.818, −0.917]
ABBYY 12	2000	−0.854	≤0.001	[−0.841, −0.865]
Stratified by year group				
1631–1882*	517	−0.919	≤0.001	[−0.904, −0.931]
1883–1947*	1885	−0.786	≤0.001	[−0.768, −0.802]
1948–1995*	1046	−0.760	≤0.001	[−0.733, −0.785]
Stratified by microfilm				
microfilm*	841	−0.765	≤0.001	[−0.735, −0.791]
no microfilm*	2607	−0.834	≤0.001	[−0.821, −0.845]

There is some slight variation between the various ABBYY Versions. ABBYY 9 has the highest correlation (−0.934), but also the lowest number of pages. For the year groups, the oldest newspapers (1631–1882) had the strongest correlation (−0.919), while the most modern newspapers had the lowest correlation (−0.760). When looking at the presence of microfilm, the strongest correlation is on the group that does not have a microfilm (−0.834).

4.2 Exploring a Proxy Measure

For the exploration of a proxy measure, we used a subset of the total dataset including only ABBYY version 12.

This set was split into a train (1400 pages) and a test set (600 pages) and the three approaches (3.3) were performed on the train set. Scatterplots with trendlines for the three approaches are shown in Fig. 2. The mean square error and R^2 were calculated to evaluate the various models and are presented in Table 3.

Of the three methods, the polynomial formula has the best fit to the data and is best suited to determine the quality based on the word confidence score,

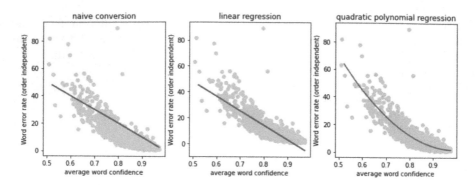

Fig. 2. Comparing proxy measures

Table 3. Comparing proxy measures

	Mean Square Error	R^2
Naive conversion	68.49	0.203
Linear regression	25.5	0.703
Quadratic polynomial regression	**17.75**	**0.793**

as it had the lowest error and the highest percentage of explained variance (17.75 and 0.793 respectively).

We therefore continued with this method, with the corresponding formula:

$$276.234x^2 - 550.55x + 275.748 \tag{4}$$

For the practical application using quality categories (Sect. 3.3) we determined the accuracy of the proxy measure test set using the polynomial regression. The results are presented in Table 4 and as a confusion matrix in Fig. 3.

We found that for ABBYY 12 the category with the lowest WER_oi, the 'Good' category, could be predicted with the highest precision and recall, followed by the 'Low' and the 'Average' categories. The total accuracy for the model is 0.83.

From the confusion matrix it is noticeable that there is almost no contamination between the two outer categories. The 'Good' quality category contains only one page (0,3%) with a true classification of 'Low' that is falsely classified as 'Good'. Similarly, the 'Low' category contains only one page (2%) with a true classification of 'Good' that is falsely classified as 'Low'. There is significantly more overlap between the 'Low' and 'Average' categories, and the 'Good' and 'Average' categories.

In most workflows the 'Low' and the 'Average' categories will likely undergo further checks as these are targets for re-OCRing or post-processing. A 'Good' that is falsely predicted as an 'Average' is therefore not as much of a problem as an 'Average' that is falsely predicted as a 'Good'. The confusion matrix shows that 28 pages with an actual 'Average' quality were falsely predicted as 'Good' quality. Whether this can be a problem depends largely on how far the actual WER_oi is from the predicted WER_oi. To establish if these 28 false positives are closer to 'Good' or to 'Low' we determined the mean of the actual WER_oi. The mean actual WER_oi of this group was 12.86, meaning that these false positives were closer to 'Good' than to 'Low'.

Table 4. Performance of proxy measure for each category

	Precision	Recall	f1-score	# pages
Good	0.93	0.88	0.90	401
Average	0.59	0.73	0.65	126
Low	0.86	0.75	0.80	73

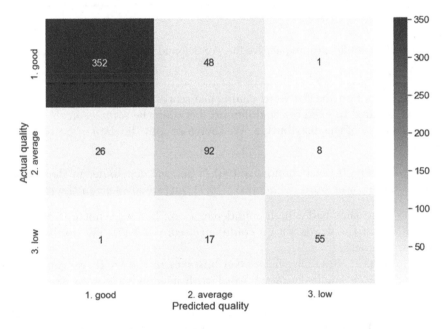

Fig. 3. Confusion matrix of proxy measure performance

4.3 Analysis on Word Level

Experiment 3a: Exploring ABBYY 12 Finereader for Producer C.
From the 2000 pages of the ABBYY 12 version from producer C, we extracted all words with their corresponding word confidence. This resulted in a list of 7,227,304 words. For each word it was determined if it could be found in the lexicon. For both groups, found or not found in the lexicon, the word confidence was plotted (Fig. 4). As can be observed from the figure, it appears that the

higher the word confidence, the higher the chance that the words can be found in a lexicon. For words that have a word confidence of 0.9 or higher, 94.8 % can be found in the lexicon indicating a strong relation between word level confidence and 'correctness' of a word.

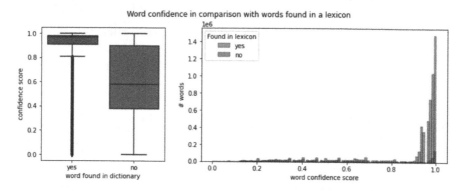

Fig. 4. Word confidence frequencies for words found or not found in a lexicon

As the histogram for the word confidence scores shows a left-skewed distribution, we decided to explore the difference between the samples in the tail and those in the body of the distribution. We therefore split the data into two specific cases:

1. a word has a high word confidence (≥ 0.7) but was not found in the lexicon;
2. a word has a lower word confidence (≤ 0.7) but was present in the lexicon.

For the cases that had a high confidence score, but were not present in the lexicon, 50.9% of the words with a confidence score of 1 (78,680 words) existed of only one character.

To determine if these may have been mistakes in the OCR, we counted the occurrence of each single-character word with a confidence score equal to or greater than 0.7 and compared these to the occurrence of the same character in the Ground Truth. From this comparison, we saw that some single character words, despite their high confidence, are likely OCR errors, while others are likely correct words. Some examples of single character words with their corresponding counts are shown in Table 5.

A negative Delta indicates that more occurrences of a character are found in the Ground Truth than in the OCR. When more single character words are found in the OCR than in the Ground Truth, the chance that these are OCR errors is high, while when there more single character words are found in the Ground Truth, these are likely correct. As we look at only a small percentage of the OCR, the difference can be quite big but it can still be used as a rough estimation that likely they where correctly detected.

Table 5. High confidence, not found in lexicon

word	words with confidence ≥ 0.7	words in GT	Δ%	likely correct
,	9,385	987	0.895	no
ʼ	2,298	31	0.987	no
:	1,544	527	0.659	no
;	1,415	210	0.852	no
f	3,263	10,328	−2.165	yes
ſ	1,935	10,205	−4.274	yes
−	3,186	6,227	−0.954	yes
a	1,771	2,100	−0.186	yes

Another noteworthy observation was that there were several small words, like articles and prepositions, with a wide variety of confidence scores. For some of these, more than half of the occurrences had a confidence score lower than 0.7. The total occurrence of these words in the OCR was close to the total of occurrences in the Ground Truth, indicating that they were likely correctly recognized, despite the variation of confidence scores. Some examples are shown in Table 6.

Table 6. Low confidence, found in lexicon

word	% with confidence ≤ 0.7	total words OCR	Total words in GT	Δ%
in	70.3%	123,822	129,013	−0.0419
en	68.6%	145,130	153,604	−0.0583
op	58.4%	64,041	63,721	0.005
de	55.7%	359,191	381,197	−0.061

Experiment 3b: Comparing ABBYY Version 10 for Producer A and B. In Sect. 4.1 we determined that the correlation between the average word confidence and the WER_oi was highly dependent on the producer. Where producer A and C have a very similar correlation, producer B had a very divergent result, with much weaker correlation between the average confidence score and the WER_oi.

To explore the influence of this producer variance, we selected all words that from both producer A and B that were digitized with the same ABBYY version (ABBYY version 10). The word confidence was analyzed for words occurring and not occurring in the lexicon. The word confidences per group are presented in a boxplot and histogram, as shown in Fig. 5 for producer A and Fig. 6 for producer B.

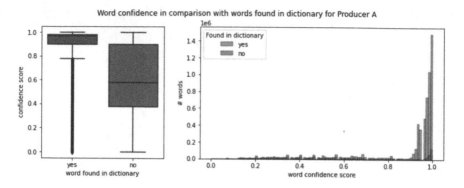

Fig. 5. Word confidence frequencies for words found or not found in a lexicon for producer A

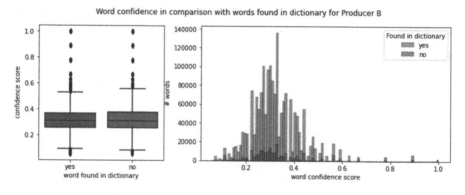

Fig. 6. Word confidence frequencies for words found or not found in a lexicon for producer B

There is a large difference in distribution between both producers. Producer A (733,4917 words) has a clear left-skewed distribution of the word confidence score, whereas producer B (1,934,017) has a more or less normally distributed word confidence score. For producer A most words that are found in the lexicon have a word confidence higher than 0.7 (80.8%), and most words that were not found in the lexicon had a word confidence lower than 0.7 (63.5%). Whereas for producer B the words 'found in the lexicon' and words 'not found in the lexicon' had similar word confidences, as can be observed from the overlapping distributions in Fig. 6. For producer B, 99.8 % of the words found in a lexicon have a word confidence lower than or equal to 0.7 against 84% of the words that were not found in a lexicon.

5   Discussion

Our results suggest that under certain conditions the confidence score can be used as a proxy measure for quality. In practice this would mean that when a small sample shows a correlation between word confidence and WER_oi, (polynomial) regression can be used to create an indication of the quality of the digitized texts. In our study polynomial regression as a proxy measure was most accurate in detecting digitized texts with the lowest WER_oi, which corresponds to the texts with the highest quality.

The results as presented in the confusion matrix show that there is almost no contamination between the 'Good' and 'Low' categories. This implies that when a page is classified as either of 'Good' or 'Low' quality, the chance of belonging to the other of these two categories is very low. As cross-contamination between these two categories undesirable, this result is very promising.

However, these formulas do not seem to generalize very well between various versions of OCR software or various producers. Meaning that with every new producer or version, the formula must be recalculated. Nonetheless, organizations can use one small, standard set of Ground Truth to re-calculate the formula when a new engine or producer is introduced, making this a quickly obtainable method that needs little in the way of resources. When the scans corresponding to this Ground Truth set are re-OCRed for every new producer or engine, it can be used as a quality control to compare the new OCR results and performance.

An interesting and unexpected result was that the year group with the highest correlation between the proxy WER measure and the WER_oi was the 1631–1882 group. In this period the Dutch language was not yet standardized and newspapers from that time had large spelling variations and generally differ more from modern Dutch than the other year groups in our study. After some inquiries, it appears that the producer added a specific historical lexicon to the OCR engine which we suspect has contributed to a higher word confidence.

A more worrying result is that the confidence scores from producer B have such a different correlation compared to the other producers. After inquiry it became apparent that the ABBYY version of producer B was a non-optimized, 'from scratch', version. The other producers trained and optimized their ABBYY versions before use. Also, we strongly suspect that provider B did not use historical lexicons, whereas the other producer included these in their workflow.

All the above combined make it very important to know what engine, producer and personalisation of the OCR was applied before drawing conclusions. The usefulness of our method is therefore strongly influenced and limited by these variables.

A remarkable result was that a large part of the high confidence words (>0.7) were in fact single character words. Initially such words would be considered incorrect, as these words were not found in the lexicon and there are only a few single letter words in Dutch. However, upon closer inspection with the Ground Truth, it appeared that a part of these single character words were identified correctly and consisted, among others, of currency marks (f and ƒ, which stands for florin, the Dutch currency from the 15th century until 2002) and parts of enumerations (the dash). In these cases the engine identified the correct 'word', even though the words did not exist as such. However, other single character words were present in much larger numbers in the OCR than in the Ground Truth, indicating that they were incorrectly identified, despite their high confidence. In addition, a large section of two character words had a low word confidence (<0.7) but could be found in the lexicon. These consisted of, among others, articles and prepositions. This may be explained by the fact that for short words it is more difficult to find the correct word when the original material is damaged or unclear. Missing one letter would mean that a large part of the word is uncertain, drastically lowering the confidence score. This shows that it is important to know what the contents of the texts are and that specific rules and conditions must be kept in mind for language, time period and type of text.

It is important to note that we used the WER_oi measure of the OCREvaluation tool [8]. As different toolings can have different results [11], it could be that our results are biased due to the choice of software for the calculation of the WER_oi. Furthermore, when looking at the results of the word level analysis, it is important to keep in mind that if a word is found in a lexicon it does not necessarily mean that the word is the correct word. Small substitutions can change a word into a different, 'correct', word that is not the word in the original text.

In this study we focused on the page level as this was most relevant for our institution. However, researchers generally prefer to know the quality at the article level as this helps them select what they need for their research. In the future we will expand the current research to include article level data. Also, We would like to replicate our findings using texts that were digitized with other engines to see how well the method generalizes.

As the success of OCR is largely dependent of the quality of the scan and the conditional of the original material, it would be interesting to determine if the confidence score does not only correlate to the quality of OCR, but also to the quality of the scan or original material. This could help institutions pinpoint where to repeat or improve scans, or where to redo the OCR.

6   Conclusion

In conclusion, when there is a correlation between word confidence and WER_oi, it is feasible to build a proxy WER_oi measure to classify texts into quality groups and use this groups to determine which pages are a good candidate for re-OCRing or post-processing or already good as it is. This can be done with a minimum of Ground Truth, making it very interesting method to quickly sort digitized texts into rough quality categories.

References

1. ABBYY: ABBYY FineReader Server (2022). www.abbyy.com/finereader-server/
2. ABBYY: FineReader Engine 12 for Windows Developer's Guide (2022). http:// help.abbyy.com/en-us/finereaderengine/12/user_guide/introduction_startpage/
3. Anderson, N., Muhlberger, G., Antonacopoulos, A.: Optical character recognition: IMPACT best practice guide. www.digitisation.eu/download/website-files/BPG/ OpticalCharacterRecognition-IBPG_01.pdf. Accessed 05 Oct 2022
4. Gupta, A., et al.: Automatic assessment of OCR quality in historical documents. Proc. AAAI Conf. Artif. Intell. 29(1) (2015). https://doi.org/10.1609/aaai.v29i1. 9487
5. Hill, M., Hengchen, S.: Quantifying the impact of dirty OCR on historical text analysis: eighteenth century collections online as a case study. Digit. Scholarsh. Hum. **34**, 825–843 (2019). https://doi.org/10.1093/llc/fqz024
6. Holley, R.: How good can it get? Analysing and improving OCR accuracy in large scale historic newspaper digitisation programs. D-Lib Mag. Mag. Digit. Libr. Forum **15** (2009)
7. Impact Centre of Competence: Confidence Level (OCR) (2018). www.digitisation. eu/glossary/confidence-level-ocr/
8. IMPACT Centre of Competence: ocrevalUAtion (2019). http://github.com/ impactcentre/ocrevalUAtion
9. Instituut voor de Nederlandse taal: INT Historische Woordenlijst (2012). http:// taalmaterialen.ivdnt.org/download/tstc-int-historische-woordenlijst/
10. Kofax: Kofax documentation. http://docshield.kofax.com/KTA/en_US/740- uc0n6j0c5s/help/SD/ScriptDocumentation/c_Welcome.html
11. Neudecker, C., Baierer, K., Gerber, M., Clausner, C., Antonacopoulos, A., Pletschacher, S.: A survey of OCR evaluation tools and metrics. In: The 6th International Workshop on Historical Document Imaging and Processing, HIP 2021, pp. 13–18. Association for Computing Machinery, New York (2021). https://doi. org/10.1145/3476887.3476888
12. Nguyen, T.T.H., Jatowt, A., Coustaty, M., Doucet, A.: Survey of post-OCR processing approaches. ACM Comput. Surv. **54**(6) (2021). https://doi.org/10.1145/ 3453476
13. OpenTaal: Nederlandse woordenlijst (2020). http://github.com/OpenTaal/ opentaal-wordlist
14. Padilla, T., Allen, L., Frost, H., Potvin, S., Russey Roke, E., Varner, S.: Final report – always already computational: collections as data, May 2019. https://doi. org/10.5281/zenodo.3152935
15. Salah, A.B., Moreux, J.P., Ragot, N., Paquet, T.: OCR performance prediction using cross-OCR alignment. In: 2015 13th International Conference on Document Analysis and Recognition (ICDAR), pp. 556–560 (2015). https://doi.org/10.1109/ ICDAR.2015.7333823
16. Smith, D., Cordell, R.: A research agenda for historical and multilingual optical character recognition (2019). http://hdl.handle.net/2047/D20298542
17. Springmann, U., Fink, F., Schulz, K.: Automatic quality evaluation and (semi-) automatic improvement of mixed models for OCR on historical documents (2016)
18. van Strien, D., Beelen, K., Ardanuy, M.C., Hosseini, K., McGillivray, B., Colavizza, G.: Assessing the impact of OCR quality on downstream NLP tasks. In: Proceedings of the 12th International Conference on Agents and Artificial Intelligence - Volume 1: ARTIDIGH, pp. 484–496. INSTICC, SciTePress (2020). https://doi. org/10.5220/0009169004840496

19. Traub, M.C., van Ossenbruggen, J., Hardman, L.: Impact analysis of OCR quality on research tasks in digital archives. In: Kapidakis, S., Mazurek, C., Werla, M. (eds.) TPDL 2015. LNCS, vol. 9316, pp. 252–263. Springer, Cham (2015). https:// doi.org/10.1007/978-3-319-24592-8_19
20. Wilms, L., Koster, T.: Historical newspaper OCR ground-truth data set (2020)

Optimizing the Performance of Text Classification Models by Improving the Isotropy of the Embeddings Using a Joint Loss Function

Joseph Attieh[1,3,4(✉)], Abraham Woubie Zewoudie[2], Vladimir Vlassov[3], Adrian Flanagan[4], and Tom Bäckström[1]

[1] Aalto University, Espoo, Finland
{joseph.attieh,tom.backstrom}@aalto.fi
[2] Silo AI, Helsinki, Finland
abraham.zewoudie@silo.ai
[3] KTH Royal Institute of Technology, Stockholm, Sweden
vladv@kth.se
[4] Huawei Technologies Oy., Helsinki, Finland
adrian.flanagan@huawei.com

Abstract. Recent studies show that the spatial distribution of the sentence representations generated from pre-trained language models is highly anisotropic. This results in a degradation in the performance of the models on the downstream task. Most methods improve the isotropy of the sentence embeddings by refining the corresponding contextual word representations, then deriving the sentence embeddings from these refined representations. In this study, we propose to improve the quality of the sentence embeddings extracted from the [CLS] token of the pre-trained language models by improving the isotropy of the embeddings. We add one feed-forward layer between the model and the downstream task layers, and we train it using a novel joint loss function. The proposed approach results in embeddings with better isotropy, that generalize better on the downstream task. Experimental results on 3 GLUE datasets with classification as the downstream task show that our proposed method is on par with the state-of-the-art, as it achieves performance gains of around 2–3% on the downstream tasks compared to the baseline.

Keywords: Text Classification · Isotropy · Embeddings · BERT · IsoScore

1 Introduction

Recent Transformer-based models have achieved significant success in various natural language processing tasks [6]. However, [3] observes that some language models including Bidirectional Encoder Representations from Transformers (BERT) produce contextualized word representations that are not isotropic.

© The Author(s), under exclusive license to Springer Nature Switzerland AG 2023
G. A. Fink et al. (Eds.): ICDAR 2023, LNCS 14191, pp. 121–136, 2023.
https://doi.org/10.1007/978-3-031-41734-4_8

In other words, the information in the embeddings is not uniformly distributed in all directions in the space. This is not desirable as these representations vary the most in top directions, which limits the expressiveness of the space. [4] referred to this problem as the representation degeneration problem. Even though researchers did not agree on the source of anisotropy, having an isotropic space is very desirable as the more isotropic the space is, the more diverse the embeddings are. Furthermore, having an isotropic space affects the optimization of the model (i.e., convergence and accuracy), and leads to improvement in the performance of the model [17].

As mentioned previously, BERT-based models suffer from the problem of having an anisotropic space. This affects the representation capacity of the embedding space and affects the accuracy of the downstream task. More specifically, [12] highlighted that the Classification ([CLS]) token representations are much more anisotropic than all representations in the fine-tuned space. The authors highlighted that this problem becomes even more dramatic after fine-tuning, as this process tends to concentrate information about the target task in the dominant directions (i.e., the top principal components). Therefore, we propose to learn an embedding transformation that renders the [CLS] embeddings more isotropic without losing the information of the target task; once improved, the embedding space should exhibit better statistical properties, which should result in a better performance on the downstream task (i.e., increase in the model performance). Thus, we proceed by freezing the parameters of the fine-tuned model and the downstream task layer, and adding an Isotropy Layer between these two models. This Isotropy Layer is one feed-forward layer, which goal is to output embeddings of better isotropy. The parameters of this layer are learned using a joint loss function that combines an isotropic loss function and the downstream task loss function.

We apply empirical methods to quantitatively measure the improvement of the models in terms of isotropy and performance on the downstream task. The improvement in isotropy is evaluated by computing two measures of isotropy, mainly the isoscore and the explained variance, while the improvement in the model performance is evaluated by computing a dataset-specific metric. Two main experiments are carried out. The first experiment compares the proposed method to the baseline (i.e., fine-tuned model with no Isotropy Layer), while the second experiment compares the method to the Isotropic Batch Normalization (IsoBN) method [22]. To the best of our knowledge, our work is the second study besides IsoBN [22] that aims to improve the isotropy of the [CLS] representations.

The main contributions of this work are as follows:

1. We provide a method to improve the isotropy of the embeddings by adding an Isotropy Layer at the output of the fine-tuned language model, and only training this layer using a joint loss function and not the entire pre-trained language model. The need to only fine-tune the Isotropy Layer while the other layers are frozen means that this approach can be applied retroactively to previously trained models.

2. As shown by [12], it is not sufficient to only improve the isotropy of the embeddings, as the embeddings need to maintain the semantics required for the downstream task. Therefore, achieving perfect isotropy of the embeddings might not ultimately lead to a better model. To achieve the right amount of isotropy needed by the model, we propose a novel joint loss function that optimizes both an isotropic loss measure as well as a downstream task loss, and results in embeddings that are more isotropic and that perform better on the downstream task. This joint loss function should encourage researchers to include unsupervised quality measures inside the loss function to enforce some statistical properties on the model.

3. We evaluate our method of improving the isotropy of embeddings on multiple datasets from the General Language Understanding Evaluation (GLUE) benchmark for several downstream tasks and compare its performance with the IsoBN method. To the best of our knowledge, our work is the second study besides IsoBN [22] that aims to improve the isotropy of the [CLS] representations (the other studies focus on improving the isotropy on the token level, rather than on the sentence level). The experiments were conducted using two bert-based models (bert-base and roberta-large). Future work will test the approach on other state-of-the-art models such as ERNIE and T5. Experimental results on 3 GLUE datasets demonstrate that our method can improve isotropy significantly, as well as improve the model performance.

4. We also provide an additional case study, which deals with abusive language detection. The experiment presented supports some of the design decisions taken in this study (i.e., freezing the pre-trained language model and the classifier, and using the joint loss).

This paper is structured as follows: Sect. 2 introduces the concept of Isotropy as well as the related work that is relevant to the study. Section 3 presents the proposed method and describes the implementation of this method. Section 4 describes the experimental evaluation as well as the results obtained from these experiments. Section 5 presents experiments conducted on a public abuse language detection dataset. Finally, Sect. 6 provides a summary of the findings and conclusions as well as the future scope of this study.

2   Isotropy

Isotropy is a geometric property that assesses the distribution of the points in space [2]. In an ideally isotropic space, the embeddings are uniformly distributed in all directions of the space, i.e., the embeddings are not biased in a specific direction.

2.1   Properties

Isotropy has been linked to multiple properties in space. For instance, in an anisotropic space, randomly sampled words tend to be highly similar to one another in terms of cosine similarity [3]. Furthermore, the representations exhibit word-frequency bias, as the high-frequency words concentrate densely in the embedding space while low-frequency words disperse sparsely in the space [7]. Multiple studies tried to explain the source of the anisotropy. [15] showed that the anisotropy in contextual models is a product of rogue dimensions of the entire embedding space that drive the similarity metrics, explaining the high similarity property between random embeddings of these spaces. Furthermore, [2] showed that embeddings do not occupy a narrow cone, but rather only appear as a cone when projected to a lower-dimensional space. The authors showed that during training, word embeddings share the same direction gradients, therefore are shifted in one dominant direction in the vector space.

2.2   Measures

There is a need to approximate the degree of isotropy of the space, i.e., the spatial utilization of the embeddings. Methods that are based on the Principle Component Analysis are the most appropriate to study the isotropy of the space. We present the two most robust PCA-based methods to quantify the isotropy, mainly the explained variance ratio and the IsoScore as highlighted by [13].

Explained Variance Ratio. The explained variance ratio, which we refer to as EV_k Score, measures how much total variance is explained by the first k principal components of the data. This metric measures the difference in variance in different directions of the space. However, computing it requires the specification of a certain number of Principle Components (PCs). Therefore, in this study, we will be numerically examining this score for the first three PCs, and graphically for the top components. Given that λ_i is the ith largest singular value of the embeddings matrix E, the variance explained ratio is computed as follows:

$$EV_k(E) = \frac{\sum_{i=1}^{k} \lambda_i^2}{\sum_{i=1}^{D} \lambda_i^2} \tag{1}$$

IsoScore: The IsoScore [13] of an embedding space can be interpreted as the fraction of dimensions uniformly used by the embedding space. This score is derived from an isotropy defect that is calculated by computing the distance between the identity matrix and the normalized covariance matrix of the PCA-reoriented data. The IsoScore scales linearly with the number of dimensions used and is stable when distributions contain highly isotropic subspaces. A high IsoScore (i.e., close to 1.0), indicates that the principal components are uniformly distributed across all dimensions of space, implying that the space is isotropic. However, a small IsoScore (i.e., close to 0.0) indicates that the first components explain almost all the variance of the data, implying a highly anisotropic space.

2.3   Related Work

In this section, we present some related work aiming to solve the representation degeneration problem and improve the isotropy of the space. We can split the studies into two: (1) studies that regularize the embeddings during the training stage and (2) studies that post-process the embeddings after the training phase.

Regularizing the Embeddings. Multiple studies applied regularization to improve the isotropy of the learned embeddings. Firstly, [4] employed cosine regularization to decrease the similarity between the embeddings and increase the representation power of the space. However, [21] proposed the Laplacian regularization as a better alternative to cosine regularization, as it minimizes the similarities between the embeddings with similar contexts (instead of applying it to all embeddings). Moreover, [17] mitigated the fast singular value decay phenomenon of anisotropic space using spectrum control. Finally, [5] learned embeddings of better isotropy by alleviating the word frequency bias of anisotropic spaces using adversarial training.

Postprocessing the Embeddings. Other studies proposed to post-process the learned embeddings to improve the isotropy of the space. Firstly, [10] introduced the All-but-the-top method that removes the common vector and dominant directions from the embeddings, rendering them more isotropic. Secondly, [11] increased the isotropy by clustering the embeddings and nulling the principal components of each cluster. Third, [8] applied a weighted removal of a selected number of dominant directions from the embedding. The weights were learned through a word similarity task applied to the embeddings. Fourth, [14] proposed a whitening operation, effectively improving the isotropy of the sentence representations and reducing their dimensionality.

Discussion. Our method of improving the isotropy of embeddings is similar to the regularization-based approaches as it improves the embeddings through a penalty term. However, the measure used in our method is unsupervised and more robust, and it directly estimates isotropy instead of computing an indicator of isotropy. Furthermore, our method is not as computationally expensive as the regularization methods, as it only trains one feed-forward layer instead of the whole language model. Our method is also a form of postprocessing of the embeddings, as the layer introduced transforms the embeddings and makes them more isotropic, without compromising the modeling power of the space.

3   Improving the Isotropy of Embeddings Using an Isotropic Layer and a Joint Loss

3.1   Overall Description

As mentioned previously, we limit the scope of our work to the embedding space of the [CLS] representations. Our proposed approach assumes that the pre-

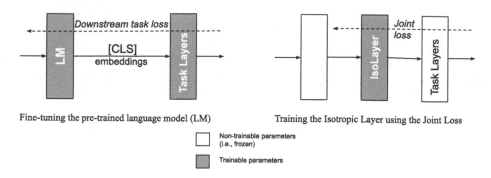

Fine-tuning the pre-trained language model (LM)   Training the Isotropic Layer using the Joint Loss

Non-trainable parameters
(i.e., frozen)

Trainable parameters

Fig. 1. Diagram summarizing the approach described in the study.

trained language model is first fine-tuned on the downstream task. This consists
of extracting the [CLS] embeddings from the language model and feeding these
embeddings to a predefined set of downstream task layers. Both the pre-trained
language model and the downstream task layers will be trained to perform the
downstream task. Furthermore, the downstream task considered in our study is
classification. Future work will test the approach on regression tasks.

In summary, the approach consists of adding a layer, referred to as the
Isotropy Layer, between the pre-trained language model and the downstream
task layers. This Isotropy Layer is responsible for transforming the [CLS] embeddings of BERT into embeddings with a better isotropic property. Since the goal
of the study is to improve the isotropy of the space to perform better on the
downstream task, we condition the learning process by freezing the parameters
of the downstream task layers as well as the language model. This choice is further justified in Sect. 5. Since only this layer is updated during training, we are
learning a clear transformation of the space; this transformation post-processes
the embeddings to improve the distribution of the embeddings in the space while
keeping the semantics needed to perform the downstream task.

The process can be summarized in Fig. 1. We present the details of the approach in the following subsections.

3.2 Freezing the Network and Adding Isotropy Layer

We insert one feed-forward layer, called the Isotropy Layer, between the fine-
tuned model and the downstream task layers. The goal of the Isotropy Layer
is to transform the [CLS] embeddings outputted by the fine-tuned model to a
new space that is more isotropic. It should be noted that this layer could be
replaced by a more complex neural network, with the only limitation that the
output of this neural network should be of the same size as its input (i.e., the
embedding vector outputted by the pre-trained model), to ensure compatibility
of the output of the network with the downstream task layers.

As mentioned previously, improving the isotropy by itself is not sufficient [12]. Therefore, the Isotropy Layer needs to maintain the semantics needed to perform the downstream task at hand. To do so, we perform the following:

- We freeze the parameters of the fine-tuned model. Freezing this fine-tuned model preserves the knowledge learned from the pre-training step, and reduces the costs of training.
- We freeze the parameters of the downstream task layers. As mentioned previously, the output of the Isotropy Layer (i.e., the transformed embeddings) needs to maintain the semantics needed to perform the downstream task. Therefore, we freeze these layers to condition the output of the Isotropy Layer to adjust to the knowledge of this layer during the training. This design choice is justified in the use case presented in Sect. 5.

3.3 Training the Isotropy Layer Using a Joint Loss

Now that the fine-tuned model and the downstream task layers are both frozen, we train the Isotropy Layer using the proposed joint loss function:

$$\mathcal{L} = \alpha \times \mathcal{L}_1 + (1 - \alpha) \times \mathcal{L}_2 \qquad (2)$$

We define the variables in the equation as follows:

1. $\mathcal{L}_1$ is the loss used to fine-tune the pre-trained model. It varies with the downstream task (i.e., CrossEntropy for the classification, Mean Squared Error for regression tasks). This measure is computed over the output of the new network (i.e., pre-trained language model followed by the Isotropy Layer and the classifier). This term ensures that the transformed embeddings maintain the semantic information required for the downstream task.
2. $\mathcal{L}_2$ quantifies the degree of the isotropy of the space of embeddings at the output of the Isotropy Layer. It is an unsupervised measure computed over a mini-batch of embeddings at a time. Intuitively, the bigger the batch of embeddings, the more accurate the isotropy measure is. This term acts as a regularizer for the new embeddings, and it pushes the weights of the Isotropy Layer to produce more isotropic embeddings.

We propose to use the IsoScore as the measure of quality used to compute $\mathcal{L}_2$. We usually desire to optimize a decreasing function. Knowing that the IsoScore increases for a better isotropic space, we propose to use the logarithmic of the inverse of the IsoScore. This decision was taken due to its sensitivity to the change in the measure used (i.e., IsoScore).

Ideally, setting α to 0.5 should result in equally acceptable isotropic property and downstream task performance. However, both losses in the joint loss operate around different scales. Therefore, the learning might be biased toward one of the losses and might optimize one of the losses at the expense of the other one. To prevent this issue, we normalize the losses using the Moving Average [19].

4 Experiments

To evaluate the proposed method, we conduct two main experiments. The first experiment evaluates the proposed method and compares it to our baseline (i.e., the text classification system without the Isotropy Layer) in terms of IsoScore, Explained Variance, and performance measurement. The second experiment compares the proposed method with the IsoBN method [22]. To our knowledge, the only study besides ours which improves the [CLS] embeddings is the IsoBN [22]. In their study, [22] first highlighted the anisotropic nature of the [CLS] embeddings. Then, they proposed to improve the isotropy of these embeddings using an isotropic variant of the batch normalization method.

4.1 Experimental Setup

Datasets: To evaluate our approach, we used multiple datasets from the GLUE benchmark [16]. The GLUE benchmark is a collection of Natural Language Understanding (NLU) tasks including question answering, sentiment analysis, and textual entailment. GLUE datasets favor models that learned to represent linguistic knowledge for sample-efficient learning and knowledge transfer across tasks [16]. Each dataset has its metric to evaluate the model performance. We selected the three specific datasets described in Table 1 because they represent the three main tasks of the GLUE benchmark, which are Inference Tasks (RTE), Single-Sentence Tasks (CoLA), and Similarity and Paraphrase Tasks (MRPC). Future work will evaluate the approach on all datasets in the GLUE benchmark. We evaluate the classification on the dev sets that were provided by these datasets[1].

Table 1. Details of the datasets used in the experimental evaluation.

Dataset ID	Dataset Name	Dataset Description	Performance Metric
RTE	Recognizing Textual Entailment	Determines whether each sentence entails a given hypothesis or not	Accuracy
CoLA	Corpus of Linguistic Acceptability	Determines whether each sentence is grammatically correct or not	Mathew's correlation coefficient
MRPC	Microsoft Research Paraphrasing Corpus	Consists of a pair of sentences and determines whether the sentences are paraphrases from one another	Accuracy

Models[2] The models were implemented using the transformers library provided by HuggingFace [18] using PyTorch. The optimizer used is AdamW [9].

---

[1] The labels of the test sets were not provided (they are only evaluated through the leaderboard at https://gluebenchmark.com/leaderboard) This setup also follows the work done by IsoBN [22].

[2] More information regarding hyper-parameter tuning is available in a technical report [1].

Early stopping was applied according to task-specific metrics on a validation set (train/validation split of 70/30). The approach is evaluated on two pre-trained language models, mainly *bert-base-cased* and *roberta-large*, as these models perform well for the English language. Since the datasets used are binary classification datasets, the downstream task layers consist of only one classification layer of 2 neurons. The activation function used is the softmax function and the loss used is the binary cross-entropy loss. The Isotropy Layer is a one-layer feed-forward neural network, with a number of neurons of the same size as the [CLS] embedding extracted from the pre-trained language model (768 in the case of BERT and 1024 in the case of RoBERTa). We could have opted out for a more complex neural network. We leave this direction for future studies.

4.2 Comparing the Proposed Approach to the Baseline

Model Performance and IsoScore. We proceed by applying our approach and evaluating the performance of the model using dataset performance metric and the isotropy of the transformed embedding space using the IsoScore. These measures have been computed over the dev set. Results are displayed in Table 2.

Table 2. Internal evaluation of the approach on 3 GLUE benchmarks. We can observe a significant increase in the isoscore and a notable increase in performance.

Dataset	Method	bert-base-case		roberta-large	
		Performance	IsoScore	Performance	IsoScore
RTE	Fine-tuned Language Model (LM)	67.87	0.0051	85.56	0.0049
	Fine-tuned LM + Isotropy Layer	71.84	0.025	87.36	0.0145
	Improvement of the approach	**+5.8%**	**+390.2%**	**+2.1%**	**+195.9%**
CoLA	Fine-tuned Language Model (LM)	61.61	0.004	67.23	0.0012
	Fine-tuned LM + Isotropy Layer	63.57	0.0255	68.77	0.0023
	Improvement of the approach	**+3.2%**	**+537.5%**	**+2.3%**	**+91.67%**
MRPC	Fine-tuned Language Model (LM)	85.29	0.0033	90.93	0.0016
	Fine-tuned LM + Isotropy Layer	87.99	0.0103	91.17	0.00245
	Improvement of the approach	**+3.17%**	**+212.12%**	**+0.26%**	**+53.13%**

Explained Variance. We examine the explained variance curve of the models trained by computing the metric over the top K=20 principal components. The results are displayed in Fig. 2. We can see that using the Isotropy Layer resulted in embeddings with a smaller explained variance compared to the baseline. This means that the singular values distribute more uniformly in the transformed space, inferring that the information is spread across more principal components uniformly (the space is more isotropic). We also notice that the proposed approach had limited improvement in explained variance for RoBERTa, compared to BERT. We provide multiple explanations in the discussion section for such behavior.

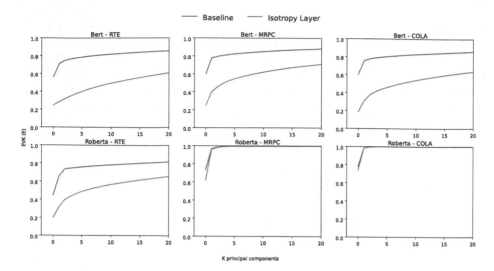

Fig. 2. The plots present the explained variance of the top principal components. These plots show that the proposed method results in a smaller explained variance compared to the baseline, which indicates that the variations of the embeddings tend to distribute equally in all directions (i.e., more isotropic space).

Table 3. Results on the dev sets of selected GLUE tasks after running 5 times with different random seeds. For the performance measures, we report the median and standard deviation over the 5 models. As for the isotropy measure, we report the explained variance of the model that exhibits median EV1.

Dataset	Method	Performance Measure		Isotropic Measure	
		IsoBN	IsoLayer	IsoBN	IsoLayer
RTE	BERT-base	67.87 (1.1)	67.72 (0.83)	0.88/0.93/0.95	0.61/0.75/0.78
	BERT-base+Isotropy	70.75 (1.6)	70.40 (1.05)	0.49/0.72/0.85	0.27/0.35/0.38
	RoBERTa-L	84.47 (1.0)	85.56 (0.8)	0.53/0.66/0.70	0.44/0.65/0.73
	RoBERTa-L+Isotropy	87.00 (1.3)	87.36 (0.6)	0.15/0.29/0.37	0.19/0.31/0.39
CoLA	BERT-base	60.72 (1.4)	60.89 (0.81)	0.49/0.58/0.64	0.60/0.75/0.77
	BERT-base+Isotropy	61.59 (1.6)	62.82 (0.85)	0.25/0.37/0.48	0.18/0.30/0.32
	RoBERTa-L	68.25 (1.1)	67.33 (0.9)	0.83/0.88/0.90	0.66/0.87/0.91
	RoBERTa-L+Isotropy	69.70 (0.8)	68.77 (0.8)	0.21/0.38/0.49	0.41/0.63/0.76
MRPC	BERT-base	85.29 (0.9)	86.64 (1.09)	0.76/0.87/0.89	0.63/0.80/0.81
	BERT-base+Isotropy	87.5 (0.6)	87.5 (0.42)	0.37/0.68/0.77	0.43/0.49/0.52
	RoBERTa-L	90.68 (0.9)	90.93 (0.2)	0.86/0.90/0.91	0.73/0.96/0.98
	RoBERTa-L+Isotropy	91.42 (0.8)	91.17 (0.3)	0.18/0.36/0.43	0.61/0.96/0.98

4.3   Comparing the Proposed Approach with IsoBN

We compare the proposed approach to IsoBN, the only approach in the literature besides ours that aims to improve the isotropy of the [CLS] embeddings. To prove that the improvements incurred by our approach are consistent, we run the approach on each dataset with 5 random seeds. Furthermore, we measure the isotropy by examining the explained variance of the top 3 principal components. Table 3 presents the results. As we can see, our approach provides improvements in both performance and isotropy. We notice that our approach is on par with IsoBN in terms of model performance. As for the isotropy, we can see that our approach results in better explained variance for all BERT models, while IsoBN results in better explained variance for the RoBERTa models. This is interpreted in the discussion section.

4.4 Discussion

BERT vs RoBERTa: We notice that the improvements on BERT were higher than the improvements in RoBERTa in terms of performance and isotropy. Investigating this phenomenon is left for future work. However, we provide some hypotheses that could explain the observed behavior. One source of this discrepancy can be attributed to the highly anisotropic nature of the RoBERTa embedding space; Fig. 2 shows that most of the variance is concentrated in the top 5 principal components (unlike BERT, where the variance is more distributed among the principal components). Furthermore, RoBERTa is more complex than BERT (larger architecture). Therefore, a solution worth exploring is to increase the complexity of the Isotropy Layer (i.e., replacing the one feed-forward neural network with a deeper network that can learn a more complex transformation resulting in better embeddings). Another source of the discrepancy observed could be due to the strict constraint imposed by the downstream task layers (that are frozen). In other terms, the downstream-task layers might rely heavily on the information encoded in the top principal components. A solution worth exploring in future work is to retrain the Isotropy Layer on the improved embeddings, fine-tune the downstream task layers on these transformed embeddings, and repeat both steps until convergence.

IsoLayer vs IsoBN: We pinpoint an interesting analogy between both approaches; the IsoBN approach employs an isotropic batch normalization to regularize the embeddings, while our method learns a transformation that adds an isotropic penalty term to regularize the embeddings. We should note that the proposed method trains only one feed-forward neural network, while the IsoBN method performs the training for the whole network. Even though both methods are on par in terms of performance, our approach can be applied to existing models and only requires training the single layer. A disadvantage of our method is that the performance is highly constrained by the downstream task layers. Perhaps, a more isotropic embedding space with better semantic properties can be reached with a different downstream task network.

5 Case Study - Abusive Language Detection

5.1 Overview

Abusive language (also known as hate speech) has seen a significant rise in the last decade. Consequently, researchers have worked on limiting the impact of abusive statements by building Automatic Abusive Language Detection models to solve this problem. Abusive Language Detection can be modeled as a text classification problem, where the text is inputted into the model and the degree of abusiveness is outputted by the model.

5.2 Experimental Setup

In this use case, we employ the Offensive Language Identification Dataset, known as OLID [20]. This dataset has been introduced in the OffensEval 2019 challenge, in which participants were asked to perform three tasks. We apply the method proposed in this paper on this dataset. We apply the approach using *bert-base-case*, with one Isotropy Layer and one downstream-task layer. The metric used to evaluate the performance of the model is the F1 Score, while the metric used to evaluate the isotropy of the embeddings is the IsoScore. We investigate and confirm some of the design decisions in our approach for the case of abusive language detection. We define different setups, where every setup represents a variant of the proposed method, differing by the modules that we choose to train as well as the loss function used. We compute the IsoScore on the embeddings outputted by the Isotropy Layer. If the Isotropy Layer is not present, the IsoScore is computed on the embeddings outputted by the pre-trained language model.

5.3 Results

Table 4 presents the different setups that we experimented with. Let's analyze the results:

- **Setups 1, 2, and 3:** Setup 1 corresponds to the baseline (i.e., the pre-trained language model and the classifier layer). We freeze the parameters of BERT from setup 1, then add the isotropy and the classification layers, and train these layers using the joint loss function (setup 2). Afterward, we freeze the parameters of both BERT and the classification layer from setup 1 and train the Isotropy Layer using the joint loss function (setup 3). Both setups 2 and 3 achieve a better performance in terms of IsoScore and F1 score compared to setup 1. We can observe that setup 2 generates embeddings of better isotropy, while setup 3 performs a more accurate classification. We can infer that freezing the classification layer and training the Isotropy Layer result in embeddings that are well tailored toward the downstream task.
- **Setups 4, 5, and 6:** In setups 4, 5, and 6, we train different modules in the system using the cross-entropy loss as shown in Table 4. The three setups generate embeddings that have a low IsoScore and do not achieve significant

performance gain compared with the baseline (setup 1). This experiment confirms that the improvements observed in the approach proposed in this paper are a result of improving the isotropy of the embeddings using the novel joint loss function.

- **Setups 7, 8, and 9:** In setups 7, 8, and 9, we train different modules in the system using the joint loss (when possible) as shown in Table 4. This experiment shows that the improvements incurred by these variants are not as significant as the improvements incurred by our approach. We can clearly see that it is better to fine-tune BERT using the cross-entropy loss before adding the Isotropy Layer.

Table 4. Results of the different setups considered. We use ✓ to indicate that the module is trained, − to indicate that the module is absent from the system, and (setup number) to indicate that the module was taken from a certain setup and its parameters were frozen. The losses that we experimented with are the cross-entropy loss (CE) and the joint loss.

Setup	BERT	Isotropy Layer	Class. Layer	Loss	IsoScore	F1 Score
1	✓	–	✓	CE	0.0043	73.87
2	(1)	✓	✓	Joint	**0.0445**	74.94
3	(1)	✓	(1)	Joint	0.0101	**75.05**
4	✓	✓	✓	CE	0.0036	72.87
5	(4)	✓	✓	CE	0.0030	73.79
	(4)	(5)	✓	CE	0.0030	74.06
7	✓	✓	✓	Joint	0.0045	73.86
	(7)	✓	✓	Joint	0.0064	73.97
	(7)	(8)	✓	CE	0.0064	73.66

6   Conclusions and Future Work

As mentioned in the previous sections, pre-trained language models and especially transformer-based models suffer from the problem of having an anisotropic embedding space. This affects the representation capacity of the embedding space and the accuracy of the downstream tasks. In our work, we proposed to learn an embedding transformation that improves the isotropy of the [CLS] embeddings by adding an Isotropy Layer at the output of the fine-tuned language model and only training this layer using a joint loss function. Once trained, the layer will output transformed embeddings of better statistical properties that result in a better performance on the downstream task. We applied empirical methods to quantitatively measure the improvement of the models in terms of isotropy and performance on the downstream task. The experimental results on

3 GLUE datasets showed that our proposed method is on par with the state-of-the-art, as it achieves performance gains of around 2-3% on the downstream tasks compared to the baseline. A promising direction would be to understand the impact of our solution on the semantics of the model. To do so, we propose to employ tools that allow us to navigate the embedding space, giving us insights into the distribution of the concepts in the embedding space (i.e., reach more interpretable results). Since this property is not supported by default, we leave this direction for future work.

Reproducibility Statement

As mentioned in the previous sections, all datasets used in this study are part of the GLUE benchmark (https://gluebenchmark.com/). Furthermore, all seeds have been fixed to ensure that the results are reproducible. More information regarding the details of the training process (including hyper-parameter tuning) is available in a separate technical report [1]. The source code of the model is available on GitHub at https://github.com/josephattieh/IsoLayer .

Acknowledgments. This work was supported by the Terminal Cloud Service Competence Center of Huawei Technologies Oy. in Helsinki, Finland in the context of the master thesis of the first author.

References

1. Attieh, J.: Optimizing the Performance of Text Classification Models by Improving the Isotropy of the Embeddings using a Joint Loss Function. Master's thesis, Aalto University. School of Science (2022). http://urn.fi/URN:NBN:fi:aalto-202209255727
2. Biś, D., Podkorytov, M., Liu, X.: Too much in common: Shifting of embeddings in transformer language models and its implications. In: Proceedings of the 2021 Conference of the North American Chapter of the Association for Computational Linguistics: Human Language Technologies. pp. 5117–5130. Association for Computational Linguistics, Online (Jun 2021). https://doi.org/10.18653/v1/2021.naacl-main.403, http://aclanthology.org/2021.naacl-main.403
3. Ethayarajh, K.: How contextual are contextualized word representations? comparing the geometry of BERT, ELMO, and GPT-2 embeddings. vol. abs/1909.00512 (2019). arXiv: abs/1909.00512
4. Gao, J., He, D., Tan, X., Qin, T., Wang, L., Liu, T.: Representation degeneration problem in training natural language generation models. In: International Conference on Learning Representations (2019). http://openreview.net/forum?id=SkEYojRqtm
5. Gong, C., He, D., Tan, X., Qin, T., Wang, L., Liu, T.Y.: Frage: Frequency-agnostic word representation. ArXiv arXiv:1809.06858 (2018)
6. Kalyan, K.S., Rajasekharan, A., Sangeetha, S.: Ammus : A survey of transformer-based pretrained models in natural language processing (2021). https://doi.org/10.48550/ARXIV.2108.05542,

7. Li, B., Zhou, H., He, J., Wang, M., Yang, Y., Li, L.: On the sentence embeddings from pre-trained language models. In: Proceedings of the 2020 Conference on Empirical Methods in Natural Language Processing (EMNLP), pp. 9119–9130. Association for Computational Linguistics, Online (Nov 2020). https://doi.org/10.18653/v1/2020.emnlp-main.733

8. Liang, Y., Cao, R., Zheng, J., Ren, J., Gao, L.: Learning to remove: Towards isotropic pre-trained Bert embedding. In: Artificial Neural Networks and Machine Learning - ICANN 2021: 30th International Conference on Artificial Neural Networks, Bratislava, Slovakia, September 14–17, 2021, Proceedings, Part V, p. 448–459. Springer-Verlag, Berlin, Heidelberg (2021). https://doi.org/10.1007/978-3-030-86383-8_36

9. Loshchilov, I., Hutter, F.: Decoupled weight decay regularization (2017). https://doi.org/10.48550/ARXIV.1711.05101

10. Mu, J., Viswanath, P.: All-but-the-top: Simple and effective post-processing for word representations (2018), publisher Copyright: © Learning Representations, ICLR 2018 - Conference Track Proceedings. All right reserved.; 6th International Conference on Learning Representations, ICLR 2018; Conference date: 30-04-2018 Through 03-05-2018

11. Rajaee, S., Pilehvar, M.T.: A cluster-based approach for improving isotropy in contextual embedding space. In: ACL (2021)

12. Rajaee, S., Pilehvar, M.T.: How does fine-tuning affect the geometry of embedding space: a case study on isotropy. In: EMNLP (2021)

13. Rudman, W., Gillman, N., Rayne, T., Eickhoff, C.: IsoScore: Measuring the uniformity of embedding space utilization. In: Findings of the Association for Computational Linguistics: ACL 2022, pp. 3325–3339. Association for Computational Linguistics, Dublin, Ireland (May 2022). https://doi.org/10.18653/v1/2022.findings-acl.262, http://aclanthology.org/2022.findings-acl.262

14. Su, J., Cao, J., Liu, W., Ou, Y.: Whitening sentence representations for better semantics and faster retrieval (2021). https://doi.org/10.48550/ARXIV.2103.15316, http://arxiv.org/abs/2103.15316

15. Timkey, W., van Schijndel, M.: All bark and no bite: Rogue dimensions in transformer language models obscure representational quality. In: Proceedings of the 2021 Conference on Empirical Methods in Natural Language Processing, pp. 4527–4546. Association for Computational Linguistics, Online and Punta Cana, Dominican Republic (Nov 2021). https://doi.org/10.18653/v1/2021.emnlp-main.372, http://aclanthology.org/2021.emnlp-main.372

16. Wang, A., Singh, A., Michael, J., Hill, F., Levy, O., Bowman, S.: GLUE: A multitask benchmark and analysis platform for natural language understanding. In: Proceedings of the 2018 EMNLP Workshop BlackboxNLP: Analyzing and Interpreting Neural Networks for NLP, pp. 353–355. Association for Computational Linguistics, Brussels, Belgium (Nov 2018). https://doi.org/10.18653/v1/W18-5446, http://aclanthology.org/W18-5446

17. Wang, L., Huang, J., Huang, K., Hu, Z., Wang, G., Gu, Q.: Improving neural language generation with spectrum control. In: International Conference on Learning Representations (2020). http://openreview.net/forum?id=ByxY8CNtvr

18. Wolf, T., et al.: Huggingface's transformers: State-of-the-art natural language processing (2019). https://doi.org/10.48550/ARXIV.1910.03771

19. Zabihzadeh, D.: Ensemble of loss functions to improve generalizability of deep metric learning methods. arXiv:2107.01130 (2021)

20. Zampieri, M., Malmasi, S., Nakov, P., Rosenthal, S., Farra, N., Kumar, R.: Predicting the Type and Target of Offensive Posts in Social Media. In: Proceedings of NAACL (2019)
21. Zhang, Z., Gao, C., Xu, C., Miao, R., Yang, Q., Shao, J.: Revisiting representation degeneration problem in language modeling. In: Findings of the Association for Computational Linguistics: EMNLP 2020, pp. 518–527. Association for Computational Linguistics, Online (Nov 2020). https://doi.org/10.18653/v1/2020.findings-emnlp.46, http://aclanthology.org/2020.findings-emnlp.46
22. Zhou, W., Lin, B.Y., Ren, X.: Isobn: Fine-tuning bert with isotropic batch normalization. Proc. AAAI Conf. Artif. Intell. 35(16), 14621–14629 (May 2021), http://ojs.aaai.org/index.php/AAAI/article/view/17718

FTDNet: Joint Semantic Learning for Scene Text Detection in Adverse Weather Conditions

Jiakun Tian, Gang Zhou$^{(\boxtimes)}$, Yangxin Liu, En Deng, and Zhenhong Jia

Key Laboratory of Signal Detection and Processing, School of Information Science and Engineering, Xinjiang University, Ürümqi, China
gangzhou_xju@126.com, jzhh@xju.edu.cn

Abstract. In recent years, convolutional neural network (CNN)-based scene text detection methods have been extensively studied and obtained successful results in public datasets. However, scene text detection in adverse weather conditions suffers from poor visibility. In this paper, we use a multi-task learning approach to resolve this issue. We construct a foggy text detection network (FTDNet) composed of dual subnetworks: a text detection subnetwork and a visibility enhancement subnetwork. We employ DBNet as the text detection subnetwork, which shares the feature extraction layers for both two subnetworks. And we design a feature visibility enhancement (FVE) module for visibility enhancement subnetwork. In order to enable joint learning of multi-task networks, a novelty loss function called the mask dehazing loss is applied. This method achieved state-of-the-art results in terms of detection on both synthetic datasets and real-to-world datasets.

Keywords: Scene Text Detection · Dual Subnetworks · Joint learning · Multi-task Learning · Mask Dehazing Loss

1 Introduction

As a key premise for scene text understanding, scene text detection aims to achieve precise localization of text in the scene image. Due to its practical applications in areas such as scene interpretation, driving navigation, and photo translation, it has been a vibrant research area for a long time. Although existing mainstream methods such as PSE [34], PAN [36], DB [16], etc. perform excellently on clear images, their performance usually shows a significant decline under adverse weather conditions. For example, in foggy weather, a portion of the spectra between the camera lens and objects are scattered and absorbed by suspended particles, increasing the difficulty of extracting features for text detection from these images. As illustrated in Fig. 1, the detection performance

Supported by National Natural Science Foundation of China under grant No. 62166040, 62261053, 62137002, Natural Science Foundation of Xinjiang Autonomous Region under grant No. 2021D01C057, and the National Key R&D Program of China under grant No. 2021ZD0113601.

G. A. Fink et al. (Eds.): ICDAR 2023, LNCS 14191, pp. 137–154, 2023.
https://doi.org/10.1007/978-3-031-41734-4_9

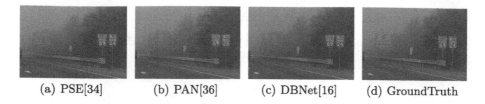

(a) PSE[34]   (b) PAN[36]   (c) DBNet[16]   (d) GroundTruth

Fig. 1. The performance of several algorithms in hazy condition.

of several text detection algorithms has been significantly impacted by the presence of foggy weather conditions. While the text on the nearby blue road signs are accurately detected by several algorithms, the text on the distant green road signs located in the fog are not effectively detected.

In general cognition, the processing of image dehazing can enhance human visual perception. Therefore, it seems reasonable to use image dehazing as a preprocessing step to enhance the detection performance in hazy conditions. However, in the field of object detection, [14] has shown that using dehazed images as input for a target detection model does not always result in an enhancement of detection performance. Additionally, this approach of performing detection and dehazing in series leads to more complex than the original detection model, resulting in a reduction in detection speed.

We present a foggy text detection method based on multi-task learning [2] with dual subnetworks, which address the text detection in poor visibility. To achieve this goal, we adopt DBNet [16] as the text detection subnetwork, and propose the FVE module on the basis of this architecture. The FVE module combines the mask dehazing loss to construct a visibility enhancement subnetwork to enhance feature visibility. We will introduce the network in detail in Sect. 3. The network concurrently optimizes the enhancement of feature visibility and text localization, and is trained in an end-to-end manner. By means of joint learning, it can share the clean features generated from the input hazy image by the visibility enhancement subnetwork, making the text detection subnetwork acquire stronger text localization ability and thus improving the network's text detection ability under foggy conditions.

Our primary contributions are:

– We propose a multi-task learning approach, which improves the network's generalization ability in foggy conditions by jointly optimizing feature visibility enhancement and text localization ability.
– The proposed FVE module, under the guidance of the mask dehazing loss, learns cooperatively with the text detection subnetwork, improving the performance of the text detection in poor visibility.
– We carried out a series of experiments on text detection problem in foggy conditions, and also constructed a natural foggy text dataset to verify the validity of our pattern.

This paper assesses the proposed network on natural and synthetic foggy text datasets. The evaluation results indicate that our proposed model has advantages in accuracy and speed compared to current state-of-the-art text detection models and the cascaded model for dehazing and detection.

2   Related Work

In this section, we introduce some methods for scene text detection, image dehazing, and multi-task learning that have made notable accomplishments in their respective domains.

2.1   Scene Text Detection

Scene text detection based on CNN has received extensive attention and can be divided into three categories: methods based on regression, methods based on segmentation, and hybrid methods.

Methods based on regression treat text as an object, aiming to directly predict the text bounding boxes. For instance, TextBoxes [15] uses larger aspect ratios of default boxes and convolutional kernels on the basis of the object detection method SSD [8] to achieve detection of long texts. Ma et al. [21] proposed a rotating region proposal network based on Faster-RCNN [26] to detect text with arbitrary orientations in scene images. Dai et al. [3] proposed a contour information aggregation method and localization mechanism that suppresses the impact of redundant and noisy contour nodes, it achieves more accurate localization of texts of any shape.

Methods based on segmentation transform text detection into a semantic segmentation problem by classifying pixels in an image as text or non-text, and using some post-processing steps to obtain text region bounding boxes. For example, Wang et al. [34] proposed a progressive scale expansion algorithm that can effectively differentiate similar text instances and has excellent detection performance for text instances of any shape. PAN [36] generates an efficient text detector through the utilization of a lightweight feature extraction network, a computationally inexpensive segmentation head, and several post-processing stages. CT [28] decomposes text instances into a combination of text kernels and centroid offsets, it utilizes centroid offset to achieve pixel aggregation, connects the boundary text pixels to the internal text area and achieves text detection. Liao et al. [16] proposed differentiable binarization, making binarization end-to-end trainable in CNNs, and separates text area into two parts for prediction, considering both detection accuracy and speed.

Hybrid methods use the segmentation-based approach to categorize text/non-text pixels while obtaining the bounding box of text through regression. EAST [39] uses a fully convolutional network to predict a rotated rectangle or quadrilateral for each point within a text region, and then employs non-maximum suppression to filter out the final words or text lines. Liao et al. [17] proposed a rotation-sensitive regression method for scene text detection.

By actively rotating the convolution kernel, it fully utilizes rotation-invariant features and better detects inclined text. Du et al. [4] proposed a text instance-level cooperative learning module to address the problem of incomplete detection and inaccurate detection caused by different backgrounds in arbitrary shape text detection based on the Mask-RCNN [5].

All the above text detection algorithms have achieved excellent results under normal weather conditions. However, there is little research on text detection for scenes in adverse weather conditions. As a consequence, adverse weather conditions continue to adversely affect practical applications.

2.2 Image Dehazing

Formation mechanism of haze images explained by the atmospheric scattering model [22] is frequently used in the form of Eq. 1 after being proposed. It has been utilized by methods such as DehazeNet [1], MSCNN [27], and DCPDN [38] in single image dehazing, and good results have been obtained.

$$I(x) = J(x)t(x) + A(1 - t(x)) \tag{1}$$

where $I(x)$ is the blurred image captured by the camera, $J(x)$ is considered as a haze-free image, A is the global atmospheric light value, $t(x)$ is the medium transmittance:

$$t(x) = e^{-\beta d(x)} \tag{2}$$

where, β is the atmospheric scattering coefficient, $d(x)$ is the scene depth.

Recently, considering the remarkable performance of CNNs in computer vision tasks, some methods have combined CNNs with atmospheric scattering models to improve performance in dehazing. DehazeNet [1] utilizes an end-to-end CNN to generate the transmission map from the image and then applies it to an atmospheric scattering model to produce the dehazed image, thus restoring the visibility of a hazy image. Light-DehazeNet [32] processes the recovered images generated by the atmospheric scattering model through color correction, making the recovery image closer to the real world. There are also some methods that are not based on the atmospheric scattering model, such as FFA-Net [25], utilizes a basic module composed of pixel attention and channel attention to adaptively learn feature weights, assigning higher weights to important features, thus achieving end-to-end image dehazing. These methods have shown excellent dehazing capability. However, the above methods are all aimed at restoring hazy images to haze-free images at the pixel level as much as possible, without considering adaptability for advanced tasks. Li et al. [13] discussed the relationship between image dehazing and target detection, and trains AOD-Net with Faster-RCNN [26] jointly. Compared to the separate training of dehazing and detection models, the results obtained show a significant improvement. However, when compared to methods that directly retrain on synthetic hazy images, the observed improvement is limited in magnitude.

2.3 Multi-task Learning

Multi-task learning refers to learning multiple connected tasks, which is learned from one task can enhance the learning of other tasks. Mask-RCNN [5] combines object detection and image segmentation, introducing this approach to the field of object detection. YOLO-R [33] combines object detection with multi-label classification and feature embedding, while DSNet [9] focuses its attention on the problem of object detection in haze weather conditions. DSNet uses RetinaNet [19] as the detection subnetwork and also proposed the feature recovery (FR) module that shares part of the backbone with the detection network. By jointly learning the multiple tasks, the shared part obtains relevant information regarding object detection and visibility enhancement, ultimately improving the object detection results under haze weather conditions.

Unlike the above methods, our FTDNet jointly learns the tasks of text localization and visibility enhancement. Compared to common targets such as pedestrians, vehicles, and animals, text in natural conditions is more diverse in terms of color, font, scale, and direction, and scene text is also easily affected by some similar background factors. Additionally, text detection is typically an upstream task of text recognition, and errors in detection can negatively impact recognition. Therefore, scene text localization should be more accurate than other objects localization. In order to effectively address the text detection problem in degraded images, FTDNet combines learnable weights to achieve a greater utilization of multi-scale information compared to DSNet. In addition, through a specialized loss design, we enhance the visibility of text regions in the image.

3 Proposed Method

In this section, we introduce the FTDNet, which can effectively enhance the performance of text detection while maintaining speed in adverse weather conditions. We achieve this objective by jointly learning the tasks of visibility enhancement and text localization. FTDNet comprises two subnetworks: a text detection subnetwork and a visibility enhancement subnetwork. The text detection subnetwork adopts the complete DBNet [16], which is responsible for text localization, and both it and the visibility enhancement subnetwork start with a common block (CB) module. The design method for the visibility enhancement subnetwork is to attach the feature visibility enhancement (FVE) module to the CB module to enhance visibility as depicted in Fig. 2. Both subnetworks commence with the CB module, ensuring that the generated clear features can be utilized simultaneously in both subnetworks during the network training process.

3.1 Text Detection Subnetwork

As previously introduced, there are currently many text detection algorithms with excellent detection performance. Among these methods, DBNet introduces a differentiable binarization that modifies the conventional binarization process.

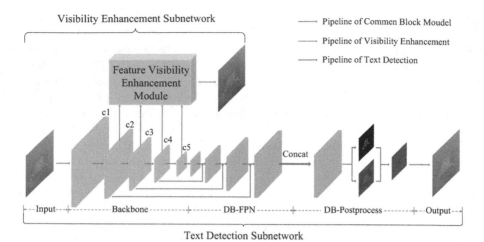

Fig. 2. Overall structure of FTDNet.

This enables the network to learn the threshold during the training process, replacing the manual setting of the threshold with an adaptive learning method and ultimately improving the detection performance. DBNet does not have complex post-processing, therefore its detection speed is relatively fast. On the network architecture, DBNet adopts a ResNet [7] as the backbone, and uses a FPN [18] to provide top-down paths and lateral connections, fully utilizing semantic information of different scales to improve the performance of detecting text of different scales. For these reasons, we adopt DBNet as the Baseline and make it share the CB module with the visibility enhancement sub-network. It completes text detection through joint learning with the sub-network during the training process. Given many available DBNet models, we chose to use the ResNet-50 with a training scale of 640 in our network.

The following processing procedure has been adopted for the text detection subnetwork. Firstly, a ResNet-50 is adopted as the feature extraction network. Then, an FPN structure is built on top of ResNet-50 to handle features at different scales. Finally, the network uses the outputs from the FPN structure to predict binary maps for both the regions surrounding text boundaries and the internal regions, and then generate binary maps for text regions with differentiable binarization, thus achieving text detection.

We utilized the loss function L_{DB} of DBNet as the loss function of the text detection subnetwork without modification.

3.2   Visibility Enhancement Subnetwork

The visibility enhancement subnetwork is responsible for generating clear features of text area to improve the performance of the network in adverse weather conditions. As depicted in Fig. 3, the subnetwork is composed of two parts: the

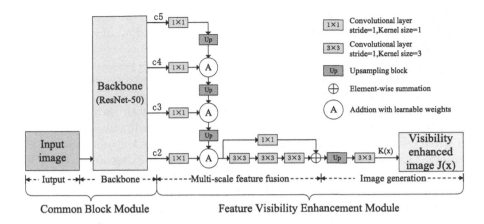

Fig. 3. Structure of the visibility enhancement subnetwork.

CB module and the FVE module. We can obtain the haze-free image $J(x)$ after dehazing of the text region through Eq. 1:

$$J(x) = \frac{1}{t(x)}I(x) - A\frac{1}{t(x)} + A \tag{3}$$

Inspired by AOD-Net [13], in order to reduce the error accumulation caused by separately estimating α and $t(x)$, we merge them into one variable in the image restoration process. Therefore, Eq. 3 can be transformed into:

$$J(x) = K(x)I(x) - K(x) + 1 \tag{4}$$

where

$$K(x) = \frac{\frac{1}{t(x)}(I(x) - A) + (A - 1)}{I(x) - 1} \tag{5}$$

Our model's generation of $K(x)$ and $J(x)$ will be further introduced in the section of FVE Module.

CB Module. In FTDNet, features are extracted from the input image by the CB module and utilized for joint learning of visibility enhancement and text localization capabilities. It is composed of a ResNet-50. It can be seen from Fig. 2, we select the different scale feature maps produced by the four stages of ResNet-50 and input them into the FVE module. Meanwhile, the text detection subnetwork, following the structure of DBNet, uses the feature maps produced by ResNet-50 that are fused by FPN for text detection.

FVE Module. Haze causes a degradation in the clarity of the image features extracted by the network, resulting in a decreased text detection effectiveness. In order to restore the clear features for effective joint learning of text detection, the visibility enhancement subnetwork uses the FVE module, which consists of two submodules: a multi-scale feature fusion submodule and an image generation submodule.

Multi-scale Feature Fusion Submodule. In order to effectively utilize feature information at different scales, the multi-scale feature fusion submodule combines the four scale feature maps output by the backbone with learnable weights. The features at each scale undergo a 1×1 convolution for dimensionality reduction when they enter the FVE module. We shall utilize the features generated in c5 stage as an illustration to present the subsequent processing steps. The FVE module uses bilinear interpolation to achieve upsampling, scaling the features to twice their size and then fusing them with the features generated in c4 that have undergone a 1×1 convolution process. Inspired by GridDehazeNet [20], we designed a set of learnable weights for feature fusion during multi-scale fusion, which is described as:

$$\bar{F}^i = h^i F_h^i + v^i F_v^i \tag{6}$$

where F_i^h refers to the i-th channel of the feature inputted from the horizontal direction, F_i^v refers to the i-th channel of the feature inputted from the vertical direction, h_i and v_i represent their learnable weights. $\bar{F}^i$ stands for the i-th channel of the merged feature. Experimental results indicate that the learnable channel weights effectively enhance the detection performance. After preliminary fusion of features in four scales, we further utilized three convolutional layers to reduce the channels of the feature map from 256 to 3. We added a skip connection with a 1×1 convolution to avoid excessive information loss.

Image Generation Submodule. In this submodule, we first use bilinear interpolation to upsample the feature maps to the required scale, then use a convolutional layer to generate $K(x)$, then use element-wise multiplication, addition and subtraction to calculate Eq. 4, and finally obtain the visibility enhanced image $J(x)$.

Mask Dehazing Loss. In order to make the text region features in the CB module more obvious, our visibility enhancement loss only uses the text region of the image for loss calculation. Mean square error (MSE) loss is a commonly used dehazing loss and it is a pixel-level loss. But our goal is to make the text region as a whole be more attention by the network. Thus, we added the perception loss [10], letting the network further enhance the visibility of features in a new feature space for the text regions and making the text regions of visibility enhanced images and clear images more perceptually similar. Our visibility enhancement subnetwork adopts the mask dehazing loss for guidance, which is described as:

$$L_{VE} = \frac{1}{n} \sum_{j=1}^{n} \left((\bar{y}_j - \bar{x}_j)^2 + \beta \sum_{i=1}^{5} \theta_i |G_i(\bar{y}_j) - G_i(\bar{x}_j)| \right) \tag{7}$$

$\bar{x}_j$ and $\bar{y}_j$ respectively represent the text region of the visibility enhanced image and its corresponding clear image. Through the coordinate information contained in labels, we can obtain the binary mask of text regions and thus obtain the text regions participating in the loss calculation. n represents the batch size during training. $G_i(i = 1, 2, \cdots, 5)$ represents the feature maps extracted from the i-th convolutional layer of a fixed pre-trained model, which is extracted from the VGG-19 [30] network trained on ImageNet. The parameters of VGG-19 are frozen during the training of FTDNet. θ_i is the weight coefficient of the i-th layer. In the experiments, we used the feature maps from the 1st, 3rd, 5th, 9th, 13th layers of VGG-19, and the corresponding θ_i are 1/32, 1/16, 1/8, 1/4, 1. β is set to 0.2.

The visibility enhancement subnetwork can generate dehazed images of text regions, but instead of using such images as input for the text detection subnetwork, our objective is to improve the clarity of the features generated in the shared module through learning the visibility enhancement task, thereby improving the detection performance of the network through joint optimization of visibility enhancement and text localization. Therefore, the activation of FVE module is limited to the training stage, and our network structure is consistent with DBNet during inference.

3.3   Loss Balance

The overall loss L of FTDNet is expressed as:

$$L = \lambda L_{DB} + \gamma L_{VE} \tag{8}$$

In the experiments, we found that the performance of the network is sensitive to the weight coefficients λ and γ of the loss, and a large number of experiments are required to determine their values. At the same time, when changing the composition of the CB module, the most suitable λ and γ values for the network are also changing. Therefore, in order to avoid the experimental cost generated by a substantial amount of manual tuning, we choose the dynamic loss weighting method Uncertainty [12] to balance the weights of detection and enhancement loss. Equation 8 is rewritten as Eq. 9.

$$L = \frac{1}{2\sigma_1^2} L_{DB} + \frac{1}{2\sigma_2^2} L_{VE} + \log \sigma_1 + \log \sigma_2 \tag{9}$$

where σ_1 and σ_1 are two learnable parameters, which are used to measure the output data noise.

4   Experiments

This section introduces our experimental results on text detection in hazy conditions. All of our experiments were conducted on the Pytorch 1.8.0 framework in the Ubuntu 20.04.1 system, which is equipped with an Intel(R) Xeon(R) Silver 4210R CPU @ 2.40 GHz, 64 G RAM, and three RTX 3090 GPUs.

(a) Normal weather   (b) Foggy weather with low   (c) Foggy weather with a
                                    visibility   bluish tint

Fig. 4. Images in different weather conditions.

4.1 Datasets

Synthetic Foggy Text Dataset. Currently, there is no public dataset specifically for foggy text detection. In order to conduct experiments, we selected 799 outdoor English images and 749 Chinese images from ICDAR 2015(IC15) [11], 2017 MLT [24], 2017 RCTW [29], 2019 MLT [23], 2019 LSVT [31], and utilized an atmospheric scattering model to synthesize foggy images. Each image was synthesized into 4 different foggy images. 600 images in each language were randomly selected to synthesize the training set, so both the Chinese and English training sets have 2400 images. The rest of the images were used to generate the test images for the Synthetic Foggy Text(SFT) dataset.

As described in Fig. 4, we observed that the brightness of natural foggy images is usually lower than that of normal weather, and the color of foggy images can sometimes appear different. Therefore, in synthesizing foggy images, in addition to randomly setting the fog density, the atmospheric light value and the blue channel value of the image were also randomly scaled within a certain range with a certain probability. To demonstrate the rationality of this synthesis, we also conducted comparative experiments using foggy images synthesized with other parameter settings. This will be discussed in more detail in Sect. 4.3.

Natural Foggy Text Dataset. To assess the performance of our method under natural foggy conditions, we collected 787 natural foggy text images by taking photos under natural foggy conditions and searching on China News Service Network[1], BURST[2], GRIBBLENATION[3] and the RTTS [14] dataset. After annotation, we formed the Natural Foggy Text (NFT) dataset, which is used the experiments in this paper. The dataset includes more than 8000 text instances, divided into 204 English text images and 583 Chinese text images.

---

[1] http://www.ecns.cn/.

[2] https://burst.shopify.com/.

[3] http://www.gribblenation.org/.

4.2   Implementation Details

In reality, photos may have text that appears blurry or small due to limitations in lighting, angle, distance, and camera equipment. The IC15 dataset was taken using Google Glass, and no specific measures were taken to improve the quality of the images during the shooting. In order to enhance the generalization ability of the trained model on small, blurry text, we trained the DBNet on the IC15 dataset for 1200 epochs to generate a pre-trained weight model, and then after the addition of the visibility enhancement network, we fine-tuned the model using the SFT dataset. The FVE module was initialized using the method of Kaiming [6]. In the fine-tuning stage, we used the Adam optimizer and trained for 1000 epochs. The batch size was configured as 16, with an initial learning rate of 0.001, which was adjusted according to the settings of the DBNet. The models used for the evaluation on the English synthetic and natural datasets were all trained based on this method.

4.3   The Impact of Parameter Setting on Dataset Synthesis

In Sect. 4.1, we explained that due to the variability of natural foggy images, we adjusted multiple parameters when synthesizing the dataset. In this section, we provide a detailed explanation of this point. Table 1 demonstrates the performance of DBNet and FTDNet trained with four different parameter settings for synthesized foggy images on the NFT-English dataset.

A larger β value represents a smaller transmittance and a higher concentration of synthetic fog. A larger α value represents a larger atmospheric light value and a brighter synthesized image. A larger α_b value represents a larger blue channel value of the image. In the experiments, we randomly set the α_b value with a probability of 5% based on the color distribution of the NFT dataset. As shown in Table 1, as the randomness of the synthesis parameter setting increases, the performance of the model improves. This suggests that the use of more randomized settings in synthesizing haze images will bring stronger generalization to the network, alleviate overfitting on the synthetic dataset after training, and effectively improve the network's performance in natural foggy conditions.

Table 1. The effect of different synthesis parameter settings on DBNet and FTDNet on the NFT-English dataset. Numerical range represents the corresponding parameters were uniformly randomly set within the range. The numbers in columns 2 and 3 of the table respectively represent the scaling factor of α and α_b. "R", "P", and "F" indicate Recall, Precision and F-measure respectively.

Parameter Settings			Baseline(DBNet)			FTDNet		
β	α	α_b	R	P	F	R	P	F
0.09	1.0	1.0	39.13	66.02	49.14	35.63	65.90	46.23
0.05, 0.07, 0.09, 0.11, 0.13	1.0	1.0	62.73	76.60	68.98	61.66	76.20	68.16
0.004–0.014	1.0	1.0	67.17	81.14	73.50	70.31	84.17	76.62
0.004–0.014	0.4–1.0	1.0–1.6	**68.21**	**83.06**	**74.91**	**71.76**	**86.06**	**78.26**

Table 2. Comparison results with other scene text detection algorithms on the English foggy text dataset. The pre-training of all networks in the table was performed with the IC15 dataset, followed by fine-tuning with the SFT-English training set.

Methods	NFT-English			SFT-English		
	R	P	F	R	P	F
EAST [39]	61.67	63.65	62.64	43.85	47.64	45.67
PSE [34]	67.05	82.39	73.93	59.01	73.20	65.34
PAN [36]	67.57	85.02	75.29	60.13	75.59	66.98
DBNet [16]	68.21	83.07	74.91	61.35	73.21	66.76
CT [28]	68.11	85.87	75.97	61.12	76.62	68.00
I3CL [4]	69.72	**86.13**	77.06	63.95	77.01	69.88
FTDNet	**71.76**	86.06	**78.26**	**65.21**	**77.84**	**70.97**

Table 3. Comparison results with other text detection algorithms on the Chinese foggy text dataset.

Methods	NFT-Chinese			SFT-Chinese		
	R	P	F	R	P	F
EAST [39]	56.50	57.02	56.76	30.17	59.23	39.98
PSE [34]	58.82	64.45	61.51	55.90	63.50	59.46
PAN [36]	65.46	52.15	58.05	51.92	65.42	57.89
DBNet [16]	59.17	70.95	64.53	55.16	68.70	59.30
PAN++ [35]	63.73	50.32	56.24	52.31	62.57	56.98
CT [28]	62.42	71.19	66.52	57.09	**70.06**	62.92
FTDNet	**66.14**	**73.00**	**69.40**	**61.21**	69.57	**65.12**

4.4   Comparisons with Previous Methods

In this section, we compare FTDNet with several other text detection algorithms in both English and Chinese. Table 2 and Table 3 are the comparison results in English and Chinese respectively, and Fig. 5 and Fig. 6 are the visualization comparison results of different algorithms. In foggy conditions, FTDNet has a significant advantage in Recall compared to other text detection algorithms. On the natural foggy text dataset NFT, the Recall is over 2% higher than the second place, and on the synthetic foggy text dataset SFT, it is over 1.2% higher. FTD-Net also achieves optimal results in F-measure. Compared to DBNet without adding visibility enhancement subnetwork, our method brings 3.55%, 2.99%, and 3.35% improvement in NFT respectively, and 3.86%, 4.63%, and 4.21% improvement in SFT respectively. In terms of Chinese dataset, we trained all the methods in Table 3 using 2400 Chinese synthetic foggy images in SFT-Chinese, without using any other additional datasets. Similar to the results on the English

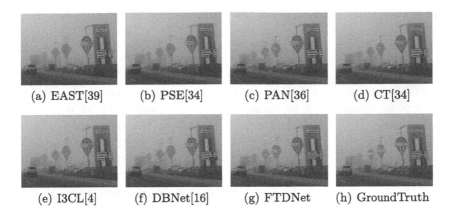

(a) EAST[39]   (b) PSE[34]   (c) PAN[36]   (d) CT[34]

(e) I3CL[4]   (f) DBNet[16]   (g) FTDNet   (h) GroundTruth

Fig. 5. Visualization results of FTDNet compared to other methods on NFT.

dataset, our method demonstrates exceptional performance on various metrics, significantly outperforming other models in Recall and F-measure.

4.5   Results Compared to Serial Structure

This section verifies the effectiveness of parallel structure for dehazing and detection, and to compare the speed of different structures. We combine DBNet with AOD-Net [13], FFA-Net [25], and AECR-Net [37] three dehazing networks. The dehazing networks are pre-trained on the RESIDE-ITS [14] dataset, while the DBNet is pre-trained on clear images corresponding to SFT-English.

The first three networks in Table 4 are two-stage structures of dehazing and detection, and the detection and dehazing networks are not jointly trained. And the dehazing networks used SFT-English for fine-tuning. The networks from the fourth to the sixth in Table 4 are composed of the dehazing network and DBNet connected together, and after pre-training, they are jointly trained on SFT-English. The training of Retrained DBNet is same as the DBNet in Table 2.

As demonstrated in Table 4, our network structure has significant advantages over other serial structures in terms of effectiveness and speed. On each of the evaluation metrics, our method has a significant improvement compared to several dehazing and detection serial structures, in which on NFT dataset, FTDNet is at least 4% higher than other methods, and on SFT dataset, it is at least 3% higher. Compared with several cascaded models, FTDNet also has a significant advantage in speed, the inference speed on RTX 3090GPU is 7 times that of FFA-DBNet. During the inference phase, no additional computational costs are incurred, thus the speed of FTDNet is comparable to that of DBNet.

Table 4. Comparison results of different structures on NFT-English and SFT-English.

Methods	NFT-English				SFT-English			
	R	P	F	FPS	R	P	F	FPS
AOD-Net [13]+DBNet	53.77	80.8	64.57	–	58.92	72.19	64.88	–
FFA-Net [25]+DBNet	50.58	80.40	62.10	–	60.12	74.08	66.37	–
AECR-Net [37]+DBNet	52.96	81.08	64.07	–	59.43	74.14	65.98	–
AOD-DBNet	64.14	79.20	70.87	10.29	61.56	72.92	66.76	8.99
FFA-DBNet	66.07	82.37	73.33	1.76	61.80	73.39	67.69	1.44
AECR-DBNet	65.10	81.73	72.47	8.27	62.04	73.11	67.12	7.65
Retrained DBNet	68.21	83.06	74.91	**12.09**	61.35	73.21	66.76	**10.37**
FTDNet	**71.76**	**86.06**	**78.26**	12.08	**65.21**	**77.84**	**70.97**	10.35

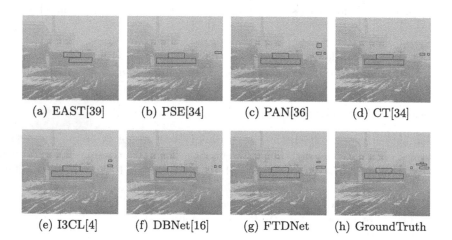

(a) EAST[39]   (b) PSE[34]   (c) PAN[36]   (d) CT[34]

(e) I3CL[4]   (f) DBNet[16]   (g) FTDNet   (h) GroundTruth

Fig. 6. Visualization results of FTDNet compared to other methods on SFT.

4.6   Ablation Study

Effect of Different Structures of CB Module. Since the FVE module requires input from four scale feature maps, we also consider incorporating it at the FPN. As shown in Table 5, using only the backbone as the CB module, all indicators have a significant improvement compared to using the Backbone+FPN as the CB module. We believe that the reason for this is that the spatial information used for visibility enhancement in the FPN is less than that of the backbone, so the improvement brought by directly connecting the FVE module to the text detection subnetwork from the FPN is also less.

Comparison with FR Module. FR module is the feature recovery module in DSNet [9]. We incorporated the FR module into the DBNet according to the

Table 5. Ablation experiments for network structure. The content in the parentheses represents the components of the CB module, Lw represents the learnable weights.

Network Structure	NFT-English			SFT-English		
	R	P	F	R	P	F
DBNet	68.21	83.06	74.91	61.35	73.21	66.76
DBNet+FR	69.82	84.78	76.58	64.81	75.24	69.64
DBNet+FVE(CB:Backbone+FPN)+Lw	70.75	85.34	77.36	64.41	76.11	69.77
DBNet+FVE(CB:Backbone)	70.46	85.97	77.45	63.65	76.89	69.65
DBNet+FVE(CB:Backbone)+Lw	**71.76**	**86.06**	**78.26**	**65.21**	**77.84**	**70.97**

Table 6. Comparison of different visibility enhancement loss functions.

Loss functions	NFT-English			SFT-English		
	R	P	F	R	P	F
MSE Loss	71.34	84.82	77.50	64.70	77.02	70.33
MSE+Perceptual Loss	71.29	85.49	77.75	64.14	**77.90**	70.36
Mask Dehazing Loss	**71.76**	**86.06**	**78.26**	**65.21**	77.84	**70.97**

settings in DSNet and compared it with our FVE module on both natural and synthetic datasets. As shown in Table 5, in foggy conditions, our FVE module is more suitable for DBNet. In F-measure, the FVE module compared to the FR module brings about 1.68% and 1.33% improvement on natural and synthetic English datasets, respectively. Huang et al. [9] think that using the c1 and c2 parts of the backbone as the CB module is more effective than the entire backbone. But in [9]'s experiments, only one scale of features was connected to the FR module from c2, c3, c4, or c5, while our FVE module makes better use of the information brought by multi-scale features, thus achieving better performance.

Impact of Learnable Weights. We compared the detection results of the network in two cases, whether to add learnable weights in the FVE module, as shown in Table 5. After adding learnable weights, FTDNet performed better on the NFT-English and SFT-English, with the F-measure metric improving by 0.81% and 1.32% respectively. The learnable weights allow the network to give more appropriate weights to different channels for different scale features.

Ablation Study of Visibility Enhancement Loss Function. We compared the effects of using MSE loss, MSE loss with perceptual loss, and mask dehazing loss on the network. As shown in Table 6, the mask dehazing loss brings about approximately 0.5% improvement over the combination of MSE with perceptual loss in terms of F-measure. We believe that the mask dehazing loss makes the FVE module pay more attention to the feature visibility enhancement of the text region, making the dehazing effect of the text area more obvious compared to the background, thus improving the detection performance.

5   Conclusion

In this paper, we have presented an end-to-end trainable CNN, called FTDNet, and demonstrated its text detection performance in adverse weather conditions. Experimental results on natural and synthetic datasets show that FTDNet outperforms other text detection networks. Compared to the serial dehazing and detection network structure, our method also has obvious advantages in accuracy and speed. This indicates that the network architecture we adopted in conjunction with the proposed FVE module and mask dehazing loss is effective. We hope to extend our method to other weather conditions, such as rain, snow, and sandstorms, and this constitutes our direction of work in the future.

References

1. Cai, B., Xu, X., Jia, K., Qing, C., Tao, D.: Dehazenet: an end-to-end system for single image haze removal. IEEE Trans. Image Process. **25**(11), 5187–5198 (2016)
2. Caruana, R.: Multitask learning. Mach. Learn. **28**(1), 41–75 (1997)
3. Dai, P., Zhang, S., Zhang, H., Cao, X.: Progressive contour regression for arbitrary-shape scene text detection. In: Proceedings of the IEEE/CVF Conference on Computer Vision and Pattern Recognition, pp. 7393–7402 (2021)
4. Du, B., Ye, J., Zhang, J., Liu, J., Tao, D.: I3cl: intra-and inter-instance collaborative learning for arbitrary-shaped scene text detection. Int. J. Comput. Vision **130**, 1961–1977 (2022)
5. He, K., Gkioxari, G., Dollár, P., Girshick, R.: Mask r-cnn. In: Proceedings of the IEEE International Conference on Computer Vision, pp. 2961–2969 (2017)
6. He, K., Zhang, X., Ren, S., Sun, J.: Delving deep into rectifiers: surpassing human-level performance on imagenet classification (2015). arXiv e-prints arXiv:1502.01852. https://doi.org/10.48550/arXiv.1502.01852
7. He, K., Zhang, X., Ren, S., Sun, J.: Deep residual learning for image recognition. In: Proceedings of the IEEE Conference on Computer Vision and Pattern Recognition, pp. 770–778 (2016)
8. He, P., Huang, W., He, T., Zhu, Q., Qiao, Y., Li, X.: Single shot text detector with regional attention. In: Proceedings of the IEEE International Conference on Computer Vision, pp. 3047–3055 (2017)
9. Huang, S.C., Le, T.H., Jaw, D.W.: Dsnet: joint semantic learning for object detection in inclement weather conditions. IEEE Trans. Pattern Anal. Mach. Intell. **43**(8), 2623–2633 (2020)
10. Johnson, J., Alahi, A., Fei-Fei, L.: Perceptual losses for real-time style transfer and super-resolution. In: Leibe, B., Matas, J., Sebe, N., Welling, M. (eds.) ECCV 2016. LNCS, vol. 9906, pp. 694–711. Springer, Cham (2016). https://doi.org/10.1007/978-3-319-46475-6_43
11. Karatzas, D., et al.: Icdar 2015 competition on robust reading. In: 2015 13th International Conference on Document Analysis and Recognition (ICDAR), pp. 1156–1160. IEEE (2015)

12. Kendall, A., Gal, Y., Cipolla, R.: Multi-task learning using uncertainty to weigh losses for scene geometry and semantics. In: Proceedings of the IEEE Conference on Computer Vision and Pattern Recognition, pp. 7482–7491 (2018)
13. Li, B., Peng, X., Wang, Z., Xu, J., Feng, D.: Aod-net: all-in-one dehazing network. In: Proceedings of the IEEE International Conference on Computer Vision, pp. 4770–4778 (2017)
14. Li, B., et al.: Benchmarking single-image dehazing and beyond. IEEE Trans. Image Process. **28**(1), 492–505 (2018)
15. Liao, M., Shi, B., Bai, X., Wang, X., Liu, W.: Textboxes: a fast text detector with a single deep neural network. In: Thirty-First AAAI Conference on Artificial Intelligence (2017)
16. Liao, M., Wan, Z., Yao, C., Chen, K., Bai, X.: Real-time scene text detection with differentiable binarization. In: Proceedings of the AAAI Conference on Artificial Intelligence, vol. 34, no. 07, pp. 11474–11481 (2020)
17. Liao, M., Zhu, Z., Shi, B., Xia, G.s., Bai, X.: Rotation-sensitive regression for oriented scene text detection. In: Proceedings of the IEEE Conference on Computer Vision and Pattern Recognition, pp. 5909–5918 (2018)
18. Lin, T.Y., Dollár, P., Girshick, R., He, K., Hariharan, B., Belongie, S.: Feature pyramid networks for object detection. In: Proceedings of the IEEE Conference on Computer Vision and Pattern Recognition, pp. 2117–2125 (2017)
19. Lin, T.Y., Goyal, P., Girshick, R., He, K., Dollár, P.: Focal loss for dense object detection. In: Proceedings of the IEEE International Conference on Computer Vision, pp. 2980–2988 (2017)
20. Liu, X., Ma, Y., Shi, Z., Chen, J.: Griddehazenet: attention-based multi-scale network for image dehazing. In: Proceedings of the IEEE/CVF International Conference on Computer Vision, pp. 7314–7323 (2019)
21. Ma, J.: Arbitrary-oriented scene text detection via rotation proposals. IEEE Trans. Multimedia **20**(11), 3111–3122 (2018)
22. McCartney, E.J.: Optics of the atmosphere: scattering by molecules and particles. New York (1976)
23. Nayef, N., et al.: Icdar 2019 robust reading challenge on multi-lingual scene text detection and recognition-rrc-mlt-2019. In: 2019 International Conference on Document Analysis and Recognition (ICDAR), pp. 1582–1587. IEEE (2019)
24. Nayef, N., et al.: Icdar 2017 robust reading challenge on multi-lingual scene text detection and script identification-rrc-mlt. In: 2017 14th IAPR International Conference on Document Analysis and Recognition (ICDAR), vol. 1, pp. 1454–1459. IEEE (2017)
25. Qin, X., Wang, Z., Bai, Y., Xie, X., Jia, H.: Ffa-net: feature fusion attention network for single image dehazing. In: Proceedings of the AAAI Conference on Artificial Intelligence, vol. 34, no. 07, pp. 11908–11915 (2020)
26. Ren, S., He, K., Girshick, R., Sun, J.: Faster r-cnn: towards real-time object detection with region proposal networks. Adv. Neural Inf. Process. Syst. **28**, 1–9 (2015)
27. Ren, W., Pan, J., Zhang, H., Cao, X., Yang, M.H.: Single image dehazing via multi-scale convolutional neural networks with holistic edges. Int. J. Comput. Vision **128**, 240–259 (2020)
28. Sheng, T., Chen, J., Lian, Z.: Centripetaltext: an efficient text instance representation for scene text detection. Adv. Neural Inf. Process. Syst. **34**, 335–346 (2021)
29. Shi, B., et al.: Icdar 2017 competition on reading Chinese text in the wild (rctw-17). In: 2017 14th IAPR International Conference on Document Analysis and Recognition (ICDAR), vol. 1, pp. 1429–1434. IEEE (2017)

30. Simonyan, K., Zisserman, A.: Very deep convolutional networks for large-scale image recognition (2014). arXiv preprint arXiv:1409.1556
31. Sun, Y., et al.: Icdar 2019 competition on large-scale street view text with partial labeling-rrc-LSVT. In: 2019 International Conference on Document Analysis and Recognition (ICDAR), pp. 1557–1562. IEEE (2019)
32. Ullah, H., et al.: Light-dehazenet: a novel lightweight cnn architecture for single image dehazing. IEEE Trans. Image Process. **30**, 8968–8982 (2021)
33. Wang, C.Y., Yeh, I.H., Liao, H.Y.M.: You only learn one representation: unified network for multiple tasks (2021). arXiv preprint arXiv:2105.04206
34. Wang, W., et al.: Shape robust text detection with progressive scale expansion network. In: Proceedings of the IEEE/CVF Conference on Computer Vision and Pattern Recognition, pp. 9336–9345 (2019)
35. Wang, W., et al.: Pan++: towards efficient and accurate end-to-end spotting of arbitrarily-shaped text. IEEE Trans. Pattern Anal. Mach. Intell. **44**(9), 5349–5367 (2021)
36. Wang, W., et al.: Efficient and accurate arbitrary-shaped text detection with pixel aggregation network. In: Proceedings of the IEEE/CVF International Conference on Computer Vision, pp. 8440–8449 (2019)
37. Wu, H., et al.: Contrastive learning for compact single image dehazing. In: Proceedings of the IEEE/CVF Conference on Computer Vision and Pattern Recognition, pp. 10551–10560 (2021)
38. Zhang, H., Patel, V.M.: Densely connected pyramid dehazing network. In: Proceedings of the IEEE Conference on Computer Vision and Pattern Recognition, pp. 3194–3203 (2018)
39. Zhou, X., et al.: East: an efficient and accurate scene text detector. In: Proceedings of the IEEE Conference on Computer Vision and Pattern Recognition, pp. 5551–5560 (2017)

DocParser: End-to-end OCR-Free Information Extraction from Visually Rich Documents

Mohamed Dhouib[1,2][✉] [iD], Ghassen Bettaieb[1] [iD], and Aymen Shabou[1] [iD]

[1] DataLab Groupe, Credit Agricole S.A, Montrouge, France
{ghassen.bettaieb,aymen.shabou}@credit-agricole-sa.fr
[2] Ecole polytechnique, Palaiseau, France
mohamed.dhouib@polytechnique.edu
https://datalab-groupe.github.io/

Abstract. Information Extraction from visually rich documents is a challenging task that has gained a lot of attention in recent years due to its importance in several document-control based applications and its widespread commercial value. The majority of the research work conducted on this topic to date follow a two-step pipeline. First, they read the text using an off-the-shelf Optical Character Recognition (OCR) engine, then, they extract the fields of interest from the obtained text. The main drawback of these approaches is their dependence on an external OCR system, which can negatively impact both performance and computational speed. Recent OCR-free methods were proposed to address the previous issues. Inspired by their promising results, we propose in this paper an OCR-free end-to-end information extraction model named DocParser. It differs from prior end-to-end approaches by its ability to better extract discriminative character features. DocParser achieves state-of-the-art results on various datasets, while still being faster than previous works.

Keywords: Information Extraction · Visually Rich Documents · OCR-free · End-to-end · DocParser

1 Introduction

Information extraction from visually rich documents (VRDs) is an important research topic that continues to be an active area of research [4,17–19,22,28,35,43,47] due to its importance in various real-world applications.

The majority of the existing information extraction from visually rich documents approaches [10,14,17,36] depend on an external deep-learning-based Optical Character Recognition (OCR) [1,2] engine. They follow a two-step pipeline: First they read the text using an off-the-shelf OCR system then they extract the fields of interest from the OCR'ed text. These two-step approaches have significant limitations due to their dependence on an external OCR engine. First

G. A. Fink et al. (Eds.): ICDAR 2023, LNCS 14191, pp. 155–172, 2023.
https://doi.org/10.1007/978-3-031-41734-4_10

of all, these approaches need positional annotations along with textual annotations for training. Also, training an OCR model requires large scale datasets and huge computational resources. Using an external pre-trained OCR model is an option, which can degrade the whole model performance in the case of a domain shift. One way to tackle this is to fine-tune these off-the-shelf OCR models which is still a delicate task. In fact, the documents full annotations are generally needed to correctly fine-tune off-the-shelf OCR models, which is time-consuming and difficult to obtain. OCR post-correction [21,38] is an option to correct some of the recognition errors. However, this brings extra computational and maintenance cost. Moreover, these two-step approaches rarely fully exploit the visual information because incorporating the textual information is already computationally expensive.

Recent end-to-end OCR-free information extraction approaches [4,12,20] were proposed to tackle some of the limitations of OCR-dependant approaches. The majority of these approaches follow an encoder-decoder scheme. However, the used encoders are either unable to effectively model global dependence when they are primarily composed of Convolutional neural network (CNN) blocks [12,22] or they don't give enough privilege to character-level features extraction when they are are primarily composed of Swin Transformer [30] blocks [5,20]. In this paper, we argue that capturing both intra-character local patterns and inter-character long-range connections is essential for the information extraction task. The former is essential for character recognition and the latter plays a role in both the recognition and the localization of the fields of interest.

Motivated by the issues mentioned above, we propose an end-to-end OCR-free information extraction model named DocParser. DocParser has been designed in a way that allows it to efficiently perceive both intra-character patterns and inter-character dependencies. Consequently, DocParser is up to two times faster than state-of-the-art methods while still achieving state-of-the-art results on various datasets.

2   Related Work

2.1   OCR-Dependant Approaches

Most of the OCR-dependant approaches simply use an off-the-shelf OCR engine and only focus on the information extraction task.

Prior to the development of deep learning techniques, earlier approaches [3,34,37] either followed a probabilistic approach, relied on rules or used manually designed features which often results in failure when applied to unfamiliar templates. The initial deep learning approaches only relied on textual information and simply used pre-trained language models [7,29]. Later, several approaches tried to take the layout information into consideration. First, [18] proposed Chargrid, a new type of text representation that preserves the 2D layout of a document by encoding each document page as a two-dimensional grid of characters. Then, [6] added context to this representation by using a BERT language model. Later, [19] improved the Chargrid model by also exploiting the visual information. Graph-based models were also proposed to exploit both textual and visual information [28,39].

To successfully model the interaction between the visual, textual and positional information, recent approaches [10,14,17,36] resorted to pre-training large models. First [46] tried to bring the success of large pre-trained language models into the multi-modal domain of document understanding and proposed LayoutLM. LayoutLMv2 [45] was later released where new pre-training tasks were introduced to better capture the cross-modality interaction in the pre-training stage. The architecture was also improved by introducing spatially biased attention and thus making the spatial information more influential. Inspired by the Vision Transformer (ViT) [17,23] modified LayoutLMv2 by using patch embeddings instead of a ResNeXt [44] Feature Pyramid Network [27] visual backbone and released LayoutLMv3. Pre-training tasks were also improved compared to previous versions. [10] proposed LAMBERT which used a modified RoBERTa [29] that also exploits the layout features obtained from an OCR system. [36] proposed TILT, a pre-trained encoder-decoder model. [14] tried to fully exploit the textual and layout information and released Bros which achieves good results without relying on the visual features. However, the efficiency and the computational cost of all the previously cited works are still hugely impacted by the used OCR system.

2.2   End-to-End Approaches

In recent years, end-to-end approaches were proposed for the information extraction task among many other Visually-Rich Document Understanding (VRDU) tasks. [12,22] both used a CNN-based encoder and a recurrent neuronal network coupled with an attention mechanism decoder. However, the accuracy of these two approaches is limited and they perform relatively badly on small datasets. [4] proposed TRIE++, a model that learns simultaneously both the text reading and the information extraction tasks via a multi-modal context block that bridges the visual and natural language processing tasks. [41] released VIES which simultaneously performs text detection, recognition and information extraction. However, both TRIE++ and VIES require the full document annotation to be trained. [20] proposed Donut, an encoder-decoder architecture that consists of a Swin Transformer [30] encoder and a Bart [25]-like decoder. [5] released Dessurt, a model that processes three streams of tokens, representing visual tokens, query tokens and the response. Cross-attention is applied across different streams to allow them to share and transfer information into the response. To process the visual tokens, Dessurt uses a modified Swin windowed attention that is allowed to attend to the query tokens. Donut and Dessurt achieved promising results, however, they don't give enough privilege to local character patterns which leads to sub-optimal results for the information extraction task.

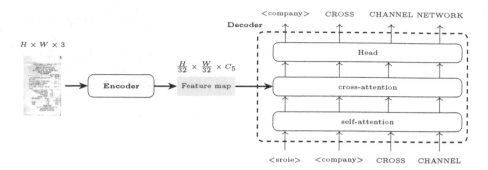

Fig. 1. An overview of DocParser's architecture. The input image of size $H \times W \times 3$ is first encoded to generate a feature map of size $\frac{H}{32} \times \frac{W}{32} \times C_5$ containing relevant visual information. The feature map is then fed to the decoder along with a task token to auto-regressively generate tokens that represent the fields of interest. For the purpose of simplification, the figure does not include residual connections and the feed-forward sub-layer.

3 Proposed Method

This section introduces DocParser, our proposed end-to-end information extraction from VRDs model.

Given a document image and a task token that determines the fields of interest, DocParser produces a series of tokens representing the extracted fields from the input image. DocParser architecture consists of a visual encoder followed by a textual decoder. An overview of DocParser's architecture is shown on Fig. 1. The encoder consists of a three-stage progressively decreased height convolutional neural network that aims to extract intra-character local patterns, followed by a three-stage progressively decreased width Swin Transformer [30] that aims to capture long-range dependencies. The decoder consists of n Transformer layers. Each layer is principally composed of a multi-head self-attention sub-layer followed by a multi-head cross-attention sub-layer and a feed-forward sub-layer as explained in [40].

3.1 Encoder

The encoder is composed of six stages. The input of the encoder is an image of size $H \times W \times 3$. It is first transformed to $\frac{H}{4} \times \frac{W}{4}$ patches of dimension C_0 via an initial patch embedding. Each patch either represents a fraction of a text character or a fraction of a non-text component of the input image. First, three stages composed of ConvNext [31] blocks are applied at different scales for character-level discriminative features extraction. Then three stages of Swin Transformer blocks are applied with varying window sizes in order to capture long-range dependencies. The output of the encoder is a feature map of size $\frac{H}{32} \times \frac{W}{32} \times C_5$ that contains multi-grained features. An overview of the encoder architecture is illustrated in Fig. 2.

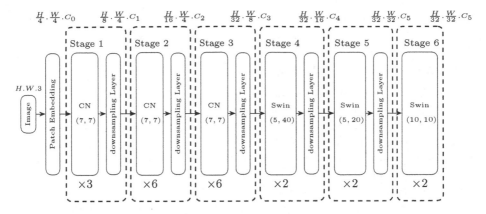

Fig. 2. The architecture of DocParser's encoder. H and W represent the height and the width of the input image. C_i, $i \in [0..5]$ represent the number of channels at different stages. The first three stages are composed of **ConvNext** (CN) blocks with a filter size of $(7,7)$. The last three stages are composed of Swin Transformer blocks with different attention window sizes. The windows' heights and widths are respectively equal to (5,40), (5,20) and (10,10).

Patch Embedding

Similar to [8], we use a progressive overlapping patch embedding. For an input image of size $W \times H \times 3$, a 3×3 convolution with stride 2 is first applied to have an output of size $\frac{W}{2} \times \frac{H}{2} \times \frac{C_0}{2}$. It is then followed by a normalization layer and another 3×3 convolution with stride 2. The size of the final output is $\frac{W}{4} \times \frac{H}{4} \times C_0$.

ConvNext-Based Stages

The first three stages of DocParser's encoder are composed of ConvNext blocks. Each stage is composed of several blocks. The kernel size is set to 7 for all stages. At the end of each stage, the height of the feature map is reduced by a factor of two and the number of channels C_i, $i \in [1, 2, 3]$ is increased to compensate for the information loss. The feature map width is also reduced by a factor of two at the end of the third stage. The role of these blocks is to capture the correlation between the different parts of each single character and to encode the non-textual parts of the image. We don't reduce the width of the feature map between these blocks in order to avoid encoding components of different characters in the same feature vector and thus allowing discriminative character features computation. We note that contrary to the first encoder stages where low-level features extraction occurs, encoding components of different characters in the same feature vector doesn't affect performance if done in the encoder last stages where high-level features are constructed. This is empirically demonstrated in Sect. 5. We chose to use convolutional blocks for the early stages mainly due to their good ability at modeling local correlation at a low computational cost.

Swin Transformer-Based Stages

The last three stages of the encoder are composed of Swin Transformer blocks. We modify Swin's window-based multi-head self-attention to be able to use rectangular attention windows. At the output of the fourth and fifth stages, the width of the feature map is reduced by a factor of two and the number of channels is increased to compensate for the information loss. The role of these layers is to capture the correlation between the different characters of the input image or between textual and non-textual components of the image. In the forth and fifth stage, the encoder focuses on capturing the correlation between characters that belong to adjacent sentences. This is accomplished through the use of horizontally wide windows, as text in documents typically has an horizontal orientation. In the last stage, the encoder focuses on capturing long-range context in both directions. This is achieved through the use of square attention windows. As a result, the output of the encoder is composed of multi-grained features that not only encode intra-character local patterns which are essential to distinguish characters but also capture the correlation between textual and non-textual components which is necessary to correctly locate the fields of interest. We note that positional embedding is added to the encoder's feature map before the encoder's forth stage.

3.2   Decoder

The decoder takes as input the encoder's output and a task token. It then outputs autoregressively several tokens that represent the fields of interest specified by the input token. The decoder consists of n^1 layers, each one is similar to a vanilla Transformer decoder layer. It consists of a multi-head self-attention sub-layer followed by a multi-head cross-attention sub-layer and a feed-forward sub-layer.

Tokenization. We use the tokenizer of the RoBERTa model [29] to transform the ground-truth text into tokens. This allows to reduce the number of generated tokens, and so the memory consumption as well as training and inference times, while not affecting the model performance as shown in Sect. 5. Similar to [20], special tokens are added to mark the start and the end of each field or group of fields. Two additional special tokens $< item >$ and $< item/ >$ are used to separate fields or group of fields appearing more than once in the ground truth.

At Training Time. When training the model, we use a teacher forcing strategy. This means that we give the decoder all the ground truth tokens as input. Each input token corresponding last hidden state is used to predict the next token. To ensure that each token only attends to previous tokens in the self-attention layer, we use a triangular attention mask that masks the following tokens.

---

[1] For our final model, we set $n = 1$.

4   Expriments and Results

4.1   Pre-training

We pre-train our model on two different steps:

Knowledge Transfer Step. Using an $L2$ Loss, we teach the ConvNext-based encoder blocks to produce the same feature map as the PP-OCR-V2 [9] recognition backbone which is an enhanced version of MobileNetV1 [15]. A pointwise convolution is applied to the output of the ConvNext-based blocks in order to obtain the same number of channels as the output of PP-OCR-V2 recognition backbone. The goal of this step is to give the encoder the ability to extract discriminative intra-character features. We use 0.2 million documents from the IIT-CDIP [24] dataset for this task. We note that even though PP-OCR-V2 recognition network was trained on text crops, the features generated by its backbone on a full image are still useful thanks to the translation equivariance of CNNs.

Masked Document Reading Step. After the knowledge transfer step, we pre-train our model on the task of document reading. In this pre-training phase, the model learns to predict the next textual token while conditioning on the previous textual tokens and the input image. To encourage joint reasoning, we mask several 32×32 blocks representing approximately fifteen percent of the input image. In fact, in order to predict the text situated within the masked regions, the model is obliged to understand its textual context. As a result, DocParser learns simultaneously to recognize characters and the underlying language knowledge. We use 1.5 million IIT-CDIP documents for this task. These documents were annotated using Donut. Regex rules were applied to identify poorly read documents, which were discarded.

4.2   Fine-Tuning

After the pre-training stage, the model is fine-tuned on the information extraction task. We fine-tune DocParser on three datasets: SROIE and CORD which are public datasets and an in-house private Information Statement Dataset.

SROIE: A public receipts dataset with 4 annotated unique fields : company, date, address, and total. It contains 626 receipts for training and 347 receipts for testing.

CORD: A public receipts dataset with 30 annotated unique fields of interest. It consists of 800 train, 100 validation and 100 test receipt images.

Information Statement Dataset (ISD): A private information statement dataset with 18 annotated unique fields of interest. It consists of 7500 train, 3250 test and 3250 eval images. The documents come from 15 different insurers, each insurer has around 4 different templates. We note that for the same template, the structure can vary depending on the available information. On Fig. 3 we show 3 samples from 3 different insurers.

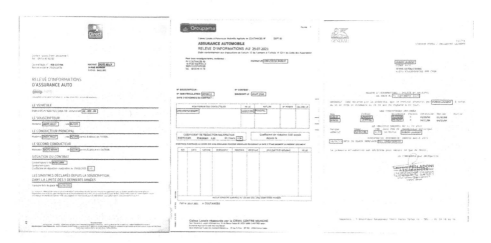

Fig. 3. Anonymized samples from our private in-house dataset. The fields of interest are located within the red boxes. (Color figure online)

4.3   Evaluation Metrics

We evaluate our model using two metrics:

Field-Level F1 Score. The field-level F1 score checks whether each extracted field corresponds exactly to its value in the ground truth. For a given field, the field-level F1 score assumes that the extraction has failed even if one single character is poorly predicted. The field-level F1 score is described using the field-level precision and recall as:

$$\text{Precision} = \frac{\text{The number of exact field matches}}{\text{The number of the detected fields}} \tag{1}$$

$$\text{Recall} = \frac{\text{The number of exact field matches}}{\text{The number of the ground truth fields}} \tag{2}$$

$$\text{F1} = \frac{2 \times \text{Precision} \times \text{Recall}}{\text{Precision} + \text{Recall}} \tag{3}$$

Document Accuracy Rate (DAR). This metric evaluates the number of documents that are completely and correctly processed by the model. If for a given document we have even one false positive or false negative, the DAR assumes that the extraction has failed. This metric is a challenging one, but requested in various industrial applications where we need to evaluate at which extent the process is fully automatable.

4.4 Setups

The dimension of the input patches and the output vectors of every stage $C_i, i \in$ [0..5] are respectively set to 64, 128, 256, 512, 768, and 1024. We set the number of decoder layers to 1. This choice is explained in Sect. 5. For both pre-training and fine-tuning we use the Cross-Entropy Loss, AdamW [33] optimizer with weight decay of 0.01 and stochastic depth [16] with a probability equal to 0.1. We also follow a light data augmentation strategy which consists of light re-scaling and rotation as well as brightness, saturation, and contrast augmentation applied to the input image. For the pre-training phase, we set the input image size to 2560 × 1920. The learning rate is set to $1e - 4$. The pre-training is done on 7 A100 GPUs with a batch size of 4 on each GPU. We use gradient accumulation of 10 iterations, leading to an effective batch size of 280. For the fine-tuning, the resolution is set to 1600 × 960 for CORD and SROIE datasets and 1600 × 1280 for the Information Statement Dataset. We pad the input image in order to maintain its aspect ratio. We also use a Cosine Annealing scheduler [32] with an initial learning rate of $3e - 5$ and a batch size of 8.

4.5 Results

Table 1. Performance comparisons on the three datasets. The field-level F1-score and the extraction time per image on an Intel Xenion W-2235 CPU are reported. In order to ensure a fair comparison, we exclude parameters related to vocabulary. Additional parameters α^* and time t^* for the OCR step should be considered for LayouLM-v3. For the ISD dataset t^* is equal to 3.6 s.

	OCR	Params(M)	SROIE		CORD		ISD	
			F1(%)	Time(s)	F1(%)	Time(s)	F1(%)	Time(s)
LayoutLM-v3	✓	$87 + \alpha^*$	77.7	$2.1 + t^*$	80.2	$2.1 + t^*$	90.8	$4.1 + t^*$
Donut		149	81.7	5.3	84	5.7	95.4	6.7
Dessurt		87	84.9	16.7	82.5	17.9	93.5	18.1
DocParser		**70**	**87.3**	3.5	**84.5**	3.7	**96.2**	**4.4**

We compare DocParser to Donut, Dessurt and LayoutLM-v3. The results are summarized in Table 1. A comparison of inference speed on an NVIDIA Quadro

RTX 6000 GPU is presented in Table 2. Per-field extraction performances on our Information Statement Dataset can be found in Table 3. DocParser achieves a new state-of-the-art on SROIE, CORD and our Information Statement Dataset with an improvement of respectively 2.4, 0.5 and 0.8 points over the previous state-of-the-art. In addition, Docparser has a significantly faster inference speed and less parameters.

Table 2. Comparison of inference speed on GPU. Extraction time (seconds) per image on an NVIDIA Quadro RTX 6000 GPU is reported. Additional time t^* for the OCR step should be considered for LayouLM-v3. For the ISD dataset t^* is equal to 0.5 s.

	SROIE	CORD	ISD
LayoutLM-v3	$0.041 + t^*$	$0.039 + t^*$	$0.065 + t^*$
Donut	0.38	0.44	0.5
Dessurt	1.2	1.37	1.39
DocParser	0.21	0.24	**0.25**

Regarding the OCR required by the LayoutLM-v3 approach, we use, for both SROIE and CORD datasets, Microsoft Form Recognizer[2] which includes a document-optimized version of Microsoft Read OCR (MS OCR) as its OCR engine. We note that we tried combining a ResNet-50 [13]-based DBNet++ [26] for text detection and an SVTR [8] model for text recognition and fine-tuned them on the fields of interest of each dataset. However, the obtained results are worse than those obtained with Microsoft Form Recognizer OCR engine. For the Information Statement Dataset, we don't use MS OCR for confidentiality purposes. Instead, we use an in-house OCR fine-tuned on this dataset to reach the best possible performances. Even though the best OCRs are used for each task, LayoutLM-v3 extraction performances are still lower than those of OCR-free models. This proves the superiority of end-to-end architectures over the OCR-dependent approaches for the information extraction task. We note that for Donut, we use the same input resolution as DocParser. For Dessurt, we use a resolution of 1152×768, which is the resolution used to pre-train the model.

5   Primary Experiments and Further Investigation

5.1   Primary Experiments

In all the experiments, the tested architectures were pre-trained on 0.5 Million synthesized documents and fine-tuned on a deskewed version of the SROIE dataset. We report the inference time on a an Intel Xenion W-2235 CPU, as we aim to provide a model suited for low resources scenarios.

---

[2] https://learn.microsoft.com/en-us/azure/applied-ai-services/form-recognizer/ concept-read?view=form-recog-3.0.0.

Table 3. Extraction performances on our Information Statement Dataset.
Per field (field-level) F1-score, field-level F1-score mean, DAR, and extraction time per image on an Intel Xenion W-2235 CPU are reported. The OCR engine inference time t^* should be considered for LayouLM-v3.

Fields	LayoutLM	DONUT	Dessurt	DocParser
Name of first driver	85.7	91.3	89.7	**92.9**
Name of second driver	82.7	90.1	89.1	**92.1**
Name of third driver	76.8	94.2	91.7	**94.7**
Bonus-Malus	96	98.5	98.1	**99**
Date of birth of first driver	97.1	97.1	97.3	**98.3**
Date of birth of second driver	95.8	96.9	96.3	**97.2**
Date of birth of third driver	91.5	**93.2**	90.8	92.8
Date of termination of contract	92	97	95.7	**97.8**
Date of the first accident	95.6	96	96.3	**97.2**
Date of the second accident	94.2	95.2	94.8	**96.1**
Subscription date	94.7	97.9	95.7	**98.5**
Date of creation of the document	95.6	97.5	96.8	**98**
Matriculation	95.2	94	94.6	**96.7**
Name of first accident's driver	85.9	85.9	85.4	**86.1**
Name of second accident's driver	84.1	**87.1**	84.9	85.2
Underwriter	92.2	92.2	92.3	**92.8**
Responsibility of the firstdriver of the accident	89.8	96.5	95.7	**97.2**
Responsibility of the seconddriver of the accident	89.5	95	94	**95.3**
Mean F1 (%)	90.8	95.4	93.5	**96.2**
DAR (%)	58.1	74	70.6	**77.5**
Time (s)	4.1+t^*	6.7	18.1	**4.4**

Table 4. Comparison of different encoder architectures. The dataset used is a deskewed version of the SROIE dataset. The field-level F1 score is reported.

Encoder architecture	F1(%)
EasyOCR-based encoder	54
PP-OCRv2-based encoder	77
Proposed encoder	81

On the Encoder's Architecture. The Table 4 shows a comparison between an EasyOCR[3]-based encoder, a PP-OCRv2 [9]-based encoder and our proposed DocParser encoder. Concerning the EasyOCR and PP-OCRv2 based encoders, each one consists of its corresponding OCR's recognition network followed by few convolutional layers that aim to further reduce the feature map size and increase the receptive field. Our proposed encoder surpasses both encoders by a large margin.

---

[3] https://github.com/JaidedAI/EasyOCR/blob/master/easyocr.

Table 5. The effect of decreasing the width of the feature map in various stages of DocParser's encoder. The dataset used is a deskewed version of the SROIE dataset. The field-level F1-score and the extraction time per image on an Intel Xenion W-2235 CPU are reported.

Encoder stages where the feature map width is reduced	Decoder	Inference time (seconds)	F1(%)
(3, 4, 5) (proposed)	Transformer	3.5	81
(3, 4, 5)	LSTM + Additive attention	3.3	58
(1, 2, 3)	Transformer	2.1	69
(1, 2, 3)	LSTM + Additive attention	1.9	52
No reduction	Transformer	6.9	81
No reduction	LSTM + Additive attention	6.6	81

On the Feature Map Width Reduction. While encoding the input image, the majority of the text recognition approaches reduce the dimensions of the feature map mainly vertically [1,8]. Intuitively, applying this approach for the information extraction task may seem relevant as it allows different characters to be encoded in different feature vectors. Our empirical results, however, show that this may not always be the case. In fact, we experimented with reducing the encoder's feature map width at different stages. As a decoder, we used both a one layer vanilla Transformer decoder and a Long Short-Term Memory (LSTM) [11] coupled with an attention mechanism that uses an additive attention scoring function [42]. Table 5 shows that reducing the width of the feature map in the early stages affects drastically the model's accuracy and that reducing the width of the feature map in the later stages achieves the best speed-accuracy trade-off. Table 5 also shows that while the LSTM-based decoder struggles with a reduced width encoder output, the performance of the vanilla Transformer-based decoder remains the same in both cases. This is probably due to the multi-head attention mechanism that makes the Transformer-based decoder more expressive than an LSTM coupled with an attention mechanism.

On the Tokenizer Choice. In addition to the RoBERTa tokenizer, we also tested a character-level tokenizer. Table 6 shows that the RoBERTa tokenizer allows faster decoding while achieving the same performance as the character-level tokenizer.

Table 6. Comparison between different tokenization techniques. The dataset used is a deskewed version of the SROIE dataset. The field-level F1-score and the decoding time per image on an Intel Xenion W-2235 CPU are reported.

Tokenizer	Decoding inference time (s)	F1(%)
RoBERTa tokenizer	0.6	81
Character-level tokenizer	1.2	81

On the Number of Decoder Layers. Table 7 shows that increasing the number of decoder layers doesn't improve DocParser's performance. Therefore, using one decoder layer is the best choice as it guarantees less computational cost.

On the Data Augmentation Strategy. Additionally to the adopted augmentation techniques, we experimented with adding different types of blur and noise to the input images for both the pre-training and the fine-tuning. We concluded that this does not improve DocParser's performance. The lack of performance improvement when using blur may be attributed to the fact that the datasets used for evaluating the model do not typically include blurred images. Additionally, it is challenging to accurately create realistic noise, thus making the technique of adding noise to the input images ineffective.

Table 7. Effect of the number of decoder layers on the performance and the decoding inference time of DocParser. The dataset used is a deskewed version of the SROIE dataset. The field-level F1-score and the decoding time per image on an Intel Xenion W-2235 CPU are reported.

Decoder layers	Decoding inference time (s)	F1(%)
1	0.6	81
2	0.8	81
4	1.2	81

Table 8. Comparison between different pre-training strategies. All the models are pre-trained for a total of 70k steps. The field-level F1-score is reported.

Pre-training tasks	SROIE	CORD	ISD
Knowledge transfer	77.5	75	89.7
Knowledge transfer + Document reading	84.7	83.7	95.6
Knowledge transfer + Masked document reading	**84.9**	**84.2**	**95.9**

5.2   Further Investigation

On the Pre-training Strategy. Table 8 presents a comparison between different pre-training strategies. To reduce compute used, all the models were pre-trained for 70k back-propagation steps, with 7k knowledge transfer steps in the case of two pre-training tasks. The results show that masking text regions during the document reading pre-training task does effectively lead to an increase in performance on all three datasets. It also confirms, as demonstrated in [20] and [5], that document reading, despite its simplicity, is an effective pre-training task.

On the Input Resolution. Figure 4 shows the effect of the input resolution on the performance of DocParser on the SROIE dataset. DocParser shows satisfying results even with a low-resolution input. It achieves 83.1 field-level F1 score with a 960×640 input resolution. The inference time for this resolution on an Intel Xenion W-2235 CPU is only 1.7 s. So, even at this resolution, DocParser still surpasses Donut and LayoutLM-v3 on SROIE while being more than three times faster. However, if the input resolution is set to 640×640 or below, the model's performance shows a drastic drop. This may be due to the fact that the characters start to be illegible at such a low resolution.

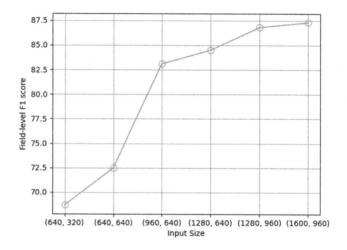

Fig. 4. The impact of the input resolution on DocParser's performance on the SROIE dataset. The field-level F1 score is reported.

6   Conclusion

We have introduced DocParser, a fast end-to-end approach for information extraction from visually rich documents. Contrary to previously proposed end-to-end models, DocParser's encoder is specifically designed to capture both intra-character local patterns and inter-character long-range dependencies. Experiments on both public and private datasets showed that DocParser achieves state-of-the-art results in terms of both speed and accuracy which makes it perfectly suitable for real-world applications.

Acknowledgments. The authors wish to convey their genuine appreciation to Prof. Davide Buscaldi and Prof. Sonia Vanier for providing them with valuable guidance. Furthermore, the authors would like to express their gratitude to Paul Wassermann and Arnaud Paran for their assistance in proofreading previous versions of the manuscript.

References

1. Baek, J., et al.: What is wrong with scene text recognition model comparisons? dataset and model analysis. In: Proceedings of the IEEE/CVF International Conference on Computer Vision, pp. 4715–4723 (2019)
2. Baek, Y., Lee, B., Han, D., Yun, S., Lee, H.: Character region awareness for text detection. In: Proceedings of the IEEE/CVF Conference on Computer Vision and Pattern Recognition, pp. 9365–9374 (2019)
3. Cesarini, F., Francesconi, E., Gori, M., Soda, G.: Analysis and understanding of multi-class invoices. Doc. Anal. Recogn. **6**, 102–114 (2003)
4. Cheng, Z., et al.: Trie++: towards end-to-end information extraction from visually rich documents (2022). arXiv preprint arXiv: arXiv:2207.06744
5. Davis, B., Morse, B., Price, B., Tensmeyer, C., Wigington, C., Morariu, V.: End-to-end document recognition and understanding with dessurt. In: Karlinsky, L., Michaeli, T., Nishino, K. (eds.) Computer Vision-ECCV 2022 Workshops, Tel Aviv, Israel, 23–27 October 2022, Proceedings, Part IV, pp. 280–296. Springer, Heidelberg (2023). https://doi.org/10.1007/978-3-031-25069-9_19
6. Denk, T.I., Reisswig, C.: Bertgrid: contextualized embedding for 2D document representation and understanding. In: Workshop on Document Intelligence at NeurIPS (2019)
7. Devlin, J., Chang, M.W., Lee, K., Toutanova, K.: BERT: pre-training of deep bidirectional transformers for language understanding. In: Proceedings of the 2019 Conference of the North American Chapter of the Association for Computational Linguistics: Human Language Technologies, vol. 1 (Long and Short Papers), pp. 4171–4186. Association for Computational Linguistics, Minneapolis (2019). https://doi.org/10.18653/v1/N19-1423
8. Du, Y., et al.: Svtr: Scene text recognition with a single visual model. In: Raedt, L.D. (ed.) Proceedings of the Thirty-First International Joint Conference on Artificial Intelligence, IJCAI-2022, International Joint Conferences on Artificial Intelligence Organization, pp. 884–890 (2022). https://doi.org/10.24963/ijcai.2022/124, main Track
9. Du, Y., et al.: Pp-ocrv2: bag of tricks for ultra lightweight ocr system (2021). ArXiv arXiv:2109.03144
10. Garncarek, Ł, et al.: LAMBERT: layout-aware language modeling for information extraction. In: Lladós, J., Lopresti, D., Uchida, S. (eds.) ICDAR 2021. LNCS, vol. 12821, pp. 532–547. Springer, Cham (2021). https://doi.org/10.1007/978-3-030-86549-8_34
11. Graves, A., Graves, A.: Long short-term memory. In: Supervised Sequence Labelling with Recurrent Neural Networks, pp. 37–45 (2012)
12. Guo, H., Qin, X., Liu, J., Han, J., Liu, J., Ding, E.: Eaten: entity-aware attention for single shot visual text extraction. In: 2019 International Conference on Document Analysis and Recognition (ICDAR), pp. 254–259 (2019). https://doi.org/10.1109/ICDAR.2019.00049
13. He, K., Zhang, X., Ren, S., Sun, J.: Deep residual learning for image recognition. In: Proceedings of the IEEE Conference on Computer Vision and Pattern Recognition, pp. 770–778 (2016)
14. Hong, T., Kim, D., Ji, M., Hwang, W., Nam, D., Park, S.: Bros: a pre-trained language model focusing on text and layout for better key information extraction from documents. In: Proceedings of the AAAI Conference on Artificial Intelligence, vol. 36, pp. 10767–10775 (2022)

15. Howard, A.G., et al.: Mobilenets: efficient convolutional neural networks for mobile vision applications (2017)

16. Huang, G., Sun, Yu., Liu, Z., Sedra, D., Weinberger, K.Q.: Deep networks with stochastic depth. In: Leibe, B., Matas, J., Sebe, N., Welling, M. (eds.) ECCV 2016. LNCS, vol. 9908, pp. 646–661. Springer, Cham (2016). https://doi.org/10.1007/978-3-319-46493-0_39

17. Huang, Y., Lv, T., Cui, L., Lu, Y., Wei, F.: Layoutlmv3: pre-training for document AI with unified text and image masking. In: Proceedings of the 30th ACM International Conference on Multimedia, MM 2022, pp. 4083–4091. Association for Computing Machinery, New York (2022). https://doi.org/10.1145/3503161.3548112

18. Katti, A.R., et al.: Chargrid: towards understanding 2D documents. In: Proceedings of the 2018 Conference on Empirical Methods in Natural Language Processing, pp. 4459–4469. Association for Computational Linguistics, Brussels (2018). https://doi.org/10.18653/v1/D18-1476

19. Kerroumi, M., Sayem, O., Shabou, A.: VisualWordGrid: information extraction from scanned documents using a multimodal approach. In: Barney Smith, E.H., Pal, U. (eds.) ICDAR 2021. LNCS, vol. 12917, pp. 389–402. Springer, Cham (2021). https://doi.org/10.1007/978-3-030-86159-9_28

20. Kim, G., et al.: Ocr-free document understanding transformer. In: Avidan, S., Brostow, G., Cissé, M., Farinella, G.M., Hassner, T. (eds.) ECCV 2022. LNCS, vol. 13688, pp. 498–517. Springer, Cham (2022). https://doi.org/10.1007/978-3-031-19815-1_29

21. Kissos, I., Dershowitz, N.: Ocr error correction using character correction and feature-based word classification. In: 2016 12th IAPR Workshop on Document Analysis Systems (DAS), pp. 198–203 (2016). https://doi.org/10.1109/DAS.2016.44

22. Klaiman, S., Lehne, M.: DocReader: bounding-box free training of a document information extraction model. In: Lladós, J., Lopresti, D., Uchida, S. (eds.) ICDAR 2021. LNCS, vol. 12821, pp. 451–465. Springer, Cham (2021). https://doi.org/10.1007/978-3-030-86549-8_29

23. Kolesnikov, A., et al.: An image is worth 16×16 words: transformers for image recognition at scale (2021)

24. Lewis, D., Agam, G., Argamon, S., Frieder, O., Grossman, D., Heard, J.: Building a test collection for complex document information processing. In: Proceedings of the 29th Annual International ACM SIGIR Conference on Research and Development in Information Retrieval, SIGIR 2006, pp. 665–666. Association for Computing Machinery, New York (2006). https://doi.org/10.1145/1148170.1148307

25. Lewis, M., et al.: BART: denoising sequence-to-sequence pre-training for natural language generation, translation, and comprehension. In: Proceedings of the 58th Annual Meeting of the Association for Computational Linguistics, pp. 7871–7880. Association for Computational Linguistics (2020). https://doi.org/10.18653/v1/2020.acl-main.703

26. Liao, M., Zou, Z., Wan, Z., Yao, C., Bai, X.: Real-time scene text detection with differentiable binarization and adaptive scale fusion. IEEE Trans. Pattern Anal. Mach. Intell. **45**(1), 919–931 (2022)

27. Lin, T.Y., Dollár, P., Girshick, R., He, K., Hariharan, B., Belongie, S.: Feature pyramid networks for object detection. In: Proceedings of the IEEE Conference on Computer Vision and Pattern Recognition, pp. 2117–2125 (2017)

28. Liu, X., Gao, F., Zhang, Q., Zhao, H.: Graph convolution for multimodal information extraction from visually rich documents. In: Proceedings of the 2019 Conference of the North American Chapter of the Association for Computational Linguistics: Human Language Technologies, vol. 2 (Industry Papers), pp. 32–39. Association for Computational Linguistics, Minneapolis (2019). https://doi.org/10.18653/v1/N19-2005

29. Liu, Y., et al.: Roberta: a robustly optimized bert pretraining approach (2019). arXiv preprint arXiv: arXiv:1907.11692

30. Liu, Z., et al.: Swin transformer: hierarchical vision transformer using shifted windows. In: Proceedings of the IEEE/CVF International Conference on Computer Vision, pp. 10012–10022 (2021)

31. Liu, Z., Mao, H., Wu, C.Y., Feichtenhofer, C., Darrell, T., Xie, S.: A convnet for the 2020s. In: Proceedings of the IEEE/CVF Conference on Computer Vision and Pattern Recognition, pp. 11976–11986 (2022)

32. Loshchilov, I., Hutter, F.: SGDR: stochastic gradient descent with warm restarts. In: ICLR 2017 (2016). arXiv preprint arXiv: arXiv:1608.03983

33. Loshchilov, I., Hutter, F.: Decoupled weight decay regularization. In: International Conference on Learning Representations (2017)

34. Medvet, E., Bartoli, A., Davanzo, G.: A probabilistic approach to printed document understanding. Int. J. Doc. Anal. Recogn. **14**(4), 335–347 (2011). https://doi.org/10.1007/s10032-010-0137-1

35. Palm, R.B., Winther, O., Laws, F.: Cloudscan-a configuration-free invoice analysis system using recurrent neural networks. In: 2017 14th IAPR International Conference on Document Analysis and Recognition (ICDAR), vol. 1, pp. 406–413. IEEE (2017)

36. Powalski, R., Borchmann, Ł, Jurkiewicz, D., Dwojak, T., Pietruszka, M., Pałka, G.: Going Full-TILT boogie on document understanding with text-image-layout transformer. In: Lladós, J., Lopresti, D., Uchida, S. (eds.) ICDAR 2021. LNCS, vol. 12822, pp. 732–747. Springer, Cham (2021). https://doi.org/10.1007/978-3-030-86331-9_47

37. Rusiñol, M., Benkhelfallah, T., dAndecy, V.P.: Field extraction from administrative documents by incremental structural templates. In: 2013 12th International Conference on Document Analysis and Recognition, pp. 1100–1104 (2013). https://doi.org/10.1109/ICDAR.2013.223

38. Schaefer, R., Neudecker, C.: A two-step approach for automatic OCR post-correction. In: Proceedings of the The 4th Joint SIGHUM Workshop on Computational Linguistics for Cultural Heritage, Social Sciences, Humanities and Literature, pp. 52–57. International Committee on Computational Linguistics (2020)

39. Sun, H., Kuang, Z., Yue, X., Lin, C., Zhang, W.: Spatial dual-modality graph reasoning for key information extraction (2021). arXiv preprint arXiv: arXiv:2103.14470

40. Vaswani, A., et al.: Attention is all you need. Adv. Neural Inf. Process. Syst. **30** (2017)

41. Wang, J., et al.: Towards robust visual information extraction in real world: new dataset and novel solution. In: Proceedings of the AAAI Conference on Artificial Intelligence, vol. 35, pp. 2738–2745 (2021)

42. Wang, W., Yang, N., Wei, F., Chang, B., Zhou, M.: Gated self-matching networks for reading comprehension and question answering. In: Proceedings of the 55th Annual Meeting of the Association for Computational Linguistics, vol. 1: Long Papers, pp. 189–198. Association for Computational Linguistics, Vancouver (2017). https://doi.org/10.18653/v1/P17-1018

43. Wei, M., He, Y., Zhang, Q.: Robust layout-aware ie for visually rich documents with pre-trained language models. In: Proceedings of the 43rd International ACM SIGIR Conference on Research and Development in Information Retrieval, pp. 2367–2376 (2020)
44. Xie, S., Girshick, R., Dollár, P., Tu, Z., He, K.: Aggregated residual transformations for deep neural networks. In: Proceedings of the IEEE Conference on Computer Vision and Pattern Recognition, pp. 1492–1500 (2017)
45. Xu, Y., et al.: LayoutLMv2: multi-modal pre-training for visually-rich document understanding. In: Proceedings of the 59th Annual Meeting of the Association for Computational Linguistics and the 11th International Joint Conference on Natural Language Processing, vol. 1: Long Papers, pp. 2579–2591. Association for Computational Linguistics (2021). https://doi.org/10.18653/v1/2021.acl-long.201
46. Xu, Y., Li, M., Cui, L., Huang, S., Wei, F., Zhou, M.: LayoutLM: pre-training of text and layout for document image understanding. In: Proceedings of the 26th ACM SIGKDD International Conference on Knowledge Discovery. ACM (2020). https://doi.org/10.1145/3394486.3403172
47. Zhao, X., Niu, E., Wu, Z., Wang, X.: Cutie: learning to understand documents with convolutional universal text information extractor (2019). arXiv preprint arXiv: arXiv:1903.12363

MUGS: A Multiple Granularity Semi-supervised Method for Text Recognition

Qi Song[1] (ID), Qianyi Jiang[2(✉)] (ID), Lei Wang[2] (ID), Lingling Zhao[2] (ID),
and Rui Zhang[2] (ID)

[1] Yonyou Network Technology Co., Ltd., Beijing, China
songq1@yonyou.com
[2] Meituan, Beijing, China
jiangqianyi02@meituan.com, zhangrui96@tsinghua.org.cn

Abstract. Most text recognition methods are trained on large amounts of labeled data. Although text images are easily accessible, labeling them is costly. Thus how to utilize the unlabeled data is worth studying. In this paper, we propose a MUltiple Granularity Semi-supervised (MUGS) method using both labeled and unlabeled data for text recognition. Inspired by the hierarchical structure (sentence-word-character) of text, we apply semi-supervised learning at both word-level and character-level. Specifically, a Dynamic Aggregated Self-training (DAS) framework is introduced to generate pseudo-labels from unlabeled data at word-level. To ensure the quality and stability of the pseudo-labeling procedure, the pseudo-labels are aggregated from one dynamic model queue which keeps updating in the whole semi-supervised training process. At the character-level, a novel module named WTC (Word To Character) that can convert sequential features to character representations is invented. Next, contrastive learning is applied to these character representations for better fine-grained visual modeling. The characters from various images that share the same classes are pulled together and the ones in different classes are set far apart in the representation space. With the combination of supervisions in different granularities, more information can be exploited from the unlabeled data. The effectiveness and robustness of the model are enhanced by a large margin. Comprehensive experiments on several public benchmarks validate our method and competitive performance is achieved with much fewer labeled data.

Keywords: Text Recognition · Semi-Supervised Learning · Multiple Granularity

1 Introduction

Text recognition has attracted great interest from both academia and industry in recent years due to its various practical applications, such as photo transcription, document processing, and product searching. Despite many efforts have

G. A. Fink et al. (Eds.): ICDAR 2023, LNCS 14191, pp. 173–188, 2023.
https://doi.org/10.1007/978-3-031-41734-4_11

been made and remarkable success has been achieved [5, 7, 10, 14, 19, 36], it is still a challenge to recognize texts in difficult conditions, i.e. diverse font style, occlusion, uneven illumination, noisy background, and random layout. Most prior works are based on deep learning and abundant image data is necessary for training robust models. To obtain the training data, human annotating and synthesizing are two prevailing approaches. While the former is very expensive and text images generated by the latter method are different from the real world data in distribution. Some examples of real and synthesized images are illustrated in Fig. 1.

Fig. 1. Examples of real images (left) and synthesized images (right). It can be observed that the real images are more complicated than the synthesized ones in diversity.

On the other hand, text is ubiquitous in our daily life, and text images without annotations can be collected easily. It is natural to adopt Semi-Supervised Learning (SSL) for text recognition which can leverage the unlabeled data. Unfortunately, most existing SSL approaches are designed for image classification, which can not be applied to text recognition directly, since sequence recognition is involved. Only recently some related studies have appeared [1, 33, 36, 45]. In these works, techniques like self-training, reinforcement learning, and contrastive learning are utilized for learning visual and semantic information from text images without annotations. Although these methods have made impressive progress, some drawbacks still exist:

1) Self-training utilized in some approaches [36, 45] is not appropriate for text recognition. Different from classification, sequence prediction like text recognition may encounter more types of mistakes, such as redundancy, absence, or repeats of characters. The word-level pseudo-labels generated by self-training may contain mistakes and harm the SSL performance. If a strict filtering procedure is adopted to get rid of these mistakes, many text images and their corresponding pseudo-labels will be discarded. Then a mass of valuable data is wasted.

2) Consistency regularization and contrastive learning which are often used in semi- and self-supervised learning methods are hard to achieve satisfactory results in text recognition. This is because these techniques are simply transferred without considering the inherent structure of text. For example, some commonly used data augmentations in semi- and self-supervised learning,

such as random cropping, shifting, and rotation are too drastic for text images. The performance for text recognition will be degraded.

3) The character-level SSL for text recognition is not fully exploited. Compared with the word-level SSL which places more emphasis on semantic information, character-level SSL can improve the visual modeling ability of characters. While few related approaches are proposed.

In this paper, we propose a novel **MU**ltiple **G**ranularity **S**emi-supervised (MUGS) approach for text recognition, which can achieve comparative performance with fully-supervised learning but using much less labeled data. Considering the drawbacks mentioned above, we improve the self-training procedure at word-level and employ character-level SSL in text recognition, as shown in Fig. 2.

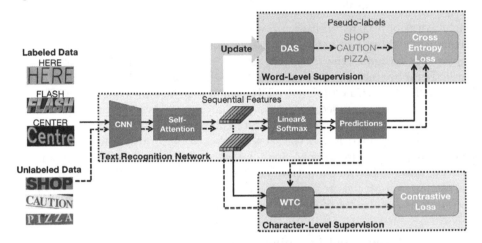

Fig. 2. The overall framework of our proposed method. The solid and dashed lines represent the flow of labeled and unlabeled data, respectively.

In detail, a novel framework named **D**ynamic **A**ggregated **S**elf-training (DAS) especially tailored for text recognition is designed for pseudo-labeling at word-level. To ensure stability, the pseudo-labels of unlabeled text images are aggregated and filtered by a dynamic model queue, instead of by one static model. During training, new models will be pushed in and old models will be popped out in this queue according to our designed strategy, which further ensures the stability of the pseudo-labels.

As for the SSL at character-level, contrastive learning is introduced in a different way from previous works. we propose a **W**ord **T**o **C**haracter (WTC) module which can convert sequential features extracted from word images into character representations. Instead of applying contrastive learning between the original text images and their augmented ones, we adopt contrastive learning on the character representations. In this way, the characters that share the same class in various images or styles are pulled together and the ones of different classes are set far apart in a representation space.

In summary, our main contributions are three-fold:

1) A robust SSL method for text recognition that utilizes self-training and contrastive learning in different granularities, namely word-level and character-level.
2) A **D**ynamic **A**ggregated **S**elf-training (DAS) framework for word-level SSL in text recognition. And a **W**ord **T**o **C**haracter (WTC) module which facilitates the implementation of character-level contrastive learning.
3) Extensive experiments validate the robustness and effectiveness of our method and competitive results are achieved.

2   Related Work

2.1   Text Recognition

Early works of text recognition are segmentation-based [12,13,31] and the performance of these methods is limited. With the rise of deep learning, recent approaches recognize the target string as a whole. These methods can be categorized into two mainstreams by decoding techniques, *i.e.*, Connectionist Temporal Classification (CTC) [2] and ATTentional sequence-to-sequence (ATT) [10]. CTC is originally proposed by Graves *et al.* for speech recognition which removes the need for pre-segmenting training data and post-processing outputs. The ATT is first proposed in [10] and the basic idea is to perform feature selection and generate characters iteratively. Some latest researches employ the combination of CTC and ATT, and show the benefits of joint learning [25,32,36,46]. These state-of-art approaches require a large number of labeled images for training, and manual annotating of such large-scale data is costly.

2.2   Semi-Supervised Learning (SSL)

To alleviate the dependence on labeled data in deep learning, SSL has become a hot topic and a large majority of the research is focused on classification. Pseudo-labeling [24] and consistency regularization are two prevailing approaches. Pseudo-labeling utilizes the model itself to generate artificial labels online or offline for unlabeled data. The generalization performance of the model can be improved when trained by both labeled and pseudo-labeled data. Self-training [39,42] and Co-training [15,37] are two most popular techniques for pseudo-labeling. Consistency regularization aims at enforcing the model to be insensitive to the noise and gradually propagating label information from labeled examples to unlabeled ones. Several recent studies present promising results using the contrastive learning [8,9,34,38].

2.3   Usage of Unlabeled Data in Text Recognition

Despite the numerous applications of SSL mentioned above, the usage of unlabeled data in text recognition remains a challenging task as sequential recognition is involved. Jiang *et al.* [33] propose a heterogeneous network for SSL in scene text recognition. Two networks predict independently from different perspectives and a discriminator is introduced for pseudo-labeling. Gao *et al.* [45] design a hybrid loss function to mix the traditional cross-entropy loss and a word-level reward-based loss. The training process takes several stages and reinforcement learning is adopted, as the edit reward is non-differentiable. Inspired by UDA, Zheng *et al.* [49] introduces an SSL framework with consistency training regularizing the outputs of two different augmentation approaches to be invariant. Recently, some self-supervised methods have also been proposed to make use of the unlabeled data. Aberdam *et al.* [1] introduces an instance-mapping function to capture atomic elements from a sequence feature and formulates the contrastive learning for unlabeled data. The data augmentation applied in this method is sophisticated which must ensure the maintenance of sequential structure in text images. Luo *et al.* [27] applies self-supervised learning of scene text in a generative manner on the assumption of the style consistency of neighboring patches among one text image. Yang   [43] investigates self-supervised learning with integration of contrastive learning and masked image modeling to learn discrimination and generation. Zhang *et al.* [48] integrates contrastive loss by pulling together clusters of identical characters within various contexts and pushing apart clusters of different characters in embedding space.

3   Method

The detailed architecture of our proposed method is shown in Fig. 2. There are three main components, namely the text recognition network, the word-level supervision, and the character-level supervision. In training, both labeled and unlabeled data are sent into the text recognition network to generate sequential features and corresponding predictions. Then in the word-level supervision, the predictions from labeled and unlabeled data are supervised by ground truths and pseudo-labels generated by DAS separately. In the character-level supervision, the sequential features and predictions are transformed into representation-label pairs by WTC. Then contrastive learning is applied. In inference, only the text recognition network is kept for predicting and the two supervisions are removed. In the following sections, we will introduce every component in detail.

3.1   Text Recognition Network

Inspired by [14,28,36], we design a simple but effective text recognition network that is composed of CNNs, self-attentions [40] and fully-connected layers, as illustrated in Fig. 2. Given an input text image $X \in \mathbb{R}^{H \times W \times 3}$, we adopt ResNet34 [20] as the CNN backbone to extract visual features $F_v \in \mathbb{R}^{T \times D}$, where T is the feature length and D is the channel number. Then two stacked self-attention modules are applied to obtain F_a from F_v. After getting F_a, we use it to predict the sequential probabilities $P \in \mathbb{R}^{T \times C}$ in parallel by calculating the Softmax of the fully-connected layer, where C is the charset size. Here we give a brief review of the self-attention, which consists of multi-head attention and feed-forward network. First, the single-head attention can be defined as follows,

$$\alpha = softmax(\frac{QK^T}{\sqrt{d_k}}), \tag{1}$$

$$Attn(Q, K, V) = \alpha V, \tag{2}$$

where Q, K, and V are the query, key, and value vectors respectively. d_k refers to the dimension of the key vector. In our implementation, the dimension of query and value vectors is identical with d_k, namely $d_k = d_q = d_v$. α is the attention weights calculated from Q, K and d_k. Then the multi-head attention is the concatenation of H single-head attention layers:

$$MultiHead(Q, K, V) = Concat(a_1, ..., a_H),$$
$$where \; a_i = Attn(Q_i, K_i, V_i) \tag{3}$$

Moreover, the Feed-Forward Network (FFN) is composed of two fully-connected layers and a ReLU activation layer, and is calculated as follows,

$$FFN(x) = W_2 ReLU(W_1 x + b_1) + b_2, \tag{4}$$

where x is the input. W_1, W_2, b_1, b_2 are learnable parameters in fully-connected layers.

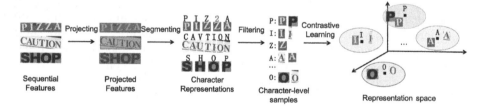

Fig. 3. The WTC module for character-level supervision. It consists of three steps, namely projecting, segmenting and filtering. Then contrastive learning is employed on the outputted character-level samples.

3.2   Word-Level Supervision

For text recognition, the predictions are sequences of characters and misreading may happen at any position. Thus, one key point is how to obtain highly credible pseudo-labels. Previous related works [36, 45] often utilize one static model trained on labeled data to generate pseudo-labels for unlabeled data. Then the model is re-trained on both labeled and pseudo labeled data. In these methods, there are two reasons which result in unsatisfactory performance. The first is placing too much trust in the ability of one single model. The second is the model for pseudo-labeling lags far behind the current training model as training progresses. To tackle these problems, in our DAS framework, a dynamic model queue with a max capacity of N is maintained. At the beginning of SSL, this queue is initialized with only one model which is pretrained by labeled data. And in the SSL procedure, the models of the queue will update dynamically, which means the latest model will be pushed in and the oldest model will be popped out to keep the capacity no more than N. In one training iteration of SSL, pseudo-labels are weighted aggregated from the models in the queue, where the weights are determined by the models' accuracies on the validation set. Then these pseudo-labels are filtered by a given threshold. The Algorithm 1 gives a detailed description of one iteration in training. Finally, the word-level supervision loss l_w is computed using ground truths and pseudo-labels on both labeled and unlabeled data.

Algorithm 1. One iteration in the DAS framework

Input:
  Unlabeled data batch $\mathcal{U}$
  Labeled data batch $\mathcal{L}$
  Model queue $\mathcal{Q}$ and current size n
  Model queue maximum capacity N
  Filtering threshold η
  Model update period ρ
  Current step number s
  Current model Θ
Output:
  The updated model $\hat{\Theta}$
1: **Aggregation:** Compute the weights $\beta_i = \frac{a_i - a_{min}}{\sum_{j=1}^{n} a_j - a_{min}}$, where a_i is the i-th model accuracy and a_{min} is the lowest model accuracy in $\mathcal{Q}$ on the validation set. The aggregated results from $\mathcal{Q}$ is computed as $\overline{P}(\mathcal{U}_t) = \sum_{i=1}^{n} \beta_i P_i(\mathcal{U}_t)$, where $P_i(\mathcal{U}_t)$ is the probability distribution of the t-th unlabeled data outputted by the i-th model.
2: **Filtering:** As $\overline{P}(\mathcal{U}_t) = [\overline{p}_1(\mathcal{U}_t), \overline{p}_2(\mathcal{U}_t), ..., \overline{p}_T(\mathcal{U}_t)]$, compute the confidence score $c_t = \min_{1 \leq t \leq T} \overline{p}(\mathcal{U}_t)$. If $c_t \leq \eta$, keep it, otherwise discard it. Then get $\overline{P}_{filtered}(\mathcal{U})$ and $\mathcal{U}_{filtered}$.
3: **Update:** Compute cross entropy loss l_w on $\mathcal{L}$, $\mathcal{U}_{filtered}$ and $\overline{P}_{filtered}(\mathcal{U})$. Update Θ to $\hat{\Theta}$ by back propagation from l_w. If $s\%\rho = 0$, push $\hat{\Theta}$ into $\mathcal{Q}$. If $n > N$, pop out the oldest model in $\mathcal{Q}$.

3.3 Character-Level Supervision

Previous works [4,36,45] mainly concentrate on word-level SSL and neglect the smallest unit that can be recognized, namely the character. There are two attractive points for applying SSL at character-level. Compared with words, the set of characters in one language is closed and relatively small. The appearances of characters are diverse across text images. Thus we can enhance the visual modeling ability by character-level SSL. However, the problem is how to obtain the character-level representations and utilize them without annotations. To solve this problem, we design a module named WTC (Word To Characters) for extracting character-level representations and then applying contrastive learning to them. As shown in Fig. 3, in WTC we first project the features F_a obtained after self-attention into a representation space,

$$F_p = Proj(F_a),$$

$$(5)$$

where $Proj$ is a non-linear projection function. Then for every F_p and its corresponding probability matrix P, we split them along the axis of feature length T. And we can construct T character pairs $G(i) = (F_p(i), P(i))$, where $F_p(i)$ and $P(i)$ are the i-th feature and its probability vector. To remove noisy data, two types of pairs will be discarded:

1) The pairs which are after the left-most pair whose top probability character is End Of String (EOS).
2) The pairs whose top probability is below a given threshold ϵ.

The valid character pairs G_v can be gathered after the filtering procedure and their character-level pseudo-labels are set to the characters with top probabilities. Finally, the contrastive loss is applied to these representations and labels to pull all samples in the same class closer and push others away. To tackle noisy labels and outliers, the Proxy Anchor loss [23] is employed as the contrastive loss. In addition, we have also conducted experiments on some other types of contrastive losses, and results are reported in Sect. 4.3. Given a set of all proxies R and a batch of valid character pairs G_v, the Proxy Anchor loss can be formulated as,

$$\ell(G_v) = \frac{1}{|R^+|} \sum_{r \in R^+} \log \left(1 + \sum_{g_v \in G_v(r+)} e^{-\gamma(z(g_v, r) - \delta)} \right)$$

$$+ \frac{1}{|R|} \sum_{r \in R} \log \left(1 + \sum_{g_v \in G_v(r-)} e^{\gamma(z(g_v, r) + \delta)} \right)$$

$$(6)$$

where $\delta > 0$ and $\gamma > 0$ are a margin and a scaling factor respectively. R^+ denotes the set of positive proxies for a specified class, and for each proxy r, the valid character pairs are divided into two sets: the set of positive pairs $G_v(r^+)$ and the set of negative pairs $G_v(r^-)$. $z()$ is the similarity computing function and the cosine distance is chosen. Finally, the character-level loss l_c is computed over both labeled and unlabeled data. For the labeled data, it should be noted that the ground-truths are utilized instead of pseudo-labels.

3.4   Objective Function

After being pretrained on labeled data, the text recognition network is finetuned on both labeled and unlabeled data. The final objective function is defined in

Eq. 7. ℓ_w and ℓ_c are the word-level and character-level supervision losses respectively. λ is used to balance the two losses and set to 0.1 empirically in all the experiments.

$$\ell_{total} = \ell_w + \lambda\ell_c. \tag{7}$$

4   Experiments

4.1   Datasets

Following the setup of [45], we use the synthetic dataset SynthText (ST) [16] as our main training data. We also conduct complemantary experiments on the union of three synthetic datasets which are MJSynth (MJ) [29], Synth-Text (ST) [16] and SynthAdd (SA) [19]. We evaluate our method on eight public scene text recognition benchmarks, including IIIT5K [30], Street View Text (SVT) [41], ICDAR2003 (IC03) [26], ICDAR2013 (IC13) [22], ICDAR2015 (IC15) [21], Street View Text Perspective (SVTP) [35], CUTE [3] and TOTAL-TEXT(TTEXT) [11].

Synthetic Datasets. MJ is a synthetic dataset with random transformations and effects, which contains 9 million text images from a lexicon of 90K English words. ST contains 800,000 images with 6 million synthetic text instances. SA contains 1.2 million text instance images with more non-alphanumeric characters.
ICDAR2003 (IC03) [26] contains 1110 images for evaluation. Following Wang *et al.* [41], we filter some samples and reduce the image number to 867.
ICDAR2013 (IC13) [22] has a testset with 1095 images. After removing words with non-alphanumeric characters, the filtered testset contains 1015 images for evaluation.
ICDAR2015 (IC15) [21] contains images captured using Google Glass. It has two versions of testset including 1811 and 2077 images respectively. We use the latter.
IIIT5K-Words (IIIT5K) [30] contains a testset of 3000 images for evaluation gathered from the Internet.
Street View Text (SVT) [41] consists of 647 word images for evaluation, most of which are horizontal text lines but blurred or of bad quality.
Street View Text Perspective (SVTP) [35] is a collection of 645 images like SVT. But most of the images are difficult to recognize because of large perspective distortions.
CUTE [3] consists of 288 images for evaluation and many of them are curved text instances.
TOTAL-TEXT (TTEXT) [11] contains 11,459 text instance images. The images have more than three different orientations, including horizontal, multi-oriented, and curved.

Table 1. The experiments of max model capacity N in DAS. The initial model is pre-trained on 5% data in ST. Only the DAS module and word-level supervision is introduced. "Avg" represents the weighted average of accuracies. Bold represents the best accuracy.

Model num.	Regular Text				Irregular Text				Avg.	Gains
	IIIT5K	SVT	IC03	IC13	IC15	SVTP	CUTE	TTEXT		
Baseline	76.5	74.0	85.7	81.5	57.8	62.2	59.0	50.7	66.4	–
1	82.1	80.5	89.9	86.0	60.8	71.3	63.2	55.6	72.3	+5.9
2	82.5	81.3	**90.9**	85.4	60.2	71.0	67.4	58.4	73.0	+6.6
3	82.5	80.2	90.2	86.0	60.7	71.0	67.0	57.7	73.0	+6.6
4	82.2	81.8	90.7	85.8	61.0	**71.9**	**69.4**	**58.7**	73.2	+6.8
5	**83.0**	**82.1**	89.3	**86.4**	**61.2**	71.2	**69.4**	57.5	**73.3**	**+6.9**

4.2   Implementation Details

The input images are resized to 32×100. Data augmentations including geometry transformation, deterioration, and color jittering are applied only to the labeled data. The pooling strides of the last two blocks in ResNet34 are changed to 2×1 following CRNN [6]. Then the length T of F_v, F_a, F_p and P is 25. In self-attention, the head number of multi-head attention is 8, and the dimension of Q, K, and V is 1024. The input and output feature dimension of the feedforward layer is 1024, and the inner-layer feature dimension is 2048. In DAS, the update frequency ρ is 0.25 epoch and the filtering threshold η for character-level pseudo-labels is set to 0.8. In WTC, the projection head consists of three fully-connected layers. The dimensions of these three layers are 1024,1024 and 128. Batch normalization and ReLU are applied after every layer. The filtering threshold ϵ for word-level pseudo-labels is set to 0.8. In training, the batch size of both labeled and unlabeled data is 128. The ADADELTA [47] is applied for optimization and the initial learning rate is 1.0, then adjust to 0.1 and 0.01 after 4 and 6 epochs. The proposed network is implemented with Pytorch and 4 NVIDIA Tesla V100 GPUs are used for training. The total number of recognition classes is 38, including 10 digits, 26 case-insensitive letters, 1 represents all punctuation symbols and 1 EOS (End Of String) symbol. No lexicons are applied after predicting. We only use greedy searching and no model ensembles are adopted during inference. Our model has similar parameters and inference speed as the small version of BCN [36]. The parameter scale is 39.0×10^6 and inference time for a batch is 17.2 ms on NVIDIA Tesla V100.

4.3   Ablation Study

In this section, we explore the effect of every important module and find out the best settings.

Table 2. The experiment about the effect of contrastive learning loss type. For all loss types, we use cosine similarity as the distance measurement. For NT-Xent loss, the temperature parameter is set to 0.07 following [18]. For circle loss, relaxation factor and gamma are 0.4 and 80 following [44]. For Proxy Anchor loss, δ and γ are 0.1 and 32 as described in [23]. Bold represents the best accuracy.

Loss types	Regular Text				Irregular Text				Avg.	Gains
	IIIT5K	SVT	IC03	IC13	IC15	SVTP	CUTE	TTEXT		
Baseline	76.5	74.0	85.7	81.5	57.8	62.2	59.0	50.7	66.4	–
Circle loss	76.4	75.3	86.3	79.5	55.5	60.9	55.9	49.7	65.6	–0.4
Contrastive loss	79.3	79.0	88.7	83.9	62.2	67.9	63.2	54.0	69.9	+3.5
NT-Xent loss	**80.2**	79.0	88.2	83.8	**64.7**	**68.5**	61.1	53.1	70.3	+3.9
Proxy Anchor loss	79.2	**79.9**	**89.2**	**84.1**	62.3	67.8	**65.3**	**55.7**	**70.4**	**+4.0**

Max Model capacity in DAS. In our DAS module, to analyze the effect of max model capacity N, we vary it from 1 to 5. To eliminate the influence of other modules, only the DAS module is employed in the experiments. As can be observed in Table 1, the best result is achieved with $N = 5$ and the accuracy improvement almost saturates when $N = 2$. Through aggregating more models, pseudo-labeling tends to be more stable and robust. While as N grows, so does the computation and memory cost during training. Given the same batch size, it takes 390 ms and 1070 ms per batch for training using $N = 1$ and $N = 5$. To make a trade-off between accuracy and efficiency, we set $N = 2$ for the following experiments. It is worth noting that the model number only affects training, the inference speed is always the same as depicted in 4.2.

Contrastive Loss. In the WTC module, we map visual sequential features into character-level representations. Then contrastive learning operates on the representations of both labeled and unlabeled data at character-level. As the contrastive loss plays an important role, we run experiments on 4 different widely used types, namely Circle loss [44], Contrastive loss [17], NT-Xent loss [9] and Proxy Anchor loss [23]. In these experiments, only WTC module is implemented and the results are shown in Table 2. Most of the loss types provide considerable accuracy improvement (vary from +3.5% to +4.0%), except for Circle loss, which decreases the accuracy by −0.4%. Proxy Anchor loss achieves the best performance (+4.0%) and is selected as the contrastive loss in our following experiments. The Proxy Anchor loss is robust against noisy labels and outliers which are very common in the SSL of text recognition. And its speed of convergence is acceptable.

Supervision. In this part, we investigate the significance of multiple granularity supervisions in our framework, namely word-level and character-level supervisions. In the experiments, we use 5% of the ST dataset as the labeled data and the rest is regarded as unlabeled data. The baseline is trained only on the

Table 3. The experiment results of using different granularity supervisions. The model capacity N in DAS is 2 and the Proxy Anchor loss is employed in WTC. The baseline model is trained only on 5% labeled data. "Avg" represents the weighted average of accuracies. Bold represents the best accuracy.

Method	Regular Text				Irregular Text				Avg.	Gains
	IIIT5K	SVT	IC03	IC13	IC15	SVTP	CUTE	TTEXT		
Baseline	76.5	74.0	85.7	81.5	57.8	62.2	59.0	50.7	66.4	–
WTC	79.2	79.9	89.2	84.1	62.3	67.8	65.3	55.7	70.4	+4.0
DAS	82.5	**81.3**	**90.9**	**85.4**	65.2	71.0	67.4	58.4	73.0	+6.6
DAS + WTC	**82.9**	80.2	90.2	85.3	**66.8**	**72.4**	**69.4**	**58.9**	**73.5**	**+7.1**

labeled data and others are trained on both labeled and unlabeled data. We set $N = 2$ for DAS and use Proxy Anchor loss for WTC. The experiment results are summarized in Table 3. Word-level and character-level supervisions marginally improve the accuracy by 6.6% and 4.0% separately. Then the combination of them achieves the best performance with an boost of 7.1%, which verifies the effectiveness of our method. In this way, both visual and semantic information in text images can be exploited effectively.

Table 4. Experiment results on the union of ST, MJ and SA with identical settings in 4.4. The baseline model is trained on 5% labeled data. "Fully" represents training on all the labeled data. "Avg" represents the weighted average of accuracies.

Method	Regular Text				Irregular Text				Avg.	Gains
	IIIT5K	SVT	IC03	IC13	IC15	SVTP	CUTE	TTEXT		
Baseline	90.1	84.1	92.7	88.9	65.8	71.5	79.2	65.2	78.6	–
Fully	93.4	88.3	94.1	92.5	73.6	76.7	79.8	70.3	83.1	+4.5
Ours	93.0	88.6	92.8	92.1	73.2	76.3	82.0	70.4	82.9	+4.3

4.4   Comparison with State-of-the-Arts

There are only a few works using unlabeled data for text recognition. In these approaches, different text recognizers, datasets, and settings are adopted in their experiments. For a fair and comprehensive comparison, we review the related works and introduce their key modules into our framework with the same datasets and settings. For contrastive learning, We reproduce the sequence-level [9] SimCLR and frame-level [1] SeqCLR to compare with our character-level module (WTC). The classic self-training method (TS) [42] is also implemented as an alternative to the DAS module. The backbone and optimizing strategy as described in Sect. 4.2 keep unchanged in all the combinations. Moreover, the

Table 5. Comparison with related SSL methods. SimCLR, SeqCLR, TS are reproduced with the same framework and optimizing strategy as described in Sect. 4.2. The results of SSL-STR [45] are excerpted from the paper. "Fully" represents the fully-supervised training on the complete ST dataset.

Method	IIIT5K		SVT		IC03		IC13		IC15		SVTP		CUTE	
	1%	5%	1%	5%	1%	5%	1%	5%	1%	5%	1%	5%	1%	5%
Baseline	66.0	76.5	66.2	74.0	78.9	85.7	70.8	81.5	44.0	57.8	51.5	62.2	39.9	59.0
SimCLR [9]	68.5	77.0	71.6	76.7	81.1	87.9	75.7	82.2	47.3	55.1	55.3	64.0	45.8	55.6
SeqCLR [1]	67.8	79.3	65.2	76.8	80.4	88.8	72.2	82.1	47.5	55.4	55.2	65.3	55.9	66.0
TS [42]	73.9	80.7	75.3	78.8	84.2	90.4	80.4	84.6	53.5	57.5	63.7	66.8	46.5	65.6
SSL-STR [45]	68.3	72.0	72.2	76.2	78.7	81.6	76.7	79.9	46.0	52.9	–	–	–	–
Ours	**79.2**	**82.9**	**77.7**	**80.2**	**88.4**	**90.2**	**81.7**	**85.3**	**55.9**	**61.7**	**65.3**	**72.4**	**63.5**	**69.4**
Fully	82.1		84.9		92.0		87.3		63.6		72.7		64.2	

SSL-STR [45] which employs a different framework is introduced in our comparison. Table 5 illustrates the comparison between our methods and others. The 1% and 5% in the table represent the data ratio of ST used as labeled data, which follows the setup in [45]. As one can see, our method achieves the best performance and significantly outperforms the others on all the testsets including both regular and irregular ones. Moreover, our method presents highly competitive or even better performances with notably fewer labeled images compared with the fully-supervised version. For example, On IIIT5K and CUTE, we outperform the fully-supervised one by a margin of 0.8% and 5.2%. We speculate that the reason for this phenomenon, especially on CUTE, is due to the WTC module. With the help of character-level contrastive learning, the fine-grained visual modeling ability is enhanced, resulting in better performance for low-quality and irregular text images. This conjecture can also be supported by the results of the last two rows in Table 3. After adding character-level supervision, the accuracy of irregular datasets is improved more significantly. We also explore the effect of different amounts of labeled images on accuracy improvement. As the ratio of labeled data increases from 1% to 5%, the accuracy gradually increases, but the margin of improvement decreases. This suggests that we can make use of more real images without annotation given less labeled images.

4.5   Complementary Experiments

To further validate the effectiveness and robustness of our method, we conduct more experiments on a larger trainset. We use the union of ST [16], MJ [29] and SA [19] which contains around 16 million text line images, are utilized as the training data. 5% text images in the trainset are regarded as the labeled data and the rest are unlabeled ones. As shown in Table 4, our method presents a gain of 4.3% on a weighted average for all the testsets and achieves almost the same performance on average as the fully-supervised version. This suggests that our method is effective and robust on a larger scale of training data. Similar to the results in 4.4, our method outperforms the fully supervised results on

SVT, TTEXT, and CUTE. Figure 4 shows some challenging examples that are not correctly recognized by full supervision but are correctly recognized by our method.

Image	Source	MUGS result	Fully result
	CUTE	CUISINE	CUIGINE
	CUTE	QUOT	OUOT
	SVT	BURBANK	BURBAM
	SVT	MUZEO	MUZBO

Fig. 4. Some examples of low quality and irregular layout images in CUTE and SVT that are not correctly recognized by full supervision but are correctly recognized by our method.

5   Conclusion

In this paper, a novel semi-supervised approach is introduced for text recognition which can achieve competitive results using only a very small scale of labeled data. Both self-training and contrastive learning are employed in multiple granularities. A Dynamic Aggregated Self-training (DAS) framework specially designed for text recognition is proposed to generate word-level pseudo-labels from unlabeled data. To adopt contrastive learning at character-level, a module named WTC (Word To Character) which can transform sequential features into character representations is invented. Due to the multiple granularity supervisions, both the visual and semantic information can be well learned by the text recognizer. Then the effectiveness and robustness are proved on several public benchmarks. In the future, we will explore methods for combining self-training and contrastive learning at both word-level and character-level.

References

1. Aberdam, A., et al.: Sequence-to-sequence contrastive learning for text recognition. In: CVPR (2021)
2. Alex, G., Santiago, F., Faustino, G., Jürgen, S.: Connectionist temporal classification: labelling unsegmented sequence data with recurrent neural networks. In: ICML (2006)

3. Anhar, R., Palaiahankote, S., Seng, C.C., Lim, T.C.: A robust arbitrary text detection system for natural scene images. Expert Syst. Appl. **41**, 8027–8048 (2014)
4. Baek, J., Matsui, Y., Aizawa, K.: What if we only use real datasets for scene text recognition? toward scene text recognition with fewer labels. In: CVPR (2021)
5. Shi, B., Yang, M., Wang, X., Lyu, P., Yao, C., Bai, X.: Aster: an attentional scene text recognizer with flexible rectification. TPAMI **41**, 2035–2048 (2018)
6. Shi, B., Bai, X., Yao, C.: An end-to-end trainable neural network for image-based sequence recognition and its application to scene text recognition. TPAMI **39**, 2298–2304 (2016)
7. Shi, B., Bai, X., Yao, C.: An end-to-end trainable neural network for image-based sequence recognition and its application to scene text recognition. TPAMI **39**, 2298–2304 (2017)
8. Berthelot, D., Carlini, N., Goodfellow, I., Papernot, N., Oliver, A., Raffel, C.A.: Mixmatch: a holistic approach to semi-supervised learning. In: NIPS (2019)
9. Chen, T., Kornblith, S., Norouzi, M., Hinton, G.: A simple framework for contrastive learning of visual representations. In: ICML (2020)
10. Chen-Yu, L., Simon, O.: Recursive recurrent nets with attention modeling for ocr in the wild. In: CVPR (2016)
11. Ch'ng, C.K., Chan, C.S., Liu, C.: Total-text: towards orientation robustness in scene text detection. IJDAR **23**, 31–52 (2020)
12. Yao, C., Bai, X., Shi, B., Liu, W.: Strokelets: a learned multi-scale representation for scene text recognition. In: CVPR (2014)
13. Yao, C., Bai, X., Liu, W.: A unified framework for multioriented text detection and recognition. TIP **23**, 4737–4749 (2014)
14. Yu, D., et al.: Towards accurate scene text recognition with semantic reasoning networks. In: CVPR (2020)
15. Dong-Dong, C., Wei, W., Wei, G., ZhiHua, Z.: Tri-net for semi-supervised deep learning. In: IJCAI (2018)
16. Gupta, A., Vedaldi, A., Zisserman, A.: Synthetic data for text localisation in natural images. In: CVPR (2016)
17. Hadsell, R., Chopra, S., LeCun, Y.: Dimensionality reduction by learning an invariant mapping. In: CVPR (2006)
18. He, K., Fan, H., Wu, Y., Xie, S., Girshick, R.: Momentum contrast for unsupervised visual representation learning. In: CVPR (2020)
19. Li, H., Wang, P., Shen, C., Zhang, G.: Show, attend and read: a simple and strong baseline for irregular text recognition. In: AAAI (2019)
20. He, K., Zhang, X., Ren, S., Sun, J.: Deep residual learning for image recognition. In: CVPR (2016)
21. Karatzas, D., et al.: Icdar 2015 competition on robust reading. In: ICDAR (2015)
22. Karatzas, D., et al.: Icdar 2013 robust reading competition. In: ICDAR (2013)
23. Kim, S., Kim, D., Cho, M., Kwak, S.: Proxy anchor loss for deep metric learning. In: CVPR (2020)
24. Lee, D.H.: Pseudo-label: the simple and efficient semi-supervised learning method for deep neural networks. In: ICML (2013)
25. Litman, R., Anschel, O., Tsiper, S., Litman, R., Mazor, S., Manmatha, R.: Scatter: selective context attentional scene text recognizer. In: CVPR (2020)
26. Lucas, S.M., et al.: ICDAR 2003 robust reading competitions: entries, results and future directions. In: IJDAR (2005)
27. Luo, C., Jin, L., Chen, J.: Siman: exploring self-supervised representation learning of scene text via similarity-aware normalization. In: Proceedings of the IEEE/CVF Conference on Computer Vision and Pattern Recognition, pp. 1039–1048 (2022)

28. Lyu, P., Yang, Z., Leng, X., Wu, X., Li, R., Shen, X.: 2D attentional irregular scene text recognizer (2019). CoRR abs/1906.05708. http://arxiv.org/abs/1906.05708
29. Max, J., Karen, S., Andrea, V., Andrew, Z.: Synthetic data and artificial neural networks for natural scene text recognition (2014). CoRR. http://arxiv.org/abs/1406.2227
30. Mishra, A., Alahari, K., Jawahar, C.: Scene text recognition using higher order language priors. In: BMVC (2012)
31. Neumann, L., Matas, J.: Real-time scene text localization and recognition. In: CVPR (2012)
32. Song, Q., Jiang, Q., Li, N., Zhang, R., Wei, X.: Reads: a rectified attentional double supervised network for scene text recognition. In: ICPR (2020)
33. Jiang, Q., Song, Q., Li, N., Zhang, R., Wei, X.: Heterogeneous network based semi-supervised learning for scene text recognition. In: Lladós, J., Lopresti, D., Uchida, S. (eds.) ICDAR 2021. LNCS, vol. 12824, pp. 64–78. Springer, Cham (2021). https://doi.org/10.1007/978-3-030-86337-1_5
34. Xie, Q., Dai, Z., Hovy, E., Luong, T., Le, Q.: Unsupervised data augmentation for consistency training. In: NIPS (2020)
35. Phan, T.Q., Shivakumara, P., Tian, S., Tan, C.L.: Recognizing text with perspective distortion in natural scenes. In: ICCV (2013)
36. Fang, S., Xie, H., Wang, Y., Mao, Z., Zhang, Y.: Read like humans: autonomous, bidirectional and iterative language modeling for scene text recognition. In: CVPR (2021)
37. Qiao, S., Shen, W., Zhang, Z., Wang, B., Yuille, A.: Deep co-training for semi-supervised image recognition. In: ECCV (2018)
38. Sohn, K., et al.: Fixmatch: simplifying semi-supervised learning with consistency and confidence. In: NIPS (2020)
39. Triguero, I., García, S., Herrera, F.: Self-labeled techniques for semi-supervised learning: taxonomy, software and empirical study. Knowl. Inf. Syst. **42**, 245–284 (2015)
40. Vaswani, A., et al.: Attention is all you need. In: NIPS (2017)
41. Wang, K., Babenko, B., Belongie, S.: End-to-end scene text recognition. In: ICCV (2011)
42. Yalniz, I.Z., Jégou, H., Chen, K., Paluri, M., Mahajan, D.: Billion-scale semi-supervised learning for image classification (2019). CoRR abs/1905.00546, http://arxiv.org/abs/1905.00546
43. Yang, M., et al.: Reading and writing: discriminative and generative modeling for self-supervised text recognition. In: Proceedings of the 30th ACM International Conference on Multimedia, pp. 4214–4223 (2022)
44. Sun, Y., et al.: Circle loss: a unified perspective of pair similarity optimization. In: CVPR (2020)
45. Gao, Y., Chen, Y., Wang, J., Lu, H.: Semi-supervised scene text recognition. TIP **30**, 3005–3016 (2021)
46. Gao, Y., Huang, Z., Dai, Y., Xu, C., Chen, K., Tuo, J.: Double supervised network with attention mechanism for scene text recognition (2018). CoRR abs/1808.00677, http://arxiv.org/abs/1808.00677
47. Zeiler, M.D.: Adadelta: an adaptive learning rate method. Comput. Sci. (2012)
48. Zhang, X., Zhu, B., Yao, X., Sun, Q., Li, R., Yu, B.: Context-based contrastive learning for scene text recognition. In: AAAI (2022)
49. Zheng, C., et al.: Pushing the performance limit of scene text recognizer without human annotation. In: Proceedings of the IEEE/CVF Conference on Computer Vision and Pattern Recognition, pp. 14116–14125 (2022)

A Hybrid Approach to Document Layout Analysis for Heterogeneous Document Images

Zhuoyao Zhong[1(✉)], Jiawei Wang[1,2], Haiqing Sun[1,3], Kai Hu[1,2],
Erhan Zhang[1,3], Lei Sun[1], and Qiang Huo[1]

[1] Microsoft Research Asia, Beijing, China
zhuoyao.zhong@gmail.com, {wangjiawei,hk970213}@mail.ustc.edu.cn
[2] Department of EEIS, University of Science and Technology of China, Hefei, China
[3] School of Software and Microelectronics, Peking University, Beijing, China
{sunhq5,zhangeh-ss}@stu.pku.edu.cn

Abstract. We present a new hybrid document layout analysis approach to simultaneously detecting graphical page objects, group text-lines into text regions according to reading order, and recognize the logical roles of text regions from heterogeneous document images. For graphical page object detection, we leverage a state-of-the-art Transformer-based object detection model, namely DINO, as a new graphical page object detector to detect tables, figures, and (displayed) formulas in a top-down manner. Furthermore, we introduce a new bottom-up text region detection model to group text-lines located outside graphical page objects into text regions according to reading order and recognize the logical role of each text region by using both visual and textual features. Experimental results show that our bottom-up text region detection model achieves higher localization and logical role classification accuracy than previous top-down methods. Moreover, in addition to the locations of text regions, our approach can also output the reading order of text-lines in each text region directly. The state-of-the-art results obtained on DocLayNet and PubLayNet demonstrate the effectiveness of our approach.

Keywords: Document layout analysis · Graphical page object detection · Text region detection · Reading order prediction

1 Introduction

Document layout analysis is the process of recovering document physical and/or logical structures from document images, including physical layout analysis and logical layout analysis [10]. Given input document images, physical layout analysis aims at identifying physical homogeneous regions of interest (also called

J. Wang, H. Sun, K. Hu and E. Zhang—This work was done when Jiawei Wang, Haiqing Sun, Kai Hu and Erhan Zhang were interns in MMI Group, Microsoft Research Asia, Beijing, China.

page objects), such as graphical page objects like tables, figures and formulas, and different types of text regions. Logical layout analysis aims at assigning a logical role to the identified regions (e.g., title, section heading, header, footer, paragraph) and determining their logical relationships (e.g., reading order relationships and key-value pair relationships). Document layout analysis plays an important role in document understanding, which can enable a wide range of applications, such as document digitization, conversion, archiving, and retrieval. However, owing to the diverse contents and complex layouts of documents, large variability in region scales and aspect ratios, and similar visual textures between different types of text regions, document layout analysis is still a challenging problem.

In recent years, many deep learning based document layout analysis approaches have emerged [17,22,36,46,47,51,53] and substantially outperformed traditional rule based or handcrafted feature based methods in terms of both accuracy and capability [3]. These methods usually borrow existing object segmentation and detection models, like FCN [29], Faster R-CNN [37], Mask R-CNN [15], Cascade R-CNN [5], SOLOv2 [44], Deformable DETR [54], to detect target page objects from document images. Although they have achieved superior results on graphical page object detection, their performance on text region detection is still unsatisfactory. First, these methods cannot detect small-scale text regions that only span one or two text-lines (e.g., header, footer, and section headings) with high localization accuracy. For example, the detection accuracy of DINO, which is a state-of-the-art Transformer-based detection model [49], for these small-scale text regions drops more than 30% when the Intersection-over-Union (IoU) threshold is increased from 0.5 to 0.75 on DocLayNet [36] based on our experimental results. Second, when two different types of text regions have similar visual textures, e.g., paragraphs and list items, paragraphs and section headings, and section headings and titles, these methods cannot distinguish them robustly. Moreover, these methods only extract the boundaries or masks of text regions and cannot output the reading order of text-lines in text regions, which makes the outputs of these methods hard to be consumed by some important downstream applications such as translation and information extraction.

To address these issues, we present a new hybrid document layout analysis approach to simultaneously detecting graphical page objects, group text-lines into text regions according to reading order, and recognize the logical roles of text regions from heterogeneous document images. For graphical page object detection, we propose to leverage the DINO [49] as a new graphical page object detector to detect tables, figures, and (displayed) formulas in a top-down manner. Furthermore, we introduce a new bottom-up text region detection model to group text-lines located outside graphical page objects into text regions according to reading order and recognize the logical role of each text region by using both visual and textual features. The DINO-based graphical page object detection model and the new bottom-up text region detection model share the same CNN backbone network so that the whole network can be trained in an end-to-end manner. Experimental results demonstrate that this new bottom-up text region detection model can achieve higher localization accuracy for small-scale text regions and better logical role classification accuracy than previous top-down text region detection approaches. Moreover, in addition to the locations of text regions, our approach can also output the reading order of text-lines in each text region directly. State-of-the-art results obtained on two large-scale document layout analysis datasets (i.e., DocLayNet [36] and PubLayNet [53]) demonstrate the effectiveness and superiority of our approach. Especially on DocLayNet, our approach outperforms the previous best-performing model by 4.2% in terms of mean Average Precision (mAP). Although PubLayNet has been well-tuned by many previous techniques, our approach still achieves the highest mAP of 96.5% on this dataset.

2 Related Work

A comprehensive survey of traditional document layout analysis methods has been given in [3]. In this section, we focus on reviewing recent deep learning based approaches that are closely related to this work. These approaches can be roughly divided into three categories: object detection based methods, semantic segmentation based methods, and graph-based methods.

Object Detection Based Methods. These methods leverage state-of-the-art top-down object detection or instance segmentation frameworks to address the document layout analysis problem. Yi et al. [48] and Oliveira et al. [35] first adapted R-CNN [12] to locate and recognize page objects of interest from document images, while their performance was limited by the traditional region proposal generation strategies. Later, more advanced object detectors, like Fast R-CNN [11], Faster R-CNN [37], Mask R-CNN [15], Cascade R-CNN [5], SOLOv2 [44], YOLOv5 [18], Deformable DETR [54], were explored by Vo et al. [42], Zhong et al. [53], Saha et al. [38], Li et al. [20], Biswas et al. [4], Pfitzmann et al. [36], and Yang et al. [46], respectively. Meanwhile, some effective techniques were also proposed to further improve the performance of these detectors. For instance, Zhang et al. [51] proposed a multi-modal Faster/Mask R-CNN model to detect page objects, in which visual feature maps extracted by CNN and two 2D text embedding maps with sentence and character embeddings were fused together to construct multi-modal feature maps and a graph neural network (GNN) based relation module was introduced to model the interactions between page object candidates. Bi et al. [2] also proposed to leverage GNN to model the interactions between page object candidates to improve page object detection accuracy. Naik et al. [34] incorporated the scale-aware, spatial-aware, and task-aware attention mechanisms proposed in DynamicHead [8] into the CNN backbone network to improve the accuracy of Faster R-CNN and Sparse R-CNN [41] for page object detection. Shi et al. [40] proposed a new lateral feature enhancement backbone network and Yang et al. [46] leveraged Swin Transformer [28] as a stronger backbone network to push the performance of Mask R-CNN and Deformable DETR on page object detection tasks, respectively. Recently, Gu et al. [13], Li et al.

[20] and Huang et al. [17] improved the performance of Faster R-CNN, Mask R-CNN, and Cascade R-CNN based page object detectors by pre-training the vision backbone networks on large-scale document images with self-supervised learning algorithms. Although these methods have achieved state-of-the-art performance on several benchmark datasets, they still struggle with small-scale text region detection and cannot output the reading order of text-lines in text regions directly.

Semantic Segmentation Based Methods. These methods (e.g., [14,23,24, 39,47]) usually use existing semantic segmentation frameworks like FCN [29] to predict a pixel-level segmentation mask first, and then merge pixels into different types of page objects. Yang et al. [47] proposed a multi-modal FCN for page object segmentation, where visual feature maps and 2D text embedding maps with sentence embeddings were combined to improve pixel-wise classification accuracy. He et al. [14] proposed a multi-scale multi-task FCN to simultaneously predict a region segmentation mask and a contour segmentation mask. After being refined by a conditional random field (CRF) model, these two segmentation masks are then input to a post-processing module to get final prediction results. Li et al. [23] incorporated label pyramids and deep watershed transformation into the vanilla FCN to avoid merging nearby page objects together. The performance of existing semantic segmentation based methods is still inferior to the other two types of methods on recent document layout analysis benchmarks.

Graph-Based Methods. These methods (e.g., [21,22,32,43]) model each document page as a graph whose nodes represent primitive page objects (e.g., words, text-lines, connected components) and edges represent relationships between neighboring primitive page objects, and then formulate document layout analysis as a graph labeling problem. Li et al. [21] used image processing techniques to generate line regions first, and then applied two CRF models to classify them into different types and predict whether each pair of line regions belong to the same instance based on visual features extracted by CNNs, respectively. Based on these prediction results, line regions belonging to the same class and the same instance were merged to get page objects. In their subsequent work [22], they used connected components to replace line regions as nodes and adopted a graph attention network (GAT) to enhance the visual features of both nodes and edges. Luo et al. [32] focused on the logical role classification task and proposed to leverage syntactic, semantic, density, and appearance features with multi-aspect graph convolutional networks (GCNs) to recognize the logical role of each page object. Recently, Wang et al. [43] focused on the paragraph identification task and proposed a GCN-based approach to grouping text-lines into paragraphs. Liu et al. [27], Long et al. [30] and Xue et al. [45] further proposed a unified framework for text detection and paragraph (text-block) identification.

Unlike these works, our unified layout analysis approach can detect page objects, predict the reading order of text-lines in text regions and recognize the logical roles of text regions from document images simultaneously.

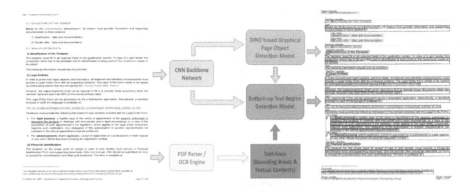

Fig. 1. The overall architecture of our hybrid document layout analysis approach.

3 Methodology

Our approach is composed of three key components: 1) A shared CNN backbone network to extract multi-scale feature maps from input document images; 2) A DINO based graphical page object detection model to detect tables, figures, and displayed formulas; 3) A bottom-up text region detection model to group text-lines located outside graphical page objects into text regions according to reading order and recognize the logical role of each text region. These three components are jointly trained in an end-to-end manner. The overall architecture of our approach is depicted in Fig. 1. The details of these components are described in the following subsections.

3.1 Shared CNN Backbone Network

Given an input document image $I \in \mathbb{R}^{H \times W \times 3}$, we adopt a ResNet-50 network [16] as the backbone network to generate multi-scale feature maps $\{C_3, C_4, C_5\}$, which represent the output feature maps of the last residual block in Conv3, Conv4, and Conv5, respectively. C_6 is obtained via a 3×3 convolutional layout with stride 2 on C_5. The resolutions of $\{C_3, C_4, C_5, C_6\}$ are $1/8, 1/16, 1/32, 1/64$ of the original document image. Then, a 1×1 convolutional layer is performed on each feature map for channel reduction. After that, all feature maps have 256 channels, which are input to the following DINO based graphical page object detection model and bottom-up text region detection model.

3.2 DINO Based Graphical Page Object Detection Model

Recently, Transformer-based object detection methods such as DETR [6], Deformable DETR [54], DAB-DETR [26], DN-DETR [19] and DINO [49] have become popular as they can achieve better performance than previous

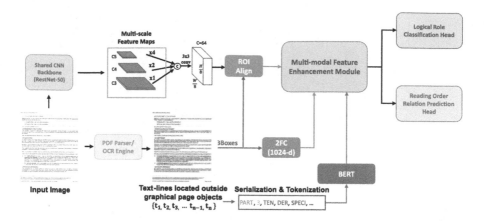

Fig. 2. A schematic view of the proposed bottom-up text region detection model.

Faster/Mask R-CNN based models without relying on manually designed components like non-maximum suppression (NMS). So, we leverage the latest state-of-the-art object detection model, DINO, as a new graphical page object detector to detect tables, figures, and displayed formulas from document images.

Our DINO based graphical page object detection model consists of a Transformer encoder and a Transformer decoder. The Transformer encoder takes multi-scale feature maps $\{C_3, C_4, C_5, C_6\}$ output by the shared CNN backbone as input and generates a manageable number of region proposals for graphical page objects, whose bounding boxes are used to initialize the positional embeddings of object queries. The Transformer decoder takes object queries as input and outputs the final set of predictions in parallel. To reduce computation cost, the deformable attention mechanism [54] is adopted in the self attention layers in the encoder and cross attention layers in the decoder, respectively. To speed up model convergence, a contrastive denoising based training method for object queries is used. We refer readers to [49] for more details. Experimental results demonstrate that this new model can achieve superior graphical page object detection performance on PubLayNet and DocLayNet benchmark datasets.

3.3 Bottom-Up Text Region Detection Model

A text region is a semantic unit of writing consisting of a group of text-lines arranged in natural reading order and associated with a logical label, such as paragraph, list/list item, title, section heading, header, footer, footnote, and caption. Given a document image D composed of n text-lines $[t_1, t_2, ..., t_n]$, the goal of our bottom-up text region detection model is to group these text-lines into different text regions according to reading order and recognize the logical role of each text region. In this work, we assume the bounding boxes and textual contents of text-lines have already been given by a PDF parser or OCR engine. Based on the detection results of our DINO based graphical page object detection model, we first filter out those text-lines located inside graphical page objects and then take the remaining text-lines as input. As shown in Fig. 2, our bottom-up text region detection model consists of a multi-modal feature extraction module, a multi-modal feature enhancement module, and two prediction heads, i.e., a reading order relation prediction head and a logical role classification head. The detailed illustrations of the multi-modal feature enhancement module and two prediction heads could be found in Fig. 3.

Multi-modal Feature Extraction Module. In this module, we extract the visual embedding, text embedding, and 2D position embedding for each text-line.

Visual Embedding. As shown in Fig. 2, we first resize C_4 and C_5 to the size of C_3 and then concatenate these three feature maps along the channel axis, which are fed into a 3×3 convolutional layer to generate a feature map C_{fuse} with 64 channels. For each text-line t_i, we adopt RoIAlign algorithm [15] to extract 7×7 features from C_{fuse} based on its bounding box $b_i = (x_i^1, y_i^1, x_i^2, y_i^2)$, where (x_i^1, y_i^1), (x_i^2, y_i^2) represent the coordinates of its top-left and bottom-right corners respectively. The final visual embedding V_i of t_i can be represented as:

$$V_i = LN(ReLU(FC(ROIAlign(C_{fuse}, b_i)))), \tag{1}$$

where FC is a fully-connected layer with 1,024 nodes and LN represents Layer Normalization [1].

Text Embedding. We leverage the pre-trained language model BERT [9] to extract the text embedding of each text-line. Specifically, we first serialize all the text-lines in a document image into a 1D sequence by reading them in a top-left to bottom-right order and tokenize the text-line sequence into a sub-word token sequence, which is then fed into BERT to get the embedding of each token. After that, we average the embeddings of all the tokens in each text-line t_i to obtain its text embedding T_i, followed by a fully-connected layer with 1,024 nodes to make the dimension the same as that of V_i:

$$T_i = LN(ReLU(FC(T_i))) . \tag{2}$$

2D Position Embedding. For each text-line t_i, we encode its bounding box and size information as its 2D position embedding B_i:

$$B_i = LN(MLP(x_i^1/W, y_i^1/H, x_i^2/W, y_i^2/H, w_i/W, h_i/H)), \tag{3}$$

where (w_i, h_i) and (W, H) represent the width and height of b_i and the input image, respectively. MLP consists of 2 fully-connected layers with 1,024 nodes, each of which is followed by ReLU.

For each text-line t_i, we concatenate its visual embeddings V_i, text embeddings T_i, and 2D position embeddings B_i to obtain its **multi-modal representation U_i.**

$$U_i = FC(Concat(V_i, T_i, B_i)), \tag{4}$$

where FC is a fully-connected layers with 1,024 nodes.

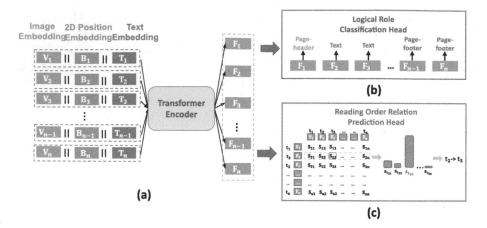

Fig. 3. Illustration of (a): Multi-modal Feature Enhancement Module; (b): Logical Role Classification Head; (c): Reading Order Relation Prediction Head.

Multi-modal Feature Enhancement Module. As shown in Fig. 3, we use a lightweight Transformer encoder to further enhance the multi-modal representations of text-lines by modeling their interactions with the self-attention mechanism. Each text-line is treated as a token of the Transformer encoder and its multi-modal representation is taken as the input embedding:

$$F = TransformerEncoder(U) \tag{5}$$

where $U = [U_1, U_2, ..., U_n]$ and $F = [F_1, F_2, ..., F_n]$ are the input and output embeddings of the Transformer encoder, n is the number of the input text-lines. To save computation, here we only use a 1-layer Transformer encoder, where the head number, dimension of hidden state, and the dimension of feedforward network are set as 12, 768, and 2048, respectively.

Reading Order Relation Prediction Head. We propose to use a relation prediction head to predict reading order relationships between text-lines. Given a text-line t_i, if a text-line t_j is its succeeding text-line in the same text region, we define that there exists a reading order relationship $(t_i \rightarrow t_j)$ pointing from text-line t_i to text-line t_j. If text-line t_i is the last (or only) text-line in a text region, its succeeding text-line is considered to be itself. Unlike many previous methods that consider relation prediction as a binary classification task [21,43], we treat relation prediction as a dependency parsing task and use a softmax cross entropy loss to replace the standard binary cross entropy loss during optimization by following [52]. Moreover, we adopt a spatial compatibility feature introduced in [50] to effectively model the spatial interactions between text-lines for relation prediction.

Specifically, we use a multi-class (i.e., n-class) classifier to calculate a score s_{ij} to estimate how likely t_j is the succeeding text-line of t_i as follows:

$$f_{ij} = FC_q(F_i) \circ FC_k(F_j) + MLP(r_{b_i,b_j}), \tag{6}$$

$$s_{ij} = \frac{\exp(f_{ij})}{\sum_N \exp(f_{ij})}, \tag{7}$$

where each of FC_q and FC_k is a single fully-connected layer with 2,048 nodes to map F_i and F_j into different feature spaces; $\circ$ denotes dot product operation; MLP consists of 2 fully-connected layers with 1,024 nodes and 1 node respectively; r_{b_i,b_j} is a spatial compatibility feature vector between b_i and b_j, which is a concatenation of three 6-d vectors:

$$r_{b_i,b_j} = (\Delta(b_i, b_j), \Delta(b_i, b_{ij}), \Delta(b_j, b_{ij})), \tag{8}$$

where b_{ij} is the union bounding box of b_i and b_j; $\Delta(.,.)$ represents the box delta between any two bounding boxes. Taking $\Delta(b_i, b_j)$ as an example, $\Delta(b_i, b_j) = (d_{ij}^{x_{ctr}}, d_{ij}^{y_{ctr}}, d_{ij}^w, d_{ij}^h, d_{ji}^{x_{ctr}}, d_{ji}^{y_{ctr}})$, where each dimension is given by:

$$\begin{aligned}
d_{ij}^{x_{ctr}} &= (x_i^{ctr} - x_j^{ctr})/w_i, & d_{ij}^{y_{ctr}} &= (y_i^{ctr} - y_j^{ctr})/h_i, \\
d_{ij}^w &= \log(w_i/w_j), & d_{ij}^h &= \log(h_i/h_j), \\
d_{ji}^{x_{ctr}} &= (x_j^{ctr} - x_i^{ctr})/w_j, & d_{ji}^{y_{ctr}} &= (y_j^{ctr} - y_i^{ctr})/h_j,
\end{aligned} \tag{9}$$

where (x_i^{ctr}, y_i^{ctr}) and (x_j^{ctr}, y_j^{ctr}) are the center coordinates of b_i and b_j, respectively.

We select the highest score from scores $[s_{ij}, k = 1, 2, ..., n]$ and output the corresponding text-line as the succeeding text-line of t_i. To achieve a higher relation prediction accuracy, we use another relation prediction head to identify the preceding text-line for each text-line further. The relation prediction results from both heads are combined to obtain the final results.

Logical Role Classification Head. Given the enhanced multi-modal representations of text-lines $F = [F_1, F_2, ..., F_n]$, we add a multi-class classifier to predict a logical role label for each text-line.

3.4   Optimization

Loss for DINO-Based Graphical Page Object Detection Model. The loss function of our DINO-based graphical page object detection model $L_{graphical}$ is exactly the same as L_{DINO} used in DINO [49], which is composed of multiple bounding box regression losses and classification losses derived from prediction heads and denoising heads. The bounding box regression loss is a linear combination of the L_1 loss and the GIoU loss [7], while the classification loss is the focal loss [25]. We refer readers to [49] for more details.

Loss for Bottom-up Text Region Detection Model. There are two prediction heads in our bottom-up text region detection model, i.e., a reading order relation prediction head and a logical role classification head. For reading order relation prediction, we adopt a softmax cross-entropy loss as follows:

$$L_{relation} = \frac{1}{N} \sum_i L_{CE}\left(\boldsymbol{s}_i, \boldsymbol{s}_i^*\right) \tag{10}$$

where $s_i = [s_{i1}, s_{i2}, ..., s_{iN}]$ is the predicted relation score vector calculated by Eqs. (6)–(7) and s_i^* is the target label.

We also adopt a softmax cross-entropy loss for logical role classification, which can be defined as

$$L_{role} = \frac{1}{N} \sum_i L_{CE}\left(\boldsymbol{c}_i, \boldsymbol{c}_i^*\right) \tag{11}$$

where c_i is the predicted label of the ith text-line output by a softmax function and c_i^* is the corresponding ground-truth label.

Overall Loss. All the components in our approach are jointly trained in an end-to-end manner. The overall loss is the sum of $L_{graphical}$, $L_{relation}$ and L_{role}:

$$L_{overall} = L_{graphical} + L_{relation} + L_{role} . \tag{12}$$

4   Experiments

4.1   Datasets and Evaluation Protocols

We conduct experiments on two representative benchmark datasets, i.e., PubLayNet [53] and DocLayNet [36] to verify the effectiveness of our approach.

PubLayNet [53] is a large-scale dataset for document layout analysis released by IBM, which contains 340,391, 11,858, and 11,983 document pages for training, validation, and testing, respectively. All the documents in this dataset are scientific papers publicly available on PubMed Central and all the ground-truths are automatically generated by matching the XML representations and the content of corresponding PDF files. It pre-defines 5 types of page objects, including Text (i.e., Paragraph), Title, List, Figure, and Table. The summary of this dataset is shown in the left part of Table 1. Because ground-truths of the testing set are not publicly available, we evaluate our approach on the validation dataset.

DocLayNet [36] is a challenging human-annotated document layout analysis dataset newly released by IBM, which contains 69,375, 6,489, and 4,999 document pages for training, testing, and validation, respectively. It covers a variety of document categories, including Financial reports, Patents, Manuals, Laws, Tenders, and Scientific Papers. It pre-defines 11 types of page objects, including Caption, Footnote, Formula, List-item, Page-footer, Page-header, Picture, Section-header, Table, Text (i.e., Paragraph), and Title. The summary of this dataset is shown in the right part of Table 1.

Table 1. Summary of PubLayNet and DocLayNet datasets.

PublayNet			DocLayNet			
Page Object	Training	Validation	Page Object	Training	Testing	Validation
Text	2,343,356	88,625	Text	431,223	49,186	29,939
Title	627,125	18,801	Title	4,424	299	332
List	80,759	4239	List-item	161,779	13,320	10,525
Figure	109,292	4327	Picture	39,621	2,775	3,533
Table	102,514	4769	Table	300,116	2,269	2,395
–	–	–	Caption	19,199	1,763	1,544
–	–	–	Footnote	5,647	312	386
–	–	–	Formula	21,175	1,894	1,969
–	–	–	Page-footer	61,267	5,571	3,992
–	–	–	Page-header	47,997	6,683	3,366
–	–	–	Section-header	118,581	15,744	8,549
Image Page	340,391	11,858	Image Page	69,375	6,489	4,999

In addition to document images, these two datasets also provide corresponding original PDF files. Therefore, we can directly use a PDF parser (e.g., PDFMiner) to obtain the bounding boxes, text contents, and reading order of textlines for exploring our bottom-up text region detection approach. The evaluation metric of these two datasets is the COCO-style mean average precision (mAP) at multiple intersection over union (IoU) thresholds between 0.50 and 0.95 with a step of 0.05.

4.2 Implementation Details

We implement our approach based on Pytorch v1.10 and all experiments are conducted on a workstation with 8 Nvidia Tesla V100 GPUs (32 GB memory). Note that, on PubLayNet, a list refers to a whole list object consisting of multiple list items, whose label is not consistent with that of a text or a title. To reduce ambiguity, we consider all lists as specific graphical page objects and use our DINO based graphical page object detection model to detect them. In training, the parameters of the CNN backbone network are initialized with a ResNet-50 model [16] pre-trained on the ImageNet classification task, while the parameters of the text embedding extractor in our bottom-up text region detection model are initialized with the pre-trained BERT$_{BASE}$ model [9]. The parameters of the newly added layers are initialized by using random weights with a Gaussian distribution of mean 0 and standard deviation 0.01. The models are optimized by AdamW [31] algorithm with a batch size of 16 and trained for 12 epochs on PubLayNet and 24 epochs on DocLayNet. The learning rate and weight decay are set as 1e-5 and 1e-4 for the CNN backbone network, 2e-5 and 1e-2 for BERT$_{BASE}$, and 1e-4 and 1e-4 for the newly added layers, respectively. The

Table 2. Ablation studies on DocLayNet testing set (in %).

	DINO	Hybrid (V)	Hybrid (V+BERT-3L)	Hybrid (V+BERT-12L)
Caption	85.5	83.2	81.9	83.2
Footnote	69.2	67.8	68.5	69.7
Formula	63.8	63.9	64.2	63.4
List-item	80.9	86.9	87.5	**88.6**
Page-footer	54.2	89.9	86.1	90.0
Page-header	63.7	70.4	**81.3**	76.3
Picture	84.1	82.0	81.8	81.6
Section-header	64.3	86.2	84.2	83.2
Table	85.7	84.4	85.4	84.8
Text	83.3	86.1	85.6	84.8
Title	82.8	75.3	82.3	**84.9**
mAP	74.3	79.6	80.8	**81.0**

Table 3. Ablation studies on PubLayNet validation set (in %).

Method	Text	Title	List	Table	Figure	mAP
DINO	94.9	91.4	96.0	98.0	97.3	95.52
Hybrid (V)	97.0	92.8	**96.4**	98.1	**97.4**	96.34
Hybrid (V+BERT-3L)	**97.7**	93.1	96.3	98.1	97.2	96.48
Hybrid (V+BERT-12L)	97.4	**93.5**	**96.4**	**98.2**	97.2	**96.54**

learning rate is divided by 10 at the 11th epoch for PubLayNet and 20th epoch for DocLayNet. The other hyper-parameters of AdamW including betas and epsilon are set as (0.9, 0.999) and 1e-8. We also adopt a multi-scale training strategy. The shorter side of each image is randomly rescaled to a length selected from [512, 640, 768], while the longer side should not exceed 800.

In the testing phase, we set the shorter side of the input image as 640. We group text-lines into text regions based on predicted reading order relationships by using the Union-Find algorithm. The logical role of a text region is determined by majority voting, and the bounding box is the union bounding box of all its constituent text-lines.

4.3   Ablation Studies

Effectiveness of Hybrid Strategy. In this section, we evaluate the effectiveness of the proposed hybrid strategy. To this end, we train two baseline models: 1) A DINO baseline to detect both graphical page objects and text regions; 2) A hybrid model (denoted as Hybrid(V)) that only uses visual and 2D position features for bottom-up text region detection. As shown in the first two columns of Table 2, compared with the DINO model, the Hybrid(V) model can achieve comparable graphical page object detection results but much higher text region detection accuracy on DocLayNet, leading to a 5.3% improvement in terms of mAP. In particular, the Hybrid(V) model can significantly improve small-scale text region detection performance, e.g., 89.9% vs. 54.2% for Page-footer, 70.4% vs. 63.7% for Page-header and 86.2% vs. 64.3% for Section-header. We observe that this accuracy improvement is mainly owing to the higher localization accuracy. Experimental results on PubLayNet are listed in the first two rows of Table 3. We can see that the Hybrid(V) model improves AP by 2.1% for Text and 1.4% for Title, leading to a 0.82% improvement in terms of mAP on PubLayNet. These experimental results clearly demonstrate the effectiveness of the proposed hybrid strategy that combines the best of both top-down and bottom-up methods.

Effectiveness of Using Textual Features. In order to evaluate the effectiveness of using textual features, we compare the performance of three hybrid models, i.e., Hybrid(V), Hybrid(V+BERT-3L) and Hybrid(V+BERT-12L). The bottom-up text region detection model of Hybrid(V) does not use textual features, while the models of Hybrid(V+BERT-3L) and Hybrid(V+BERT-12L) use the first 3 and 12 Transformer blocks of BERT$_{BASE}$ to extract text embeddings for text-lines, respectively. Experimental results on DocLayNet and PubLayNet are listed in the second and third parts of Table 2 and Table 3. We can see that both Hybrid(V+BERT-3L) and Hybrid(V+BERT-12L) models are consistently better than Hybrid(V), and Hybrid(V+BERT-12L) achieves the best results. The large performance improvements of these two models mainly come from the categories of Title, Page-header and List-item on DocLayNet and the categories of Text and Title on PubLayNet, respectively.

Table 4. Performance comparisons on DocLayNet testing set (in %). The results of Mask R-CNN, Faster R-CNN, and YOLOv5 are obtained from [36].

	Human	Mask R-CNN	Faster R-CNN	YOLOv5	DINO	Proposed
Caption	84-89	71.5	70.1	77.7	**85.5**	83.2
Footnote	83-91	71.8	73.7	**77.2**	69.2	69.7
Formula	83-85	63.4	63.5	**66.2**	63.8	63.4
List-item	87-88	80.8	81.0	86.2	80.9	**88.6**
Page-footer	93-94	59.3	58.9	61.1	54.2	**90.0**
Page-header	85-89	70.0	72.0	67.9	63.7	**76.3**
Picture	69-71	72.7	72.0	77.1	**84.1**	81.6
Section-header	83-84	69.3	68.4	74.6	64.3	83.2
Table	77-81	82.9	82.2	**86.3**	85.7	84.8
Text	84-86	85.8	85.4	**88.1**	83.3	84.8
Title	60-72	80.4	79.9	82.7	82.8	**84.9**
mAP	82-83	73.5	73.4	76.8	74.3	**81.0**

Table 5. Performance comparisons on PubLayNet validation set (in %). Vision and Text stands for using visual and textual features, respectively.

Method	Modality	Text	Title	List	Table	Figure	mAP
Faster R-CNN [53]	Vision	91.0	82.6	88.3	95.4	93.7	90.2
Mask R-CNN [53]		91.6	84.0	88.6	96.0	94.9	91.0
Naik et al. [34]		94.3	88.7	94.3	97.6	96.1	94.2
Minouei et al. [33]		94.4	90.8	94.0	97.4	96.6	94.6
DiT-L [20]		94.4	89.3	96.0	97.8	**97.2**	94.9
SRRV [2]		95.8	90.1	95.0	97.6	96.7	95.0
DINO [49]		94.9	91.4	96.0	98.0	97.3	95.5
TRDLU [46]		95.8	92.1	97.6	97.6	96.6	96.0
UDoc [13]	Vision+Text	93.9	88.5	93.7	97.3	96.4	93.9
LayoutLMv3 [17]		94.5	90.6	95.5	97.9	97.0	95.1
VSR [51]		96.7	93.1	94.7	97.4	96.4	95.7
Proposed	Vision	97.0	92.8	96.4	98.1	**97.4**	96.3
Proposed	Vision+Text	**97.4**	**93.5**	**96.4**	**98.2**	97.2	**96.5**

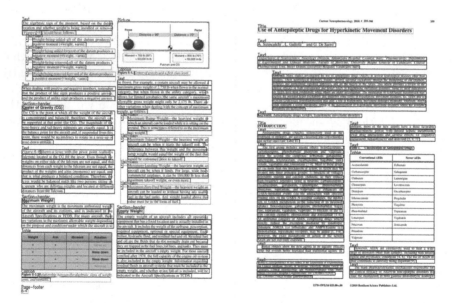

Fig. 4. Qualitative results of our approach: DocLayNet (Left); PubLayNet (Right).

4.4 Comparisons with Prior Arts

DocLayNet. We compare our approach with other most competitive methods, including Mask R-CNN, Faster R-CNN, YOLOv5, and DINO on DocLayNet. As shown in Table 4, our approach outperforms the closest method YOLOv5 substantially by improving mAP from 76.8% to 81.0%. Considering that DocLayNet is an extremely challenging dataset that covers a variety of document scenarios and contains a large number of text regions with fine-grained logical roles, the superior performance achieved by our proposed approach can demonstrate the advantage of our approach.

PubLayNet. We compare our approach with several state-of-the-art vision-based and multi-modal methods on PubLayNet. Experimental results are presented in Table 5. We can see that our approach outperforms all these methods no matter whether textual features are used in our bottom-up text region detection model or not.

Qualitative Results. The state-of-the-art performance achieved on these two datasets demonstrates the effectiveness and robustness of our approach. Furthermore, our approach provides a new capability of outputting the reading order of text-lines in each text region. Some qualitative results are depicted in Fig. 4.

5   Summary

In this paper, we propose a new hybrid document layout analysis approach, which consists of a new DINO based graphical page object detection model to detect tables, figures, and formulas in a top-down manner and a new bottom-up text region detection model to group text-lines located outside graphical page objects into text regions according to reading order and recognize the logical role of each text region by using both visual and textual features. The state-of-the-art results obtained on DocLayNet and PubLayNet demonstrate the effectiveness and superiority of our approach. Furthermore, in addition to bounding boxes, our approach can also output the reading order of text-lines in each text region directly, which is crucial to various downstream tasks such as translation and information extraction. In future work, we will explore how to use a unified model to solve various physical and logical layout analysis tasks, including page object detection, inter-object relation prediction, list parsing, table of contents generation, and so on.

References

1. Ba, J.L., Kiros, J.R., Hinton, G.E.: Layer normalization. arXiv preprint arXiv:1607.06450 (2016)
2. Bi, H., et al.: Srrv: A novel document object detector based on spatial-related relation and vision. IEEE Transactions on Multimedia (2022)
3. Binmakhashen, G.M., Mahmoud, S.A.: Document layout analysis: a comprehensive survey. ACM Comput. Surv. (CSUR) **52**(6), 1–36 (2019)
4. Biswas, S., Banerjee, A., Lladós, J., Pal, U.: Docsegtr: an instance-level end-to-end document image segmentation transformer. arXiv preprint arXiv:2201.11438 (2022)

5. Cai, Z., Vasconcelos, N.: Cascade r-cnn: high quality object detection and instance segmentation. IEEE Trans. Pattern Anal. Mach. Intell. **43**(5), 1483–1498 (2019)

6. Carion, N., Massa, F., Synnaeve, G., Usunier, N., Kirillov, A., Zagoruyko, S.: End-to-end object detection with transformers. In: Proceedings of the European Conference on Computer Vision, pp. 213–229 (2020)

7. Carion, N., Massa, F., Synnaeve, G., Usunier, N., Kirillov, A., Zagoruyko, S.: End-to-end object detection with transformers. In: Proceedings of the European Conference on Computer Vision, pp. 213–229 (2020)

8. Dai, X., et al.: Dynamic head: Unifying object detection heads with attentions. In: Proceedings of the IEEE Conference on Computer Vision and Pattern Recognition, pp. 7373–7382 (2021)

9. Devlin, J., Chang, M.W., Lee, K., Toutanova, K.: Bert: Pre-training of deep bidirectional transformers for language understanding. In: Proceedings of the Conference of the North American Chapter of the Association for Computational Linguistics: Human Language Technologies, pp. 4171–4186 (2019)

10. Doermann, D., Tombre, K. (eds.): Handbook of Document Image Processing and Recognition. Springer, London (2014). https://doi.org/10.1007/978-0-85729-859-1

11. Girshick, R.: Fast r-cnn. In: Proceedings of the International Conference on Computer Vision, pp. 1440–1448 (2015)

12. Girshick, R., Donahue, J., Darrell, T., Malik, J.: Rich feature hierarchies for accurate object detection and semantic segmentation. In: Proceedings of the IEEE Conference on Computer Vision and Pattern Recognition, pp. 580–587 (2014)

13. Gu, J., et al.: Unified pretraining framework for document understanding. arXiv preprint arXiv:2204.10939 (2022)

14. He, D., Cohen, S., Price, B., Kifer, D., Giles, C.L.: Multi-scale multi-task fcn for semantic page segmentation and table detection. In: Proceedings of the International Conference on Document Analysis and Recognition. vol. 1, pp. 254–261 (2017)

15. He, K., Gkioxari, G., Dollár, P., Girshick, R.: Mask r-CNN. In: Proceedings of the International Conference on Computer Visio, pp. 2961–2969 (2017)

16. He, K., Zhang, X., Ren, S., Sun, J.: Deep residual learning for image recognition. In: Proceedings of the IEEE Conference on Computer Vision and Pattern Recognition, pp. 770–778 (2016)

17. Huang, Y., Lv, T., Cui, L., Lu, Y., Wei, F.: Layoutlmv3: Pre-training for document ai with unified text and image masking. In: Proceedings of the ACM International Conference on Multimedia, pp. 4083–4091 (2022)

18. Jocher, G., et al.: ultralytics/yolov5: v5.0 - YOLOv5-P6 1280 models, AWS, Supervise.ly and YouTube integrations (Apr 2021)

19. Li, F., Zhang, H., Liu, S., Guo, J., Ni, L.M., Zhang, L.: Dn-detr: Accelerate detr training by introducing query denoising. arXiv preprint arXiv:2203.01305 (2022)

20. Li, J., Xu, Y., Lv, T., Cui, L., Zhang, C., Wei, F.: Dit: Self-supervised pre-training for document image transformer. In: Proceedings of the ACM International Conference on Multimedia. pp. 3530–3539 (2022)

21. Li, X.H., Yin, F., Liu, C.L.: Page object detection from pdf document images by deep structured prediction and supervised clustering. In: Proceedings of the International Conference on Pattern Recognition, pp. 3627–3632 (2018)

22. Li, X.H., Yin, F., Liu, C.L.: Page segmentation using convolutional neural network and graphical model. In: Proceedings of the International Workshop on Document Analysis Systems, pp. 231–245 (2020)

23. Li, X.H., et al.: Instance aware document image segmentation using label pyramid networks and deep watershed transformation. In: Proceedings of the International Conference on Document Analysis and Recognition, pp. 514–519 (2019)

24. Li, Y., Zou, Y., Ma, J.: Deeplayout: A semantic segmentation approach to page layout analysis. In: Proceedings of the International Conference on Intelligent Computing Methodologies, pp. 266–277 (2018)

25. Lin, T.Y., Goyal, P., Girshick, R., He, K., Dollár, P.: Focal loss for dense object detection. In: Proceedings of the International Conference on Computer Vision, pp. 2980–2988 (2017)

26. Liu, S., et al.: Dab-detr: Dynamic anchor boxes are better queries for detr. arXiv preprint arXiv:2201.12329 (2022)

27. Liu, S., Wang, R., Raptis, M., Fujii, Y.: Unified line and paragraph detection by graph convolutional networks. In: Proceedings of the International Workshop on Document Analysis Systems, pp. 33–47 (2022)

28. Liu, Z., et al.: Swin transformer: Hierarchical vision transformer using shifted windows. In: Proceedings of the IEEE Conference on Computer Vision and Pattern Recognition, pp. 10012–10022 (2021)

29. Long, J., Shelhamer, E., Darrell, T.: Fully convolutional networks for semantic segmentation. In: Proceedings of the IEEE Conference on Computer Vision and Pattern Recognition, pp. 3431–3440 (2015)

30. Long, S., Qin, S., Panteleev, D., Bissacco, A., Fujii, Y., Raptis, M.: Towards end-to-end unified scene text detection and layout analysis. In: Proceedings of the IEEE Conference on Computer Vision and Pattern Recognition, pp. 1049–1059 (2022)

31. Loshchilov, I., Hutter, F.: Decoupled weight decay regularization. arXiv preprint arXiv:1711.05101 (2017)

32. Luo, S., Ding, Y., Long, S., Han, S.C., Poon, J.: Doc-gcn: Heterogeneous graph convolutional networks for document layout analysis. arXiv preprint arXiv:2208.10970 (2022)

33. Minouei, M., Soheili, M.R., Stricker, D.: Document layout analysis with an enhanced object detector. In: Proceedings of the International Conference on Pattern Recognition and Image Analysis, pp. 1–5 (2021)

34. Naik, S., Hashmi, K.A., Pagani, A., Liwicki, M., Stricker, D., Afzal, M.Z.: Investigating attention mechanism for page object detection in document images. Appl. Sci. 12(15), 7486 (2022)

35. Oliveira, D.A.B., Viana, M.P.: Fast cnn-based document layout analysis. In: Proceedings of the International Conference on Computer Vision Workshops, pp. 1173–1180 (2017)

36. Pfitzmann, B., Auer, C., Dolfi, M., Nassar, A.S., Staar, P.W.: Doclaynet: A large human-annotated dataset for document-layout analysis. arXiv preprint arXiv:2206.01062 (2022)

37. Ren, S., He, K., Girshick, R., Sun, J.: Faster r-cnn: Towards real-time object detection with region proposal networks. In: Proceedings of the Advances in Neural Information Processing Systems, pp. 91–99 (2015)

38. Saha, R., Mondal, A., Jawahar, C.: Graphical object detection in document images. In: Proceedings of the International Conference on Document Analysis and Recognition, pp. 51–58 (2019)

39. Sang, Y., Zeng, Y., Liu, R., Yang, F., Yao, Z., Pan, Y.: Exploiting spatial attention and contextual information for document image segmentation. In: Proceedings of the Advances in Knowledge Discovery and Data Mining, pp. 261–274 (2022)

40. Shi, C., Xu, C., Bi, H., Cheng, Y., Li, Y., Zhang, H.: Lateral feature enhancement network for page object detection. IEEE Trans. Instrum. Meas. 71, 1–10 (2022)

41. Sun, P., et al.: Sparse r-cnn: End-to-end object detection with learnable proposals. In: Proceedings of the IEEE Conference on Computer Vision and Pattern Recognition, pp. 14454–14463 (2021)
42. Vo, N.D., Nguyen, K., Nguyen, T.V., Nguyen, K.: Ensemble of deep object detectors for page object detection. In: Proceedings of the International Conference on Ubiquitous Information Management and Communicatio, pp. 1–6 (2018)
43. Wang, R., Fujii, Y., Popat, A.C.: Post-ocr paragraph recognition by graph convolutional networks. In: Proceedings of the IEEE Winter Conference on Applications of Computer Vision, pp. 493–502 (2022)
44. Wang, X., Zhang, R., Kong, T., Li, L., Shen, C.: Solov2: Dynamic and fast instance segmentation. In: Proceedings of the Advances in Neural information processing systems. vol. 33, pp. 17721–17732 (2020)
45. Xue, C., Huang, J., Zhang, W., Lu, S., Wang, C., Bai, S.: Contextual text block detection towards scene text understanding. In: Proceedings of the European Conference on Computer Vision, pp. 374–391 (2022)
46. Yang, H., Hsu, W.: Transformer-based approach for document layout understanding. In: Proceedings of the International Conference on Image Processing, pp. 4043–4047 (2022)
47. Yang, X., Yumer, E., Asente, P., Kraley, M., Kifer, D., Lee Giles, C.: Learning to extract semantic structure from documents using multimodal fully convolutional neural networks. In: Proceedings of the IEEE Conference on Computer Vision and Pattern Recognition, pp. 5315–5324 (2017)
48. Yi, X., Gao, L., Liao, Y., Zhang, X., Liu, R., Jiang, Z.: Cnn based page object detection in document images. In: Proceedings of the International Conference on Document Analysis and Recognition. vol. 1, pp. 230–235 (2017)
49. Zhang, H., et al.: Dino: Detr with improved denoising anchor boxes for end-to-end object detection. arXiv preprint arXiv:2203.03605 (2022)
50. Zhang, J., Elhoseiny, M., Cohen, S., Chang, W., Elgammal, A.: Relationship proposal networks. In: Proceedings of the IEEE Conference on Computer Vision and Pattern Recognition, pp. 5678–5686 (2017)
51. Zhang, P., et al.: Vsr: a unified framework for document layout analysis combining vision, semantics and relations. In: Proceedings of the International Conference on Document Analysis and Recognition, pp. 115–130 (2021)
52. Zhang, Y., Bo, Z., Wang, R., Cao, J., Li, C., Bao, Z.: Entity relation extraction as dependency parsing in visually rich documents. In: Proceedings of the Conference on Empirical Methods in Natural Language Processing, pp. 2759–2768 (2021)
53. Zhong, X., Tang, J., Yepes, A.J.: Publaynet: largest dataset ever for document layout analysis. In: Proceedings of the International Conference on Document Analysis and Recognition, pp. 1015–1022 (2019)
54. Zhu, X., Su, W., Lu, L., Li, B., Wang, X., Dai, J.: Deformable detr: Deformable transformers for end-to-end object detection. In: Proceedings of the International Conference on Learning Representations (2021)

ColDBin: Cold Diffusion for Document Image Binarization

Saifullah Saifullah[1,2](✉) (iD), Stefan Agne[1,3] (iD), Andreas Dengel[1,2] (iD),
and Sheraz Ahmed[1,3] (iD)

[1] Smart Data and Knowledge Services (SDS), German Research Center for Artificial
Intelligence GmbH (DFKI), Trippstadter Straße 122, 67663 Kaiserslautern, Germany
`{Saifullah.Saifullah,Stefan.Agne,Andreas.Dengel,Sheraz.Ahmed}@dfki.de`
[2] Department of Computer Science, RPTU Kaiserslautern-Landau,
Erwin-Schrödinger-Straße 52, 67663 Kaiserslautern, Germany
[3] DeepReader GmbH, 67663 Kaiserlautern, Germany

Abstract. Document images, when captured in real-world settings, either modern or historical, frequently exhibit various forms of degradation such as ink stains, smudges, faded text, and uneven illumination, which can significantly impede the performance of deep learning-based approaches for document processing. In this paper, we propose a novel end-to-end framework for binarization of degraded document images based on cold diffusion. In particular, our approach involves training a diffusion model with the objective of generating a binarized document image directly from a degraded input image. To the best of the authors' knowledge, this is the first work that investigates diffusion models for the task of document binarization. In order to assess the effectiveness of our approach, we evaluate it on 9 different benchmark datasets for document binarization. The results of our experiments show that our proposed approach outperforms several existing state-of-the-art approaches, including complex approaches utilizing generative adversarial networks (GANs) and variational auto-encoders (VAEs), on 7 of the datasets, while achieving comparable performance on the remaining 2 datasets. Our findings suggest that diffusion models can be an effective tool for document binarization tasks and pave the way for future research on diffusion models for document image enhancement tasks. The implementation code of our framework is publicly available at: https://github.com/saifullah3396/coldbin.

Keywords: Document Binarization · Document Image Enhancement · Diffusion Models · Cold Diffusion · Document Image Analysis

1 Introduction

In the era of automation, accurate and efficient automated processing of documents is of the utmost importance for streamlining modern business workflows

---

This work was supported by the BMBF projects SensAI (BMBF Grant 01IW20007).

[1–3]. At the same time, it has vast applications in the preservation of historical scriptures [4–6] that contain valuable information about ancient cultural heritages and scientific contributions. Deep learning (DL) has recently emerged as a powerful tool for handling a wide variety of document processing tasks, showing remarkable results in areas such as document classification [1,7], optical character recognition (OCR) [8], and named entity recognition (NER) [2,9]. However, it remains challenging to apply DL-based models to real-world documents due to a variety of distortions and degradations that frequently occur in these documents. Document image enhancement (DIE) is a core research area in document analysis that focuses on recovering clean and improved images of documents from their degraded counterparts. Depending on the severity of the degradation, a document may display wrinkles, stains, smears, or bleed-through effects [10–12]. Additionally, distortions may result from scanning documents with a smartphone, which may introduce shadows [13], blurriness [14], or uneven illumination. Such degradations, which are particularly prevalent in historical documents, can significantly deteriorate the performance of deep learning models on downstream document processing tasks [15]. Therefore, it is essential that prior to applying these models, there be a pre-processing step that performs denoising and recovers a clean version of the degraded document image.

Over the past few decades, DIE has been the subject of several research efforts, including both classical [16,17] and deep learning-based studies [6,13,18, 19]. Lately, generative models such as deep variational autoencoders (VAEs) [20] and generative adversarial networks (GANs) [21] have gained popularity in this domain, owing to their remarkable success in natural image generation [21,22] and restoration tasks [23–25]. Generative models have attracted considerable attention due to their ability to accurately capture the underlying distribution of the training data, which allows them not only to generate highly realistic and diverse samples [22], but also to generate missing data when necessary [26]. As a result, a number of GAN and VAE based approaches have been recently proposed for DIE tasks, such as binarization [6,18,27], deblurring [6,19], and watermark removal [6].

Diffusion models [28] are a new class of generative models inspired by the process of diffusion in non-equilibrium thermodynamics. In the context of image generation, the underlying mechanism of diffusion models involves a fixed forward process of gradually adding Gaussian noise to the image, and a learnable reverse process to denoise and recover the clean image, utilizing a Markov chain structure. Diffusion models have been shown to have several advantages over GANs and VAEs such as their high training stability [28–30], diverse and realistic image synthesis [31,32], and better generalization to out-of-distribution data [33]. Additionally, conditional diffusion models have been employed to perform image synthesis with an additional input, such as class labels, text, or source image and have been successfully adapted for various natural image restoration tasks, including super-resolution [34], deblurring [35], and JPEG restoration [36]. Despite their growing popularity, however, there is no existing literature that has explored their potential in the context of document image enhancement.

In this study, we investigate the potential of diffusion models for the task of document image binarization, and introduce a novel approach for restoring clean binarized images from degraded document images using cold diffusion. We conduct a comprehensive evaluation of our proposed approach on multiple publicly available benchmark datasets for document binarization, demonstrating the effectiveness of our methodology in producing high-quality binarized images from degraded document images. The main contributions of this paper are two-fold:

- To the best of the authors' knowledge, this is the first work that presents a flexible end-to-end document image binarization framework based on diffusion models.
- We evaluate the performance of our approach on 9 different benchmark datasets for document binarization which include DIBCO '9 [37], H-DIBCO '10 [11], DIBCO '11 [38], H-DIBCO '12 [39], DIBCO '13 [12], H-DIBCO '14 [40], H-DIBCO '16 [41], DIBCO '17 [42], and H-DIBCO '18 [43].
- Through a comprehensive quantitative and qualitative evaluation, we demonstrate that our approach outperforms several classical approaches as well as the existing state-of-the-art on 7 of the datasets, while achieving competitive performance on the remaining 2 datasets.

2 Related Work

2.1 Document Image Enhancement

Document image enhancement (DIE) has been extensively studied in the literature over the past few decades [5,16,44–46]. Classical approaches to DIE were primarily based on global thresholding [16], local thresholding [44,47] or their hybrids [48]. These approaches were based on determining threshold values to segment the image pixels of a document into foreground or background. In a different direction, energy-based segmentation approaches such as Markov random fields (MRFs) [49] and conditional random fields (CRFs) [50] and classical machine learning-based approaches such as support vector machines (SVMs) [17,51] have also been widely explored in the past.

In recent years, there has been a burgeoning interest in the application of deep learning-based techniques for the enhancement of document images [4,52–54]. The earliest work in this area was majorly focused on utilizing convolutional neural networks (CNNs) [4,5,52,55]. One notable example of this is the work of Pastor-Pellicer et al. [56], who proposed a CNN-based classifier in conjunction with a sliding window approach for segmenting images into foreground and background regions. Building upon this, Tensmeyer et al. [52] presented a more advanced methodology that entailed feeding raw grayscale images, along with relative darkness features, into a multi-scale CNN, and training the network using a pseudo F-measure loss. Another approach was proposed by Calvo-Zaragoza et al. [55], in which they utilized a CNN-based auto-encoder (AE) to train the model to map degraded images to clean ones in an end-to-end fashion. A similar approach was presented by Kang *et al.* [5], who employed a pre-trained U-Net based auto-encoder model for binarization, with minimal training data requirements. Since then, a number of AE-based approaches have been proposed for DIE tasks [53,54]. In a slightly different direction, Castellanos *et al.* [57] has also investigated domain adaptation in conjunction with deep neural networks for the task of document binarization.

Generative Adversarial Networks (GANs) have also been extensively explored in this field to generate clean images by conditioning on degraded versions [6,19,27,46]. These methods typically consist of a generative model that generates a clean binarized version of the image, along with a discriminator that assesses the results of the binarization. Zhao *et al.* [27] proposed a cascaded GAN-based approach for the task of document image binarization and demonstrated excellent performance on a variety of benchmark datasets. Jemni *et al.* [58] recently presented a multi-task GAN-based approach which incorporates a text recognition network in combination with the discriminator to further improve text readability along with binarization. Similarly, Yu *et al.* [46] proposed a multi-stage GAN-based approach to document binarization that first applies discrete wavelet transform to the images to perform enhancement, and then trains a separate GAN for each channel of the document image. Besides GANs and CNN-based auto-encoders, the recent success of transformers in natural language processing (NLP) [9] and vision [59] has also sparked interest in transformers for the enhancement of document images. In a recent study, Souibgui *et al.* [45] proposed a transformer-based auto-encoder model that demonstrated state-of-the-art performance on several document binarization datasets.

3   ColDBin: The Proposed Approach

This section presents the details of our proposed approach and explains its relationship to standard diffusion [28]. The overall workflow of our approach is illustrated in Fig. 1. Primarily inspired by cold diffusion [60], our approach involves training a deep diffusion network for document binarization in two steps: a forward diffusion step and a reverse restoration step. As shown, in the forward diffusion step, a clean ground-truth document image is degraded to a specified severity level based on a given type of input degradation. In the reverse restoration step, a neural network is tasked with undoing the forward diffusion process in order to generate a clean ground-truth image from an intermediary degraded image. These forward and reverse steps are repeated in a cycle, and the neural network is trained for the binarization task by applying image reconstruction loss to its output. In the following sections, we provide a more detailed explanation of the forward and reverse steps of our approach.

3.1   Forward Diffusion

In the context of document binarization, let $P = \{(x, x_0) \sim (\mathcal{X}, \mathcal{X}_0)\}_{n=1}^{N}$ define a training set consisting of pairs of degraded document images x and their

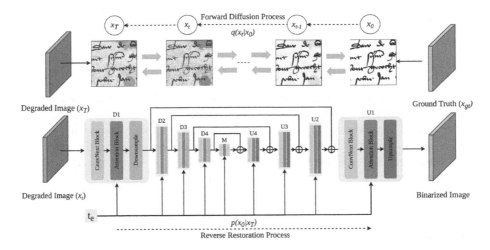

Fig. 1. Demonstration of the forward diffusion and reverse restoration processes of our approach. The forward diffusion process incrementally degrades a clean ground-truth image into its degraded counterpart. Whereas, the reverse restoration process, defined by a neural network, generates a clean binary image from a degraded input image

corresponding binarized ground-truth images x_0. Let $\mathbb{D}(x_0, t)$ be a diffusion operator that adds degradation to a clean ground-truth image x_0 proportional to the severity $t \in \{0, 1, \ldots, T\}$, T being the maximum severity permitted, then the degraded image at any given severity t can be derived as follows:

$$x_t = \mathbb{D}(x_0, t) \tag{1}$$

Consequently, the following constraint must be satisfied:

$$\mathbb{D}(x_0, 0) - x_0 \tag{2}$$

Generally in standard diffusion [28], this forward diffusion operator $\mathbb{D}(x_0, t)$ is defined as a fixed Markov process that gradually adds Gaussian noise ϵ to the image using a variance schedule specified by $\beta_1 \ldots \beta_T$. In particular, it is defined as the posterior $q(x_1, \ldots, x_T | x_0)$ that converts the data distribution $q(x_0)$ to the latent distribution $q(x_T)$ as follows:

$$q(x_1, \ldots, x_T | x_0) := \prod_{t=1}^{T} q(x_t | x_{t-1})$$

$$q(x_t | x_{t-1}) := \mathcal{N}(x_t; \sqrt{\beta_t} x_{t-1}, (1 - \beta_t)\mathbf{I})$$

where β_t is a hyper-parameter that defines the severity of degradation at each severity level t. An important property of the above forward process is that it allows sampling x_t at any arbitrary severity t in closed form: using the notation $\alpha_t := 1 - \beta_t$ and $\hat{\alpha}_t := \Pi_{s=1}^{t} \alpha_s$, we have

$$q(x_t | x_0) := \mathcal{N}(x_t; \sqrt{\hat{\alpha}_t} x_{t-1}, (1 - \hat{\alpha}_t)\mathbf{I}) \tag{3}$$

Which results in the following the diffusion operator $\mathbb{D}(x_0, t)$:

$$x_t = \mathbb{D}(x_0, t) = \sqrt{\hat{\alpha}_t} x_0 + \sqrt{1 - \hat{\alpha}_t}\epsilon, \quad \epsilon \sim \mathcal{N}(\mathbf{0}, \mathbf{I}) \tag{4}$$

Our approach maintains the same forward process as standard diffusion, except that Gaussian noise ϵ is not used to define the diffusion operator $\mathbb{D}(x_0, t)$ (hot diffusion). Rather, we define it as a cold diffusion operation that interpolates between the binarized ground-truth image x_0 and its degraded counterpart image x based on the noise schedule $\beta_1 \ldots \beta_T$. More formally, given a fully degraded input image x and its respective binarized ground-truth image x_0, an intermediate degraded image x_t at severity t is then defined as follows:

$$x_t = \mathbb{D}(x_0, x, t) = \sqrt{\hat{\alpha}_t} x_0 + \sqrt{1 - \hat{\alpha}_t} x, \quad x_0 \sim \mathcal{X}_0, x \sim \mathcal{X} \tag{5}$$

Note that this procedure is essentially the same as adding Gaussian noise ϵ in standard diffusion, except that here we are adding a progressively higher weighted degraded image to the clean ground-truth image to generate an intermediary noisy image. In addition, our diffusion operator for binarization is slightly modified $\mathbb{D}(x_0, x, t)$ and requires both the ground-truth image x_0 and the target degraded image x for forward the process.

3.2   Reverse Restoration

Let $\mathbb{R}(x_t, t)$ define the reverse restoration operator that restores any degraded image x_t at severity t to its clean binarized form x_0:

$$\mathbb{R}(x_t, t) \approx x_0 \tag{6}$$

In standard diffusion [28], generally this restoration operator $\mathbb{R}(x_t, t)$ is defined as a reverse Markov process $p(x_0, \ldots, x_{T-1} | x_T)$ that transforms the data from the latent variable distribution $p_\theta(x_T)$ to the data distribution $p_\theta(x_0)$ parameterized by θ; the process generally starting from $p(x_T) = \mathcal{N}(x_T; \mathbf{0}, \mathbf{I})$:

$$p(x_0, \ldots, x_{T-1} | x_T) := \prod_{t=1}^{T} p_\theta(x_{t-1} | x_t)$$

$$p_\theta(x_{t-1} | x_t) := \mathcal{N}(x_{t-1}; \mu_\theta(x_t, t), \sigma_\theta(x_t, t)^2 \mathbf{I})$$

Our approach uses the same reverse restoration process as the standard diffusion [28], with the exception that it begins with a degraded input image $x_T \sim \mathcal{X}$ instead of Gaussian noise $x_T \sim \mathcal{N}(x_T; \mathbf{0}, \mathbf{I})$. In practice, $\mathbb{R}(x_t, t)$ is generally implemented as a neural network $\mathbb{R}_\theta(x_t, t)$ parameterized by θ which is trained to perform the reverse restoration task. In our approach, the restoration network $\mathbb{R}_\theta(x_t, t)$ is trained by minimizing the following loss:

$$\min_\theta \mathbb{E}_{x \sim \mathcal{X}, x_0 \sim \mathcal{X}_0} \| \mathbb{R}_\theta(D(x_0, x, t), t) - x_0 \| \tag{7}$$

Algorithm 1. Training

1: **Input**: Ground truth image x_0 and its corresponding degraded image x pairs
   $P = \{(x_0, x)\}_{k=1}^{K}$, and total diffusion steps T
2: **Initialize**: Randomly initialize the restoration network $\mathbb{R}_\theta(x_t, t)$
3: **repeat**
4:     Sample $(x_0, x) \sim P$, and $t \sim \text{Uniform}(\{1, \ldots, T\})$
5:     Take the gradient step on
6:         $\nabla_\theta \|\mathbb{R}_\theta(x_t, t) - x_0\|, x_t = \sqrt{\hat{\alpha}_t} x_0 + \sqrt{1 - \hat{\alpha}} x$
7: **until** converged

---

where $\|\cdot\|$ defines a norm, which we took as standard ℓ_2 norm in this work. The overall training process of the restoration network is given in Algorithm 1. As shown, the restoration network $\mathbb{R}_\theta(x_t, t)$ is initialized with a maximum severity level of T. In each training iteration, a mini-batch of degraded images x and their corresponding binarized ground-truth images x_0 is randomly sampled from the training set P, and the degradation severity is randomly sampled from the integer set $\{1, \ldots, T\}$ The severity value t is then used in combination with the ground-truth x_0 and degraded image x pairs to compute the intermediate interpolated images x_t using Eq. 5 (line 6). The restoration network $\mathbb{R}_\theta(x_t, t)$ is then used to recover a binarized image from the interpolated image x_t. Finally, the network is optimized in each step by taking the gradient step on Eq. 7 (line 6).

3.3   Restoration Network

The complete architecture of the restoration network $\mathbb{R}_\theta(x_t, t)$ used in our approach is illustrated in Fig. 1. As shown, we used a U-Net [61] inspired architecture as the restoration network which takes as input the degraded image x_t and the diffusion severity $t \in 1, 2, \ldots, T$ and generates a binarized image as the output. The input severity level t is transformed into a severity embedding t_e based on sinusoidal positional encoding as proposed in [62]. The embedded severity and the image are then passed through multiple downsampling blocks, a middle processing block and then multiple upsampling blocks to generate the output image. Each downsampling and upsampling block is characterized by two ConvNeXt [63] blocks, a residual block with a linear attention layer, and a downsampling layer. The middle block consists of a ConvNeXt block followed by an attention module and another ConvNeXt block and is inserted between the downsampling and upsampling phases.

3.4   Inference Strategies

We investigated two different inference strategies for restoring images from their degraded counterparts: direct restoration and cold diffusion sampling. Direct restoration simply applies the restoration operator $\mathbb{R}_\theta(x_t, t)$ to a degraded input image x with degradation severity t set to T. On the other hand, cold diffusion sampling as proposed in [60] iteratively performs the reverse restoration process over T steps as described in Algorithm 2. Although a number of sampling strategies have been proposed previously for diffusion models [28,64], Bansal *et al.* [60] demonstrated in their work that this sampling strategy performs better than standard sampling [28] for cold diffusion processes, and therefore it has been investigated in this study.

4   Experiments and Results

In this section, we first describe the experimental setup, including datasets, evaluation metrics, and the training process. Subsequently, we present a comprehensive quantitative and qualitative analysis of our results.

4.1   Experimental Setup

Datasets. 9 different DIBCO document image binarization datasets were used to assess the performance of our proposed approach. These datasets include DIBCO '9 [37], DIBCO '11 [38], DIBCO '13 [12], and DIBCO '17 [42], as well as H-DIBCO '10 [11], H-DIBCO '12 [39], H-DIBCO '14 [40], H-DIBCO '16 [41], and H-DIBCO '18 [43]. A variety of degraded printed and handwritten documents are included in these datasets, which exhibit various degradations such as ink bleed through, smudges, faded text strokes, stain marks, background texture, and artifacts.

Evaluation Metrics. Several evaluation methods have been commonly used in the literature for evaluating the binarization of document images, including FM (F-Measure), pFM (pseudo-F-Measure), PSNR (Peak Signal-to-Noise Ratio), and DRD (Distance Reciprocal Distortion), which have been adopted in this study. A higher value indicates better binarization performance for the first three metrics, while the opposite is true for DRD. Due to space constraints, detailed definitions of these metrics are omitted here and can be found in [11,12].

Data Preprocessing. To train the restoration model on a specific DIBCO dataset, all the images from other DIBCO and H-DIBCO datasets as well as the Palm Leaf dataset [65] were used. The training set was prepared by splitting each degraded image and its corresponding ground truth image into overlapped patches

---

Algorithm 2. Cold Diffusion Sampling Strategy [60]

1: **Input**: A degraded sample x
2: **for** s=t,t-1,...,1 **do**
3:     $\hat{x}_0 = R(x_s, s)$
4:     $x_{s-1} = x_s - D(\hat{x}_0, s) + D(\hat{x}_0, s-1)$
5: **end for**

Table 1. The size of the training and test sets for all DIBCO datasets is provided.

	DIBCO '9 [37]	H-DIBCO '10 [11]	DIBCO '11 [38]	H-DIBCO '12 [39]	DIBCO '13 [12]	H-DIBCO '14 [40]	H-DIBCO '16 [41]	DIBCO '17 [42]	H-DIBCO '18 [43]
Train	17716	17669	17448	16885	16231	17492	17300	15983	17202
Test (256)	135	150	217	356	542	182	256	610	266
Test (512)	40	46	66	104	166	55	74	184	74

of size $384 \times 384 \times 3$. Table 1 shows the total number of training set samples that were generated for each DIBCO dataset as a result of using the above strategy. During training, a random crop of size $256 \times 256 \times 3$ was extracted from each image and then fed to the model. Additionally, a number of data augmentations were used such as horizontal flipping, vertical flipping, color jitter, grayscale conversion, and Gaussian blur, all of which were randomly applied to the images. A specific augmentation we used in our approach was to randomly colorize the degraded image using the inverted ground truth image as a mask. This augmentation was necessary to prevent the models from overfitting to black-color text since most of the images in the DIBCO datasets consisted of black-color text on various backgrounds. Furthermore, we used ImageNet normalization with per-channel means of $\mu_{RGB} = \{0.485, 0.456, 0.406\}$ and standard deviations of $\sigma_{RGB} = \{0.229, 0.224, 0.225\}$ to normalize each image before feeding it to the model.

Training Hyperparameters. We initialized our restoration networks with maximum diffusion severity T set to 200 and severity embedding set to 64. For the forward diffusion process, we used a cosine beta noise schedule $\beta_1, \ldots, \beta_T$ as described in [66]. We trained our networks for 400k iterations with a batch size of 128, Adam optimizer, and a fixed learning rate of $2e - 5$ on 4–8 NVIDIA A100 GPUs.

Evaluation Hyperparameters. To evaluate our approach, we divided each image into patches of fixed input size, restored them using the inference strategies outlined in Sect. 3.4, and then reassembled them to produce the final binarized image. Depending on the size of the input patch, binarization performance can be greatly affected, since smaller patches provide less context for the model, whereas larger patches provide more context. In this work, we examined two different patch sizes at test time, which were 256×256 and 512×512. It should be noted that we trained the models solely on 256×256 input images, and used images of size 512×512 only during evaluation.

Table 2. Comparison of different evaluation strategies on DIBCO '9 [37], H-DIBCO '12 [39], and DIBCO '17 [42] datasets. The top strategy for each metric is bolded.

	DIBCO '9 [37]				H-DIBCO '12 [39]				DIBCO '17 [42]			
Strategy / Patch size	FM↑	p-FM↑	PSNR↑	DRD↓	FM↑	p-FM↑	PSNR↑	DRD↓	FM↑	p-FM↑	PSNR↑	DRD↓
Direct Restoration / 256	93.83	96.25	20.34	2.75	96.09	97.16	23.07	1.42	92.21	94.33	18.93	2.80
Direct Restoration / 512	**94.19**	**96.52**	**20.65**	**2.58**	**96.37**	**97.41**	**23.40**	1.28	**93.04**	**95.12**	**19.32**	**2.29**
Cold Sampling / 256	93.55	96.05	20.03	2.71	95.70	96.68	22.69	1.46	89.57	91.66	18.18	3.61
Cold Sampling / 512	93.69	96.08	20.21	2.69	96.10	97.09	23.07	1.28	90.81	92.86	18.59	2.99

FM = F-Measure, p-FM = pseudo F-Measure PSNR = Peak Signal-to-Noise Ratio, DRD = Distance Reciprocal Distortion

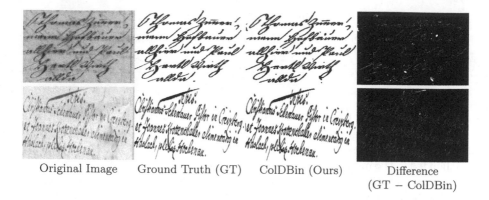

Original Image　　Ground Truth (GT)　　ColDBin (Ours)　　Difference
(GT − ColDBin)

Fig. 2. Binarization results for images 1 (top) and 10 (bottom) of the H-DIBCO '16 [41] dataset using our approach. The difference between the ground truth image and the binarized output of our proposed approach is shown to emphasize that our model produces slightly thicker strokes for this dataset

4.2　Choosing the Best Evaluation Strategy

In this section, we compare the results of direct restoration and cold diffusion sampling strategies with varying patch sizes on three different datasets DIBCO '9 [37], H-DIBCO '12 [39], and DIBCO '17 [42] as shown in Table 2. It is evident from the table that direct restoration performed significantly better than cold diffusion sampling for binarization with both patch sizes of 256 and 512. While diffusion models are well known for providing better reconstruction/image generation performance when using sampling as compared to direct inference over T steps, sampling resulted in poorer FM, p-FM, PSNR, and DRD values than direct restoration in our case. Moreover, sampling is a very computationally intensive process, requiring multiple forward and reverse diffusion steps, whereas direct inference requires only a single step and, therefore, is extremely fast. Also evident from the table is that the model performed better with patch sizes of 512 × 512 as opposed to 256 × 256. This was the case for the majority of DIBCO datasets we examined. However, we observed that the 256 × 256 patch size provided better performance for some datasets such as H-DIBCO '16 [41] and H-DIBCO '18 [43]. This raises the question of whether it is possible to develop a more effective evaluation approach that is able to accommodate images of different sizes and text resolutions within those images. However, we leave those questions to future research. Since 512 × 512 patch size with direct restoration offered the best performance for most datasets, we present only the results from this evaluation strategy when doing a performance comparison in the subsequent sections.

4.3　Performance Comparison

In this section, we present a quantitative comparison of our approach against a variety of other approaches, including classical approaches [16,44,67,68, 72], CNN-based VAEs [52,69,70], GAN-based approaches [27,46,58,71], and

Table 3. Performance evaluation of different methods for document binarization on all the DIBCO/H-DIBCO evaluation datasets. For each metric, the top **1st**, *2nd*, and <u>3rd</u> methods are **bolded**, *italicized*, and <u>underlined</u>, respectively. The results presented here were generated using the **Direction Restoration / 512** evaluation strategy.

		Otsu[16]	Sauvola[44]	Lu[67]	Su[68]	Tensmeyer[52]	Vo[69]	He[70]	Zhao[27]	Suh[71]	Xiong[72]	Soulbgui[45]	Jemni[58]	Yu[46]	Ours
		1979	2000	2010	2013	2017	2018	2019	2019	2020	2021	2022	2022	2022	2023
		Thres.	Thres.	CV	CV	CNN-AE	CNN-AE	CNN-AE	GAN	GAN	SVM	Tr-VAE	Multitask-GAN	Multiple GANs	Diffusion
Datasets	**Metrics**														
DIBCO '9 [37]	FM↑	78.72	85.41	91.24	<u>93.50</u>	89.76	-	-	*94.10*	-	93.46	-	-	-	**94.19**
	p-FM↑	-	-	-	-	*92.50*	-	-	*95.86*	-	-	-	-	-	**96.52**
	PSNR↑	15.34	16.39	18.66	19.65	18.43	-	-	*20.30*	-	<u>20.01</u>	-	-	-	**20.65**
	DRD↓	-	-	-	-	<u>4.89</u>	-	-	**1.82**	-	-	-	-	-	*2.58*
H-DIBCO '10 [11]	FM↑	85.27	75.30	86.41	92.03	*94.89*	-	-	<u>94.03</u>	-	93.73	-	-	-	**95.29**
	p-FM↑	90.83	84.22	88.25	94.85	*97.65*	-	-	<u>95.39</u>	-	95.18	-	-	-	**96.67**
	PSNR↑	17.51	15.96	18.14	20.12	*21.84*	-	-	<u>21.12</u>	-	20.97	-	-	-	**22.06**
	DRD↓	-	-	-	-	1.26	-	-	<u>1.55</u>	-	-	-	-	-	*1.36*
DIBCO '11 [38]	FM↑	82.10	82.35	81.67	87.80	93.60	92.58	91.92	92.62	93.57	90.72	*94.37*	-	<u>94.08</u>	**95.23**
	p-FM↑	85.96	88.63	-	-	97.70	94.67	95.82	95.38	95.93	-	96.15	-	*97.08*	<u>96.93</u>
	PSNR↑	15.72	15.73	15.59	17.56	20.11	19.16	19.49	19.58	20.22	18.85	*20.81*	-	<u>20.51</u>	**21.53**
	DRD↓	8.95	7.86	11.24	4.84	1.85	2.38	2.37	2.55	1.99	4.47	*1.63*	-	<u>1.75</u>	**1.44**
H-DIBCO '12 [39]	FM↑	80.18	82.89	-	-	92.53	-	-	94.96	-	94.26	*95.31*	<u>95.18</u>	-	**96.37**
	p-FM↑	82.65	87.95	-	-	*96.67*	-	-	96.15	-	95.15	<u>95.22</u>	94.63	-	**97.41**
	PSNR↑	15.03	16.71	-	-	20.00	-	-	21.91	-	21.68	*22.29*	<u>22.00</u>	-	**23.40**
	DRD↓	26.46	6.50	-	-	2.48	-	-	*1.55*	-	2.08	<u>1.6</u>	1.62	-	**1.28**
DIBCO '13 [12]	FM↑	80.04	82.73	-	-	93.17	93.43	93.36	93.86	<u>95.01</u>	93.51	-	-	*95.24*	**96.62**
	p-FM↑	83.43	88.37	-	-	96.81	95.34	<u>96.70</u>	96.47	96.49	94.54	-	-	*97.51*	97.17
	PSNR↑	16.63	16.98	-	-	20.71	20.82	20.88	21.53	<u>21.99</u>	21.32	-	-	*22.27*	**23.98**
	DRD↓	10.98	7.34	-	-	2.21	2.26	2.15	2.32	<u>1.76</u>	2.77	-	-	*1.59*	2.29
H-DIBCO '14 [40]	FM↑	91.82	83.72	-	-	91.96	95.97	95.95	96.09	96.36	*96.77*	-	-	<u>96.65</u>	**97.80**
	p-FM↑	95.69	87.49	-	-	94.78	97.42	**98.76**	*98.85*	97.87	97.73	-	-	<u>98.19</u>	98.10
	PSNR↑	18.72	17.48	-	-	20.76	21.49	21.60	21.88	21.96	*22.47*	-	-	<u>22.27</u>	**24.38**
	DRD↓	2.65	5.05	-	-	2.72	1.09	1.12	1.20	1.07	*0.95*	-	-	<u>0.96</u>	**0.66**
H-DIBCO '16 [41]	FM↑	86.59	84.27	-	-	89.52	90.01	91.19	<u>91.06</u>	*92.24*	89.64	-	**94.05**	91.46	89.50
	p-FM↑	89.92	89.10	-	-	93.76	93.44	<u>95.74</u>	94.58	*95.95*	93.56	-	94.55	**96.32**	93.73
	PSNR↑	17.79	17.15	-	-	18.67	18.74	19.51	19.64	19.93	18.69	-	21.85	<u>19.66</u>	18.71
	DRD↓	5.50	6.00	-	-	3.76	3.91	3.02	2.89	*2.77*	4.03	-	1.56	<u>2.94</u>	3.84
DIBCO '17 [42]	FM↑	77.73	77.11	-	-	-	-	-	90.73	-	89.37	*92.53*	89.80	<u>90.25</u>	**93.04**
	p-FM↑	77.89	84.10	-	-	-	-	-	92.58	-	90.8	**95.15**	89.95	<u>93.72</u>	*95.12*
	PSNR↑	13.85	14.25	-	-	-	-	-	17.83	-	17.99	*19.11*	17.45	<u>18.57</u>	**19.32**
	DRD↓	15.54	8.85	-	-	-	-	-	3.58	-	5.51	*2.37*	4.03	<u>2.94</u>	2.29
H-DIBCO '18 [43]	FM↑	51.45	67.81	-	-	-	-	-	87.73	-	88.34	89.21	92.41	91.66	<u>89.71</u>
	p-FM↑	53.05	74.08	-	-	-	-	-	90.60	-	90.37	92.54	*94.35*	**95.53**	<u>93.00</u>
	PSNR↑	9.74	13.78	-	-	-	-	-	18.37	-	19.11	19.47	20.18	*20.02*	<u>19.53</u>
	DRD↓	59.07	17.69	-	-	-	-	-	4.58	-	4.93	3.96	2.60	*2.81*	<u>3.82</u>

FM=F-Measure, p-FM=pseudo F-Measure PSNR=Peak Signal-to-Noise Ratio, DRD=Distance Reciprocal Distortion

Transformer-based autoencoders [45]. The results of our evaluation are summarized in Table 3, where FM, p-FM, PSNR, and DRD of each method are compared for different DIBCO/H-DIBCO datasets, with the top three approaches for each dataset **bolded**, *italicized* and <u>underlined</u>, respectively. As shown, our approach outperforms existing classical and state-of-the-art (SotA) approaches on 7 datasets, including DIBCO '9 [37], H-DIBCO '10 [11], DIBCO '11 [38], H-DIBCO '12 [39], DIBCO '13 [12], H-DIBCO '14 [40], and DIBCO '17 [42], ranking first on the majority of metrics, while performing competitively on the remaining 2 datasets H-DIBCO '16 [41] and H-DIBCO '18 [43]. It is worth mentioning that a number of recent SotA binarization techniques, including those presented by Yu *et al.* [46] and Jemni *et al.* [58], utilize several training stages, networks, or target objectives in order to achieve the reported results. Comparatively, our approach employs only a single diffusion network in an end-to-end fashion, and is able to outperform these methods across multiple datasets.

On DIBCO '9 [37] dataset, our approach scored the highest on all metrics except DRD, on which it ranked second. Furthermore, it demonstrated signifi-

Original Image	Ground Truth	ColDBin (Ours)

Fig. 3. Qualitative results of our proposed method for the restoration of a few samples from the DIBCO and H-DIBCO datasets. These images are arranged in columns as follows: Left: original image, Middle: ground truth image, Right: binarized image using our proposed method

cant improvements in FM and PSNR on the H-DIBCO '10 [11] and DIBCO '11 [38] datasets in comparison to existing methods. We also observed a particularly noticeable improvement in PSNR with our approach on the H-DIBCO '12 [39], DIBCO '13 [12], and H-DIBCO '14 [40] datasets, with increases of 1.11, 1.71, and 1.91 compared to the previous state-of-the-art method, respectively. Similarly, despite lower DRD values on some datasets, it was significantly improved for these three datasets, with values of 1.28, 1.20, and 0.66, respectively. Similar performance improvements were observed on the DIBCO '17 [42] dataset as well, where our approach ranked first on FM, PSNR, and DRD, and ranked second on p-FM. On H-DIBCO '18 [43], our approach placed third; however, it is evident from the results that our model demonstrated comparable performance to the top approaches.

Despite the high performance achieved on other datasets, our approach failed to achieve satisfactory results on the H-DIBCO '16 [41] dataset. Interestingly, upon inspecting the binarization outputs, we found that our approach was, in fact, quite capable of producing high quality binarization results for this dataset. The approach, however, had the tendency to generate slightly thicker text strokes compared to the ground truth images, which may explain why it did not produce the best quantitative results on this dataset. Figure 2 illustrates this effect by presenting two samples from the H-DIBCO '16 [41] dataset along with their corresponding ground truth images, binarized images derived from our method, and their difference. As can be seen from the difference image, our proposed approach produces binarized outputs very similar to the ground truth but with slightly thicker strokes in comparison. Overall, we observed that our approach demonstrated relatively consistent performance across the majority of DIBCO datasets and provided the highest FM and PSNR.

4.4 Qualitative Evaluation

This section presents a qualitative analysis of the binarization performance of our approach. In Fig. 3, we compare the binarization results of our approach with the ground truth for a few randomly selected samples from the different DIBCO and H-DIBCO datasets. As evident from the figure, our approach was highly effective at removing various types of noise, such as stains, smears, faded

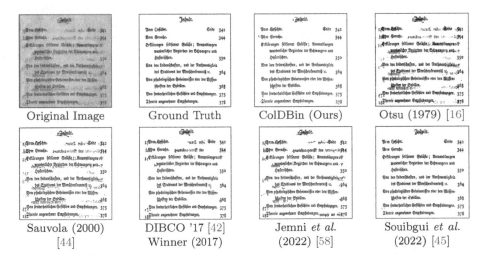

Fig. 4. Document binarization results for the input image 12 of DIBCO '17 [42] by different methods

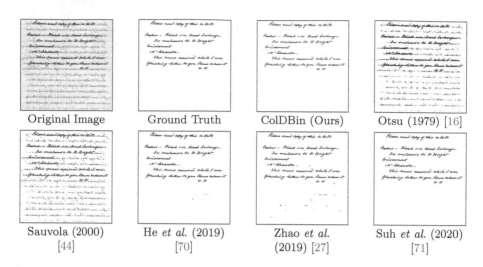

Fig. 5. Document binarization results for the input image HW5 of DIBCO '13 [12] by different methods

text, and background texture from a number of degraded document images. Moreover, it was able to produce high-quality binarized images that were visually comparable to the corresponding ground truth images, reflecting the exceptional quantitative performance discussed in the previous section.

Aside from comparisons with ground truth, we also compare the results of our approach to both classical and existing state-of-the-art (SotA) approaches. Figure 4 illustrates the binarization performance of various approaches, including ours, on sample 12 of the DIBCO '17 [42] dataset. The results demonstrate that our approach was successful in restoring a highly degraded document sample that many other approaches, including the multi-task GAN approach by Jemni et al. [58], failed to sufficiently restore. Interestingly, our results for this sample were visually similar to those obtained by Souibgui et al. [45], who used an encoder-decoder Transformer architecture for binarization. In Fig. 5, we compare the binarization performance of various approaches on another sample, namely, the HW5 from the DIBCO '13 [12] dataset. As can be seen, our approach was successful in restoring the image entirely, with the resulting image looking strikingly identical to the ground truth image. Additionally, we observed that our results for this sample were similar but slightly better than those of Suh et al. [71] and Yu et al. [46], who employed two-stage and three-stage GAN-based approaches for binarization, respectively.

4.5   Runtime Evaluation

In this section, we briefly analyze the runtime of our approach and compare it with other approaches. Since binarization speed depends on the size of input images, we evaluate the runtime in terms of secs/megapixel (MP) as used in

Table 4. Average runtimes for different binarization methods.

Runtime of different methods (secs/megapixel (MP))													
Otsu [16]	Sauvola [44]	Niblack [47]	Lu [67]	Su [68]	Bhowmik [51]	Tensemeyer [52]	Vo [69]	Zhao [27]	Xiong [72]	Ours (D-256)	Ours (D-512)	Ours (S-256)	Ours (S-512)
Thres.	Thres.	Thres.	CV	CV	Game Theory	CNN-AE	CNN-AE	GAN	SVM	Diffusion	Diffusion	Diffusion	Diffusion
0.042	0.092	0.106	12.839	7.372	80.845	6.436	3.043	0.9819	19.306	0.9679	0.9918	135.47	193.49

D-256= Direct Reconstruction (256×256), **D-512** = Direct Reconstruction (512×512), **S-256** = Cold Sampling with T=200 (256×256), **S-512** = Cold Sampling with T=200 (512×512)

prior works [27,72]. Both direct reconstruction and cold sampling were evaluated using a single NVIDIA GTX 1080Ti GPU with batch sizes of 4 and 32 for 512 × 512 and 256 × 256 image resolutions, respectively. The evaluation runtimes for other approaches were obtained directly from two papers [27,72], which may have used different resources for evaluation and therefore we are only able to make a rough comparison. As shown in Table 4, with direct reconstruction, our approach had a runtime of ∼1 sec/MP for both input image resolutions, which is comparable to the approach developed Zhao *et al.* [27], and is much lower than other computer vision methods [67,68] and deep learning approaches [52, 69]. In contrast, the runtime for cold sampling scaled proportionally with the number of diffusion steps T. With T=200 in our experiments, for a 256 × 256 input resolution, sampling took ∼135× more time than direct reconstruction, and for a 512 × 512 input resolution, it took ∼193× more time. Thus, direct reconstruction was not only effective quantitatively and qualitatively, but also time-efficient in comparison with sampled reconstruction. It is worth noting that the problem of unreasonably high sampling times in diffusion models is well-known, and different sampling strategies [64,73] have been proposed recently to overcome this problem.

5   Conclusion

This paper presents an end-to-end approach for binarization of document images using cold diffusion, which involves gradually transforming clean images into their degraded counterparts and then training a diffusion model that learns to reverse that process. The proposed approach was evaluated on 9 different DIBCO document benchmark datasets, and our results demonstrate that it outperforms traditional and state-of-the-art methods on a majority of datasets and does equally well on others. Despite its promising potential for document binarization, we believe it is also pertinent to discuss its limitations. As is the case with deep networks generally, the reliability of our models was quite dependent on the availability of data. While training datasets (DIBCO and Palm Leaf combined) have quite a lot of diversity in terms of sample distribution, the intra-class variance of samples was rather low, which necessitated training the models for a large number of iterations with various data augmentations in order to achieve the reported results. Therefore, to further enhance the performance of deep network-based approaches in the future, it may be worthwhile to invest resources in the creation of a large independent and diverse training dataset (whether synthetic or not) for binarization. We also observed a significant correlation between patch size and binarization performance with our approach. To address this issue in the future, it may be worthwhile to investigate the possibility of conditioning the output of our model on the surrounding context of each image patch.

References

1. Afzal, M.Z., Kolsch, A., Ahmed, S., Liwicki, M.: Cutting the error by half: investigation of very deep CNN and advanced training strategies for document image classification. In: Proceedings of the International Conference on Document Analysis and Recognition, ICDAR, vol. 1, pp. 883–888 (2017)
2. Xu, Y., Li, M., Cui, L., Huang, S., Wei, F., Zhou, M.: LayoutLM: pre-training of text and layout for document image understanding. In: Proceedings of the ACM SIGKDD International Conference on Knowledge Discovery and Data Mining, vol. 20, pp. 1192–1200 (2020)
3. Li, P., et al.: SelfDoc: self-supervised document representation learning (2021). https://arxiv.org/abs/2106.03331
4. Hradiš, M., Kotera, J., Zemcık, P., Šroubek, F.: Convolutional neural networks for direct text deblurring. In: Proceedings of BMVC, vol. 10, no. 2 (2015)
5. Kang, S., Iwana, B.K., Uchida, S.: Complex image processing with less data-document image binarization by integrating multiple pre-trained U-net modules. Pattern Recogn. **109**, 107577 (2021)
6. Souibgui, M.A., Kessentini, Y.: DE-GAN: a conditional generative adversarial network for document enhancement. IEEE Trans. Pattern Anal. Mach. Intell. (2020)
7. Saifullah, S., Agne, S., Dengel, A., Ahmed, S.: DocXClassifier: towards an interpretable deep convolutional neural network for document image classification **9** (2022). https://doi.org/10.36227/techrxiv.19310489.v4
8. Subramani, N., Matton, A., Greaves, M., Lam, A.: A survey of deep learning approaches for OCR and document understanding. ArXiv, abs/2011.13534 (2020)
9. Devlin, J., Chang, M.-W., Lee, K., Toutanova, K.: BERT: pre-training of deep bidirectional transformers for language understanding. In: NAACL HLT 2019–2019 Conference of the North American Chapter of the Association for Computational Linguistics: Human Language Technologies - Proceedings of the Conference, vol. 1, pp. 4171–4186 (2018). https://arxiv.org/abs/1810.04805v2
10. Sulaiman, A., Omar, K., Nasrudin, M.F.: Degraded historical document binarization: a review on issues, challenges, techniques, and future directions. J. Imag. **5**(4) (2019). https://www.mdpi.com/2313-433X/5/4/48
11. Pratikakis, I., Gatos, B., Ntirogiannis, K.: H-DIBCO 2010 - handwritten document image binarization competition. In: 2010 12th International Conference on Frontiers in Handwriting Recognition, pp. 727–732 (2010)
12. Pratikakis, I., Gatos, B., Ntirogiannis, K.: ICDAR 2013 document image binarization contest (DIBCO 2013). In: 2013 12th International Conference on Document Analysis and Recognition, pp. 1471–1476 (2013)
13. Bako, S., Darabi, S., Shechtman, E., Wang, J., Sunkavalli, K., Sen, P.: Removing shadows from images of documents. In: Asian Conference on Computer Vision (ACCV 2016) (2016)
14. Chen, X., He, X., Yang, J., Wu, Q.: An effective document image deblurring algorithm. In: CVPR 2011, pp. 369–376 (2011)

15. Saifullah, S., Siddiqui, S.A., Agne, S., Dengel, A., Ahmed, S.: Are deep models robust against real distortions? A case study on document image classification. In: 2022 26th International Conference on Pattern Recognition (ICPR), pp. 1628–1635 (2022)
16. Otsu, N.: A threshold selection method from gray level histograms. IEEE Trans. Syst. Man Cybern. **9**, 62–66 (1979)
17. Xiong, W., Xu, J., Xiong, Z., Wang, J., Liu, M.: Degraded historical document image binarization using local features and support vector machine (SVM). Optik **164**, 218–223 (2018)
18. Bhunia, A.K., Bhunia, A.K., Sain, A., Roy, P.P.: Improving document binarization via adversarial noise-texture augmentation. In: IEEE International Conference on Image Processing (ICIP) 2019, pp. 2721–2725 (2019)
19. Neji, H., Halima, M.B., Hamdani, T.M., Nogueras-Iso, J., Alimi, A.M.: Blur2Sharp: a GAN-based model for document image deblurring. Int. J. Comput. Intell. Syst. **14**, 1315–1321 (2021). https://doi.org/10.2991/ijcis.d.210407.001
20. Kingma, D.P., Welling, M.: An introduction to variational autoencoders. Foundations Trends® Mach. Learn. **12**(4), 307–392 (2019). https://doi.org/10.15612F2200000056
21. Goodfellow, I., et al.: Generative adversarial networks. Commun. ACM **63**(11), 139–144 (2020)
22. Karras, T., Aila, T., Laine, S., Lehtinen, J.: Progressive growing of GANs for improved quality, stability, and variation (2017). https://arxiv.org/abs/1710.10196
23. Mao, X., Shen, C., Yang, Y.-B.: Image restoration using very deep convolutional encoder-decoder networks with symmetric skip connections. In: Lee, D., Sugiyama, M., Luxburg, U., Guyon, I., Garnett, R. (eds.) Advances in Neural Information Processing Systems, vol. 29. Curran Associates Inc. (2016). https://proceedings.neurips.cc/paper/2016/file/0ed9422357395a0d4879191c66f4faa2-Paper.pdf
24. Dong, C., Loy, C.C., He, K., Tang, X.: Image super-resolution using deep convolutional networks (2015). https://arxiv.org/abs/1501.00092
25. Isola, P., Zhu, J.-Y., Zhou, T., Efros, A.A.: Image-to-image translation with conditional adversarial networks (2016). https://arxiv.org/abs/1611.07004
26. Yu, J., Lin, Z., Yang, J., Shen, X., Lu, X., Huang, T.S.: Generative image inpainting with contextual attention (2018). https://arxiv.org/abs/1801.07892
27. Zhao, J., Shi, C., Jia, F., Wang, Y., Xiao, B.: Document image binarization with cascaded generators of conditional generative adversarial networks. Pattern Recogn. **96**, 106968 (2019). https://www.sciencedirect.com/science/article/pii/S0031320319302717
28. Ho, J., Jain, A., Abbeel, P.: Denoising diffusion probabilistic models. In: Larochelle, H., Ranzato, M., Hadsell, R., Balcan, M., Lin, H. (eds.) Advances in Neural Information Processing Systems, vol. 33, pp. 6840–6851. Curran Associates Inc. (2020). https://proceedings.neurips.cc/paper/2020/file/4c5bcfec8584af0d967f1ab10179ca4b-Paper.pdf
29. Dhariwal, P., Nichol, A.: Diffusion models beat GANs on image synthesis (2021). https://arxiv.org/abs/2105.05233
30. Karras, T., Aittala, M., Aila, T., Laine, S.: Elucidating the design space of diffusion-based generative models (2022). https://arxiv.org/abs/2206.00364
31. Saharia, C., et al.: Photorealistic text-to-image diffusion models with deep language understanding (2022). https://arxiv.org/abs/2205.11487
32. Ramesh, A., Dhariwal, P., Nichol, A., Chu, C., Chen, M.: Hierarchical text-conditional image generation with CLIP Latents (2022). https://arxiv.org/abs/2204.06125

33. Kawar, B., Elad, M., Ermon, S., Song, J.: Denoising diffusion restoration models (2022). https://arxiv.org/abs/2201.11793
34. Saharia, C., Ho, J., Chan, W., Salimans, T., Fleet, D.J., Norouzi, M.: Image super-resolution via iterative refinement. IEEE Trans. Pattern Anal. Mach. Intell., 1–14 (2022)
35. Whang, J., Delbracio, M., Talebi, H., Saharia, C., Dimakis, A.G., Milanfar, P.: Deblurring via stochastic refinement. In: 2022 IEEE/CVF Conference on Computer Vision and Pattern Recognition (CVPR), pp. 16 272–16 282 (2022)
36. Kawar, B., Song, J., Ermon, S., Elad, M.: Jpeg artifact correction using denoising diffusion restoration models (2022). https://arxiv.org/abs/2209.11888
37. Gatos, B., Ntirogiannis, K., Pratikakis, I.: DIBCO 2009: document image binarization contest. IJDAR **14**, 35–44 (2011)
38. Pratikakis, I., Gatos, B., Ntirogiannis, K.: ICDAR 2011 document image binarization contest (DIBCO 2011). In: International Conference on Document Analysis and Recognition 2011, pp. 1506–1510 (2011)
39. Pratikakis, I., Gatos, B., Ntirogiannis, K.: ICFHR 2012 competition on handwritten document image binarization (H-DIBCO 2012). In: International Conference on Frontiers in Handwriting Recognition 2012, pp. 817–822 (2012)
40. Ntirogiannis, K., Gatos, B., Pratikakis, I.: ICFHR2014 competition on handwritten document image binarization (H-DIBCO 2014). In: 2014 14th International Conference on Frontiers in Handwriting Recognition, pp. 809–813 (2014)
41. Pratikakis, I., Zagoris, K., Barlas, G., Gatos, B.: ICFHR2016 handwritten document image binarization contest (H-DIBCO 2016). In: 2016 15th International Conference on Frontiers in Handwriting Recognition (ICFHR), pp. 619–623 (2016)
42. Pratikakis, I., Zagoris, K., Barlas, G., Gatos, B.: ICDAR2017 competition on document image binarization (DIBCO 2017). In: 2017 14th IAPR International Conference on Document Analysis and Recognition (ICDAR), vol. 01, pp. 1395–1403 (2017)
43. Pratikakis, I., Zagori, K., Kaddas, P., Gatos, B.: ICFHR 2018 competition on handwritten document image binarization (H-DIBCO 2018). In: 2018 16th International Conference on Frontiers in Handwriting Recognition (ICFHR), pp. 489–493 (2018)
44. Sauvola, J., Pietikäinen, M.: Adaptive document image binarization. Pattern Recogn. **33**(2), 225–236 (2000). https://www.sciencedirect.com/science/article/pii/S0031320399000552
45. Souibgui, M.A.: DocEnTr: an end-to-end document image enhancement transformer. In: 2022 26th International Conference on Pattern Recognition (ICPR) (2022)
46. Lin, Y.-S., Ju, R.-Y., Chen, C.-C., Lin, T.-Y., Chiang, J.-S.: Three-stage binarization of color document images based on discrete wavelet transform and generative adversarial networks (2022). https://arxiv.org/abs/2211.16098
47. Niblack, W.: An Introduction to Digital Image Processing. Strandberg Publishing Company, DNK (1985)
48. Ntirogiannis, K., Gatos, B., Pratikakis, I.: A combined approach for the binarization of handwritten document images. Pattern Recogn. Lett. **35**, 3–15 (2014). Frontiers in Handwriting Processing. https://www.sciencedirect.com/science/article/pii/S016786551200311X
49. Pinto, T., Rebelo, A., Giraldi, G.A., Cardoso, J.S.: Music score binarization based on domain knowledge. In: Iberian Conference on Pattern Recognition and Image Analysis (2011)

50. Ahmadi, E., Azimifar, Z., Shams, M., Famouri, M., Shafiee, M.J.: Document image binarization using a discriminative structural classifier. Pattern Recogn. Lett. **63**(C), 36–42 (2015). https://doi.org/10.1016/j.patrec.2015.06.008
51. Bhowmik, S., Sarkar, R., Das, B., Doermann, D.S.: GIB: a game theory inspired binarization technique for degraded document images. IEEE Trans. Image Process. (2019)
52. Tensmeyer, C., Martinez, T.: Document image binarization with fully convolutional neural networks (2017). https://arxiv.org/abs/1708.03276
53. Akbari, Y., Al-Maadeed, S., Adam, K.: Binarization of degraded document images using convolutional neural networks and wavelet-based multichannel images. IEEE Access **8**, 153 517–153 534 (2020)
54. Lore, K.G., Akintayo, A., Sarkar, S.: LLNet: a deep autoencoder approach to natural low-light image enhancement (2015). https://arxiv.org/abs/1511.03995
55. Calvo-Zaragoza, J., Gallego, A.-J.: A selectional auto-encoder approach for document image binarization. Pattern Recogn. **86**, 37–47 (2019). https://www.sciencedirect.com/science/article/pii/S0031320318303091
56. Pastor-Pellicer, J., Boquera, S.E., Zamora-Martínez, F., Afzal, M.Z., Bleda, M.J.C.: Insights on the use of convolutional neural networks for document image binarization. In: International Work-Conference on Artificial and Natural Neural Networks (2015)
57. Castellanos, F.J., Gallego, A.-J., Calvo-Zaragoza, J.: Unsupervised neural domain adaptation for document image binarization. Pattern Recogn. **119**, 108099 (2021)
58. Jemni, S.K., Souibgui, M.A., Kessentini, Y., Fornés, A.: Enhance to read better: a multi-task adversarial network for handwritten document image enhancement. Pattern Recogn. **123**, 108370 (2022). https://doi.org/10.1016%2Fj.patcog.2021.108370
59. Dosovitskiy, A.: An image is worth 16x16 words: transformers for image recognition at scale (2020). https://arxiv.org/abs/2010.11929
60. Bansal, A., et al.: Cold diffusion: inverting arbitrary image transforms without noise (2022). https://arxiv.org/abs/2208.09392
61. Ronneberger, O., Fischer, P., Brox, T.: U-net: convolutional networks for biomedical image segmentation (2015). https://arxiv.org/abs/1505.04597
62. Vaswani, A.: Attention is all you need (2017). https://arxiv.org/abs/1706.03762
63. Liu, Z., Mao, H., Wu, C.-Y., Feichtenhofer, C., Darrell, T., Xie, S.: A convnet for the 2020s (2022). https://arxiv.org/abs/2201.03545
64. Song, J., Meng, C., Ermon, S.: Denoising diffusion implicit models (2020). https://arxiv.org/abs/2010.02502
65. Suryani, M., Paulus, E., Hadi, S., Darsa, U.A., Burie, J.-C.: The handwritten Sundanese palm leaf manuscript dataset from 15th century. In: 2017 14th IAPR International Conference on Document Analysis and Recognition (ICDAR), vol. 01, pp. 796–800 (2017)
66. Nichol, A.Q., Dhariwal, P.: Improved denoising diffusion probabilistic models. In: Meila, M., Zhang, T. (eds.) Proceedings of the 38th International Conference on Machine Learning, ser. Proceedings of Machine Learning Research, vol. 139. PMLR, 18–24 July 2021, pp. 8162–8171 (2021). https://proceedings.mlr.press/v139/nichol21a.html
67. Lu, S., Su, B., Tan, C.L.: Document image binarization using background estimation and stroke edges. Int. J. Doc. Anal. Recogn. (IJDAR) **13**(4), 303–314 (2010)
68. Su, B., Lu, S., Tan, C.L.: Robust document image binarization technique for degraded document images. IEEE Trans. Image Process. **22**(4), 1408–1417 (2013)

69. Vo, Q.N., Kim, S., Yang, H.-J., Lee, G.: Binarization of degraded document images based on hierarchical deep supervised network. Pattern Recognit. **74**, 568–586 (2018)
70. He, S., Schomaker, L.: DeepOtsu: document enhancement and binarization using iterative deep learning. Pattern Recogn. **91**, 379–390 (2019). https://doi.org/10. 1016%2Fj.patcog.2019.01.025
71. Suh, S., Kim, J., Lukowicz, P., Lee, Y.O.: Two-stage generative adversarial networks for document image binarization with color noise and background removal (2020). https://arxiv.org/abs/2010.10103
72. Xiong, W., Zhou, L., Yue, L., Li, L., Wang, S.: An enhanced binarization framework for degraded historical document images. J. Image Video Process. **2021**(1) (2021). https://doi.org/10.1186/s13640-021-00556-4
73. Kong, Z., Ping, W.: On fast sampling of diffusion probabilistic models (2021)

You Only Look for a Symbol Once: An Object Detector for Symbols and Regions in Documents

William A. P. Smith[(✉)] and Toby Pillatt

University of York, York, UK
{william.smith,toby.pillatt}@york.ac.uk

Abstract. We present YOLSO, a single stage object detector specialised for the detection of fixed size, non-uniform (e.g. hand-drawn or stamped) symbols in maps and other historical documents. Like YOLO, a single convolutional neural network predicts class probabilities and bounding boxes over a grid that exploits context surrounding an object of interest. However, our specialised approach differs from YOLO in several ways. We can assume symbols of a fixed scale and so need only predict bounding box centres, not dimensions. We can design the grid size and receptive field of a grid cell to be appropriate to the known scale of the symbols. Since maps have no meaningful boundary, we use a fully convolutional architecture applicable to any resolution and avoid introducing unwanted boundary dependency by using no padding. We extend the method to also perform coarse segmentation of regions indicated by symbols using the same single architecture. We evaluate our approach on the task of detecting symbols denoting free-standing trees and wooded regions in first edition Ordnance Survey maps and make the corresponding dataset as well as our implementation publicly available.

Keywords: non-uniform symbol detection · object detection · map digitisation

1 Introduction

Historical maps and documents often contain an array of symbols and markings. Recognising and localising these symbols is useful for many reasons, including digitisation of historic documents and georeferencing of map contents. However, while the symbols may be (part-)standardised, to modern eyes (and computer vision) they are non-uniform, being either hand-drawn or reproduced inconsistently. For example, although draftsmen and printers often used stamps to speed up reproduction and adhere to drawing conventions, it was not uncommon for different workers to use different stamps for the same symbol, introducing both subtle and stark variations between maps and documents. Even for the same draftsman using a single stamp, small disparities in the way the stamp was applied or in the other processes of printmaking—during etching, engraving or inking, for example—could result in subtle differences in symbol appearance.

© The Author(s), under exclusive license to Springer Nature Switzerland AG 2023
G. A. Fink et al. (Eds.): ICDAR 2023, LNCS 14191, pp. 227–243, 2023.
https://doi.org/10.1007/978-3-031-41734-4_14

This could be the case even within a single map sheet or page. A more contemporary problem is that the digitisation of an original printed map or document can introduce distortions and corruptions that result in further variability, increasing difficulty of automated detection and all but ruling out simple template matching. Furthermore, unlike handwriting recognition, where the symbols are limited to alphanumeric characters and where words and sentences provide strong context to help disambiguate characters, more general symbols can be much more varied, may have far fewer examples per class, have less constraining local context and can partially overlap with each other.

Object detection is a widely researched topic with many high performing methods based on CNNs [9, 26] or, very recently, transformer-based architectures [34]. However, we argue that the task of non-uniform symbol detection is subtly different to general object recognition in photos and that significant performance gains can be achieved by specifically tuning an architecture to exploit these differences. Concretely, the main differences are:

1. **Scale**: Symbols are drawn at a fixed scale so we do not need to predict bounding box dimensions. Instead, it is enough to predict the box centre and symbol class. We can then use a fixed size bounding box for each class.
2. **Regions**: Some symbols indicate that a bounded area forms a region of a particular class (e.g. woodland or orchard), rather than indicating a precise spatial location with particular meaning. This means that the symbol detection problem also encompasses a partial semantic segmentation problem.
3. **Context**: Local context is important. For example, in the tree detection task in our experiments, useful local context includes knowing that trees cannot appear on a road or inside a building or that trees commonly grow at the boundaries of fields. For region segmentation, sometimes a segmentation decision must be based only on local context when a grid cell contains no symbols indicating it is inside a region. However, beyond a certain locality, context becomes unimportant so we do not want to consider global context.
4. **Boundaries and absolute position**: The absolute position of a grid cell conveys no information about whether that cell contains a particular symbol. Therefore we do not want our network to learn spatial reasoning, for example from distance to boundary.
5. **Large scale inference**: Map and document datasets can easily scale to extreme size. For example, at the resolution we work at in our experiments, a national scale map would be of terrapixel size. Therefore, efficient methods are required that can process very large volumes of data.
6. **Symbol sparsity**: Since symbol occurrence is relatively sparse within a map, hard negative mining and weighting is essential for good performance.

In this paper, we propose a method that we call 'You Only Look for a Symbol Once' (YOLSO) that takes inspiration from the single-shot detector methods that build upon YOLO [26]. It is specifically adapted to exploit the properties and tackle the challenges of the symbol detection problem given above. We apply our method to a new dataset of first edition Ordnance Survey maps for the task of detecting symbols denoting free-standing trees and wooded regions. We show that the specialisation of YOLSO to this task leads to significant performance improvements compared to applying generic object detection pipelines to the task of map symbol detection.

2 Related Work

Object Detection. Object detection has benefited significantly from developments in deep learning and we only mention learning-based methods here. Early attempts to apply deep neural networks to the problem of object detection separated the task into two phases: object proposal and object recognition. Region-based CNN (R-CNN) [9] used Selective Search [31] to propose regions of interest (ROIs). Each ROI is then cropped and nonuniformly resized to a fixed size for input to a feature extraction CNN. These features are then passed to a support vector machine for classification as either background or one of a set of possible object classes. Since there may be thousands of object proposals per image, this method is extremely slow. Fast R-CNN [8] partially addressed this by passing the input image through the feature extraction network only once and then cropping and resizing the ROI features from the feature volume. Faster R-CNN [29] further improved speed and unified the object proposal and recognition stages by using the same network for both ROI generation and feature extraction. Since all of these methods pass cropped input to the object classifier, they cannot exploit wider context within the image. Classifying each region independently also remains a performance bottleneck.

"You Only Look Once" (YOLO) [26] was the first of what are now called *single shot detectors*. With a single forward pass through a CNN, these methods tackle the whole object detection problem in one go. The underlying idea is to discretise the output space of possible bounding boxes to one or more grids, each cell of which can make a fixed number of bounding box predictions. This eliminates the need for object proposals entirely, shares extraction of overlapping features between grid cells for efficiency and can exploit context according to the receptive field of the output grid cell. YOLO used a network that included fully connected layers. This enables it to exploit global context but this is unhelpful for symbol detection where useful context is relatively local to a symbol. The Single Shot Detector (SSD) [21] used multiscale grids for detection of objects at different sizes, used anchor boxes so that each scale has a notion of default box shapes and sizes, and replaced the fully connected layers with a fully convolutional architecture. YOLOv2 [27] and YOLOv3 [28] also switched to a fully convolutional architecture and incorporated many minor improvements that increased performance while retaining the same underlying approach. All of these approaches use architectures such that the receptive field of each grid cell is very large (potentially larger than the image) and therefore includes a lot of content outside the image which must be filled with padding. This introduces a boundary dependency such that the networks can exploit positional context (for example, learning that a face often occurs near the centre of the image). This is unhelpful for symbol detection in documents without meaningful boundaries. In addition, the very large receptive fields make recognition and precise localisation of small symbols very challenging. Our approach is inspired by these single shot techniques but is specialised to the problem of symbol detection.

Non-uniform Symbol Detection. Some of the earliest and best known work in CNN-based computer vision, e.g. LeCun et al. [19] tackled the problem of handwritten character *recognition* (i.e. where the symbol has already been located and cropped from the document). There have been relatively few attempts to tackle the specific problem of non-uniform symbol detection within documents and the majority of these applied existing methods without adaptation. Early work was based on classical template matching combined with multiple classifiers and explicit handling of rotation invariance [16]. Julca-Aguilar and Hirata [15] address the symbol detection problem in the context of online handwriting recognition using Faster R-CNN. Elyan et al. [6] apply YOLO to the problem of symbol detection in engineering drawings. In both cases, they use fixed sized small crops from the overall document such that the symbols are relatively large within the crop. Adorno et al. [1] take a different approach, treating the problem as one of dense semantic segmentation and post-processing to extract symbol bounding boxes. Recently, attention-based architectures have been applied to symbol detection, for example for recognising mathematical expressions [20].

Historic Maps. Historic maps are important sources for reconstructing social and landscape history. As collections have been digitised access has been significantly improved, but the conversion of scanned map rasters to labelled vector data remains a laborious task when performed manually. Early attempts at map vectorisation used image processing techniques [2,4] but more recently deep learning has been applied to both object detection and segmentation tasks. For example, Maxwell et al. [23] use a UNet architecture for semantic segmentation to identify a single feature class - historic mine extents - from a standardised series of polychrome topographic maps, while Petitpierre [25] experiments with using UNet architecture with a ResNet encoder for multiclass segmentation across whole corpuses of cadastral maps produced to different styles and conventions, and Garcia-Molsosa et al. [7] use a variety of detectors built on the UNet-based Picterra[1] platform to identify and segment the different, non-uniform symbols used to denote archaeological sites on multiple early twentieth-century colonial map series. Perhaps ostensibly most similar to the approach we describe is Groom et al.'s [10] use of CNN symbol detection to classify segments produced using colour-based image object analyses. However, unlike our method, this is predicated on polychrome maps and is not reliant on particularly high object detection accuracy.

Across the above examples, it is notable that while computer vision has developed markedly in recent years, the richness of map data, overlapping features, and imprecise georeferencing have continued to complicate analyses. Moreover, a particular limitation of deep learning models is the significant time and resource

---

[1] https://picterra.ch/.

investment required to train them. Consequently, an ongoing strand of research, pioneered by the 'Linked Maps' project[2] [5], is on ways of creating training data from more recent mapping, sometimes by directly applying modern labelling to corresponding regions on old maps [30], or sometimes with intermediary steps, such as by using Generative Adversarial Networks to produce synthetic maps images that imitate historical map styles but preserve spatial patterning and labelling from modern maps [33]. While it is conceivable that some applications of YOLSO would benefit from such an approach, for our example the Ordnance Survey stopped mapping individual trees in the early twentieth century meaning there is no direct analogue that can be used to generate training data.

The need to invest time and resources in training means that deep learning tasks tend to be targeted towards big datasets, where any efficiencies to gained from such an approach are realised. When working with maps, this generally means working across multiple mapsheets encompassing large areas. The Living With Machines[3] project recently developed a series of useful tools for working with very large collections of historical Ordnance Survey maps [11,12]. Instead of a pixel-level segmentation model, they train a binary classification model using labelled 'patches' that can be of any size, enabling them to explore urbanisation and railway expansion across the whole of nineteeth-century Great Britain. However, the resolution of the analysis is determined by the patch size, and multiple iterations of labelling, training and analysis are required if one needs to drill down to examine groups of features at different resolutions. Conversely, an advantage of our method is that you only look for a symbol once.

Tree Mapping. Detailed and accurate data concerning the location of trees are an important prerequisite of tree risk management, and form the basis of economic and ecosystem service-based valuations [22]. A significant area of research is in using airborne remote sensing, including through LiDAR, photogrammetry and multispectral image analysis, to map urban and managed forest environments. These can be combined with computer vision analyses of street-level imagery, such as from Google Street View, to produce detailed tree maps [3,18]. The fact that many of these types of dataset are now available spanning multiple epochs opens up the prospect of studies of treescape change over time. However, where trees were in the more distant past is understood only in very broad terms. This is in spite of the fact that trees were often routinely plotted on old maps and plans, reflecting their value as sources of food, fuel and materials [32]. Identifying and geolocating trees on historical OS maps will present a baseline from which to explore changes to Britain's treed environment throughout the 20th century.

3   Method

The YOLSO architecture is a CNN that outputs a grid, in which each cell contains the symbol detection result for a corresponding region in the input image

---

[2] https://usc-isi-i2.github.io/linked-maps/.
[3] https://livingwithmachines.ac.uk/.

(see Fig. 1). The network is fully convolutional, meaning that we do not need to choose and fix the dimensions or aspect ratio of the input image and the output grid dimension is determined by the input image dimensions. The entire symbol detection result is computed with a single forward pass through a small CNN, meaning it is computationally efficient and therefore applicable to very large datasets such as national-scale maps.

3.1 Architecture

The network is formed of blocks comprising: a convolution layer with no padding, ReLU activation, batch normalisation [13] and max pooling. The input image size, grid cell resolution, layer hyperparameters and number of blocks must be carefully chosen to ensure appropriately sized output.

For an input of spatial dimension $H \times W$, the output is a grid of spatial dimension $S_H \times S_W$ where $S_H = (H - 2P)/R$ and $S_W = (W - 2P)/R$. Each output grid cell contains the symbol detection result for the corresponding $R \times R$ pixel region in the input image. This means that R must be a factor of both $H - 2P$ and $W - 2P$. The P pixel boundary is required to ensure all grid cells can benefit from local context and to avoid using padding in the convolutional layers which would introduce unwanted boundary dependency (see below).

We use max pooling layers with 2×2 kernel for downsampling. Hence, to reduce an $R \times R$ region to a 1×1 grid cell, our network must contain $p = \log_2 R$ pooling layers and hence the grid cell resolution must be a power of 2, $R = 2^p$.

3.2 Padding and Receptive Field

We use no padding in our convolutional layers for two reasons. First, padding provides a network with a means to learn boundary dependence [14,17]. Since we do not expect symbols to be any more likely to appear in any particular location within a random map crop, positional dependence is undesirable. Second, padding introduces incorrect local context. Zero or replication padding introduces featureless regions. Reflection padding introduces reversed symbols that will not occur in real data.

Using no padding means that each convolution layer reduces the spatial size of the feature map. To compensate for this, we include a P pixel boundary in the input image. Each output grid cell has a receptive field in the original image of size $2P + R \times 2P + R$ (see Fig. 1, red dashed square) and this boundary region ensures that the receptive field of all output grid cells lies within the image.

The appropriate value for P depends on the parameters of the convolutional layers and the number of pooling layers. Suppose that the parameters of the convolution layers are stored in the set $\mathcal{L}$. Each layer comprises the pair (k, d) where $d \in \{0, \ldots, p\}$ is the depth, i.e. the number of pooling layers that precede it, and k is the kernel size. The required padding can be computed as:

$$P = \sum_{(k,d)\in\mathcal{L}} \frac{k-1}{2} \cdot 2^d. \tag{1}$$

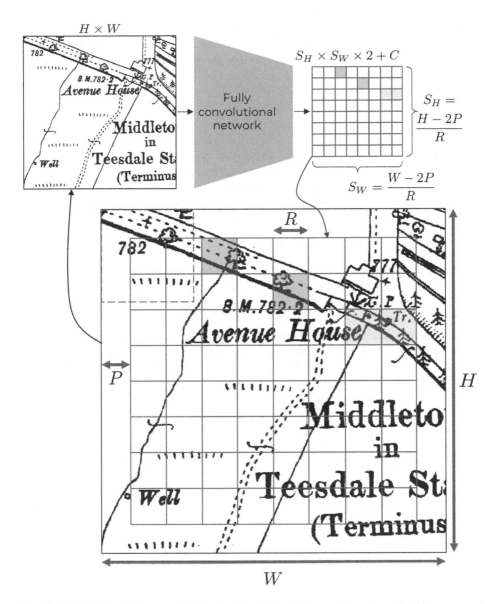

Fig. 1. YOLSO is a fully convolutional network that takes as input an $H \times W$ image and outputs a $(H-2P)/R \times (W-2P)/R$ grid. Each cell in the output grid indicates whether the corresponding $R \times R$ region in the input image contains a symbol (class indicated by coloured fill) and, if so, the offset of the bounding box centre. The convolution layers use no padding and, hence, the receptive field of each grid cell (shown in red for the top left cell) is contained entirely within the original input image. To enable this, the input image must include a border of width P which ensures all output grid cells benefit from local context and no boundary dependency is introduced. (Color figure online)

At training time, it is important that the boundary region contains typical content, such that the local context around a grid cell is meaningful. For this reason, we use $(H - 2P)/R \times (W - 2P)/R$ crops from the training images. At inference time, in order to detect symbols for the whole of any input image of arbitrary size, we pad images by P on the left and top, $P + T_W$ on the right and $P + T_H$ on the bottom where $T_W = R(1 - W/R + \lfloor W/R \rfloor)$ and $T_H = R(1 - H/R + \lfloor H/R \rfloor)$ ensure that the grid covers all of the input image.

For our task of symbol detection, it is important that the local context (and hence receptive field of the output grid cells) is not too large. While local context is helpful, beyond some modest distance, map content becomes unrelated to the symbol within the grid cell and unhelpful to the detection process. This means that the depth of the network must be kept relatively modest. We find that a very small network is adequate for the symbol detection task on our dataset.

3.3   Bounding Box Regression

We can choose the grid cell resolution R appropriately for our data to ensure that only one symbol will lie within any grid cell. This means that we only need to detect a single bounding box per grid cell, simplifying our output and loss functions (contrast this with YOLO [26] which uses relatively large grid cells and must allow multiple detections per cell). In order to regress bounding box coordinates and class labels, the output tensor (including channel dimension) is of size $S_H \times S_W \times 2 + C$, where C is the number of classes (including one for background) and hence logits that we output per cell. We supervise the classification output with cross entropy loss (see below).

The additional two outputs predict bounding box centre offsets from the centre of the grid cell. We apply sigmoid activation to these spatial outputs such that they represent normalised coordinates within a grid cell. We compute an L1 loss over predicted bounding box centre offsets for those cells that contain a symbol:

$$\mathcal{L}_{\text{coord}} = \frac{1}{\sum_{i=1}^{S_W} \sum_{j=1}^{S_H} \mathbb{1}_{ij}^{\text{sym}}} \sum_{i=1}^{S_W} \sum_{j=1}^{S_H} \mathbb{1}_{ij}^{\text{sym}} |x_{i,j} - \hat{x}_{i,j}|, \tag{2}$$

where $\mathbb{1}_{ij}^{\text{sym}} \in \{0,1\}$ indicates whether grid cell (i,j) contains a symbol, $x_{i,j}$ is the ground truth bounding box centre offset and $\hat{x}_{i,j}$ the estimated one.

3.4   Region Segmentation

Our approach and network can naturally be extended to perform coarse semantic segmentation of regions simultaneously with detection of individual symbols. In the context of maps, regions are usually represented by a number of symbols (for example indicating woodland of a particular type) enclosed by symbols indicating boundaries such as walls, hedges and fences. Here, context is crucially important as the input region corresponding to a grid cell might be empty but the cell belongs to a region because it is surrounded by symbols indicating that

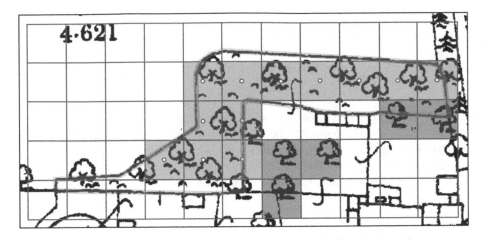

Fig. 2. Coarse semantic segmentation with the YOLSO architecture. A region annotation (shown in green) is converted to a semantic segmentation at grid resolution by applying a point-in-polygon test to the centre of each grid cell (shown as red dots for cells with centres inside the region polygon). Cells labelled as inside the region are shaded orange while cells containing single objects can still be detected as normal (shaded blue here). (Color figure online)

region. In addition, the network must be able to reason about which side of a boundary symbol a cell is.

We label segments by drawing arbitrarily shaped polygons. Then, for each grid cell, we perform a point-in-polygon test for the coordinate of the centre of the cell. For any cells with centres lying inside the polygon, we label that cell with the class of the segment. This requires no modification to training for object detection. The number of output classes C is simply increased to include both the symbol and region classes and we define $\mathbb{1}_{ij}^{\text{sym}} = 0$ if grid cell (i, j) contains a region class (i.e. we do not compute a coordinate loss for region cells).

3.5 Hard Negative Mining

In many datasets, including our map data, symbols and regions are relatively sparse. This means that the vast majority of grid cells are labelled as background. Naively training directly on this data leads to a very high false negative rate. For this reason, hard negative mining is essential for good performance. We do not ignore easy negatives entirely since this leads to misclassifying many background cells but instead downweight them as follows:

$$\mathcal{L}_{\text{class}} = \frac{1}{S_W S_H} \sum_{i=1}^{S_W} \sum_{j=1}^{S_H} -w_{ij} \log \frac{\exp z_{ijc_{ij}}}{\sum_{k=1}^{C} \exp z_{ijk}}, \tag{3}$$

where z_{ijk} is the kth logit for grid cell (i, j), c_{ij} is the ground truth class for grid cell (i, j) and w_{ij} downweights easy background cells as follows:

$$w_{ij} = \begin{cases} W & \text{if } c_{ij} = 1 \wedge z_{ij1} = \max_k z_{ijk} \\ 1 & \text{otherwise} \end{cases}. \tag{4}$$

An easy background cell is defined as one which has ground truth background label ($c_{ij} = 1$) and the logit for the background class is maximal over all logits for that cell ($z_{ij1} = \max_k z_{ijk}$), i.e. the network currently correctly labels it as background. We use a weight of $W = 0.1$ in all our experiments. In general, this weight should be set similarly to the proportion of non background symbols in the dataset.

3.6   Implementation

The symbols in our dataset originally had bounding boxes of size 48×48. We choose the next smallest power of two, $R = 32$, as our grid cell resolution. Since no bounding box centres occur closer then 32 pixels, this ensures that our assumption of only one object detection per grid cell is valid.

We use five blocks of convolution/batchnorm/ReLU/max pooling. The first convolution layer has kernel size 7×7, the following four have size 3×3 and the number of output channels per convolution layer are: 64, 64, 128, 256, 512. The final part of the network reasons at grid cell scale in order to compute final detection output. We apply a convolution/batchnorm/ReLU with kernel size 3×3 and 512 output channels in order to share information between adjacent grid cells before a final 1×1 convolution layer with $C + 2$ output channels. This architecture requires a boundary of width $P = 65$ pixels on the input images. This means that the receptive field of each output grid cell (corresponding to a 32×32 region in the input images) is 162×162. We find that this provides sufficient context for both symbol detection and region segmentation.

4   Experiments

4.1   Dataset

Our work focuses on the Ordnance Survey First Edition 1:2500 County Series, produced in the latter half of the nineteenth century and covering most of Great Britain. Only uncultivated areas in highest uplands and lowest lowlands were excluded [24]. These historic maps are unusual in that they attempt to show all freestanding trees. More modern maps tend to dispense with the gargantuan task of surveying so many trees, limiting themselves to only showing groups of trees and woodlands. Our objective is to develop a method of quickly detecting and georeferencing both freestanding trees and areas of trees on the historic maps, thereby creating new digital 'National Historic Tree Map', which can be used a baseline for researching landscape-scale environmental change throughout the

pxpx

Fig. 3. A range of different symbols are used to denote freestanding broadleaved trees.

twentieth century and beyond. However, this is not a straightforward task. The symbols denoting trees are not consistent across the whole First Edition County Series (Fig. 3), they can vary as a result of the mapmaking and digitisation processes, and they can be associated with other symbols and features that make them hard to detect or segment (Fig. 4).

Historic Ordnance Survey maps are available from a number of third party data providers, libraries and archives. We elected to use digitised versions of the historic maps produced by the Landmark Information Group and made available through the online Edina Digimap platform[4]. One advantage of using this dataset is that the original mapsheets, which were surveyed using the Cassini Projection, have been re-projected to the modern National Grid and stitched together. The maps are thus served as individual 1 km^2 tiles in GeoTIFF format, scanned at two resolutions such that they are either 4724 or 6000px square. For our dataset, we randomly selected 135 individual map tiles from the within the historic county of Yorkshire (encompassing the North, West and East Ridings). This comprises approximately 1% of the total land area of the historic county. Each tile was manually labelled using GIS software, and point coordinates denoting individual trees and polygons marking out groups of trees were stored in shapefiles relating to each map tile.

782 regions (wooded areas) were labelled manually by defining a polygon and assigning one of four classes (deciduous woodland, coniferous woodland, mixed woodland, orchard). Point locations for individual tree symbols were labelled semi-automatically. 23,807 approximate locations were labelled manually. These were refined or discarded and one of two classes (broadleaf and conifer) chosen by applying template matching within a 96px square centred on the manually plotted point. The templates were of size 48px and, within each target area, the location of the highest template match was used to derive a pixel coordinate centre point for that tree symbol. Poor matches were excluded, such as those not reaching a certain confidence threshold and those that would purport to

---

[4] 1:2500 County Series 1st Edition [TIFF geospatial data], Scale 1:2500, Updated: 30 November 2010, Historic, Using: EDINA Historic Digimap Service, https://digimap. edina.ac.uk. Downloaded: 2015–2022. 1' Crown Copyright and Landmark Information Group Limited 2023. All rights reserved. 1890–1893.

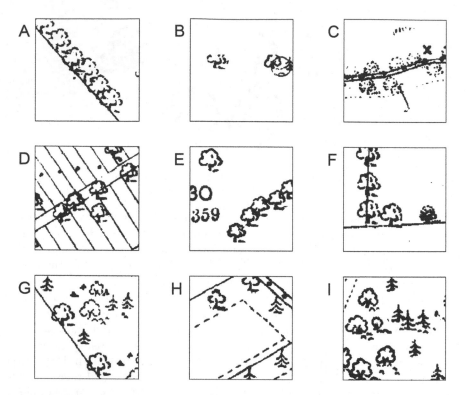

Fig. 4. Characteristics of the map images that make tree symbols difficult to detect and segment: **(A)** Touching or overlapping tree symbols. **(B)** Symbols partially truncated at the edges of map tiles. **(C)** Poorly reproduced symbols. **(D)** Symbols partially occluded by other map features. **(E)** Ostensibly similar symbols of different sizes. **(F)** Different symbols for the same class of tree within a single map tile - this is a result of different map sheets from different areas, sometimes produced at different times, being stitched together to produce a single contiguous map. **(G)** Individual tree symbols shown in close proximity to or overlapping with areas of woodland. **(H)** Areas where it is unclear whether individual trees or groups of trees are being depicted. **(I)** Areas of woodland that change abruptly from one class to another (here from deciduous broadleaved to coniferous) with no intervening boundary.

match with symbols that exceed the bounds of the 96px square target area. As a result, the final labelled dataset comprised 22,882 point features (96.1% of those initially manually labelled).

4.2   Results

In Fig. 5 we show some qualitative results from our method. We choose two test regions which illustrate performance mainly on point symbols (top row) and on region segmentation (bottom row). In the top row, our approach is able to accurately label point symbols even with heavy distractors from overlaid boundary

Estimated   Actual

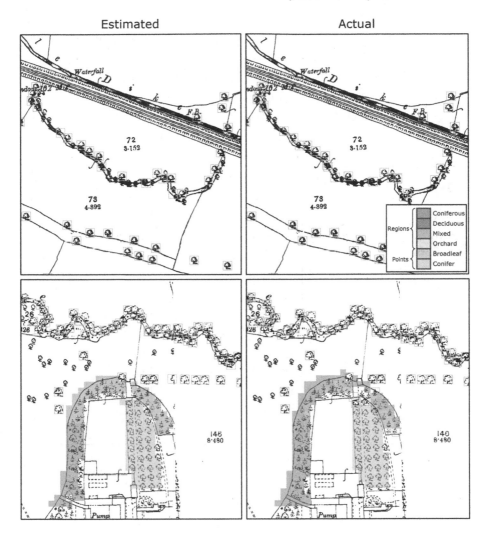

Fig. 5. Qualitative object detection and region segmentation results. On the first row we show a crop in which many point labelled individual trees are close together and obscured by boundary markings. In the bottom row we show a mix of point and region segmentations in which two different region types are adjacent.

symbols. In the bottom row, our approach accurately recovers the shape and changing class of adjacent regions and correctly ignores symbols that do not indicate freestanding trees outside of the regions.

We compare our approach against YOLOv3 [28] and Faster R-CNN [29]. During training, we use different strategies for the different methods for cropping training images from the 4k × 4k images. For YOLSO, we train on a 16 × 16 grid, which, including the necessary boundary, means that we crop 642 × 642 images. The crops are random which has the same effect as translation augmentation -

Table 1. Quantitative object detection (middle column) and region segmentation (right) results. YOLOv3 [28] and Faster R-CNN [29] do not perform segmentation.

Method	Point mAP@0.5	Region mAP
YOLSO	97.9%	94.6%
YOLOv3 [28]	43.6%	n/a
Faster R-CNN [29]	70.2%	n/a

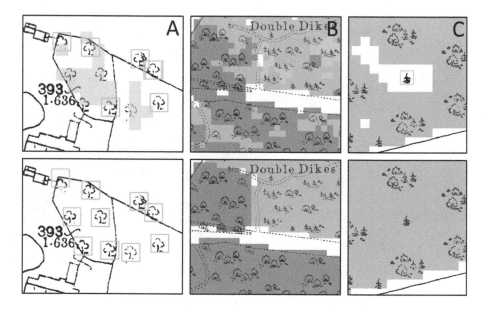

Fig. 6. Illustrative failure cases (top row: estimated, bottom row: ground truth). A: misinterpreting point symbols as regions. B: confusion between region classes. C: false negatives in empty regions where context is too distant.

i.e. every symbol may be observed at every possible position within a cell during training. While our approach can handle training crops with very few positive samples (due to our easy negative down-weighting strategy and the fact that regions provide additional positive samples) for YOLOv3 and Faster R-CNN we found that performance was improved by excluding crops that contained too few positive samples and also shifting the crop window such that partial symbols at the boundary were avoided. We use 512 × 512 crops for Faster R-CNN and 416 × 416 for YOLOv3. At test time, our method can process an entire tile in one go (using the padding strategy described in Sect. 3.2). Whereas for YOLOv3 and Faster R-CNN, we retain the image size used for training and process a tile by sliding a window over the full image.

We provide quantitative results in Table 1. For point symbols, we define a correct prediction as one with correct class and IOU>0.5. Our approach significantly outperforms the two standard methods. Note that both comparison methods often estimate bounding boxes of completely the wrong aspect ratio or scale since they do not incorporate the constraint of known size that our approach does. Both comparison methods suffer more from false negatives than false positives, likely due to the extreme symbol sparseness and small size of the symbols compared to the objects these methods are usually used to detect.

In Fig. 6 we illustrate the most common failure cases of our method. In A, symbols are interpreted as indicating a region due to the context provided by the boundary. However, despite the similarity, these symbols actually indicate freestanding trees and should be detected as points. In B, a mistake is made between the mixed and deciduous classes. This is challenging since the deciduous symbols are used in mixed regions with the context of other symbols (potentially quite distance) providing the disambiguation. In C, parts of a region are labelled as background due to lack of any symbol context in the surrounding region. Note also that a region symbol is misinterpreted as a freestanding conifer - again, the symbols are very similar.

5 Conclusions

We have proposed a variant of the YOLO object detection framework that is specialised for the detection of non-uniform symbols and regions depicted by symbols in documents such as historic maps. By accounting for the specific properties of the task, namely that symbol scale is fixed, local context is important but global context and spatial dependence should be ignored, we arrive at a model that is lightweight and efficient but significantly outperforms the application of existing generic object detection methods to the problem. In addition, our approach additionally computes coarse semantic segmentation using the same single network. We have shown that a relatively small network performs well on this well constrained task and perhaps even aids performance by avoiding the receptive field becoming too large. We believe that performance can be further improved by additional tuning of this trade off, particularly for featureless areas inside regions which require larger context. We did nothing to handle the imbalance between classes nor between region and point symbols and believe additional performance gains could be achieved here.

Acknowledgments. This research was conducted as part of the Future Of UK Treescapes project 'Branching Out: New routes to valuing urban treescapes', funded by UK Research and Innovation [Grant Number: NE/V020846/1].

References

1. Adorno, W., Yi, A., Durieux, M., Brown, D.: Hand-drawn symbol recognition of surgical flowsheet graphs with deep image segmentation. In: 2020 IEEE 20th International Conference on Bioinformatics and Bioengineering (BIBE), pp. 295–302. IEEE (2020)

2. Baily, B.: The extraction of digital vector data from historic land use maps of great britain using image processing techniques. E-perimetron **2**(4), 209–223 (2007)
3. Branson, S., Wegner, J.D., Hall, D., Lang, N., Schindler, K., Perona, P.: From google maps to a fine-grained catalog of street trees. ISPRS J. Photogramm. Remote. Sens. **135**, 13–30 (2018)
4. Budig, B.: Extracting spatial information from historical maps: algorithms and interaction. Würzburg University Press (2018)
5. Chiang, Y.-Y., Duan, W., Leyk, S., Uhl, J.H., Knoblock, C.A.: Using Historical Maps in Scientific Studies. SG, Springer, Cham (2020). https://doi.org/10.1007/978-3-319-66908-3
6. Elyan, E., Jamieson, L., Ali-Gombe, A.: Deep learning for symbols detection and classification in engineering drawings. Neural Netw. **129**, 91–102 (2020)
7. Garcia-Molsosa, A., Orengo, H.A., Lawrence, D., Philip, G., Hopper, K., Petrie, C.A.: Potential of deep learning segmentation for the extraction of archaeological features from historical map series. Archaeol. Prospect. **28**(2), 187–199 (2021)
8. Girshick, R.: Fast R-CNN. In: Proceedings of the IEEE International Conference On Computer Vision, pp. 1440–1448 (2015)
9. Girshick, R., Donahue, J., Darrell, T., Malik, J.: Rich feature hierarchies for accurate object detection and semantic segmentation. In: Proceedings of the IEEE Conference on Computer Vision and Pattern Recognition, pp. 580–587 (2014)
10. Groom, G.B., Levin, G., Svenningsen, S.R., Perner, M.L.: Historical maps machine learning helps us over the map vectorisation crux. In: Automatic Vectorisation of Historical Maps: International workshop organized by the ICA Commission on Cartographic Heritage into the Digital, pp. 89–98. Department of Cartography and Geoinformatics, ELTE Eötvös Loránd University (2020)
11. Hosseini, K., McDonough, K., van Strien, D., Vane, O., Wilson, D.C.: Maps of a nation? the digitized ordnance survey for new historical research. J. Vic. Cult. **26**(2), 284–299 (2021)
12. Hosseini, K., Wilson, D.C., Beelen, K., McDonough, K.: Mapreader: a computer vision pipeline for the semantic exploration of maps at scale. In: Proceedings of the 6th ACM SIGSPATIAL International Workshop on Geospatial Humanities, pp. 8–19 (2022)
13. Ioffe, S., Szegedy, C.: Batch normalization: Accelerating deep network training by reducing internal covariate shift. In: International conference on machine learning, pp. 448–456. PMLR (2015)
14. Islam, M.A., Jia, S., Bruce, N.D.: How much position information do convolutional neural networks encode? In: International Conference on Learning Representations (2019)
15. Julca-Aguilar, F.D., Hirata, N.S.: Symbol detection in online handwritten graphics using faster R-CNN. In: 2018 13th IAPR International Workshop on Document Analysis Systems (DAS), pp. 151–156. IEEE (2018)
16. Kara, L.B., Stahovich, T.F.: An image-based, trainable symbol recognizer for hand-drawn sketches. Comput. Graph. **29**(4), 501–517 (2005)
17. Kayhan, O.S., van Gemert, J.C.: On translation invariance in CNNs: Convolutional layers can exploit absolute spatial location. In: Proceedings of the IEEE/CVF Conference on Computer Vision and Pattern Recognition, pp. 14274–14285 (2020)
18. Laumer, D., Lang, N., van Doorn, N., Mac Aodha, O., Perona, P., Wegner, J.D.: Geocoding of trees from street addresses and street-level images. ISPRS J. Photogramm. Remote. Sens. **162**, 125–136 (2020)
19. LeCun, Y., Bottou, L., Bengio, Y., Haffner, P.: Gradient-based learning applied to document recognition. Proc. IEEE **86**(11), 2278–2324 (1998)

20. Li, Z., Jin, L., Lai, S., Zhu, Y.: Improving attention-based handwritten mathematical expression recognition with scale augmentation and drop attention. In: 2020 17th International Conference on Frontiers in Handwriting Recognition (ICFHR), pp. 175–180. IEEE (2020)

21. Liu, W., et al.: SSD: single shot multibox detector. In: Leibe, B., Matas, J., Sebe, N., Welling, M. (eds.) ECCV 2016. LNCS, vol. 9905, pp. 21–37. Springer, Cham (2016). https://doi.org/10.1007/978-3-319-46448-0_2

22. Ltd, B.I.: National tree map (Nov 2022). https://bluesky-world.com/ntm/

23. Maxwell, A.E.: Semantic segmentation deep learning for extracting surface mine extents from historic topographic maps. Remote Sensing **12**(24), 4145 (2020)

24. Oliver, R.: Ordnance Survey Maps: a concise guide for historians. Charles Close Society (1993)

25. Petitpierre, R.: Neural networks for semantic segmentation of historical city maps: Cross-cultural performance and the impact of figurative diversity. arXiv:abs/2101.12478

26. Redmon, J., Divvala, S., Girshick, R., Farhadi, A.: You only look once: Unified, real-time object detection. In: Proceedings of the IEEE Conference on Computer Vision and Pattern Recognition (CVPR), pp. 779–788 (2016)

27. Redmon, J., Farhadi, A.: YOLO9000: better, faster, stronger. In: Proceedings of the IEEE Conference on Computer Vision and Pattern Recognition, pp. 7263–7271 (2017)

28. Redmon, J., Farhadi, A.: YOLOv3: An incremental improvement. arXiv preprint arXiv:1804.02767 (2018)

29. Ren, S., He, K., Girshick, R., Sun, J.: Faster R-CNN: Towards real-time object detection with region proposal networks. In: Advances in Neural Information Processing Systems, vol. 28 (2015)

30. Uhl, J.H., Leyk, S., Chiang, Y.Y., Knoblock, C.A.: Towards the automated large-scale reconstruction of past road networks from historical maps. Comput. Environ. Urban Syst. **94**, 101794 (2022)

31. Uijlings, J.R., Van De Sande, K.E., Gevers, T., Smeulders, A.W.: Selective search for object recognition. Int. J. Comput. Vision **104**(2), 154–171 (2013)

32. Williamson, T., Barnes, G., Pillatt, T.: Trees in England: management and disease since 1600. University of Hertfordshire Press (2017)

33. Wong, C.S., Liao, H.M., Tsai, R.T.H., Chang, M.C.: Semi-supervised learning for topographic map analysis over time: a study of bridge segmentation. Sci. Rep. **12**(1), 18997 (2022)

34. Zhu, X., Su, W., Lu, L., Li, B., Wang, X., Dai, J.: Deformable DETR: Deformable transformers for end-to-end object detection. In: International Conference on Learning Representations (2020)

SAN: Structure-Aware Network for Complex and Long-Tailed Chinese Text Recognition

Junyi Zhang, Chang Liu, and Chun Yang[✉]

School of Computer and Communication Engineering,
University of Science and Technology Beijing, Beijing, China
lasercat@gmx.us, chunyang@ustb.edu.cn

Abstract. In text recognition, complex glyphs and tail classes have always been factors affecting model performance. Specifically for Chinese text recognition, the lack of shape-awareness can lead to confusion among close complex characters. Since such characters are often tail classes that appear less frequently in the training-set, making it harder for the model to capture its shape information. Hence in this work, we propose a structure-aware network utilizing the hierarchical composition information to improve the recognition performance of complex characters. Implementation-wise, we first propose an auxiliary radical branch and integrate it into the base recognition network as a regularization term, which distills hierarchical composition information into the feature extractor. A Tree-Similarity-based weighting mechanism is then proposed to further utilize the depth information in the hierarchical representation. Experiments demonstrate that the proposed approach can significantly improve the performances of complex characters and tail characters, yielding a better overall performance. Code is available at https://github.com/Levi-ZJY/SAN.

Keywords: Structure awareness · Text recognition · Radical · Tree Similarity

1 Introduction

Chinese text recognition plays an important role in the field of text recognition due to its huge audience and market. Most current text recognition methods, including Chinese text recognition methods, are character based, where characters are the basic elements of the prediction. Specifically, most methods fit into the framework formulated by Beak et al. [1], which includes an optional rectifier (Trans.), a feature extractor (Feat.), a sequence modeler (Seq.), and a character classifier (Pred.). Many typical Chinese text recognition methods [9,25,29], also fall into this category, where the feature extractor generally takes the form of a Convolutional Neural Network, and the classifier part is mostly implemented as a linear classifier decoding input features into predicted character probabilities.

G. A. Fink et al. (Eds.): ICDAR 2023, LNCS 14191, pp. 244–258, 2023.
https://doi.org/10.1007/978-3-031-41734-4_15

However, the naive classification strategy has limited performance on Chinese samples, due to the large character set, severely unbalanced character frequency, and the complexity of Chinese glyphs.

To address the frequency skew, compositional learning strategies are widely used in low-shot Chinese character recognition tasks [6,12,31]. For compositional information exploited, the majority of implementations [2,20,21,24,31] utilize the radical representation, where the components and the structural information of each character are hierarchically modeled as a tree. Specifically, the basic components serve as leaf nodes and the structural information (the spatial placement of each component) serves as non-leaf nodes. Some methods are also seen to utilize stroke [6] or Wubi [12] representations. Besides the Chinese language, characters in many other languages can be similarly decomposed into basic components [3,4,6,16]. These methods somewhat improve text recognition performance under low-shot scenarios. However, compositional-based methods are rarely seen in regular recognition methods due to their complexity and less satisfactory performance. This limitation is solved by PRAB [7], which proposes to use radical information as a plug-in regularization term. The method decodes the character feature at each timestamp to its corresponding radical sequence and witnesses significant overall performance improvement on several SOTA baselines. However, the method still has two major limitations. First, PRAB [7] only applies to text recognition methods with explicitly aligned character features [13,18] and does not apply to implicitly aligned CTC-based methods like CRNN [17]. Furthermore, PRAB and most of the aforementioned radical method treats large parts and small parts alike, ignoring the depth information of the hierarchical representation.

To alleviate the limitations, we propose a Structural-Aware Network (SAN) which distills hierarchical composition information into the feature extractor with the proposed alignment-agnostic and depth-aware Auxilary Radical Branch (ARB). The ARB serves as a regularization term which directly refines feature maps extracted by the Feat. stage to preserve the hierarchical composition information without explicitly aligning the feature map to each character. The module allows the model to focus more on local features and learn more structured visual features, which significantly improves complex character recognition accuracy. As basic components are shared among head and tail classes alike, it also improves the tail-classes performance by explicitly exploiting their connections with head classes. Furthermore, we proposed a novel Tree Similarity (TreeSim) semimetric serves as a more fine-grained measure of the visual similarity between two characters. The proposed TreeSim semimetric further allows us to exploit the depth information of the hierarchical representation, which is implemented by weighting the tree nodes accordingly in ARB. The suggested method substantially enhances the accuracy of complex character recognition. As fundamental elements are shared between head and tail classes, it also boosts the performance of tail classes by leveraging their relationships with head classes.

Experiments demonstrate the effectiveness of ARB in optimizing the recognition performance of complex characters and long-tailed characters, and it also improves the overall recognition accuracy of Chinese text. The contributions of this work are as follows:

- We propose a SAN for complex character recognition by utilizing the hierarchical components information of the character.
- ARB based on the tree modeling of the label is introduced, which enhances the structure awareness of visual features. ARB shows promising improvement in complex character and long-tailed character recognition and it also improves the overall recognition accuracy of Chinese text.
- We propose a novel TreeSim method to measure the similarity of two characters, and propose a TreeSim-based weighting mechanism for ARB to further utilize the depth information in the hierarchical representation.

2   Related Work

2.1   Character-Based Approaches

In Chinese text recognition, early works are often character-based. Some works are based on CNN model [9,25,29] to design improved or integrated methods. For example, MCDNN [9] integrates multiple models including CNN, which shows advantageous performance in handwritten characters recognition. ART-CNN [25] alternatively trains a relaxation CNN to regularize the neural network during the training procedure and achieves state-of-the-art accuracy. Later, DNN-HMM [10] sequentially models the text line and adopts DNN to model the posterior probability of all HMM states, which significantly outperforms the best over-segmentation-based approach [19]. The SOTA text recognition model, ABINet [11], recommends blocking the gradient flow between the vision and language models, and introduces an iterative correction approach for the language model. These strategies promote explicit language modeling and effectively mitigate the influence of noisy input. However, these methods did not put forward effective countermeasures against the difficult problems of Chinese text recognition like many complex characters, insufficient training samples, and so on, so the performance improvement of these models is greatly limited.

2.2   Chinese Enhanced Character-Based Approaches

Several methods attempt to design targeted optimization strategies according to the characteristics of Chinese text [8,22,23,26–28]. Wu et al [26]. use MLCA recognition framework and new writing-style-aware image synthesis method to overcome large character sets and great insufficient training samples problems. In [8], the authors apply Maximum Entropy Regularization to regularize the training process to optimize the large amount of fine-grained Chinese characters and the great imbalance over class problems. Wang et al. [23] utilize the similarity of Chinese characters to reduce the total number of HMM states and model the tied states more accurately. These methods pay attention to the particularity of Chinese characters and give targeted optimization. However, they are still character-based, which makes it difficult to further explore the deep features of Chinese characters.

2.3   Radical-Based Approaches

In recent years, radical-based approaches have shown outstanding advantages in Chinese recognition [2,20,21,24,31]. RAN [31] employs an attention mechanism to extract radicals from Chinese characters and to detect spatial structures among the radicals, which reduce the vocabulary size and can recognize unseen characters. FewShotRAN [20] maps each radical to a latent space and uses CAD to analyze the radical representation. HDE [2] designs an embedding vector for each character and learns both radicals and structures of characters via a semantic vector, which achieves superior performance in both traditional HCCR and zero-shot HCCR tasks. Inspired by these works, our method emphasizes the role of structural information of radical in visual perception and proposes to utilize the common local and structural traits between characters to optimize the recognition performance in complex and long-tailed characters.

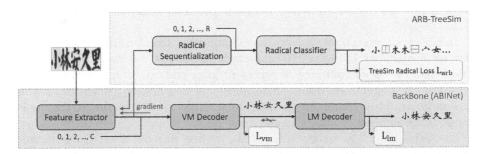

Fig. 1. The Structure-Aware Network (SAN). The orange square frame is the ARB-TreeSim and the blue square frame is the base network. The gradient flow of the ARB-TreeSim and the VM decoder influence the feature extractor together. (Color figure online)

3   Our Method

3.1   Label Modeling

As characters can be decomposed into basic components, we model it as a tree composed of various components hierarchically and model the label as a forest composed of character trees, which gives the label a structural representation.

As shown in Fig. 2(a), in Chinese, one popular modeling method is to decompose characters into radicals and structures [7]. Radicals are the basic components of characters, and structures describe the spatial composition of the radicals in each character. One structure is often associated with two or more radicals, and the structure is always the root node of a subtree. In this way, all Chinese characters can be modeled as a tree, in which leaf nodes are radicals, and

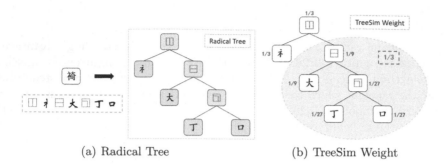

(a) Radical Tree   (b) TreeSim Weight

Fig. 2. In figure (a), the red nodes are called structure and the blue nodes are called radical. In figure (b), the yellow circle indicates that the total weight of the subtree is 1/3. (Color figure online)

non-leaf nodes are structures. For simplicity, we call such trees "radical trees" in this paper.

In this work, we propose a novel Tree Similarity (TreeSim) metric to measure the visual similarity between two characters represented with radical trees. In a radial tree, the upper components represent large and overall information, while the lower components represent small and local information. When comparing the similarity of two characters, humans tend to pay more attention to the upper components and less attention to the lower ones.

Correspondingly, as shown in Fig. 2(b), we first propose a weighting method for TreeSim. The method includes the following three characteristics: First, upper components in the radical tree have greater penalty weight. Second, every subtree is regarded as an independent individual, the root node and its subtrees have equal weight. Third, the total weight of the tree is 1. The calculation procedure is depicted in Algorithm 1, where $Node$ is the pointer of the root node and w_{sub} is the weight of the subtree. The weight of each node can be calculated by the recursion method.

Based on this weighting method, the TreeSim calculation sample is shown in Fig. 3. The calculation process is as follows: First, select any one of the two radical trees, and traverse every node in preorder. Second, judge whether it is matching for each node. The matching rules include: 1)Every ancestor of this node is matching. 2)This location in the other tree also has a node and the two nodes are the same. Last, TreeSim is equal to the sum of the weights of all matching nodes.

The proposed TreeSim is hierarchical, bidirectional, and deep-independent. Hierarchical means TreeSim pays different attention to different levels of the tree. Bidirectional means no matter select which of the two trees, the calculation result is the same. Depth-independent means the weight of nodes at a certain level is independent of the depth of the tree.

Based on this label modeling method, we proposed two loss design strategies. Considering that a rigorous calculation for the loss of two forests may be

---

Algorithm 1. TreeSim Weight Algorithm

Require: *Node*: Pointer of the root node;
$\qquad w_{sub}$: Weight of the subtree, which equals to 1 in the initial call;
  1: **function** GETTREESIMWEIGHT($Node, w_{sub}$)
  2:   $n \leftarrow$ number of children of this node
  3:   **if** $n == 0$ **then**
  4:       weight of this node $\leftarrow w_{sub}$
  5:       **return**
  6:   **else**
  7:       weight of this node $\leftarrow w_{sub}/(n+1)$
  8:       **for** $c \leftarrow Node.Children$ **do**
  9:           GETTREESIMWEIGHT($c, w_{sub}/(n+1)$)
 10:       **end for**
 11:   **end if**
 12:   **return**
 13: **end function**

---

complex and nonlinear, we introduce two linear approximate calculation methods in Sect. sec3.2 to supervise the loss calculation.

3.2 Auxiliary Radical Branch

Inspired by the radical-based methods [7,20], we first propose an Auxiliary Radical Branch (ARB) module to enhance the structure-awareness of visual features by using the hierarchical representation as a regularization term. Unlike the PRAB [7] method, the proposed ARB module directly decodes the feature map extracted from the Feat. stage, thus do not need individual character features and can be theoretically applied to most methods with a Feat. stage [1]. As shown in Fig. 1, ARB includes the following characteristics: 1) The label representation in ARB is hierarchical, which contains component and structural information. 2) ARB is regarded as an independent visual perception optimization branch, which takes feature as input, decodes according to structured representation, and finally refines the visual feature extraction.

The role of ARB is reflected in two aspects: One is that ARB can use hierarchical label representation to supply and refine the feature extraction, improving the structure-awareness of the model. The other is that by taking advantage of the common component combination traits between Chinese characters, ARB could exploit the component information connections between tail classes and head classes, which can optimize long-tailed character recognition.

In ARB, we propose two linear loss design strategies, which reflect the hierarchical label representation while simplifying the design and calculation.

Sequence Modeling. Considering that the proposed label modeling method is top-down and root-first, in loss design, we intend to perpetuate this design mode to match them. We propose using the sequence by traversing the radical

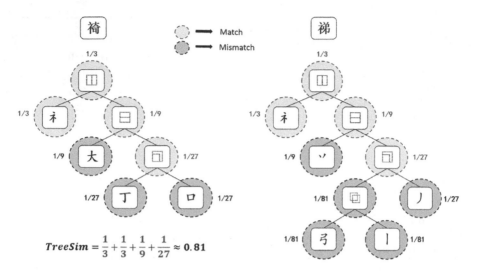

$$TreeSim = \frac{1}{3} + \frac{1}{3} + \frac{1}{9} + \frac{1}{27} \approx 0.81$$

Fig. 3. TreeSim Calculation. Nodes with a red circle are the match modes and nodes with a black circle are the mismatch nodes. The calculation results of the two radical trees are the same. (Color figure online)

tree in preorder to model each tree in the label, so we use the linear radical structure sequence [7]. This sequence follows the root-first design strategy and retains the hierarchical structure information of the radical tree to a certain extent. By connecting each radical structure sequence in the label, we get the sequence modeling label.

The ARB prediction is supervised with the ARB loss L_{arb}, which is a weighted cross-entropy loss of each element.

$$L_{arb} = \sum_{i}^{l_{rad}} w_{rad_{(i)}} * log P(r_{(i)}^*), \tag{1}$$

where r_i^* is the ground truth of the i-th radical, l_{rad} is the total radical length of **all** characters in the sample.

The radical structure sequence contains the component and structural information of the label, so this modeling method implements the introduction of hierarchical information and effectively improves the vision model performance.

Naive Weighting. We first use a naive way to set equal weights to radicals in each character, i.e., $\mathbf{w}_{rad} = \mathbf{w}_{naive} = \mathbf{1}$. Albeit the Naive method demonstrates some extent of performance improvement, it treats all components as equal importance, which may lead to over-focusing on minor details.

TreeSim Enhanced Weighting. In sequence modeling, the structural information is implicit, we want to make it more explicit. So we propose TreeSim enhanced Weighting method to explicitly strengthen the structural information of label representation.

We add the TreeSim weight (Fig. 2(b)) to the naive weight as a regularization term and get the final radical weight.

$$\mathbf{w}_{rad} = \mathbf{w}_{naive} + \lambda_{TreeSim}\mathbf{w}_{TreeSim} \tag{2}$$

where $\lambda_{TreeSim}$ is set to 1. This method enhanced the structural information in loss explicitly, which shows better performance on the vision model than the Naive method.

3.3   Optimization

The training gradient flow by two branches is superimposed and updates the weight of the feature extractor together, which allows the radical level prediction model to supplement and refine the visual feature extraction of the character prediction model in the training process. We define the loss as follows

$$L_{overall} = L_{base} + L_{arb}, \tag{3}$$

where L_{base} is the loss of base model, which is the loss from ABINet in this work, i.e.,

$$L_{base} = L_{vm} + L_{lm}. \tag{4}$$

Both the character prediction branch and radical structure prediction branch affect the feature extraction. For fairness, we give them the same weight of influence on visual feature extraction, to ensure that the performance of both parties is fully reflected.

4   Experiment

4.1   Implementation Details

The datasets used in our experiments are Web Dataset and Scene Dataset [7]. The Web Dataset contains 20,000 Chinese and English web text images from 17 different categories on the Taobao website and the Scene Dataset contains 636,455 text images from several competitions, papers, and projects. The number of radical classes is 960. We set the value of R (Fig. 1) to be 33 for the web dataset and 39 for the scene dataset, covering a substantial 95% of the training samples from both datasets.

We implement our method with PyTorch and conduct experiments on three NVIDIA RTX 3060 GPUs. Each input Image is resized to 32×128 with data augmentation. The batch size is set to 96 and ADAM optimizer is adopted with the initial learning rate of $1e^{-4}$, which is decayed to $1e^{-5}$ after 12 epochs.

Table 1. Ablation study of ARB. The dataset used is Web dataset [7].

Model	Naive Radical loss	TreeSim weighing loss	Accuracy	1-NED
VM-Base			59.1	0.768
VM-Naive	✓		60.7	0.779
VM-TreeSim	✓	✓	**61.3**	**0.786**
ABINet-Base			65.6	0.803
ABINet-Naive	✓		66.8	0.812
ABINet-TreeSim (SAN)	✓	✓	**67.3**	**0.817**

4.2   Ablative Study

We discuss the performance of the proposed approaches (ARB and TreeSim weighting) with two base-model configurations: Vision Model and ABINet. Experiment results are recorded in Table 1.

First, the proposed ARB is proved useful, by significantly improving the accuracy of both base-metods. Specifically, the VM-Naive outperforms VM-Base by 1.6% accuracy and 0.011 1-NED, ABINet-Naive outperforms ABINet-Base by 1.2% accuracy and 0.009 1-NED. Second, TreeSim enhanced weighting also achieves expected improvements. VM-TreeSim boosts accuracy by 0.6% and 1-NED by 0.007 than VM-Naive, SAN boosts accuracy by 0.5% and 1-NED by 0.005 than ABINet-Naive.

The above observations suggest that the hierarchical components information is useful to the feature extractor, which can significantly improve the performance of the vision model. The proposed approaches also yields significant improvement against full ABINet indicating hierarchical components information still plays a significant part, even when the language models can, to some extent, alleviate the confusions caused by insufficiency shape awareness.

4.3   Comparative Experiments

Compared with Chinese text recognition benchmarks and recent SOTA works that are trained on web and scene datasets, SAN also shows impressive performance (Table 2). We can see from the comparison, our SAN outperforms ABINet with 1.7%, 1.8% on Web and Scene datasets respectively. Also, SAN achieves the best 1-NED on both datasets. Some successful recognition examples are shown in Fig. 4, which shows the complex characters and long-tailed characters that predict failure using the base model (ABINet) while predicted successfully by the full model.

4.4   Property Analysis

To validate the optimization of ARB on complex character recognition and long-tailed character recognition, we divide them into different groups according to

Table 2. Performance on Chinese text recognition benchmarks [7]. † indicates results reported by [7], * indicates results from our experiments.

Method	Web		Scene	
	Accuracy	1-NED	Accuracy	1-NED
CRNN† [17]	54.5	0.736	53.4	0.734
ASTER† [18]	52.3	0.689	54.5	0.695
MORAN† [14]	49.9	0.682	51.8	0.686
SAR† [13]	54.3	0.725	62.5	0.785
SRN† [30]	52.3	0.706	60.1	0.778
SEED† [15]	46.3	0.637	49.6	0.661
TransOCR† [5]	62.7	0.782	67.8	0.817
TransOCR-PRAB† [7]	63.8	0.794	71.0	0.843
ABINet* [11]	65.6	0.803	71.8	0.853
SAN (Ours)	**67.3**	**0.817**	**73.6**	**0.863**

	(a) Complex	(b) Complex	(c) Long-tailed	(d) Long-tailed
ABINet	韩用	博雅圆書城	将记	郑县红光晶宝店
SAN	藥用	博雅圖書城	奂记	郓县红光晶宝店
GT	藥用	博雅圖書城	奂记	郓县红光晶宝店

Fig. 4. Successful recognition examples using ARB. (a) (b) are complex character examples. (c) (d) are long-tailed character examples. The text strings are ABINet prediction, SAN prediction and Ground Truth.

their complexity and frequency. For each category, to give a better understanding of the differences in recognizing performances, the character prediction samples by SAN and ABINet are studied in more detail. Specifically, we calculated the accuracy of character prediction results and the average TreeSim between the predicted results and the ground truth. TreeSim serves as a more fine-grained metric compared to accuracy, as it can indicate the awareness of structural information.

Experiments on Different Complexity Characters. We observe that the complexity of the character structure is proportional to its radical structure sequence length (RSSL), so we use RSSL to represent the complexity of each character. According to observations, we denote characters with complexity of RSSL5 (characters that RSSL is equal to 5) and RSSL6 as medium complexity characters, characters with longer RSSL are denoted as complex and characters with shorter RSSL are considered as simple. Accodingly, we divide character classes into three parts: RSSL $\leq$ 4 (average 34% in web and 30% in scene),

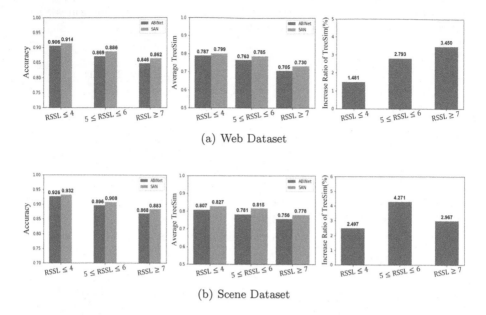

(a) Web Dataset

(b) Scene Dataset

Fig. 5. The accuracy (left), the average TreeSim (middle) and the increase ratio of TreeSim (right) divided by RSSL on character prediction samples by SAN and ABINet. RSSL represents the length of the radical structure sequence of a character.

$5 \leq \text{RSSL} \leq 6$ (average 38% in web and 37% in scene), $\text{RSSL} \geq 7$ (average 28% in web and 33% in scene), and call them simple characters, sub-complex characters, and complex characters respectively.

We calculate the accuracy and the average TreeSim between character prediction samples and ground truth values within each of these three parts (Fig. 5). In terms of accuracy, we note that in both the web dataset (Fig. 5(a)) and the scene dataset (Fig. 5(b)), there is a consistent trend of higher improvement in accuracy for sub-complex ($5 \leq \text{RSSL} \leq 6$) and complex ($\text{RSSL} \geq 7$) characters compared to simple characters ($\text{RSSL} \leq 4$). Regarding TreeSim, the average TreeSim in all three parts consistently increases. In the web dataset, the rising trend becomes more pronounced as RSSL increases, with the growth of complex characters being particularly noticeable. In the scene dataset, the increase in sub-complex characters is extremely significant, and the growth rates of both sub-complex and complex characters surpass that of simple characters.

The experiments show that ARB can prominently improve the similarity between the recognition results and ground trues of complex characters, indicating that ARB has a more distinct perception of complex characters, which is due to the introduction of hierarchical structure information improving the structure-awareness of the vision model and making it easier for models to distinguish different components in complex characters.

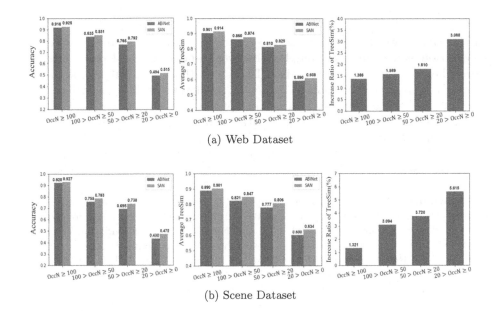

(a) Web Dataset

(b) Scene Dataset

Fig. 6. The accuracy (left), the average TreeSim (middle) and the increase ratio of TreeSim (right) divided by OccN on character prediction samples by SAN and ABINet. OccN represents the occurrence number of a character in the training dataset.

Experiments on Long-Tailed Characters. To validate the feasibility of our model on long-tailed characters, we sort character prediction samples in descending according to their occurrence number (OccN) in the training dataset, and divide them into four parts: $OccN \geq 100$ (average 22% in web and 39% in scene), $100 > OccN \geq 50$ (average 12% in web and 10% in scene), $50 > OccN \geq 20$ (average 19% in web and 15% in scene), $20 > OccN \geq 0$ (average 47% in web and 36% in scene).

We calculate the accuracy and the average TreeSim between the character prediction samples and their corresponding ground truth within each of these parts (Fig. 6). In terms of accuracy, we observe that for both the web dataset (Fig. 6(a)) and the scene dataset (Fig. 6(b)), the improvement in accuracy for infrequent characters ($OccN < 50$) is consistently higher than that for frequent characters ($OccN \geq 50$). Regarding TreeSim, the rising trend becomes more pronounced as the occurrence number decreases. The average TreeSim of tail classes ($20 > OccN \geq 0$) exhibits the most noticeable increase in both datasets.

These results demonstrate that the ARB can make the recognition results of long-tailed characters more similar to the ground truths, indicating that ARB can learn more features of long-tailed characters, which is because of the common components combination traits shared between characters, ARB can take advantage of the traits learning on head classes to optimize the tail classes recognition.

5   Conclusion

In this paper, we propose a Structure-Aware Network (SAN) to optimize the recognition performance of complex characters and long-tailed characters, by using the proposed Auxiliary Radical Branch (ARB) which utilizes the hierarchical components information of characters. Besides, we propose using Tree Similarity (TreeSim) to measure the similarity of two characters and using TreeSim weight to enhance the structural information of label representation. Experiment results demonstrate the superiority of ARB on complex and long-tailed character recognition and validate that our method outperforms standard benchmarks and recent SOTA works.

Acknowledgement. The research is supported by National Key Research and Development Program of China (2020AAA0109700), National Science Fund for Distinguished Young Scholars (62125601), National Natural Science Foundation of China (62076024, 62006018), Interdisciplinary Research Project for Young Teachers of USTB (Fundamental Research Funds for the Central Universities)(FRF-IDRY-21-018).

References

1. Baek, J., et al.: What is wrong with scene text recognition model comparisons? Dataset and model analysis. In: ICCV, pp. 4714–4722 (2019)
2. Cao, Z., Lu, J., Cui, S., Zhang, C.: Zero-shot handwritten Chinese character recognition with hierarchical decomposition embedding. Pattern Recognit. **107**, 107488 (2020)
3. Chanda, S., Baas, J., Haitink, D., Hamel, S., Stutzmann, D., Schomaker, L.: Zero-shot learning based approach for medieval word recognition using deep-learned features. In: 16th International Conference on Frontiers in Handwriting Recognition, ICFHR 2018, Niagara Falls, 5–8 August 2018, pp. 345–350. IEEE Computer Society (2018). https://doi.org/10.1109/ICFHR-2018.2018.00067
4. Chanda, S., Haitink, D., Prasad, P.K., Baas, J., Pal, U., Schomaker, L.: Recognizing bengali word images - a zero-shot learning perspective. In: 2020 25th International Conference on Pattern Recognition (ICPR), pp. 5603–5610 (2021). https://doi.org/10.1109/ICPR48806.2021.9412607
5. Chen, J., Li, B., Xue, X.: Scene text telescope: text-focused scene image super-resolution. In: 2021 IEEE/CVF Conference on Computer Vision and Pattern Recognition (CVPR), pp. 12021–12030 (2021)
6. Chen, J., Li, B., Xue, X.: Zero-shot Chinese character recognition with stroke-level decomposition. In: IJCAI, pp. 615–621 (2021)
7. Chen, J., et al.: Benchmarking Chinese text recognition: datasets, baselines, and an empirical study. arXiv preprint arXiv:2112.15093 (2021)
8. Cheng, C., Xu, W., Bai, X., Feng, B., Liu, W.: Maximum entropy regularization and Chinese text recognition. arXiv preprint arXiv:2007.04651 (2020)
9. Ciresan, D.C., Meier, U.: Multi-column deep neural networks for offline handwritten chinese character classification. In: 2015 International Joint Conference on Neural Networks (IJCNN), pp. 1–6 (2013)
10. Du, J., Wang, Z., Zhai, J.F., Hu, J.: Deep neural network based hidden Markov model for offline handwritten chinese text recognition. In: 2016 23rd International Conference on Pattern Recognition (ICPR), pp. 3428–3433 (2016)

11. Fang, S., Xie, H., Wang, Y., Mao, Z., Zhang, Y.: Read like humans: autonomous, bidirectional and iterative language modeling for scene text recognition. In: CVPR, pp. 7098–7107 (2021)
12. He, S., Schomaker, L.: Open set Chinese character recognition using multi-typed attributes. arXiv preprint arXiv:1808.08993 (2018)
13. Li, H., Wang, P., Shen, C., Zhang, G.: Show, attend and read: a simple and strong baseline for irregular text recognition. In: AAAI, pp. 8610–8617 (2019)
14. Luo, C., Jin, L., Sun, Z.: MORAN: a multi-object rectified attention network for scene text recognition. Pattern Recognit. **90**, 109–118 (2019)
15. Qiao, Z., Zhou, Y., Yang, D., Zhou, Y., Wang, W.: SEED: semantics enhanced encoder-decoder framework for scene text recognition. In: CVPR, pp. 13525–13534 (2020)
16. Rai, A., Krishnan, N.C., Chanda, S.: Pho(sc)net: an approach towards zero-shot word image recognition in historical documents. arXiv preprint arXiv:2105.15093 (2021)
17. Shi, B., Bai, X., Yao, C.: An end-to-end trainable neural network for image-based sequence recognition and its application to scene text recognition. IEEE Trans. Pattern Anal. Mach. Intell. **39**(11), 2298–2304 (2017)
18. Shi, B., Yang, M., Wang, X., Lyu, P., Yao, C., Bai, X.: ASTER: an attentional scene text recognizer with flexible rectification. IEEE Trans. Pattern Anal. Mach. Intell. **41**(9), 2035–2048 (2019)
19. Wang, Q.F., Yin, F., Liu, C.L.: Handwritten Chinese text recognition by integrating multiple contexts. IEEE Trans. Pattern Anal. Mach. Intell. **34**, 1469–1481 (2012)
20. Wang, T., Xie, Z., Li, Z., Jin, L., Chen, X.: Radical aggregation network for few-shot offline handwritten Chinese character recognition. Pattern Recognit. Lett. **125**, 821–827 (2019)
21. Wang, W., Shu Zhang, J., Du, J., Wang, Z., Zhu, Y.: Denseran for offline handwritten Chinese character recognition. In: 2018 16th International Conference on Frontiers in Handwriting Recognition (ICFHR), pp. 104–109 (2018)
22. Wang, Z., Du, J.: Joint architecture and knowledge distillation in CNN for Chinese text recognition. Pattern Recognit. **111**, 107722 (2019)
23. Wang, Z., Du, J., Wang, J.: Writer-aware CNN for parsimonious hmm-based offline handwritten Chinese text recognition. arXiv preprint arXiv:1812.09809 (2018)
24. Wu, C.J., Wang, Z., Du, J., Shu Zhang, J., Wang, J.: Joint spatial and radical analysis network for distorted Chinese character recognition. In: 2019 International Conference on Document Analysis and Recognition Workshops (ICDARW), vol. 5, pp. 122–127 (2019)
25. Wu, C., Liang Fan, W., He, Y., Sun, J., Naoi, S.: Handwritten character recognition by alternately trained relaxation convolutional neural network. In: 2014 14th International Conference on Frontiers in Handwriting Recognition, pp. 291–296 (2014)
26. Wu, Y., Hu, X.: From textline to paragraph: a promising practice for Chinese text recognition. In: Arai, K., Kapoor, S., Bhatia, R. (eds.) FTC 2020. AISC, vol. 1288, pp. 618–633. Springer, Cham (2021). https://doi.org/10.1007/978-3-030-63128-4_48
27. Xiao, X., Jin, L., Yang, Y., Yang, W., Sun, J., Chang, T.: Building fast and compact convolutional neural networks for offline handwritten Chinese character recognition. Pattern Recognit. **72**, 72–81 (2017)

28. Xiao, Y., Meng, D., Lu, C., Tang, C.K.: Template-instance loss for offline handwritten Chinese character recognition. In: 2019 International Conference on Document Analysis and Recognition (ICDAR), pp. 315–322 (2019)
29. Yin, F., Wang, Q.F., Zhang, X.Y., Liu, C.L.: ICDAR 2013 chinese handwriting recognition competition. In: 2013 12th International Conference on Document Analysis and Recognition, pp. 1464–1470 (2013)
30. Yu, D., et al.: Towards accurate scene text recognition with semantic reasoning networks. In: CVPR, pp. 12110–12119 (2020)
31. Shu Zhang, J., Zhu, Y., Du, J., Dai, L.: Ran: radical analysis networks for zero-shot learning of chinese characters. arXiv preprint arXiv:1711.01889 (2017)

TACTFUL: A Framework for Targeted Active Learning for Document Analysis

Venkatapathy Subramanian[1,2]([envelope]) [iD], Sagar Poudel[1] [iD], Parag Chaudhuri[1] [iD], and Ganesh Ramakrishnan[1] [iD]

[1] Indian Institute of Technology Bombay, Mumbai 400076, Maharashtra, India
{sagar.poudel,paragc,ganesh}@cse.iitb.ac.in
[2] Anaadi Rural AI Center, Dindigul, India
venkatapathy@cse.iitb.ac.in

Abstract. Document Layout Parsing is an important step in an OCR pipeline, and several research attempts toward supervised, and semi-supervised deep learning methods are proposed for accurately identifying the complex structure of a document. These deep models require a large amount of data to get promising results. Creating such data requires considerable effort and annotation costs. To minimize both cost and effort, Active learning (AL) approaches are proposed. We propose a framework TACTFUL for Targeted Active Learning for Document Layout Analysis. Our contributions include (i) a framework that makes effective use of the AL paradigm and Submodular Mutual Information (SMI) functions to tackle object-level class imbalance, given a very small set of labeled data. (ii) an approach that decouples object detection from feature selection for subset selection that improves the targeted selection by a considerable margin against the current state-of-the-art and is computationally effective. (iii) A new dataset for legacy Sanskrit books on which we demonstrate the effectiveness of our approach, in addition to reporting improvements over state-of-the-art approaches on other benchmark datasets.

Keywords: Active Learning · Submodular Functions · Balancing Dataset · Document Layout Analysis

1 Introduction

Digitization of scanned documents such as historical books, papers, reports, contracts, *etc.*, is one of the essential tasks required in this information age. A typical digitization workflow consists of different steps such as pre-processing, page layout segmentation, object detection, Optical Character Recognition (OCR), post-processing, and storage. Though OCR is the most important step in this pipeline, the preceding steps of page layout segmentation and object detection play a crucial role. This is especially so when there is a requirement to preserve the layout of the document beyond OCR. For over two decades, the scientific community has proposed various techniques [2] for document layout analysis, yet recent deep-learning methods have attained improved performance by leaps and bounds. Several supervised, and semi-supervised deep learning methods [16–18] can accurately identify the complex structure of a document. This performance improvement, though, comes at a cost. Deep models require a significantly large amount of data to yield promising results. Creating such data requires considerable annotation effort and cost. To minimize both cost and effort, active learning (AL) approaches [19] are proposed with a constraint on the annotation budget. AL can help iteratively select a small amount of data for annotation, from a large unlabeled pool of data, on which the model can be trained at each iteration. Though such approaches work, sometimes page-level AL techniques may be biased and miss out on rare classes while selecting images. It is especially true in the case of document layout analysis where document objects (such as titles, images, equations, *etc.*) are complex, dense, and diverse. It can be somewhat challenging to select pages that lead to balance across classes in the training set and specially balanced performance across all. Most state-of-the-art AL approaches tend to decrease the models' performance on rare classes in the pursuit of overall accuracy. Summarily what we wanted is an effective AL technique that can help address the class imbalance problem while selecting page images. Towards this, we propose a framework for **T**argeted **ACT**ive Learning **F**or DocUment Layout AnaLysis (TACTFUL). The proposed framework uses sub-modular mutual information (SMI) functions in its active learning strategies to select relevant data points from a large pool of unlabeled datasets. The SMI (Submodular Mutual Information) functions are useful in two complementary ways: 1) By taking advantage of the natural diminishing returns property of sub-modular functions, the framework maximizes the utility of subsets selected during each AL round. 2) By quantifying the mutual dependence between two random variables (a known rare class and an unknown object of interest in our case), we can maximize the mutual information function to get relevant objects and through them the page images for annotation. Our contribution, through the TACTFUL framework, is as follows:

1. We propose an end-to-end AL framework that can tackle object-level class imbalance while acquiring page images for labeling, given a very small set of labeled data. Within this framework, we make effective use of two complementing paradigms, *viz.*, i) AL paradigm that aims to select a subset of samples with the highest value from a large set, to construct the training samples, and ii) Submodular functions that have higher marginal utility in adding a new element to a subset than adding the same element to a superset. Within submodular functions, we use SMI functions [4, 7] that can model the selection of subsets similar to a smaller query set from rare classes thereby avoiding severe data imbalance.
2. For subset selection, we decouple the object detection and feature selection steps thereby overcoming the limitation [3] present in current object detection models. We show that the pre-trained model, without additional fine-tuning, works effectively well for representing objects. The decoupling strategy improves the targeted selection by a considerable margin against the current state-of-the-art and is computationally effective.

3. We empirically prove that our model performs well compared to the current SOTA framework having an increase of about 9.3% over the SOTA models. We also release a new dataset for document layout analysis for legacy Sanskrit books and show that our framework works for similar settings for documents in other languages and helps improve the AP by 8.6 % over the baseline.

2   Motivation

Layout detection is a continuing challenge in the field of Document Analysis. Many state-of-the-art models and datasets [1,11,23] have been created to address this problem. The recent availability of a large amount of annotated data has resulted in good-quality ML models for layout detection for English documents. There is still a dearth of good quality ground truth data for other languages. This dearth is due to the following reasons: (i) Manual labeling is time-consuming and expensive (ii) It is difficult, if not impossible, to replicate the alternative ways of creating ground truth data such as those created for the English documents. Large datasets such as Publaynet [23] and DocBank [11] are created in a weakly supervised manner by extracting the ground-truth layout information from an available parallel corpus such as scientific latex documents [5] and then manually post-editing or correcting the output. Such availability is rare in other languages. Hence the need for large annotated datasets for other languages remains unaddressed. One case study we consider is the digitization of ancient Sanskrit documents, where there is an immediate need for high-quality document layout detection methods. In this aforementioned work, we have tens of thousands of scanned images that are old manuscripts. Thus, the only way of training a good model for layout detection is to annotate a subset of the available pages. The question that subsequently arises is *'How can we identify pages to be annotated such that the model performance improves across all classes?'*

As an attempt to answer the question, we performed a retrospective study on one of the largest datasets available for document layout analysis, *viz.*, Pub-LayNet [23]. The dataset was created by automatically matching the XML representations and the content of over 1 million PDF articles publicly available on PubMed Central. It contains over 360 thousand annotated document images and the deep neural networks trained on PubLayNet achieve an Average Precision (AP) of over 90%. The dataset consists of five categories of annotations, and those include TITLE, TITLE, LIST, TABLE, and FIGURE. The statistics of the layout categories associated with this dataset are summarized in Table 1:

In Fig. 1a, we depict the share of each class amongst the total objects present in the training dataset of PubLayNet. From the chart, we can notice that there is a class imbalance for the different objects. Given this imbalanced class distribution, we initially investigate how an object detection model learns. We randomly sampled a fraction of about 2000 data points from the dataset (from the official train and test sets) while maintaining the same class imbalance as in the original set. In those 2000 pages, there were a total of 9920 TITLE objects, 900 TITLE objects, 74 LIST objects, 76 TABLE objects, and 123 FIGURE objects. We subsequently trained a Faster RCNN model from scratch. The model reached an AP

Table 1. Statistics of layout categories in the training, development, and testing sets of PubLayNet. PubLayNet is one or two orders of magnitude larger than any existing document layout dataset obtained from [23]

	Training	Development	Testing
Pages			
Plain pages	87,608	1,138	1,121
TITLEpages	46,480	$2,059^+$	$2,021^+$
Pages with LISTs	53,793	2,984	3,207
Pages with TABLEs	86,950	$3,772^+$	$3,950^+$
Pages with FIGUREs	96,656	$3,734^+$	$3,807^+$
Total	340,391	11,858	11,983
Instances			
TITLE	$2,376,702$	93,528	95,780
TITLE	633,359	19,908	20,340
LISTs	81,850	4,561	5,156
TABLEs	103,057	4,905	5,166
FIGUREs	116,692	4,913	5,333
Total	$3,311,660$	127,815	131,775

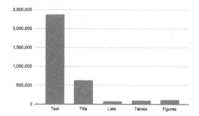

(a) Distribution of class objects

(b) Pagewise spread of different class objects

Fig. 1. Figures depicting the share of objects in the training dataset.

of 33% before plateauing. Figure 2a shows the improvement of average AP and classwise AP during the training. The AP of TITLE is the highest with 75%, followed by TITLE with 50%. The worst performing classes are also the rare classes, with the FIGURE class averaging 20%, TABLE 15%, and LIST having the worst performance with just 0.3%. It's also interesting to note that TABLE and FIGURE with counts as of LIST perform better. It could be easier to detect FIGURE and TABLE objects due to their unique shapes than the rest of the page objects. The average AP is drastically reduced by the last three tail objects of the distribution. Given the skewed distribution of objects in the dataset, this is expected, as the model might not have seen the rare class objects. We later selected just 500 data points but maintained the balance among the classes.

This time there were a total of 2583 TITLE objects, 231 TITLE objects, 14 LIST objects, 17 TABLE objects, 27 FIGURE objects. Again we trained a Faster RCNN model from scratch. Figure 2a shows the improvement of average AP and class-wise AP during the training. The model reached an AP of 48% before plateauing. The difference in AP between the model trained with 2000 data points with a severely imbalanced set and the model trained on 1/4 of the former(500 data points) but without class imbalance is large, with a balanced dataset performing better. **There is then the scope for improving the AP in the same setting if there is a way to select images so that the rare classes are covered, and the class imbalance is avoided.**

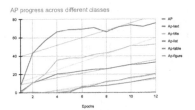 

(a) AP and classwise AP during training of 2000 data points with class imbalance

(b) Confusion Matrix for the five different classes as classified by models trained on 2000 data points

Fig. 2. Figure depicting the AP, classwise AP, and confusion matrix for different classes trained on 2000 data points

We further plotted the confusion matrix for the object detected on the initial dataset of 2000 points. The model trained on this initial dataset had good object detection accuracy, as depicted in Fig. 2b. But performs poorly in the object classification task. Empirically it is clear that training a model for both classification and object detection compromises the model's ability to learn efficient features in the embedding space [3], which can be observed in our case as well. Given these observations, our goal is to mitigate the class imbalance during image acquisition for annotations, thereby improving the model's performance for rare classes. We explain our framework in detail in the next section.

3   Our Approach

Our Active Learning (AL) paradigm [14] approach to address the problem of class imbalance discussed in Sect. 2 is visually depicted in Fig. 3. Similar to standard AL techniques [17] we train a detection model Θ to detect n layout objects of an image X_i. An object n_j consists of the bounding box b_j and its category c_j. $Y_i = \{(b_j, c_j)\}_{j=1}^n$ are the object annotations for X_i. Θ is initially trained on a small labelled dataset $\mathcal{L}_0 = \{(\mathcal{X}_i, \mathcal{Y}_i\}_{i=1}^l$ and contains a large set of unlabelled dataset $\mathcal{U}_0 = \{(\mathcal{X}_i\}_{i=l}^{u+l}\}$. Given the unlabeled set $\mathcal{U}$, the goal is to *acquire* a subset of the images $\mathcal{A} \subseteq \mathcal{U}$ of budget $k = \|\mathcal{A}\|$, iteratively, to improve the model's performance. At each iteration t, m samples $\mathcal{M}_t = \{\mathcal{X}_i\}_{i=l}^m \subseteq \mathcal{U}_{t-1}$ are queried for labelling and the corresponding labeled set $\mathcal{M}_t = \{(\mathcal{X}_i \, \mathcal{Y}_{i=l}^m)\} \subseteq \mathcal{U}_{t-1}$ is added to the existing labeled set $\mathcal{L}_t = \{\mathcal{L}_{t-1} \cup \mathcal{M}_t\}$. For the next iteration, the unlabeled set becomes $\mathcal{U}_t = \{\mathcal{U}_{t-1} \setminus \mathcal{M}\}$. Iteration is stopped when the model reaches the desired performance as described in Sect. 2, we wanted to select a subset of images from the unlabelled page image distribution (eg, from distribution as shown in Fig. 1b) such that the model performance improves for selected rare class[es].

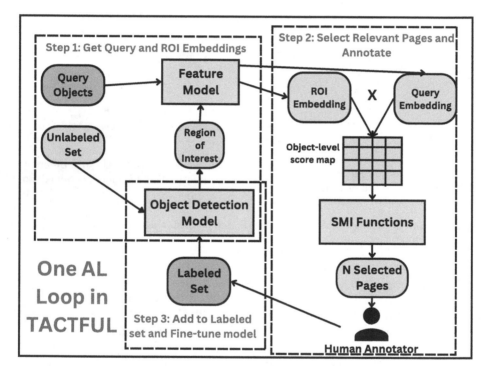

Fig. 3. Three-step approach of one AL loop of TACTFUL

To achieve this, we will use the core idea of SMI functions [8, 9, 21] as acquisition functions to acquire data points from $\mathcal{U}$. In the following section, we discuss some of the notations and preliminaries required for our work as introduced in [6, 7] and their extensions introduced in [8, 9]

3.1 Submodular Functions

$\mathcal{V}$ denotes the *ground-set* of n data points $\mathcal{V} = \{1, 2, 3, ..., n\}$ and a set function $f : 2^{\mathcal{V}} \to \mathbb{R}$. The function f is submodular [4] if it satisfies diminishing returns, namely $f(j\|\mathcal{X}) \geq f(j\|\mathcal{Y})$ for all $\mathcal{X} \subseteq \mathcal{Y} \subseteq \mathcal{V}, j \notin \mathcal{X}, \mathcal{Y}$. Facility location, graph cut, log determinants, *etc.*, are some examples [21].

3.2 Submodular Mutual Information (SMI)

Given a set of items $\mathcal{A}, \mathcal{Q} \subseteq \mathcal{V}$, the submodular mutual information (SMI) [6,7] is defined as $I_f(\mathcal{A}; \mathcal{Q}) = f(\mathcal{A}) + f(\mathcal{Q}) - f(\mathcal{A} \cup \mathcal{Q})$. Intuitively, SMI measures the similarity between $\mathcal{Q}$ and $\mathcal{A}$, and we refer to $\mathcal{Q}$ as the query set. In our context, $\mathcal{A} \subset \mathcal{U}$ is our unlabelled set of images, and the query set $\mathcal{Q}$ is the small target set containing the rare class images and page annotations for the target set. To find an optimal subset $\mathcal{M} \subseteq \mathcal{U}$, given a query set $\mathcal{Q}$ we can define $g_\mathcal{Q}(\mathcal{A}) = I_f(\mathcal{A}; \mathcal{Q})$, $\mathcal{A} \subseteq \mathcal{V}$ and maximize the same.

3.3 Specific SMI Functions Used in This Work

For any two data points $i \in \mathcal{V}$ and $j \in \mathcal{Q}$, let s_{ij} denote their similarity.

Graph Cut MI (Gcmi): The SMI instantiation of graph-cut (GCMI) is defined as: $I_{GC}(\mathcal{A}; \mathcal{Q}) = 2 \sum_{i \in \mathcal{A}} \sum_{j \in \mathcal{Q}} s_{ij}$. Since maximizing GCMI maximizes the joint pairwise sum with the query set, it will lead to a summary similar to the query set Q. Specific instantiations of GCMI have been used for query-focused summarization for videos [20] and documents [10,12].

Facility Location MI (Flmi): We consider two variants of FLMI. The first variant is defined over $\mathcal{V}$(FLVMI), the SMI instantiation can be defined as: $I_{FLV}(\mathcal{A}; \mathcal{Q}) = \sum_{i \in \mathcal{V}} \min(\max_{j \in \mathcal{A}} s_{ij}, \max_{j \in \mathcal{Q}} s_{ij})$. The first term in the min(.) of FLVMI models diversity, and the second term models query relevance.

For the second variant, which is defined over $\mathcal{Q}$ (FLQMI), the SMI instantiation can be defined as: $I_{FLQ}(\mathcal{A}; \mathcal{Q}) = \sum_{i \in \mathcal{Q}} \max_{j \in \mathcal{A}} s_{ij} + \sum_{i \in \mathcal{A}} \max_{j \in \mathcal{Q}} s_{ij}$. FLQMI is very intuitive for query relevance as well. It measures the representation of data points most relevant to the query set and vice versa.

Log Determinant MI (Logdetmi): The SMI instantiation of LOGDETMI can be defined as: $I_{LogDet}(\mathcal{A}; \mathcal{Q}) = \log \det(S_\mathcal{A}) - \log \det(S_\mathcal{A} - S_{\mathcal{A},\mathcal{Q}} S_\mathcal{Q}^{-1} S_{\mathcal{A},\mathcal{Q}}^T)$. $S_{\mathcal{A},\mathcal{Q}}$ denotes the cross-similarity matrix between the items in sets $\mathcal{A}$ and $\mathcal{Q}$.

3.4 Object-Level Feature Extraction

In [8], the similarity is calculated using the feature vector of Dimension D for T Region of Interest(ROIs) in Query images $\mathcal{Q}$ and P region proposals obtained using a Region Proposal Network(RPN) in Unlabelled images $\mathcal{U}$. Then the dot product along the feature dimension is computed to obtain pairwise similarities between T and P. One problem we encountered with this approach is that extracting the feature from the same model as the one being trained for detection did not give a good feature representation of objects(see Sect. 2 and observations for Fig. 2b). Through detailed analysis, Chen et al. [3] also show that though the object detection networks are trained with additional annotations, the resulting embeddings are significantly worse than those from classification models for image retrieval. In line with observation, we propose decoupling the process by object-level feature selection- detecting objects with object detectors and then encoding them with pre-trained classification models.

3.5   Targeted Active Learning for Document Layout Analysis (TACTFUL)

Stitching together the concepts presented in Sects. 3.1 through 3.4, we propose our approach called Targeted ACTive Learning For DocUment Layout AnaLysis (**Tactful**). The algorithm is presented in 1. In TACTFUL, a user can provide a small annotated set that can be used for initial training as well as query object selection; With a partially trained object detection model and pre-trained feature selection model, TACTFUL can be used to do targeted active learning on a new and unknown set of document pages, and can effectively avoid the curse of class imbalance and thus improving the performance of document layout analysis compared to other methods. We did extensive experiments on the proposed framework, detailed in the following section.

Algorithm 1. Targeted Active Learning for Document Layout Analysis

---

Require: $\mathcal{L}_0$: initial Labeled set, $\mathcal{U}$ unlabeled set, $\mathcal{Q}$ under-performing query set, k: selection budget, Θ Object-Detection model, Φ feature selection model(eg a Pre-Trained ResNET50)

1: Train model on labeled set $\mathcal{L}_0$ to obtain the model parameters Θ_0 { Obtain model macro AP }
2: Evaluate the model and obtain the under-performing class
3: Crop query and unlabeled set into objects set using ground truth and bounding box detection on model Θ. { $\hat{\mathcal{Q}} \leftarrow n \times \mathcal{Q}, \hat{\mathcal{U}} \leftarrow m \times \mathcal{U}$ where n and m are objects in query and unlabeled set respectively }
4: Compute the embeddings $\left\{ \nabla_\Phi \mathcal{L}\left(x_i, y_i\right), i \in \hat{\mathcal{U}} \right\}$ and $\left\{ \nabla_{\theta_\varepsilon} \mathcal{L}\left(x_i, y_i\right), i \in \hat{\mathcal{Q}} \right\}$ { Obtain vectors for computing kernel in Step 5
5: Compute the similarity kernels S and define a MI function $I_f(\hat{\mathcal{U}}; \hat{\mathcal{Q}}) = f(\hat{\mathcal{U}}) + f(\hat{\mathcal{Q}}) - f(\hat{\mathcal{U}} \cup \hat{\mathcal{Q}})$ {Instantiate Functions }
6: $\hat{A} \leftarrow \max_{A \subseteq \mathcal{U}, (A) \leq k} \left(I_f(A; \mathcal{T} \mid \mathcal{P}) \right)$ {Obtain the subset optimally matching the target }
7: Retrace the subset to original data point $\{\bar{A} \leftarrow \hat{A}\}$ and obtain the label of the element in $\bar{A}$ as $\text{Ł}(\bar{A})$
8: Train a model on the combined labeled set $\mathcal{L}_i \in \text{Ł}(\bar{A})$ and repeat steps 2 to 8 until the model reaches the desired performance

---

4   Experiments

We ran experiments for the TACTFUL framework on two datasets. First, we recreated the experiment from Sect. 2, where we randomly took a fraction of about 2000 data points from the Publaynet while maintaining the class imbalance among objects as in the original set. In the following experiment, we took a budget of 50 objects in each AL round in the Publaynet dataset and a total budget of 2000 pages. For the experiment, we used detectron2 [22] to fine-tune

Faster RCNN [15] and train the model. The experiment was done on a system with 120 GB RAM and 2 GPUs of 50 GB Nvidia RTX A6000. We performed two different types of tests. The first strategy was to dynamically update the query set, i.e. in each active learning cycle, we found the underperforming object and updated the query set Q with the underperforming objects. In the second test (Static Query set), we selected the under-performing objects after the initial training and performed subset selection. We compare TACTFUL with TALISMAN [8] as the baseline. Here we show the effectiveness of TACTFUL's object-level selection against TALISMAN's object selection which is limited for each image. We also show the effectiveness of using a different model for feature selection. To assess the scalability and generalizability of our proposed approach, we performed a series of experiments by varying the size of the initial training set and the number of sampled pages per cycle. Specifically, we utilized 6300 train data points and 10,000 unlabelled data points and allocated 200 and 2000 budgets for active learning and the total budget, respectively.

Through experiments, we show that the performance of our active learning approach is not heavily dependent on the choice of pre-trained models used for image embedding as long as the embeddings accurately capture the image content. Thus, we selected pre-trained models based on their ease of use and availability while ensuring they provide high-quality embeddings. We used a similar setting mentioned above for this experiment.

The paper introduces a new dataset called the **Sanskrit dataset**. The Sanskrit dataset is a collection of images in the Sanskrit language, which is an ancient Indian language of Hinduism and a literary and scholarly language in Buddhism, Jainism, and Sikhism. The dataset contains four types of class objects, Image, Math Table, and Text. The dataset contains 1388 training images, 88 validation images, and 82 test images. Table 2 provides the distribution of layout objects. The images in the dataset were collected from various sources, including Sanskrit textbooks, manuscripts, and art. They cover many topics, including religious texts, philosophical texts, poetry, and art.

We randomly selected 334 images from the 1388 training images to create the initial training set. The remaining 1054 training images were used as unlabelled data points. We employed an active learning strategy with a total budget of 300 images, allocating 30 for each active learning cycle. We fine-tuned a Faster R-CNN [15] model pre-trained on the COCO dataset [13] for this experiment. We conducted 10 iterations of the active learning process to evaluate our approach.

5   Results

For **Dynamic Query Set** strategy i.e. new query set in each active learning cycle experiment, We observed that model performance oscillates for each round. Figure 4a shows the progression of model performance. The reason could be constant changes in the query set for each active learning cycle. Table 3 shows the result with respect to different strategies. The GCMI strategy performed better than other strategies. In COM strategy, the FIGURE object class got less AP, maybe due to constant updates to the query set.

Table 2. Statistics of layout categories in the training, development, and testing sets of Sanskrit Dataset

	Training	Validation	Testing
Pages	1356	88	82
Instances Image	263	57	65
Math	2725	340	406
Table	63	24	31
Text	24615	1761	1309
Total	37666	2182	1811

Table 3. AP with respect to different Active learning strategy with dynamic under-performing classes

	AP	AP-TITLE	AP-TITLE	AP-LIST	AP-TABLE	AP-FIGURE
Random	50.8703	84.5026	67.2462	38.0769	36.2	28.3259
GCMI	**54.9445**	85.4826	65.5047	50.084	36.3451	**37.3062**
FL2MI	51.1594	83.9618	65.1775	36.1761	**36.6315**	33.85
COM	48.2865	**85.7209**	**67.6509**	**44.5379**	23.5566	19.966

In **Static Query Set** strategy, we selected the worst performing class after initial training, and the query set class is fixed for all AL rounds. Figure 4b shows that the model oscillates lesser than before when adding a new data point to the trained labeled set. Table 4 shows the result with respect to different strategies. In this experiment, FL2MI performed better than all other strategies. We selected the LIST category as a query set in this experiment. FL2MI gets more AP-LIST than random strategy, beating it by $\tilde{1}0\%$.

We compared TACTFUL with TALISMAN [8] as a baseline. In TALISMAN, the same feature model is used for selecting the image features. As seen from Table 5, TACTFUL has outperformed TALISMAN in all three strategies. This validates our proposal that decoupling detection and feature selection models considerably impact SMI strategies.

Figure 5 shows the cumulative sum of rare classes augmented to the trained labeled set $(\mathcal{L}_i)$ in each active learning round for the dynamic query set. It can be seen that GCMI and FLMI strategies take twice as many objects from rare classes as random ones. This shows the effectiveness of our approach to tackling class imbalance during AL data acquisition.

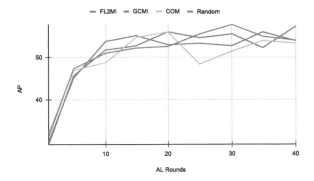

(a) TACTFUL: dynamic query set(different query set) for Publaynet dataset.The plot shows the model AP at interval 5 active learning rounds.

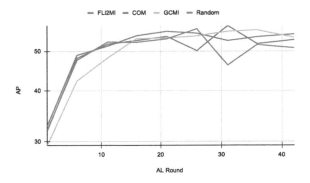

(b) TACTFUL: static query set(list) for Publaynet dataset. The plot shows the model AP at interval 5 active learning rounds. The model oscillates less compare to dynamic query list

Fig. 4. AP with respect to Dynamic Query and Static Query set

Table 4. AP with respect to different Active learning strategy with static rare classes and margin sampling

	AP	AP-TITLE	AP-TITLE	AP-LIST	AP-TABLE	AP-FIGURE
Random	50.8703	84.5026	**67.2462**	38.0769	36.2	28.3259
GCMI	53.949	83.90303	64.8621	39.76392	**45.3594**	35.8586
FL2MI	**55.2586**	84.482	66.0408	**48.9131**	35.348	**43.43945**
COM	53.26239	**84.68513**	63.2735	47.98171	37.80424	31.635
Margin	53.1542	**82.7765**	64.5724	40.57466	42.68292	35.1645

To evaluate our active learning approach, we conducted extensive experiments with varying initial training set sizes and different numbers of sampled

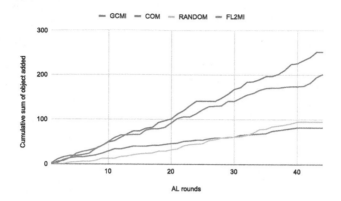

Fig. 5. Cumulative Target Object Added vs AL rounds

Table 5. AP with respect to TALISMAN [8] and TACTFUL

	FL2MI	GCMI	COM	Random
TACTFUL	56.4369	55.2120	55.3783	51.64246
Talisman	51.6316	52.8814	53.0556	51.64246

pages per cycle. In the second set of experiments, we evaluated the impact of different pre-trained models on the performance of our active learning framework. Table 6 gives the experiment's outcome, and fl2mi gives the most significant score. A similar result can be observed with multiple pre-trained models.

Table 6. AP with respect to larger train data points and multiple pre-trained model

	FL2MI	GCMI	COM	Random
RESNET101	**79.46**	78.989	78.71	78.26
RESNET50	**79.211**	78.357	78.526	78.26
RESNET18	**79.196**	78.8507	78.5441	78.26

Finally, we experimented on **Sanskrit Dataset** that contains fewer data points as compared to large corpora like Docbank [11] and Publaynet [23]. Table 7 shows the result with respect to the Sanskrit dataset. All SMI strategies [4, 7, 21] performed better than the random strategy. Among all SMI strategies, GCMI gave the best result. Figure 6a shows the training plot for the Sanskrit dataset and we can see that all SMI functions give better results than the random function.

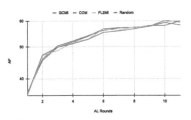
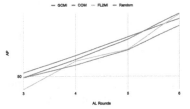

(a) TACTFULfor the Sanskrit dataset for different AL functions with static query set

(b) Zoomed graph depicting AP between 3 to 6 AL rounds. Demonstrates SMI functions work better than random functions.

Fig. 6. AP with respect to Static Query set for the Sanskirt Dataset

Table 7. AP with respect to Sanskrit dataset using TADA

	FL2MI	GCMI	COM	Random
Sanskrit Dataset	49.4122	51.7727	49.1080	48.2420

6   Related Work

In our research, we build upon the work presented in PRISM [9], which addresses two specific subset selection domains: 1) Targeted subset selection and 2) Guided Summarization. PRISM employs a submodular function to determine the similarity between the query set and the lake setting, using distinct submodular function variations to solve the two types of subset selection problems. Our contribution enhances the approach proposed in PRISM and its extension in TALSIMAN [8] by incorporating the object embedding decoupling strategy.

Another related study, by Shekhar et al. [16], focuses on learning an optimal acquisition function through deep Q-learning and models active learning as a Markov decision process. The primary distinction between our framework and theirs is that their approach necessitates pre-training with an underlying dataset. At the same time, our method can be utilized without any pre-training requirements.

Shen et al.'s OLALA [17] attempt to leverage human annotators solely in areas with the highest ambiguity in object predictions within an image. Although they operate at the object level, images are still randomly selected. We propose that our targeted selection approach can be integrated into the OLALA framework, potentially enhancing the efficiency of both methodologies. This integration, however, remains a topic for future research. Furthermore, an additional related work, "ActiveAnno: General-Purpose Document-Level Annotation Tool with Active Learning Integration" [?], presents a versatile annotation tool designed for document-level tasks, integrating active learning capabilities. This tool aims to reduce the human effort required for annotation while maintaining high-quality results. It achieves this by identifying and prioritizing the most informative instances for annotation, thus optimizing the use of human expertise during the annotation process.

Our framework could potentially benefit from incorporating elements of ActiveAnno's approach to streamline the targeted selection and annotation process further. Combining our targeted selection methodology with ActiveAnno's active learning integration may improve both systems' overall efficiency and effectiveness. This potential integration and its implications warrant further exploration and experimentation in future research.

7   Conclusion

In this paper, we propose a Targeted ACTive Learning For DocUment Layout AnaLysis (TACTFUL). That mitigates the class imbalance during image acquisition for annotations, thereby improving the model's performance for rare classes. Through different experiments, we show that our framework significantly improves the model accuracy compared to random, relative to the object level. We also decoupled the model and showed that it can perform better than TALISMAN. This approach can be used with language with fewer data points to improve accuracy.

Acknowledgements. We acknowledge the support of a grant from IRCC, IIT Bombay, and MEITY, Government of India, through the National Language Translation Mission-Bhashini project.

References

1. Abdallah, A., Berendeyev, A., Nuradin, I., Nurseitov, D.: Tncr: table net detection and classification dataset. Neurocomputing **473**, 79–97 (2022)
2. Binmakhashen, G.M., Mahmoud, S.A.: Document layout analysis: a comprehensive survey. ACM Comput. Surv. (CSUR) **52**(6), 1–36 (2019)
3. Chen, B.C., Wu, Z., Davis, L.S., Lim, S.N.: Efficient object embedding for spliced image retrieval. In: Proceedings of the IEEE/CVF Conference on Computer Vision and Pattern Recognition, pp. 14965–14975 (2021)
4. Fujishige, S.: Submodular Functions and Optimization. Elsevier, Amsterdam (2005)
5. Ginsparg, P.: Arxiv at 20. Nature **476**(7359), 145–147 (2011)
6. Gupta, A., Levin, R.: The online submodular cover problem. In: ACM-SIAM Symposium on Discrete Algorithms (2020)
7. Iyer, R., Khargoankar, N., Bilmes, J., Asnani, H.: Submodular combinatorial information measures with applications in machine learning (2020). arXiv preprint arXiv:2006.15412
8. Kothawade, S., Ghosh, S., Shekhar, S., Xiang, Y., Iyer, R.: Talisman: targeted active learning for object detection with rare classes and slices using submodular mutual information (2021). arXiv preprint arXiv:2112.00166
9. Kothawade, S., Kaushal, V., Ramakrishnan, G., Bilmes, J., Iyer, R.: Prism: a rich class of parameterized submodular information measures for guided subset selection (2021). arXiv preprint arXiv:2103.00128

10. Li, J., Li, L., Li, T.: Multi-document summarization via submodularity. Appl. Intell. **37**(3), 420–430 (2012)
11. Li, M., et al.: Docbank: a benchmark dataset for document layout analysis (2020). arXiv preprint arXiv:2006.01038
12. Lin, H.: Submodularity in natural language processing: algorithms and applications. PhD thesis (2012)
13. Lin, T.-Y., et al.: Microsoft COCO: common objects in context. In: Fleet, D., Pajdla, T., Schiele, B., Tuytelaars, T. (eds.) ECCV 2014. LNCS, vol. 8693, pp. 740–755. Springer, Cham (2014). https://doi.org/10.1007/978-3-319-10602-1_48
14. Ren, P., et al.: A survey of deep active learning. ACM Comput. Surv. (CSUR) **54**(9), 1–40 (2021)
15. Ren, S., He, K., Girshick, R., Sun, J.: Faster r-cnn: towards real-time object detection with region proposal networks. Adv. Neural Inf. Process. Syst. **28** (2015)
16. Shekhar, S., Guda, B.P.R., Chaubey, A., Jindal, I., Jain, A.: Opad: an optimized policy-based active learning framework for document content analysis. In: Proceedings of the IEEE/CVF Conference on Computer Vision and Pattern Recognition, pp. 2826–2836 (2022)
17. Shen, Z., Zhao, J., Dell, M., Yu, Y., Li, W.: Olala: object-level active learning for efficient document layout annotation. arXiv preprint arXiv:2010.01762 (2020)
18. Sun, H.Y., Zhong, Y., Wang, D.H.: Attention-based deep learning methods for document layout analysis. In: Proceedings of the 8th International Conference on Computing and Artificial Intelligence, pp. 32–37 (2022)
19. Tharwat, A., Schenck, W.: A survey on active learning: state-of-the-art, practical challenges and research directions. Mathematics **11**(4), 820 (2023)
20. Vasudevan, A.B., Gygli, M., Volokitin, A., Gool, L.V.: Query-adaptive video summarization via quality-aware relevance estimation. In: Proceedings of the 25th ACM International Conference on Multimedia, pp. 582–590 (2017)
21. Wei, K., Iyer, R., Bilmes, J.: Submodularity in data subset selection and active learning. In: International Conference on Machine Learning, pp. 1954–1963. PMLR (2015)
22. Wu, Y., Kirillov, A., Massa, F., Lo, W.Y., Girshick, R.: Detectron2 (2019). https://github.com/facebookresearch/detectron2
23. Zhong, X., Tang, J., Yepes, A.J.: Publaynet: largest dataset ever for document layout analysis. In: 2019 International Conference on Document Analysis and Recognition (ICDAR), pp. 1015–1022. IEEE (2019)

End-to-End Multi-line License Plate Recognition with Cascaded Perception

Song-Lu Chen[1], Qi Liu[1], Feng Chen[2], and Xu-Cheng Yin[1(✉)]

[1] University of Science and Technology Beijing, Beijing, China
{songluchen,xuchengyin}@ustb.edu.cn
[2] EEasy Technology Company Ltd., Zhuhai, China

Abstract. Due to the irregular layout, multi-line license plates are challenging to recognize, and previous methods cannot recognize them effectively and efficiently. In this work, we propose an end-to-end multi-line license plate recognition network, which cascades global type perception and parallel character perception to enhance recognition performance and inference speed. Specifically, we first utilize self-information mining to extract global features to perceive plate type and character layout, improving recognition performance. Then, we use the reading order to attend plate characters parallelly, strengthening inference speed. Finally, we propose extracting recognition features from shallow layers of the backbone to construct an end-to-end detection and recognition network. This way, it can reduce error accumulation and retain more plate information, such as character stroke and layout, to enhance recognition. Experiments on three datasets prove our method can achieve state-of-the-art recognition performance, and cross-dataset experiments on two datasets verify the generality of our method. Moreover, our method can achieve a breakneck inference speed of 104 FPS with a small backbone while outperforming most comparative methods in recognition.

Keywords: License plate recognition · Multi-line · End-to-end

1 Introduction

License Plate Recognition (LPR) is essential to intelligent transportation systems like traffic supervision and vehicle management. However, as shown in Fig. 1, multi-line license plates are challenging to recognize because they are irregular in character layouts compared with single-line license plates.

Fig. 1. Double-line and single-line license plates.

G. A. Fink et al. (Eds.): ICDAR 2023, LNCS 14191, pp. 274–289, 2023.
https://doi.org/10.1007/978-3-031-41734-4_17

As shown in Fig. 2, previous multi-line LPR methods can be roughly categorized into three types, (a) character segmentation, (b) line segmentation, and (c,d) segmentation-free. Firstly, like Fig. 2(a), Silva et al. [26,27,41] propose to detect each plate character independently. However, these methods require labor-intensive annotations and suffer character detection errors. Secondly, like Fig. 2(b), Cao et al. [4,11] propose to bisect the double-line LP horizontally and then splice it into a single-line LP for recognition. However, these methods cannot handle tilted license plates due to incorrect bisection. Thirdly, researchers propose to spatially attend plate characters without segmentation, including CNN-based [19,31,42] and RNN-based [17,33,38]. However, like Fig. 2(c), the CNN-based methods can only extract local features and lack the overall perception of plate type and character layout, which causes attention errors and affects recognition performance. Like Fig. 2(d), the RNN-based methods can pre-classify plate type as the prior knowledge, but they are time-consuming due to encoding features step by step. To solve the above problems, like Fig. 2(e), we propose to cascade global type perception and parallel character perception to improve multi-line LPR. **Firstly**, we use self-information mining to extract global features to pre-perceive plate type and character layout, which can improve multi-line LPR by enhancing the subsequent character attention. **Then**, we use the reading order to spatially attend plate characters, which can parallelly encode features to enhance the inference speed. **Finally**, we can accurately recognize multi-line license plates with globally and spatially enhanced features.

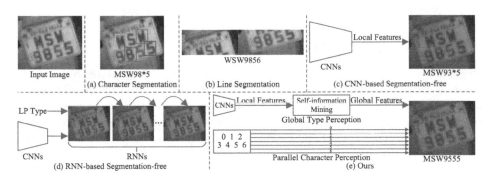

Fig. 2. Our proposed method can improve multi-line LPR by cascading global type perception and parallel character perception.

Moreover, previous methods [7,26,42] utilize separate networks for LPR, where the detected license plates are extracted from the original image for recognition. However, these methods could cause error accumulation because the subsequent recognition will inevitably fail if the previous detection fails. To reduce error accumulation, researchers propose end-to-end methods [16,24,34,35] to jointly optimize detection and recognition, where recognition features are extracted from deep layers of the backbone. However, deep-layer features are not conducive to LPR due to small-sized features after multiple down-sampling operations. To solve this problem, we propose to extract recognition features from shallow layers of the backbone because shallow-layer features retain more plate information for LPR, such as character stroke and layout.

Extensive experiments on RodoSol [13], UFPR [14], and CCPD [34] prove our method can achieve state-of-the-art recognition performance, especially for multi-line license plates. We also conduct cross-dataset experiments to verify the generality of our proposed method, i.e., training on CCPD [34] and testing on PKUData [37] and CLPD [38]. Moreover, our method can achieve the fastest inference speed of 104 FPS with a small backbone while outperforming most comparative methods in recognition performance.

2   Related Work

2.1   License Plate Detection

License plate detection (LPD) methods [6,14,16,26,34] aim to regress the bounding box and predict the confidence of the license plate. In this work, we primarily aim to verify the proposed multi-line license plate recognizer and end-to-end network. Hence, we use the prestigious SSD [20] as the baseline detector, which can sufficiently prove the capability of the proposed recognizer and network. SSD can detect the license plate based on anchor boxes, i.e., hand-picked rectangles with different sizes and aspect ratios. Moreover, we adopt CIoU loss [39] to enhance LPD because it simultaneously considers the overlap area, central point distance, and aspect ratio for better and faster bounding box regression.

2.2   License Plate Recognition

Previously, researchers [8,25,29] proposed to recognize license plates by extracting one-dimensional features. However, these methods cannot effectively recognize multi-line license plates because characters of different lines are squeezed together, causing mutual interference. To solve this problem, researchers have mainly proposed three methods, i.e., character segmentation [26,27,41], line segmentation [4,11], and segmentation-free [17,19,31,33,38,42]. The character segmentation methods regard LPR as an object detection task, i.e., locating characters and combining them into a string. However, these methods require labor-intensive character-level annotations. The line segmentation methods propose to bisect the double-line license plate horizontally into two parts from the original image [11] or feature map [4] and then splice the upper and lower parts into a single-line license plate for recognition. However, the horizontal bisection is fixed in position, which cannot accurately recognize tilted license plates due to incorrect bisection. The segmentation-free methods can spatially attend plate characters based on CNNs [19,31,36,42] or RNNs [17,33,38]. However, the CNN-based methods cannot extract global features for plate type and character layout, causing attention errors. The RNN-based methods are time-consuming due to the step-by-step encoding process. In this work, we propose cascaded perception to improve recognition performance and inference speed.

2.3   End-to-End Recognition

Previous researchers [7,26,42] propose to extract the detected license plate from the original image, then use a separate classifier to recognize it, causing error accumulation. To solve this problem, end-to-end networks [16,24,34,35] are proposed to jointly optimize detection and recognition, where recognition features are extracted from deep layers of the backbone. However, deep-layer features are not conducive to LPR due to losing much information. We propose to extract recognition features from shallow layers of the backbone, which can retain more essential features, such as character stroke and layout, to enhance recognition.

3   Methodology

Figure 3 illustrates the proposed network, including the backbone, detection head, recognition feature extraction, and recognition. **Firstly**, the backbone can extract shared features for detection and recognition. **Then**, the detection head can accurately detect the license plate using multiple deep-layer features. **Finally**, the license plate can be accurately recognized with cascaded perception, where the recognition features are extracted from the first convolutional layer of the backbone to construct an end-to-end network.

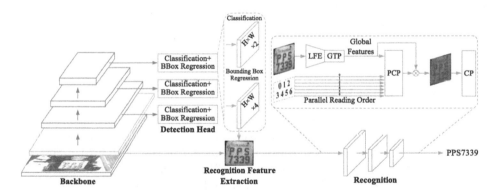

Fig. 3. Overall architecture. LFE: Local Feature Extraction; GTP: Global Type Perception; PCP: Parallel Character Perception; CP: Character Prediction.

3.1   Backbone

We use the large VGGNet [28] or small MobileNetV1 [10] as the backbone to evaluate the generality, which can extract shared features for detection and recognition. For VGGNet, we use the same network as the vanilla SSD [20] for a fair comparison. For MobileNetV1, we adopt the network implemented by the third party [1]. The input size of the backbone is set to 512×512. If not specified, VGGNet is used by default.

3.2   License Plate Detection

We use the prestigious SSD [20] as the detector and replace the bounding box regression loss from Smooth L1 loss [9] to CIoU loss [39]. This way, it can enhance detection because CIoU loss considers the overlap area, central point distance, and aspect ratio for better and faster regression. The training loss comprises the classification loss $\mathcal{L}_{cls}$ and bounding box regression loss $\mathcal{L}_{CIoU}(B, B_{gt})$.

$$\mathcal{L}_{det} = \frac{1}{N}\left(\mathcal{L}_{cls} + \alpha\mathcal{L}_{CIoU}(B, B_{gt})\right), \tag{1}$$

where $\mathcal{L}_{cls}$ is implemented by Softmax loss. N denotes the number of matched anchor boxes. α represents the loss balance term and is set to 5 by default. B and B_{gt} denote the predicted box and ground-truth box, respectively.

3.3   Recognition Feature Extraction

Generally, deep-layer features have strong semantics and a large receptive field [18], which are conducive to LPD. However, deep-layer features are not conducive to LPR because they undergo multiple down-sampling operations and lose much information, such as character stroke and layout. On the contrary, shallow-layer features have weak semantics and a large resolution, which retain the information of character stroke and layout to enhance recognition. In this work, we propose to extract recognition features from the first convolutional layer of the backbone and construct the network end-to-end. This way, it can reduce error accumulation and achieve the best recognition performance.

3.4   License Plate Recognition

After recognition feature extraction, we can get plate features $X_{96\times32}$ with the size of 96×32, which is the input of the recognition network. As shown in Fig. 3, the proposed recognizer consists of local feature extraction, global type perception (GTP), parallel character perception (PCP), and character prediction. GTP and PCP are implemented by self-attention and spatial attention, respectively.

Local Feature Extraction. The extracted recognition features $X_{96\times32}$ can generate local features $X_{24\times8}$ through multi-layer convolutions. As shown in Eq. (2), there are eighteen convolutional layers, including six convolutions of 64, 128, and 256 channels, respectively. All convolutional layers consist of Convolution, BatchNorm, and ReLU. There are two down-sampling operations after the sixth and twelfth layers, so the size of local features is 24×8.

$$X_{24\times8} = [ReLU(BN(Conv(X_{96\times32})))]^{\times18}. \tag{2}$$

Global Type Perception. However, the local features $X_{24\times8}$ cannot perceive plate type and character layout, causing character attention errors. As shown in Fig. 4, we use GTP to extract global features $Y_{24\times8}$, enhancing the subsequent character attention. The global features are generated via self-information mining, and the key, query, and value are all generated by local features. Specifically, we adopt the vanilla transformer unit [30] and stack two transformer units with attention heads $h = 4$, feed-forward dimension $d_{ff} = 512$, and output dimension $d_{model} = 256$. W_K, W_Q, W_V, and W denote learnable parameters for the key, query, value, and position-wise feed-forward network, respectively.

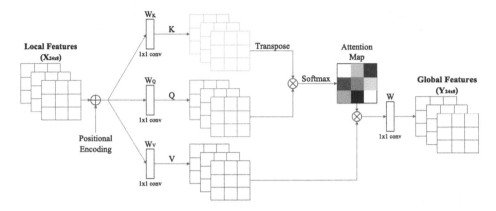

Fig. 4. The architecture of global type perception. It can extract global features to enhance the perception of plate type and character layout.

Parallel Character Perception. As shown in Fig. 5, we use the reading order to parallelly attend plate characters of different layouts, thus improving the inference speed. This way, the global features $Y_{24\times8}$ can generate spatially-attended features $Z_{maxT \times d_{model}}$, where $maxT = 8$ denotes the maximum character length of the license plate. Specifically, the key-value set is generated from global features $Y_{24\times8}$, and V is the same as $Y_{24\times8}$, without any extra operation. The query is the parallel reading order, which is first transformed into one-hot vectors and then encoded by parallel embedding with the hidden dimension $d_h = 256$. W_K, W_Q, W_E, and W represent learnable parameters for the key, query, embedding function, and output function, respectively. The embedding function can be regarded as a fully-connected layer without bias.

Character Prediction. Finally, we adopt a fully-connected layer to map the spatially-attended features $Z_{maxT \times d_{model}}$ to the class probabilities $P_{maxT \times C}$. As shown in Eq. (3), we use the cross-entropy loss as the recognition loss to maximize the prediction confidence of each character. Notably, we add blank characters at the end of the license plate for parallel training, which ensures that all license plates are aligned to the same character length of $maxT$.

$$\mathcal{L}_{rec} = -\sum_{t=1}^{maxT} \sum_{c=1}^{C} y_{tc} \log\left(p_{tc}\right), \tag{3}$$

where C denotes the number of character categories. y and p represent the ground-truth label and predicted class probability, respectively.

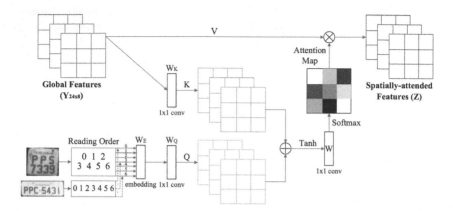

Fig. 5. The architecture of parallel character perception. It can parallelly attend plate characters to improve the inference speed.

The whole network can be trained in an end-to-end manner.

$$\mathcal{L} = \mathcal{L}_{det} + \mathcal{L}_{rec}. \tag{4}$$

4   Experiments

4.1   Datasets and Evaluation Metrics

RodoSol [13] is collected at toll booths on Brazilian highways, containing 20k images. RodoSol contains license plates of two layouts: single-line for cars and double-line for motorcycles. The number of single-line and double-line license plates is the same. We follow the official split, i.e., 8k, 4k, and 8k images for training, validation, and testing, respectively. **UFPR** [14] is collected on Brazilian roads with driving recorders, containing 4.5k images. Like RodoSol, UFPR contains single-line and double-line license plates, where the number of single-line is three times that of double-line. We randomly select 40%, 20%, and 40% images for training, validation, and testing, respectively. **CCPD** [34] is collected in roadside parking scenes of Chinese cities, including two versions: **CCPDv1** and **CCPDv2**, and all license plates are single-line. CCPDv1, containing about 250k images, is divided into subsets Base, DB, FN, Rotate, Tilt, Weather (abbr. Wea.), and Challenge (abbr. Cha.). CCPDv2 contains about 300k images with an extra subset Blur without the subset Weather. The subset Base is used for training and validation, and the other subsets are used for testing.

PKUData [37] is collected on Chinese highways under various conditions. Three subsets are used, i.e., G1 (daytime under normal conditions), G2 (daytime with sunshine glare), and G3 (nighttime), which contain a total of 2253 images.

CLPD [38] is collected from all provinces in mainland China under various real-world scenes. CLPD contains 1.2k images captured with various shooting angles, image resolutions, and backgrounds. Following the literature [32], we conduct cross-dataset experiments on PKUData and CLPD to evaluate the generality, i.e., training on CCPD-Base [34] and testing on PKUData and CLPD.

We use text recognition rate to evaluate the recognition accuracy; that is, all characters must be recognized correctly. The recognition accuracy for single-line and double-line license plates is denoted as **Single** and **Double**, respectively. For RodoSol and UFPR, we also calculate the average accuracy (abbr. **Avg**) of single-line and double-line license plates to evaluate the overall performance. For CCPD, **Avg** denotes the average recognition accuracy of all the subsets. Moreover, we use frames per second (**FPS**) to evaluate the inference speed. Notably, there is only one license plate in each image of all the above datasets.

4.2 Training Settings

When training the detection network alone, the network is trained for 60k iterations using Adam optimizer [12] with initial learning rate 10^{-4}, β_1 momentum 0.9, β_2 momentum 0.99, weight decay 5×10^{-4}, and batch size 32. The learning rate is decreased by ten times at 20k and 40k iterations. Three data augmentation strategies are adopted to enhance the network robustness, i.e., color jittering, random scaling, and random mirror. When training the end-to-end network, we set the initial learning rate to 10^{-5} for the detection branch, which can stabilize the training process. As for the recognition branch, we fix the learning rate to 5×10^{-4}. Considering that image mirroring will affect LPR, we adopt only two data augmentation strategies, i.e., color jittering and random scaling. When training the recognition network alone, we crop the ground truth from the original image and use the same training parameters as the above recognition branch. All experiments are conducted with four NVIDIA Titan Xp GPUs.

4.3 Ablation Study

As shown in Table 1, we conduct ablation experiments of different network components on RodoSol [13], UFPR [14], and CCPD [34], including DE for detection, GTP and PCP for recognition, and E2E for the end-to-end network.

Table 1. Ablation study of different network components.

Method	E2E?	RodoSol			UFPR			CCPDv2	CCPDv1	FPS
		Avg	Single	Double	Avg	Single	Double	Single	Single	
Ours	✓	**96.55**	**98.08**	**95.03**	**99.67**	**99.58**	**100.00**	**73.48**	**98.8**	37
w/o DE	✓	96.28	97.60	94.95	98.22	98.75	96.15	72.12	98.7	37
w/o GTP	✓	93.94	96.38	91.50	98.94	99.16	98.08	70.03	98.6	**38**
w/o PCP	✓	96.40	97.85	94.95	99.00	99.23	98.08	72.59	98.7	*33*
w/o E2E		93.94	97.63	90.25	98.28	98.47	97.53	67.32	98.5	37

E2E?: Is it an end-to-end license plate recognition network?

DE (Detection Enhancement) means using CIoU loss [39] for bounding box regression instead of Smooth L1 loss [9] in the vanilla SSD [20]. CIoU loss regresses the box as a whole unit and considers the overlap area, central point distance, and aspect ratio to improve detection, thus enhancing recognition.

GTP (Global Type Perception) means the global perception of plate type and character layout. Although removing GTP can improve the inference speed by 1 FPS, it will significantly reduce the recognition accuracy, especially for the double-line license plate. For RodoSol and UFPR, the recognition accuracy of single-line license plates decreases by 1.70% and 0.42%, while that of double-line license plates decreases by 3.53% and 1.92%, respectively. As shown in Fig. 6, GTP can improve multi-line LPR by enhancing character attention.

Fig. 6. Character attention with or without GTP. The character attention might be deviated or messy without GTP, affecting character recognition on respective positions.

PCP (Parallel Character Perception) means parallelly attending plate characters using the reading order. Without PCP, it will degrade to the RNN-based segmentation-free model [17,33,38], which sequentially encodes character features and is time-consuming. With PCP, our method can run 4 FPS faster due to encoding parallelly. Moreover, encoding the relationship between characters through RNNs will introduce noises because of no or minor semantics between plate characters. Our method can decouple the semantic relationship between characters by using reading orders independently, thus improving recognition.

E2E (End-to-End) means the proposed end-to-end network, i.e., extracting recognition features from the first convolutional layer of the backbone. Without E2E, it will consist of a separate detector and recognizer, which extracts the detected license plate from the original image. This way, it will significantly reduce recognition accuracy because of error accumulation, especially for double-line license plate. For RodoSol and UFPR, the recognition accuracy of single-line license plates decreases by 0.45% and 1.11%, while that of double-line license plates decreases by 4.78% and 2.47%, respectively.

As shown in Table 2, we study the effect of different recognition inputs, i.e., extracting the detected LP from the original image or convolutional layer. If from the original image, it will cause error accumulation to affect recognition; if from

Table 2. Ablation study of different recognition inputs. Conv: convolution.

Input	E2E?	Size	RodoSol			UFPR			CCPDv2	CCPDv1
			Avg	Single	Double	Avg	Single	Double	Single	Single
Image		–	93.94	97.63	90.25	98.28	98.47	97.53	67.32	98.5
1st Conv	✓	512	**96.55**	**98.08**	**95.03**	**99.67**	**99.58**	**100.00**	**73.48**	**98.8**
4th Conv	✓	256	94.36	96.05	92.68	99.28	99.09	100.00	73.31	**98.8**
7th Conv	✓	128	37.98	34.35	41.60	97.28	97.01	98.35	53.01	96.9
10th Conv	✓	64	0.00	0.00	0.00	89.78	91.57	82.69	5.08	53.2
13th Conv	✓	32	0.00	0.00	0.00	81.00	79.67	86.26	0.00	0.0

the convolutional layer, the network can be optimized jointly to improve recognition. Moreover, shallow convolutional layers are conducive to LPR because they are large-sized and retain more essential features, such as character stroke and layout. Deep-layer features are small-sized after multiple down-sampling operations, which are not conducive to LPR because of losing much information. In this work, we choose the first convolutional layer of the backbone as the recognition input, achieving the best recognition performance for all datasets.

4.4   Comparative Experiments on Multi-line License Plates

As shown in Table 3, for the dataset RodoSol, we compare our method with the character segmentation method CR-Net [27], line segmentation method [4], 1D-sequence recognition methods LPRNet [40], TRBA [3], RNN-based segmentation-free method RNN-Attention [38], and CNN-based segmentation-free method SALPR [19]. The detectors are the vanilla SSD [20], SSD(CIoU) [39], or Chen et al. [5]. SSD(CIoU) denotes using CIoU loss [39] for bounding box regression instead of Smooth L1 loss [9] in the vanilla SSD. Chen et al. can detect the multidirectional license plate by regressing its vertices. Generally, Chen et al. perform better but slower than SSD and SSD(CIoU). We use SSD(CIoU) as the detector in our end-to-end network. The comparative experiments include two types: a separate detector and recognizer and an end-to-end network. For example, SSD+LPRNet means a separate detector SSD [20] and recognizer LPRNet [40]. Ours_Small and Ours_Large are our proposed end-to-end networks. With a large backbone, our method can achieve the best recognition performance with a fast inference speed. When using a small backbone, our method can still outperform all comparative methods in recognition and achieve a breakneck inference speed of 104 FPS, proving its effectiveness and efficiency. Notably, LPRNet [40] performs poorly on double-line license plates because character features of different lines are squeezed together, causing mutual interference. Moreover, the character segmentation [27] and line segmentation methods [4] generally lag behind the segmentation-free methods [19,38] because of character detection errors or inaccurate bisection. However, SALPR [19] cannot effectively perceive plate type and character layout, causing attention errors to affect recognition. RNN-Attention [38] runs slowly due to the step-by-step process.

Table 3. Comparative experiments on RodoSol. Our method can achieve the best recognition accuracy and fastest inference speed for multi-line license plates.

Method	E2E?	RodoSol			FPS
		Avg	Single	Double	
SSD [20]+LPRNet [40]		33.24%	65.08%	1.40%	40
SSD(CIoU) [39]+LPRNet [40]		34.23%	66.88%	1.58%	40
Chen et al. [5]+LPRNet [40]		37.23%	72.66%	1.80%	35
CR-Net[†] [27]		55.80%	–	–	–
SSD [20]+Cao et al. [4]		56.35%	65.50%	47.20%	32
SSD(CIoU) [39]+Cao et al. [4]		57.53%	67.20%	47.85%	32
TRBA[†] [3]		59.60%	–	–	–
Chen et al. [5]+Cao et al. [4]		61.55%	69.85%	53.25%	27
SSD [20]+SALPR [19]		91.65%	94.15%	89.15%	36
SSD(CIoU) [39]+SALPR [19]		91.98%	93.95%	90.00%	36
Chen et al. [5]+SALPR [19]		92.04%	93.95%	90.13%	31
SSD [20]+RNN-Attention [38]		92.79%	95.10%	90.48%	33
SSD(CIoU) [39]+RNN-Attention [38]		93.48%	95.78%	91.18%	33
Chen et al. [5]+RNN-Attention [38]		93.84%	95.98%	91.70%	28
Ours_Small	✓	94.01%	95.88%	92.15%	**104**
Ours_Large	✓	**96.55%**	**98.08%**	**95.03%**	37

†: These data are from the literature [13].
Small, Large: The backbone in Fig. 3 is set to the small MobileNetV1 or large VGGNet.

As shown in Table 4, our method achieves the best recognition performance and fastest inference speed on UFPR, verifying its generality. Sighthound [22], OpenALPR [2], and Laroca et al. [14,15] are LPR methods based on character segmentation. These methods perform worse on double-line license plates because of the irregular character layouts. Notably, we do not compare the multidirectional license plate detector [5] on UFPR because of no vertex annotations.

4.5   Comparative Experiments on Single-Line License Plates

We test the dataset CCPD to verify the generality of our method, which contains only single-line license plates. As shown in Table 5, our method can achieve the best average recognition accuracy and perform best on most subsets of CCPDv2. However, our method performs worse than Chen et al. [5]+RNN-Attention [38] on the subsets Rotate and Tilt that contain highly tilted license plates. Chen et al. can detect multidirectional license plates by vertex regression, while our detector can only detect horizontal license plates. In future work, we will improve the detection and recognition of multidirectional license plates.

Table 4. Comparative experiments on UFPR. Our method can achieve the best recognition accuracy and fastest inference speed for multi-line license plates.

Method	E2E?	UFPR			FPS
		Avg	Single	Double	
SSD [20]+LPRNet [40]		32.11%	37.60%	11.81%	40
SSD(CIoU) [39]+LPRNet [40]		35.33%	42.13%	8.79%	40
Sighthound[‡] [22]		47.40%	58.40%	3.30%	–
OpenALPR[‡] [2]		50.90%	58.00%	22.80%	–
Laroca et al.[‡] [14]		64.90%	72.20%	35.60%	–
TRBA[†] [3]		72.90%	–	–	–
SSD [20]+Cao et al. [4]		74.44%	78.62%	57.97%	32
CR-Net[†] [27]		78.30%	–	–	–
SSD [20]+RNN-Attention [38]		83.56%	87.05%	69.78%	33
SSD [20]+SALPR [19]		83.72%	89.55%	60.71%	36
SSD(CIoU) [39]+Cao et al. [4]		85.83%	86.49%	83.24%	32
Laroca et al.[‡] [15]		90.00%	95.90%	66.30%	–
SSD(CIoU) [39]+RNN-Attention [38]		90.00%	90.81%	86.81%	33
SSD(CIoU) [39]+SALPR [19]		93.28%	95.47%	84.62%	36
Ours_Small	✓	94.94%	96.59%	88.46%	**104**
Ours_Large	✓	**99.67%**	**99.58%**	**100.00%**	37

‡, †: These data are from the literature [13, 15], respectively.

Moreover, as shown in Table 6, our method can achieve state-of-the-art average recognition accuracy and perform best on most subsets of CCPDv1, significantly outperforming all end-to-end methods [16, 23, 24, 34, 35]. Our method can achieve the fastest inference speed with a small backbone while outperforming most comparative methods in recognition.

5   Cross-Dataset Experiments

We conduct cross-dataset experiments on PKUData [37] and CLPD [38] with the training set CCPD-Base [34]. As shown in Table 7, our method can achieve the best cross-dataset recognition performance, verifying its generality. PKU-Data [37] has similar region code distributions with the training set. Hence, the recognition accuracy with and without region codes has little difference. However, CLPD differs from the training set in region code distributions, so the recognition accuracy significantly lags when considering the region code.

Table 5. Comparative experiments on CCPDv2. Our method achieves the best average recognition accuracy and fastest inference speed.

Method	E2E?	Avg	DB	Blur	FN	Rotate	Tilt	Cha.	FPS
SSD+LPRNet		27.74	21.59	20.59	33.57	20.19	14.34	34.57	40
SSD(CIoU)+LPRNet		28.47	22.45	21.49	34.54	20.65	14.48	35.48	40
SSD+Cao et al.		28.95	24.05	19.63	32.60	24.64	19.38	33.74	32
SSD(CIoU)+Cao et al.		29.82	25.32	20.63	34.06	25.30	20.09	34.46	32
Chen et al.+Cao et al.		41.43	26.55	17.51	40.03	69.45	53.50	34.69	27
Chen et al.+LPRNet		41.53	24.05	18.00	40.06	67.87	52.36	35.54	35
SSD+SALPR		43.34	35.50	26.67	45.78	46.15	41.18	45.29	36
SSD+HC[†]		43.42	34.47	25.83	45.24	52.82	52.04	44.62	11
SSD(CIoU)+SALPR		44.45	36.85	27.87	47.51	48.83	42.29	46.10	36
SSD+R-A		47.23	39.66	27.84	48.98	53.70	47.40	47.35	33
SSD(CIoU)+R-A		47.89	40.72	28.82	49.66	54.59	47.71	48.10	33
Chen et al.+SALPR		51.91	34.78	23.25	53.13	82.43	68.14	43.54	31
Chen et al.+R-A		57.23	38.81	24.11	58.89	**88.02**	**78.05**	46.66	28
Ours_Small	✓	70.68	57.52	**65.35**	68.76	69.32	66.42	**75.60**	**104**
Ours_Large	✓	**73.48**	**63.85**	63.00	**75.18**	75.70	73.40	74.63	37

†: It is the official benchmark method. **R-A**: It is the abbreviation of RNN-Attention.

Table 6. Comparative experiments on CCPDv1. Our method achieves the best average recognition accuracy.

Method	E2E?	Avg	Base	DB	FN	Rotate	Tilt	Wea.	Cha.	FPS
Zherzdev et al.[‡] [40]		93.0	97.8	92.2	91.9	79.4	85.8	92.0	69.8	56
TE2E[‡] [16]	✓	94.4	97.8	94.8	94.5	87.9	92.1	86.8	81.2	3
Silva et al.[‡] [26]		94.6	98.7	86.5	85.2	94.5	95.4	94.8	91.2	31
RPNet[†] [34]	✓	95.5	98.5	96.9	94.3	90.8	92.5	87.9	85.1	61
DAN[‡] [31]		96.6	98.9	96.1	96.4	91.9	93.7	95.4	83.1	19
MANGO[‡] [23]	✓	96.9	99.0	97.1	95.5	95.0	96.5	95.9	83.1	8
HomoNet[‡] [35]	✓	97.5	99.1	96.9	95.9	97.1	98.0	97.5	85.9	19
ILPRNet[‡] [42]		97.5	99.2	98.1	98.5	90.3	95.2	97.8	86.2	-
Qin et al.[‡] [24]	✓	97.5	99.5	93.3	93.7	98.2	95.9	98.9	**92.9**	26
MORAN[‡] [21]		98.3	99.5	98.1	98.6	98.1	**98.6**	97.6	86.5	55
AttentionNet[‡] [38]		98.5	99.6	98.8	**98.8**	96.4	97.6	98.5	88.9	40
Ours_Small	✓	98.1	99.3	98.3	98.0	97.3	97.5	98.2	87.6	**104**
Ours_Large	✓	**98.8**	**99.7**	**99.1**	**98.8**	**98.4**	98.5	98.8	90.0	37

†: It is the official benchmark method. ‡: These data are from existing literature.

Table 7. Cross-dataset experiments. R.C. denotes the region code, i.e., the first Chinese character of the license plate. Syn. represents using additional synthetic data.

Method	PKUData		CLPD	
	w/o R.C.	Overall	w/o R.C.	Overall
Sighthound[†] [22]	89.3%	–	85.2%	–
RPNet[†] [34]	78.4%	77.6%	78.9%	66.5%
AttentionNet[†] [38]	86.5%	84.8%	86.1%	70.8%
AttentionNet w/ Syn.[†] [38]	90.5%	88.2%	87.6%	76.8%
Ours	**92.8%**	**92.7%**	**92.4%**	**81.1%**

†: These data are from the literature [38].

6 Conclusion

In this work, we propose an end-to-end license plate recognition network, which can recognize multi-line license plates effectively and efficiently by cascading global type perception and parallel character perception. Extensive experiments on three datasets prove our method can achieve state-of-the-art recognition performance with a fast inference speed. Cross-dataset experiments on two datasets verify the generality of our proposed method. Moreover, we can achieve the fastest inference speed of 104 FPS with a small backbone while outperforming most comparative methods in recognition. In future work, we will enhance multidirectional license plate recognition, especially for highly tilted license plates.

Acknowledgement. The research is supported by National Key Research and Development Program of China (2020AAA0109700), National Science Fund for Distinguished Young Scholars (62125601), and National Natural Science Foundation of China (62076024, 62006018, U22B2055).

References

1. MobileNetV1 SSD (2023). https://github.com/qfgaohao/pytorch-ssd
2. OpenALPR Cloud API (2023). https://www.openalpr.com/carcheck-api.html
3. Baek, J., Kim, G., Lee, J., et al.: What is wrong with scene text recognition model comparisons? dataset and model analysis. In: ICCV, pp. 4714–4722 (2019)
4. Cao, Y., Fu, H., Ma, H.: An end-to-end neural network for multi-line license plate recognition. In: ICPR, pp. 3698–3703 (2018)
5. Chen, S., Tian, S., Ma, J., et al.: End-to-end trainable network for degraded license plate detection via vehicle-plate relation mining. Neurocomputing **446**, 1–10 (2021)
6. Chen, S., Yang, C., Ma, J., et al.: Simultaneous end-to-end vehicle and license plate detection with multi-branch attention neural network. IEEE Trans. Intell. Transp. Syst. **21**(9), 3686–3695 (2020)
7. Dong, M., He, D., Luo, C., et al.: A cnn-based approach for automatic license plate recognition in the wild. In: BMVC, pp. 1–12 (2017)
8. Duan, S., Hu, W., Li, R., et al.: Attention enhanced convnet-rnn for Chinese vehicle license plate recognition. In: PRCV, pp. 417–428 (2018)

9. Girshick, R.: Fast R-CNN. In: ICCV, pp. 1440–1448 (2015)
10. Howard, A., Zhu, M., Chen, B., et al.: Mobilenets: efficient convolutional neural networks for mobile vision applications (2017). arXiv:1704.04861
11. Jain, V., Sasindran, Z., Rajagopal, A., et al.: Deep automatic license plate recognition system. In: ICVGIP, pp. 1–8 (2016)
12. Kingma, D., Ba, J.: Adam: a method for stochastic optimization. In: ICLR, pp. 1–15 (2015)
13. Laroca, R., Cardoso, E., Lucio, D.: On the cross-dataset generalization in license plate recognition. In: VISIGRAPP, pp. 166–178 (2022)
14. Laroca, R., Severo, E., Zanlorensi, L., et al.: A robust real-time automatic license plate recognition based on the YOLO detector. In: IJCNN, pp. 1–10 (2018)
15. Laroca, R., Zanlorensi, L., Gonçalves, G., et al.: An efficient and layout-independent automatic license plate recognition system based on the YOLO detector. IET Intell. Transp. Syst. **15**(4), 483–503 (2021)
16. Li, H., Wang, P., Shen, C.: Towards end-to-end car license plates detection and recognition with deep neural networks. IEEE Trans. Intell. Transp. Syst. **20**(3), 1126–1136 (2019)
17. Li, H., Wang, P., Shen, C., et al.: Show, attend and read: a simple and strong baseline for irregular text recognition. In: AAAI, pp. 8610–8617 (2019)
18. Lin, T., Maji, S.: Visualizing and understanding deep texture representations. In: CVPR, pp. 2791–2799 (2016)
19. Liu, Q., Chen, S., Li, Z., et al.: Fast recognition for multidirectional and multi-type license plates with 2D spatial attention. In: ICDAR, pp. 125–139 (2021)
20. Liu, W., et al.: SSD: single shot multibox detector. In: Leibe, B., Matas, J., Sebe, N., Welling, M. (eds.) ECCV 2016. LNCS, vol. 9905, pp. 21–37. Springer, Cham (2016). https://doi.org/10.1007/978-3-319-46448-0_2
21. Luo, C., Jin, L., Sun, Z.: MORAN: a multi-object rectified attention network for scene text recognition. Pattern Recogn. **90**, 109–118 (2019)
22. Masood, S.Z., Shu, G., Dehghan, A., et al.: License plate detection and recognition using deeply learned convolutional neural networks (2017). arXiv:1703.07330
23. Qiao, L., Chen, Y., Cheng, Z., et al.: MANGO: a mask attention guided one-stage scene text spotter. In: AAAI, pp. 2467–2476 (2021)
24. Qin, S., Liu, S.: Towards end-to-end car license plate location and recognition in unconstrained scenarios. Neural Comput. Appl. **34**(24), 21551–21566 (2022)
25. Shi, B., Bai, X., Yao, C.: An end-to-end trainable neural network for image-based sequence recognition and its application to scene text recognition. IEEE Trans. Pattern Anal. Mach. Intell. **39**(11), 2298–2304 (2017)
26. Silva, S., Jung, C.: License plate detection and recognition in unconstrained scenarios. In: ECCV, pp. 593–609 (2018)
27. Silva, S., Jung, C.: Real-time license plate detection and recognition using deep convolutional neural networks. J. Vis. Commun. Image Represent. **71**, 102773 (2020)
28. Simonyan, K., Zisserman, A.: Very deep convolutional networks for large-scale image recognition. In: ICLR, pp. 1–14 (2015)
29. Spanhel, J., Sochor, J., Juránek, R., et al.: Holistic recognition of low quality license plates by CNN using track annotated data. In: AVSS, pp. 1–6 (2017)
30. Vaswani, A., Shazeer, N., Parmar, N., et al.: Attention is all you need. In: NIPS, pp. 5998–6008 (2017)
31. Wang, T., Zhu, Y., Jin, L., et al.: Decoupled attention network for text recognition. In: AAAI, pp. 12216–12224 (2020)

32. Wang, Y., Bian, Z., Zhou, Y., et al.: Rethinking and designing a high-performing automatic license plate recognition approach. IEEE Trans. Intell. Transp. Syst. **23**(7), 8868–8880 (2022)
33. Xu, H., Zhou, X., Li, Z., et al.: EILPR: toward end-to-end irregular license plate recognition based on automatic perspective alignment. IEEE Trans. Intell. Transp. Syst. **23**(3), 2586–2595 (2022)
34. Xu, Z., Yang, W., Meng, A., et al.: Towards end-to-end license plate detection and recognition: a large dataset and baseline. In: ECCV, pp. 261–277 (2018)
35. Yang, Y., Xi, W., Zhu, C., et al.: Homonet: unified license plate detection and recognition in complex scenes. In: CollaborateCom, pp. 268–282 (2020)
36. Yu, D., Li, X., Zhang, C., et al.: Towards accurate scene text recognition with semantic reasoning networks. In: CVPR, pp. 12110–12119 (2020)
37. Yuan, Y., Zou, W., Zhao, Y., et al.: A robust and efficient approach to license plate detection. IEEE Trans. Image Process. **26**(3), 1102–1114 (2017)
38. Zhang, L., Wang, P., Li, H., et al.: A robust attentional framework for license plate recognition in the wild. IEEE Trans. Intell. Transp. Syst. **22**(11), 6967–6976 (2021)
39. Zheng, Z., Wang, P., Liu, W., et al.: Distance-iou loss: faster and better learning for bounding box regression. In: AAAI, pp. 12993–13000 (2020)
40. Zherzdev, S., Gruzdev, A.: Lprnet: license plate recognition via deep neural networks (2018). arXiv:1806.10447
41. Zhuang, J., Hou, S., Wang, Z., et al.: Towards human-level license plate recognition. In: ECCV, pp. 314–329 (2018)
42. Zou, Y., Zhang, Y., Yan, J., et al.: License plate detection and recognition based on yolov3 and ILPRNET. Signal Image Video Process. **16**(2), 473–480 (2022)

Evaluating Adversarial Robustness on Document Image Classification

Timothée Fronteau[ID], Arnaud Paran[✉][ID], and Aymen Shabou[ID]

DataLab Groupe, Credit Agricole S.A, Montrouge, France
{arnaud.paran,aymen.shabou}@credit-agricole-sa.fr
https://datalab-groupe.github.io/

Abstract. Adversarial attacks and defenses have gained increasing interest on computer vision systems in recent years, but as of today, most investigations are limited to natural images. However, many artificial intelligence models actually handle documentary data, which is very different from real world images. Hence, in this work, we try to apply the adversarial attack philosophy on documentary data and to protect models against such attacks. Our methodology is to implement untargeted gradient-based, transfer-based and score-based attacks and evaluate the impact of defenses such as adversarial training, JPEG input compression and grey-scale input transformation on the robustness of ResNet50 and EfficientNetB0 model architectures. To the best of our knowledge, no such work has been conducted by the community in order to study the impact of these attacks on the document image classification task.

Keywords: Adversarial attacks · Computer vision · Deep learning

1 Introduction

The democratization of artificial intelligence (AI) and deep learning systems has raised public concern about the reliability of these new technologies, particularly in terms of safety and robustness. These considerations among others have motivated the elaboration of a new European regulation, known as the *AI Act*, which aims to regulate the use of AI systems.

In a context of substantial increase of incoming customer documents in big companies, document classification based on computer vision techniques has been found to be an effective approach for automatically classifying documents such as ID cards, invoices, tax notices, etc. Unfortunately, these techniques have demonstrated to be vulnerable to adversarial examples, which are maliciously modified inputs that lead to adversary-advantageous misclassification by a targeted model [28].

However, although robustness of image classification models against several adversarial example generation methods has been benchmarked on datasets like ImageNet and Cifar10 [7], the conclusions of these studies might not apply to

G. A. Fink et al. (Eds.): ICDAR 2023, LNCS 14191, pp. 290–304, 2023.
https://doi.org/10.1007/978-3-031-41734-4_18

document image classification. In fact, documents have different semantic information than most images: they have text, a layout template, often a light background and logos. This is why the adversarial robustness of visual models for document image classification should be evaluated in an appropriate attack setting, and on an appropriate dataset.

In this paper, we evaluate the robustness of two state-of-the-art visual models for document classification on the RVL-CDIP dataset [13]. Our contributions are as follows:

- We establish a threat model that is consistent with the document classification task we study following the taxonomy of [3], and propose a new constraint for generating adversarial examples, adapted to documents.
- We evaluate an EfficientNetB0 architecture [29] and a ResNet50 architecture [15], each one trained two times using different methods, against several adversarial attacks that are representative of the different threat scenarios we can imagine.

After giving an overview of related works on adversarial attacks and document image classification, we present our experimental data and method. Then we describe our experiments and results. Finally, we discuss our findings and conclude the work.

2   Related Work

2.1   Adversarial Attacks on Image Classification

Szegedy et al. [28] are the first ones to show that well performing deep neural networks can be vulnerable to adversarial attacks, by generating adversarial examples for an image classification task. The topic rapidly gained in popularity among the deep learning community, which led to the publishing of several surveys on the matter.

Following that work, multiple gradient based attacks emerged. Those attacks used projected gradient descent in order to maximize the loss of the model [8,12,19]. The idea behind that approach is that when the loss is at a high value, the model will misclassify the data. Carlini and Wagner [4] presented an attack which works in a targeted scenario meaning that we choose the category we want the data to be labeled as. More recently, Sriraman et al. [27] used graduated optimization and replaced the common crossentropy from the loss by a margin between the probability of the correct label and the probability of the adversarial label. That replacement leads to better results than using crossentropy.

In a black box setting, successfully training a surrogate model that will be attacked proved being a decent approach [33].

With the goal of hiding adversarial attacks in the details of pictures, Jia et al. [16] didn't attack in pixel space but rather in discrete cosine transform space on KxK square patches of the image. The idea behind that approach is that the attack will merge better with the original image.

Liu et al. [20] perfected an approach which tries to find optimal perturbations to initialize adversarial search by using adaptative methods. That initialization allows us to explore better the space of adversarial candidates.

Ensemble methods try to attack multiple models at once. One issue is that the models can be very different. To tackle that issue, SVRE [31] uses a variance reduction strategy which allows gradient descent to perform better.

The field of universal perturbations explored attacks which can fool most of the images of the dataset the attack is performed on [22].

Chakraborty et al. [5] and Machado et al. [21] categorize adversarial attacks into threat frameworks depending on the attack surface, capabilities and on the adversary's goal, and present representative methods for each attack threat. The taxonomy of the first article covers adversarial attacks in the vast domain of computer vision, while the second one focuses on image classification.

This threat framework, or threat model, is important to compare attacks. In particular, the authors distinguish white-box to black-box attacks. In the first kind of attack, the adversary has complete knowledge of the target model and its parameters. In the second kind, the attacker has partial or no knowledge of the model and of its training data. Depending on the attacker's knowledge, different attack approaches can be excluded by the threat model, which offers the possibility to compare attacks within a threat model.

Both articles also provide a state of the art of the main defense strategies. However, among the many possibilities that explain the existence of adversarial examples, the scientific community commonly accepts none and it is challenging to design effective defense strategies [21]. In order to help in this process, Carlini et al. [3] gave a thorough set of recommendations on how to evaluate attack methods and defense strategies depending on the attack threat.

Dong et al. [7] selected several adversarial example generation methods that cover the spectrum of categories listed by the previous articles, and benchmarked the robustness of defense models against these attacks on a classification task. They used the CIFAR-10 [17] and ImageNet [6] datasets of labelled pictures. They implemented 19 attack methods and 8 defense models using the PyTorch framework and published a library of example generation methods in Python and robustness evaluation pipelines on GitHub[1]. Another library, CleverHans, implements five of these attacks using the Tensorflow framework [23].

2.2   Document Image Classification

Few open datasets are accessible to evaluate and compare methods for document image classification. The RVL-CDIP dataset [14] is at the time a commonly used dataset to compare deep learning methods for classifying documents[2]. The most recent approaches that obtain state-of-the-art accuracy are Transformer-based

---

[1] Ares robustness library URL: https://github.com/thu-ml/ares.

[2] Meta AI. *Document Image Classification*. June 2020. URL: https://paperswithcode.com/task/document-image-classification.

approaches that make use of visual, textual and layout information, like Doc-Former [2] or LayoutLMv2 [32]. These methods make use of hundreds of thousands of parameters to classify one document, and require the use of an OCR as a preprocessing step to extract the text from documents. For these two reasons, their inference time is longer than deep learning models based on convolutional layers. Also, those models are very heavy compared to convolutional models that only use the visual information of documents while reaching good accuracy, although a bit lower [11]. With the notable exception of VGG16 [25], these models have an order of magnitude smaller number of parameters [1,11]. For those reasons, lightweight convolutional models will be favored unless there is a real need of the extra power of multimodal models. That's why we decided to focus on visual convolutional models as this case will cover more real life applications in several industries.

These models have been trained and optimized to obtain state-of-the-art accuracy, without consideration for their adversarial robustness. However, accurate models are not necessarily robust to adversarial examples, indeed they often perform poorly in comparison to humans. Therefore, evaluating adversarial robustness is a tool for understanding where these systems fail and make progress in deep learning [3].

3   Method and Data

In our attempt to evaluate the adversarial robustness of common document image classification methods, we carefully selected a proper dataset for document classification, a range of threat models in which we want to perform our evaluation, and a few attack methods and defense models that suit these settings.

3.1   Data

The *RVL-CDIP* dataset [14] is an open dataset containing 400,000 black-and-white document images split into 16 categories among which we can find "news article", "advertisement", "email" or "handwritten", for example. We subsampled the images into images of shape 240×240 pixels, and duplicate the pixel values into the three RGB channels so that they fit our models, which are presented below. The images are therefore of shape $240 \times 240 \times 3$.

The dataset is already split into a training set, a validation set and a test set of 360,000, 40,000 and 40,000 documents respectively, saved in TIFF image format. We randomly selected 1000 documents from the test set to perform our evaluations, containing 54 to 68 examples for each of the 16 classes.

3.2   Threat Model

In order to have correct metrics, and to assess the validity of our approach in a real-world context, it is essential that we define a threat model using a precise taxonomy. We define this threat model according to Carlini et al. [3] by defining the goals, capabilities, and knowledge of the target AI system, that the adversary has.

Let $C(x) : X \to Y$ be a classifier, and $x \in X \subset \mathbb{R}^d$ an input to the classifier. Let $y^{true} \in Y = \{1, 2, ..., N\}$ be the ground truth of the input x, i.e. the true class label of x among N classes. We call $y^{pred} = C(x)$ the predicted label for x.

In our study, we perform *untargeted attacks*, i.e. the adversary's goal is to generate an adversarial example x^{adv} for x that is misclassified by C. This goal is easier to reach than the one of a targeted attack, where x^{adv} should be classified as a predefined $y^{adv} \neq y^{true}$ to be considered adversarial. x^{adv} should also be as *optimal* as possible, which means that it fools the model with high confidence in the wrong predicted label, with an input x^{adv} that is indistinguishable from x by a human eye [21].

Formally, we can define the *capability* of the adversary by defining a *perturbation budget* ε so that $\|x - x^{adv}\|_p < \varepsilon$ where $\| \cdot \|_p$ is the L_p norm with $p \in \{2, \infty\}$. For some attack methods, the capability of the adversary can also be described with the *strength* of the attack, that we define as the number of queries an adversary can make to the model.

3.3   Attack Methods

Here we present the nine attack methods we used for generating adversarial examples. We selected them in order to cover a large spectrum of the capability settings an adversary can have. We cover white-box to black-box attacks, namely gradient-based, transfer-based and score-based attacks, under the L_∞ and the L_2 norms.

Examples of the adversarial images computed with gradient-based and score-based attacks are presented in Fig. 1.

Gradient-Based Attacks. Considering that the adversary has access to all parameters and weights of a target model, we can generate a perturbation for x by using the gradient of a loss function computed with the model parameters. The Fast Gradient Method (FGM) [12], the Basic Iterative Method (BIM) [19] and the Momentum Iterative Method (MIM) [8] are three gradient-based methods we evaluated in this paper. BIM and MIM are recursive versions of FGM, which requires only one computation step.

The three attacks have a version well suited for generating perturbations under the L_∞ constraint and another that suits the L_2 constraint, so we evaluated both versions for a total of six different attacks. When necessary, we differentiate the L_∞ and the L_2 versions of the attacks with the suffix -L_∞ and -L_2 respectively.

Transfer-Based Attacks. Adversarial images generated with gradient-based attacks on a substitute model have a chance of also leading the target model to misclassification. To evaluate the extent of this phenomenon of transferability [24], we used the perturbed examples generated with robust models to attack a target model in a transfer-based setting. The examples are generated under L_∞ constraint with FGM and MIM, which is a variation of BIM designed to improve the transferability of adversarial examples. Note that the adversary has access to the training set of the target model to train the substitute model.

Score-Based Attack. Considering that the adversary has only access to the prediction array of a target model and can query it a maximum number of times defined as the attack strengh, we can design the so-called score-based attacks. The Simultaneous Perturbation Stochastic Approximation method (SPSA) [30] is one of them. We evaluate it under L_∞ constraint.

3.4 Models and Defenses

Model Backbones. We conducted our experiments on two visual model architectures that currently perform best on RVL-CDIP with less than a hundred million of parameters, as far as we know[3]: an EfficientNetB0 and a ResNet50. According to the method of Ferrando et al. [11], who observed the best accuracy of both models on our classification task, we initialized these models with weights that have been pre-trained on the ImageNet classification task.

Reactive Defenses. We implemented preprocessing steps, that precede the model backbone, with the aim of improving their robustness. The first of these reactive defenses we implemented is the JPEG preprocessing. We compress the document image into the JPEG format, then decompress each image that is given as input of the model backbone (**JPEG** defense). This transformation has been identified by Dsiugaite et al. [10] as a factor of robustness against adversarial attacks. This transformation is all the more interesting because it is widely used for industrial applications.

The grey-scale transformation is an other low-cost preprocessing step that we evaluated. The model backbones require that we use three color channels, but document images may contain less semantic information accessible via the colors of the image. As a matter of fact, the RVL-CDIP dataset is even composed of grey-scale images. Therefore, constraining the input images before the model backbone by averaging the values of each RGB pixel (**Grey** defense) does not affect the test accuracy of the models, but might improve adversarial robustness.

Proactive Defense. Some training methods have proven to improve robustness against adversarial attacks on image classification tasks [7]. In this article, we compare the natural fine-tuning proposed in Ferrando et al. [11] with the adversarial training method suggested in Kurakin et al. [18], which has the advantage of being very easy to implement.

In this last method, the fine-tuning step is performed using the adversarial examples generated with BIM (**Adversarial** defense). One adversarial example

---

[3] https://paperswithcode.com/sota/document-image-classification-on-rvl-cdip.

is generated for each input of the training set and for each epoch of the training. We use the same learning rate scheduling as for the natural training method. This means that the attacks are performed against the training model and not only against the naturally trained model.

4   Experiments and Results

Similar to Dong et al. [7], we selected two distinct measures to assess the robustness of the models under attack, defined as follows. Given an attack method A that generates an adversarial example $x^{adv} = A(x)$ for an input x, the accuracy of a classifier C is defined as $Acc(C, A) = \frac{1}{N} \sum_{i=1}^{N} \mathbb{1}(C(A(x_i)) = y_i)$, where $\{x_i, y_i\}_{1 \leq i \leq N}$ is the test set and $\mathbb{1}(\cdot)$ is the indicator function. The attack success rate of an untargeted attack on the classifier is defined as $Asr(C, A) = \frac{1}{M} \sum_{i=1}^{N} \mathbb{1}(C(x_i) = y_i \wedge C(A(x_i)) \neq y_i)$, where $M = \Sigma_{i=1}^{N} \mathbb{1}(C(x_i) = y_i)$.

On Figs. 2, 3 and 4, we draw curves of accuracy vs. perturbation budget of each model under the nine attacks we evaluated. On the other hand, the test accuracy of each model under no attack is rendered on Table 1, and ranges from 86.1% to 90.8%.

Table 1. Test accuracy of each model. The test accuracy has been computed on the same 1000 examples as for all experiments.

Defense Method	Model Backbone	
	EfficientNet	ResNet
No Defense	**90.8**	**89.0**
Grey	**90.8**	**89.0**
JPEG	86.8	86.1
Adversarial	89.0	87.3
Grey + JPEG + Adversarial	89.3	87.2

4.1   Adversarial Robustness Under Gradient-Based Attack

We performed the gradient-based attacks under L_∞ and L_2 constraints for perturbation budgets within the ranges of 0.005 to 0.20, and of 0.5 to 20 respectively. We display the results in Fig. 2 and Fig. 3 respectively, and render them in Table 2. For BIM and MIM, we set the attack strength to 20 queries to the target model.

First we can see that the accuracy of undefended models drops drastically from more than 88% to less than 0.6% under BIM-L_∞ and MIM-L_∞ attacks with $\varepsilon = 0,02$, and to less than 14.9% under BIM-L_2 and MIM-L_2 attacks with $\varepsilon = 2$. We see that the JPEG and Grey defenses don't necessarily improve the

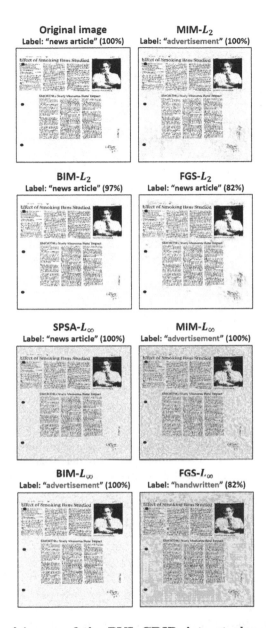

Fig. 1. An original image of the RVL-CDIP dataset along with perturbed images generated with gradient-based and score-based attacks against EfficientNet with JPEG defense. The true label of the image is "news article". We set the perturbation budget to 0.2 for the attacks under L_∞ norm, and to 20 for the others. On top of each image figures the label predicted by the model, along with the confidence of the model in its prediction. The images with red labels are successful attack images (Color figure online).

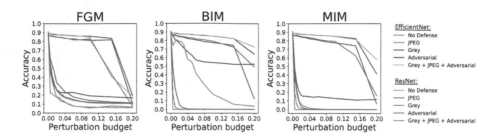

Fig. 2. The accuracy vs. perturbation budget curves of gradient-based attacks under L_∞ norm.

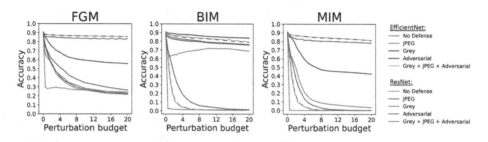

Fig. 3. The accuracy vs. perturbation budget curves of gradient-based attacks under L_2 norm.

Table 2. The accuracy models under several gradient-based and score-based attacks. The attacks have been computed with a perturbation budget of 0.02 for attacks under L_∞ constraint, and of 2 for attacks under L_2 constraint.

Backbone	Defense Method	Attack Method						
		FGM	BIM	MIM	FGM	BIM	MIM	SPSA
		X-L_∞			X-L_2			$-L_\infty$
EfficientNet	No defense	18.3	00.0	00.0	28.9	01.7	00.0	02.5
	Grey	26.2	76.9	16.8	75.2	**88.7**	74.8	–
	JPEG	17.6	78.8	06.7	54.4	65.0	49.3	79.1
	Adversarial	87.7	87.7	87.7	**87.5**	86.3	**86.6**	**88.0**
	Grey + JPEG + Adversarial	**88.1**	**87.9**	**87.9**	87.4	86.6	**86.6**	–
ResNet	No defense	25.3	00.3	00.6	54.2	12.6	14.9	39.7
	Grey	26.5	00.4	00.7	56.4	15.0	17.4	–
	JPEG	34.2	05.8	03.2	60.8	41.4	36.6	47.9
	Adversarial	85.0	84.3	84.4	85.2	82.2	82.9	**85.1**
	Grey + JPEG + Adversarial	**85.1**	**85.1**	**85.1**	**85.3**	84.1	83.5	–

adversarial robustness of target models, and when they do, the improvements are inconsistent depending on the attack and the considered model backbone. For example, with $\varepsilon = 0.02$ against BIM-L_∞, the accuracy of EfficientNet with Grey defense goes up to 76.9% from 0%, while the accuracy of ResNet only improves by 0.1 point from 0.3%, and the accuracy of EfficientNet under MIM-L_∞ goes up to 16.8% from 0%.

The adversarial training, on the other hand, improves consistently for all model backbones and under all evaluated attacks the robustness of targeted models. In fact, accuracies of adversarially trained models only drop of 1.9 points in average against attacks under L_∞ constraint with $\varepsilon = 0.02$, and of 2.9 points in average against attacks under L_2 constraint with $\varepsilon = 2$.

4.2 Adversarial Robustness Under Transfer-Based Attack

We generated perturbed examples with FGM-L_∞ and MIM-L_∞ by using the adversarially trained model backbones EfficientNet and ResNet as subtitute models to attack every other model in a transfer-based setting, and render the results in Fig. 4 and Table 3.

Table 3. The accuracy models under several transfer-based attacks under L_∞ constraint, with a perturbation budget of 0.02. Target models have been attacked with two substitute models: an adversarially trained EfficientNet and an adversarially trained ResNet. As we consider transfer attacks we didn't rerun attacks towards the same model.

Backbone	Defense Method	Substitute Model			
		EfficientNet		ResNet	
		Method			
		FGM-L_∞	MIM-L_∞	FGM-L_∞	MIM-L_∞
EfficientNet	No defense			83.6	83.9
	Grey			82.0	83.1
	JPEG			72.0	73.3
	Adversarial			88.9	88.9
	Grey + JPEG + Adversarial			**89.2**	**89.2**
ResNet	No defense	80.8	81.7		
	Grey	80.9	81.8		
	JPEG	73.8	78.1		
	Adversarial	86.4	86.3		
	Grey + JPEG + Adversarial	**86.5**	**86.7**		

We can observe that undefended models are vulnerable to adversarial images generated with the more robust substitute models (up to 36 points of accuracy decrease), while adversarially trained models stay robust (less than 4 points of accuracy decrease) for a perturbation budget $\varepsilon = 0.10$. Interestingly, the JPEG

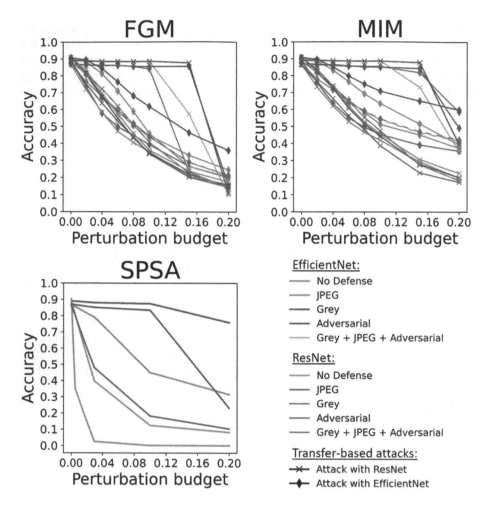

Fig. 4. The accuracy vs. perturbation budget curves of transfer-based attacks (FGM, MIM) and a score-based attack (SPSA) under L_∞ norm.

defense seems to increase the vulnerability of models to adversarial attacks: the curves of EfficientNet and ResNet with JPEG defense stay under the curves of undefended EfficientNet and ResNet when they are attacked by the same perturbed examples. The Grey defense doesn't change the accuracy of the naturally trained ResNet by more than 2 points, while the Grey defense on the naturally trained EfficientNet has inconsistent behaviour depending on the adversary's model, improving robustness against the EfficientNet substitute model but reducing it against ResNet model.

4.3   Adversarial Robustness Under Score-Based Attack

The SPSA attack requires to set several parameters. According to the taxonomy of Uesato et al. [30], we perform 20 iterations of the method as for BIM and MIM, and for each iteration, we compute a batch of 10 images with 40 pixel perturbations each. Therefore, it was 40 times longer to compute an adversarial example with SPSA than with BIM or MIM, as we could observe. With such parameters, we can reduce the robustness of every model in a similar way than with gradient-based attacks (see Fig. 4 and Table 3). In fact, in average, models with no defense have 2.5% and 39.7% of accuracy under SPSA attack with a perturbation budget $\varepsilon = 0.10$, with EfficientNet and ResNet backbones respectively. With JPEG defense, they still have 79.1% and 47.9% of accuracy respectively, and the accuracy goes up to 88.0% and 85.1% for adversarially trained models. It shows that naturally trained document classification models can be vulnerable to score-based attacks if the threat model (the attack strength in particular) allows it.

5   Conclusion and Future Works

As expected, a convolutional model such as an EfficientNetB0 or a ResNet50 trained without any strategy to improve its robustness, is very sensitive to optimal adversary examples generated with gradient-based white-box untargeted attacks, which constitute attacks performed in the best threat model for an adversary. Restraining the values of pixels so that the images are grey-scale doesn't improve much the robustness of the models. Compressing and then decompressing input images of a model using JPEG protocol improves the model robustness against adversarial images, but not consistently.

On the other hand, the adversarial training of both models using the method of Kurakin et al. [18], strongly improves robustness of both models. This training method is quite easy to implement and does not affect a lot the test accuracy of models on our classification task. Therefore, this method seems far more effective on a document classification task than against an image classification task, as we can see in Dong et al. [7].

The black-box attack we evaluated generates blurry examples that appear darker than legitimate document images. It seems to be the case with other black-box attacks, since the perturbation is generated randomly using probabilistic laws that don't take into account the fact that document images are brighter than other images [9].

There are many ways to improve the robustness of a model that would only use the visual modality of a document [5,21]. However, state-of-the-art approaches to document classification take advantage of other information modalities, such as the layout of the document, and the text it contains see footnote 3. Therefore, after this work on evaluating the robustness of visual models, it would be interesting to evaluate the transferability of the generated examples to a multimodal model such as DocFormer or LayoutLMv2, which use optical character recognition (OCR) and transformer layers. Furthermore, we could explore the possibility of designing adversarial attacks to which these models are more sensitive, for example by targeting OCR prediction errors [26] that affect the textual modality and may also affect the robustness of such models [34]. Dealing with the added modality given by text means that more approaches can be explored attacking only one modality or both.

Other defense mechanisms could be explored too, for example training a model to detect adversarial examples is a common approach. We did not favor that one as it implies using multiple models which makes inference longer but future works could explore that possibility.

On the other hand, attacks specific to document data could be thought of, for example changing the style of the text by changing the font, the size, using bold or italic characters. That would imply a much more complex pipeline to find those adversarial examples but those attacks can benefit from being able to generate perturbations of large amplitude which is a case we did not explore here as our perturbations were bounded in norm.

In conclusion, with this paper, we open up the path to adversarial attacks on documentary data as this issue is a major concern and has not been studied much in the litterature. Our first conclusions are reassuring as adversarial training works on the attacks we explored but further work is needed to explore other attack and defense scenarios.

References

1. Afzal, M.Z., Kölsch, A., Ahmed, S., Liwicki, M.: Cutting the error by half: investigation of very deep cnn and advanced training strategies for document image classification. In: 2017 14th IAPR International Conference on Document Analysis and Recognition (ICDAR), vol. 1, pp. 883–888. IEEE (2017)
2. Appalaraju, S., Jasani, B., Kota, B.U., Xie, Y., Manmatha, R.: Docformer: end-to-end transformer for document understanding. In: Proceedings of the IEEE/CVF International Conference on Computer Vision, pp. 993–1003 (2021)
3. Carlini, N., et al.: On evaluating adversarial robustness (2019). arXiv preprint arXiv:1902.06705
4. Carlini, N., Wagner, D.: Towards evaluating the robustness of neural networks. In: 2017 IEEE Symposium on Security and Privacy (SP), pp. 39–57. IEEE (2017)
5. Chakraborty, A., Alam, M., Dey, V., Chattopadhyay, A., Mukhopadhyay, D.: A survey on adversarial attacks and defences. CAAI Trans. Intell. Technol. 6(1), 25–45 (2021)
6. Deng, J., Dong, W., Socher, R., Li, L.J., Li, K., Fei-Fei, L.: Imagenet: a large-scale hierarchical image database. In: 2009 IEEE Conference on Computer Vision and Pattern Recognition, pp. 248–255. IEEE (2009)
7. Dong, Y., et al.: Benchmarking adversarial robustness on image classification. In: Proceedings of the IEEE/CVF Conference on Computer Vision and Pattern Recognition, pp. 321–331 (2020)
8. Dong, Y., et al.: Boosting adversarial attacks with momentum. In: Proceedings of the IEEE Conference on Computer Vision and Pattern Recognition, pp. 9185–9193 (2018)

9. Dong, Y., et al.: Efficient decision-based black-box adversarial attacks on face recognition. In: Proceedings of the IEEE/CVF Conference on Computer Vision and Pattern Recognition, pp. 7714–7722 (2019)

10. Dziugaite, G.K., Ghahramani, Z., Roy, D.M.: A study of the effect of jpg compression on adversarial images (2016). arXiv preprint arXiv:1608.00853

11. Ferrando, J., et al.: Improving accuracy and speeding up document image classification through parallel systems. In: Krzhizhanovskaya, V.V., et al. (eds.) ICCS 2020. LNCS, vol. 12138, pp. 387–400. Springer, Cham (2020). https://doi.org/10.1007/978-3-030-50417-5_29

12. Goodfellow, I.J., Shlens, J., Szegedy, C.: Explaining and harnessing adversarial examples (2014). arXiv preprint arXiv:1412.6572

13. Harley, A.W., Ufkes, A., Derpanis, K.G.: Evaluation of deep convolutional nets for document image classification and retrieval. In: International Conference on Document Analysis and Recognition (ICDAR) (2015)

14. Harley, A.W., Ufkes, A., Derpanis, K.G.: Evaluation of deep convolutional nets for document image classification and retrieval. In: 2015 13th International Conference on Document Analysis and Recognition (ICDAR), pp. 991–995. IEEE (2015)

15. He, K., Zhang, X., Ren, S., Sun, J.: Deep residual learning for image recognition. In: Proceedings of the IEEE Conference on Computer Vision and Pattern Recognition, pp. 770–778 (2016)

16. Jia, S., Ma, C., Yao, T., Yin, B., Ding, S., Yang, X.: Exploring frequency adversarial attacks for face forgery detection. In: Proceedings of the IEEE/CVF Conference on Computer Vision and Pattern Recognition, pp. 4103–4112 (2022)

17. Krizhevsky, A., Hinton, G.: Learning multiple layers of features from tiny images. Technical Report, University of Toronto, Toronto, Ontario (2009)

18. Kurakin, A., Goodfellow, I., Bengio, S.: Adversarial machine learning at scale (2016). arXiv preprint arXiv:1611.01236

19. Kurakin, A., Goodfellow, I.J., Bengio, S.: Adversarial examples in the physical world. In: Artificial Intelligence Safety and Security, pp. 99–112. Chapman and Hall/CRC (2018)

20. Liu, Y., Cheng, Y., Gao, L., Liu, X., Zhang, Q., Song, J.: Practical evaluation of adversarial robustness via adaptive auto attack. In: Proceedings of the IEEE/CVF Conference on Computer Vision and Pattern Recognition, pp. 15105–15114 (2022)

21. Machado, G.R., Silva, E., Goldschmidt, R.R.: Adversarial machine learning in image classification: a survey toward the defender's perspective. ACM Comput. Surv. (CSUR) 55(1), 1–38 (2021)

22. Moosavi-Dezfooli, S.M., Fawzi, A., Fawzi, O., Frossard, P.: Universal adversarial perturbations. In: Proceedings of the IEEE Conference on Computer Vision and Pattern Recognition, pp. 1765–1773 (2017)

23. Papernot, N., et al.: Technical report on the cleverhans v2.1.0 adversarial examples library (2018). arXiv preprint arXiv:1610.00768

24. Papernot, N., McDaniel, P., Goodfellow, I., Jha, S., Celik, Z.B., Swami, A.: Practical black-box attacks against deep learning systems using adversarial examples, vol. 1, no. 2, p. 3 (2016). arXiv preprint arXiv:1602.02697

25. Simonyan, K., Zisserman, A.: Very deep convolutional networks for large-scale image recognition (2014). arXiv preprint arXiv:1409.1556

26. Song, C., Shmatikov, V.: Fooling ocr systems with adversarial text images (2018). arXiv preprint arXiv:1802.05385

27. Sriramanan, G., Addepalli, S., Baburaj, A., et al.: Guided adversarial attack for evaluating and enhancing adversarial defenses. Adv. Neural Inf. Process. Syst. 33, 20297–20308 (2020)

28. Szegedy, C., et al.: Intriguing properties of neural networks (2013). arXiv preprint arXiv:1312.6199
29. Tan, M., Le, Q.: Efficientnet: rethinking model scaling for convolutional neural networks. In: International Conference on Machine Learning, pp. 6105–6114. PMLR (2019)
30. Uesato, J., O'donoghue, B., Kohli, P., Oord, A.: Adversarial risk and the dangers of evaluating against weak attacks. In: International Conference on Machine Learning, pp. 5025–5034. PMLR (2018)
31. Xiong, Y., Lin, J., Zhang, M., Hopcroft, J.E., He, K.: Stochastic variance reduced ensemble adversarial attack for boosting the adversarial transferability. In: Proceedings of the IEEE/CVF Conference on Computer Vision and Pattern Recognition, pp. 14983–14992 (2022)
32. Xu, Y., et al.: Layoutlmv2: multi-modal pre-training for visually-rich document understanding (2020). arXiv preprint arXiv:2012.14740
33. Zhang, J., et al.: Towards efficient data free black-box adversarial attack. In: Proceedings of the IEEE/CVF Conference on Computer Vision and Pattern Recognition, pp. 15115–15125 (2022)
34. Zhang, W.E., Sheng, Q.Z., Alhazmi, A., Li, C.: Adversarial attacks on deep-learning models in natural language processing: a survey. ACM Trans. Intell. Syst. Technol. (TIST) 11(3), 1–41 (2020)

UTRNet: High-Resolution Urdu Text Recognition in Printed Documents

Abdur Rahman(✉)⬤, Arjun Ghosh, and Chetan Arora

Indian Institute of Technology Delhi, Delhi, India
ch7190150@iitd.ac.in

Abstract. In this paper, we propose a novel approach to address the challenges of printed Urdu text recognition using high-resolution, multi-scale semantic feature extraction. Our proposed UTRNet architecture, a hybrid CNN-RNN model, demonstrates state-of-the-art performance on benchmark datasets. To address the limitations of previous works, which struggle to generalize to the intricacies of the Urdu script and the lack of sufficient annotated real-world data, we have introduced the UTRSet-Real, a large-scale annotated real-world dataset comprising over 11,000 lines and UTRSet-Synth, a synthetic dataset with 20,000 lines closely resembling real-world and made corrections to the ground truth of the existing IIITH dataset, making it a more reliable resource for future research. We also provide UrduDoc, a benchmark dataset for Urdu text line detection in scanned documents. Additionally, we have developed an online tool for end-to-end Urdu OCR from printed documents by integrating UTRNet with a text detection model. Our work not only addresses the current limitations of Urdu OCR but also paves the way for future research in this area and facilitates the continued advancement of Urdu OCR technology. The project page with source code, datasets, annotations, trained models, and online tool is available at abdur75648.github.io/UTRNet.

Keywords: Urdu OCR · UTRNet · UTRSet · Printed Text Recognition · High-Resolution Feature Extraction

1 Introduction

Printed text recognition, also known as optical character recognition (OCR), involves converting digital images of text into machine-readable text and is an important topic of research in the realm of document analysis with applications in a wide variety of areas [61]. While OCR has transformed the accessibility & utility of written/printed information, it has traditionally been focused on Latin languages, leaving non-Latin low-resource languages such as Urdu, Arabic, Pashto, Sindhi and Persian largely untapped. Despite recent developments in Arabic script OCR [5,18,28], research on OCR for Urdu remains limited [27,34, 35]. With over 230 million native speakers and a huge literature corpus, including

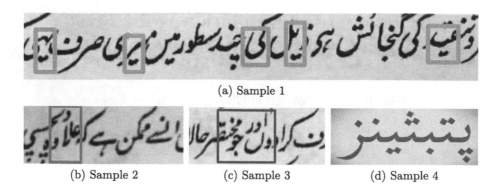

(a) Sample 1

(b) Sample 2   (c) Sample 3   (d) Sample 4

Fig. 1. Intricacies of the script. (a) All 5 characters inside the boxes are the same, but they look different as the shape of the character relies on its position and context in the ligature. The box contains 8 characters in (b) and 11 in (c), demonstrating that the script has a very high degree of overlap. (c) The word shown consists of 6 distinct characters of similar shape, which differ merely by the arrangement of little dots (called *"Nuqta"*) around them.

classical prose and poetry, newspapers, and manuscripts, Urdu is the 10th most spoken language in the world [30,43]. Hence the development of a robust OCR system for Urdu remains an open research problem and a crucial requirement for efficient storage, indexing, and consumption of its vast heritage, mainly its classical literature.

However, the intricacies of the Urdu script, which predominantly exists in the Nastaleeq style, present significant challenges. It is primarily cursive, with a wide range of variations in writing style and a high degree of overall complexity, as shown in Fig. 1. Though Arabic script is similar [56], the challenges faced in recognizing Urdu text differ significantly. Arabic text is usually printed in the Naskh style, which is mostly upright and less cursive, and has only 28 alphabets [28], in contrast to the Urdu script, which consists of 45 main alphabets, 26 punctuation marks, 8 honorific marks, and 10 Urdu digits, as well as various special characters from Persian and Arabic, English alphabets, numerals, and punctuation marks, resulting in a total of 181 distinct glyphs [30] that need to be recognized. Furthermore, the lack of large annotated real-world datasets in Urdu compounds these challenges, making it difficult to compare different models' performance accurately and to continue advancing research in the field (Sect. 4). The lack of standardization in many Urdu fonts and their rendering schemes (particularly in early Urdu literature) further complicates the generation of synthetic data that closely resembles real-world representations. This hinders experiments with more recent transformer-based OCR models that require large training datasets (Table 3B) [37,57,70]. As a result, a naive application of the methods developed for other languages does not result in a satisfactory performance for Urdu (Table 3), highlighting the need for exclusive research in OCR for Urdu.

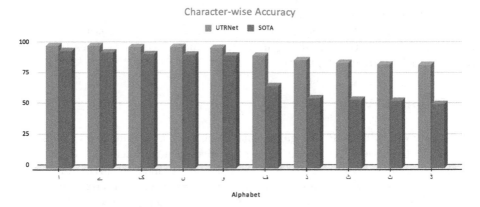

Fig. 2. Plot showing character-wise accuracies of UTRNet-Small and SOTA for Urdu OCR, presented in [30]. It can be observed that the accuracy gap is larger for the last 5 characters (right side), which differ from several other characters having the same shape only in terms of presence and arrangement of dots around them, as compared to the first 5 characters, which are simpler ones.

The purpose of our research is to address these long-standing limitations in printed Urdu text recognition through the following key contributions:

1. We propose a novel approach using high-resolution, multi-scale semantic feature extraction in our UTRNet architecture, a hybrid CNN-RNN model, that demonstrates state-of-the-art performance on benchmark datasets.
2. We create UTRSet-Real, a large-scale annotated real-world dataset comprising over 11,000 lines.
3. We have developed a robust synthetic data generation module and release UTRSet-Synth, a high-quality synthetic dataset of 20,000 lines closely resembling real-world representations of Urdu text.
4. We correct many annotation errors in one of the benchmark datasets [30] for Urdu OCR, thereby elevating its reliability as a valuable resource for future research endeavours and release the corrected annotations publicly.
5. We have curated UrduDoc, a real-world Urdu documents text line detection dataset. The dataset is a byproduct of our efforts towards UTRSet-Real, and contains line segmentation annotation for 478 pages generated from more than 130 books.
6. To make the output of our project available to a larger non-computing research community, as well as lay users, we have developed an online tool for end-to-end Urdu OCR, integrating UTRNet with a third-party text detection model.

In addition to our key contributions outlined above, we conduct a thorough comparative analysis of state-of-the-art (SOTA) Urdu OCR models under similar experimental setups, using a unifying framework introduced by [8] that encompasses feature extraction, sequence modelling and prediction stages and provides a common perspective for all the existing methods. We also examine the contributions of individual modules towards accuracy as an ablation study (Table 2). Finally, we discuss the remaining limitations and potential avenues for future research in Urdu text recognition.

2   Related Work

The study of Urdu OCR has gained attention only in recent years. While the first OCR tools were developed more than five decades back [34], the earliest machine learning based Urdu OCR was developed in 2003 [47]. Since then, the research in this field has progressed from isolated character recognition to word/ligature level recognition to line level recognition (see [27,34,35] for a detailed survey). Early approaches primarily relied on handcrafted features, and traditional machine learning techniques, such as nearest neighbour classification, PCA & HMMs [2,33,53,64], to classify characters after first segmenting individual characters/glyphs from a line image. These techniques often required extensive pre-processing and struggled to achieve satisfactory performance on large, varied datasets.

Recently segmentation-free end-to-end approaches based on CNN-RNN hybrid networks [30] have been introduced, in which a CNN [38] is used to extract low-level visual features from input data which is then fed to an RNN [63] to get contextual features for the output transcription layer. Among the current SOTA models for Urdu OCR (Table 3C), VGG networks [60] have been used for feature extraction in [30,44,45], whereas ResNet networks [25] have been utilized in [31]. For sequential modelling, BiLSTM networks [54] have been used in [30,31], while MDLSTM networks [13] have been employed in [44,45]. All of these approaches utilize a Connectionist Temporal Classification (CTC) layer [22] for output transcription. In contrast, [6] utilizes a DenseNet [26] and GRU network [16] with an attention-based decoder layer [9] for final transcription. Arabic, like Urdu, shares many similarities in the script as discussed above, and as such, the journey of OCR development has been similar [18,29]. Recent works have shown promising results in recognising Arabic text through a variety of methods, including traditional approaches [55,77], as well as DL-based approaches such as CNN-RNN hybrids [20,29], attention-based models [12,62], and a range of network architectures [23,32,36,49]. However, these approaches still struggle when it comes to recognizing Urdu text, as evident from the low accuracies achieved by SOTA Arabic OCR methods like [12,20,62] in our experimental results presented in Table 3.

While each of the methods proposed so far has claimed to have pushed the boundary of the technology, they often rely on the same methodologies used for other languages without considering the complexities specific to Urdu script and, as such, do not fully utilize the potential in this field. Additionally, a fair comparison among these approaches has been largely missing due to inconsistencies in the datasets used, as described in the dataset section below.

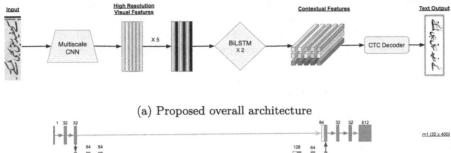

(a) Proposed overall architecture

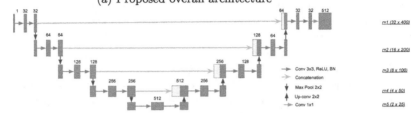

(b) Multiscale feature extraction module for UTRNet-Small based on UNet [50]

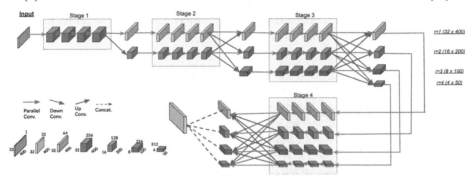

(c) Multiscale feature extraction module for UTRNet-Large based on HRNet [66]

Fig. 3. Proposed Architecture

3 Proposed Architecture

Recently, transformer-based models have achieved state-of-the-art performance on various benchmarks [24]. However, these models have the drawback of being highly data-intensive and requiring large amounts of training data, making it difficult to use them for several tasks with limited real-world data, such as printed Urdu text recognition (as discussed in Sect. 1). In light of this, we propose UTRNet (Fig. 3), a novel CNN-RNN hybrid network that offers a viable alternative. The proposed model effectively extracts high-resolution multi-scale features while capturing sequential dependencies, making it an ideal fit for Urdu text recognition in real-world scenarios. We have designed UTRNet in two versions: UTRNet-Small (10.7M parameters) and UTRNet-Large (47.3M parameters). The architecture consists of three main stages: feature extraction, sequential modelling, and decoding. The feature extraction stage makes use of convolutional layers to extract feature representations from the input image. These representations are then passed to the sequential modelling stage to learn sequential dependencies among them. Finally, the decoding module converts the sequential data feature thus obtained into the final output sequence.

3.1 High-Resolution Multiscale Feature Extraction

In our proposed method, we address the wide mismatch in the accuracy of different characters observed in existing techniques, as shown in Fig. 2. We posit that this is due to the lack of attention given to small features associated with most Urdu characters by the existing methods. These methods rely upon using standard CNNs, such as RCNN [21], VGG [60], ResNet [25], etc., as their backbone. However, a significant drawback of using these CNNs is the low resolution representation generated at the output layer. The representation lacks low-level feature information, such as the dots (called *"Nuqta"*) in Urdu characters. To overcome this limitation, in our proposed method, we propose to use a high-resolution multiscale feature extraction technique to extract features from an input image while preserving the small details of the image.

UTRNet-Small. To address the issue of computational efficiency, we propose a lighter variant of our novel UTRNet architecture, referred to as UTRNet-Small, which employs a standard U-Net model (as shown in Fig. 3b), initially proposed in [50] for biomedical image segmentation. The lighter version proposed by us addresses captures high resolution feature maps using the standard U-Net model, originally proposed in [50] for biomedical image segmentation. We first encode the low-resolution representation of input image X, which captured context from a large receptive field. We then recovers the high-resolution representation using learnable convolutions from the previous decoder layer, and skip connections from the encoder layers. For any resolution with index $r \in \{1, 2, 3, 4, 5\}$, the feature map at that resolution can be defined as: $F_r = \text{downsample}(F_{r-1})$ during downsampling. Here downsample(F_{r-1}) is the downsampled feature map from the resolution index $(r - 1)$. Similarly, $F_r = \text{concat}(\text{upsample}(F_{r+1}), M_r)$ represents feature map during upsampling, where upsample(F_{r+1}) is the upsampled feature map from the resolution index $(r + 1)$, and M_r is the feature map from the downsampling path corresponding to that resolution). This allows the model to aggregate image features from multiple image scales.

UTRNet-Large The model (Fig. 3c) maintains high-resolution representation throughout the process and captures the fine-grained details more efficiently, using an HRNet architecture [66]. The resulting network consists of 4 stages, with the Ith stage containing streams coming from $\#I$ different resolutions and giving out $\#I$ streams corresponding to the different resolutions. Each stream in the Ith stage is represented as R_r^I, where $r \in \{1, 2, \ldots, I\}$. These streams which are then inter-fused among themselves to get $\#(I+1)$ final output streams, $R_r^{I,O}$ for the next stage:

$$R_r^{I,O} = \sum_{i=1}^{I} f_{ir}(R_i^I), \qquad r \in \{1, 2, \ldots, I+1\}$$

The transform function f_{ir} is dependent on the input resolution index i and the output resolution index r. If $x = r$, then $f_{xr}(R) = R$. However, if $x < r$, then $f_{xr}(R)$ downsamples the input representation R. Similarly, if $x > r$, then $f_{xr}(R)$ upsamples the input resolution. UTRNet-Large uses repeated multi-resolution fusions which allows effective exchange of information across multiple resolutions. Thus, giving a multi-dimensional and high-resolution feature representation of the input image, which is semantically richer.

3.2 Sequential Modeling and Prediction

The output from the feature extraction stage is a feature map $V = \{v_i\}$. To prevent over-fitting, we implement a technique called Temporal Dropout [14], in which we randomly drop half of the visual features before passing them to the next stage. We do this 5 times in parallel and take the average. In order to capture the rich contextual information and temporal relationships between the features thus obtained, we pass it through 2 layers of BiLSTM [54] (DBiLSTM) [58]. Each BiLSTM layer identifies two hidden states, h_t^f and h_t^b, calculated forward and backward through time, respectively, which are combined to determine one hidden state h^t, using an FC layer. The sequence $H = \{h_t\} = \text{DBiLSTM}(V))$ thus obtained has rich contextual information from both directions, which is crucial for Urdu text recognition, especially because the shape of each character depends upon characters around it. The final prediction output $Y = \{y1, y2, \ldots\}$, a variable-length sequence of characters, is generated by the prediction module from the input sequence H. For this, we use the Connectionist temporal classification (CTC), as described in [22].

4   Current Publicly Available and Proposed Datasets

The availability of datasets for the study of Urdu Optical Character Recognition (OCR) is limited, with only a total of six datasets currently available: UPTI [51], IIITH [30], UNHD [1], CENPARMI [52], CALAM [17], and PMU-UD [3]. Of these datasets, only IIITH and UPTI contain printed text line samples, out of which only UPTI has a sufficient number of samples for training. However, the UPTI dataset is synthetic in nature, with limited diversity and simplicity in comparison to real-world images (See Fig. 4). There is currently no comprehensive real-world printed Urdu OCR dataset available publicly for researchers. As a result, different studies have used their own proprietary datasets, making it difficult to determine the extent to which proposed models improve upon existing approaches. This lack of standardisation and transparency hinders the

(a) IIITH [30]   (b) UPTI [51]

(c) Proposed UTRSet-Synth

(d) Proposed UTRSet-Real

Dataset	Training Set	Validation Set	Vocab Length	Type
IIITH [30]	NA	1,610	5,772	Real
UPTI [51]	8,051	2,012	12,054	Synthetic
UTRSet-Real	9,065	2,096	22,964	Real
UTRSet-Synth	20,000	NA	28,187	Synthetic

Fig. 4. Sample images, and statistics of the publicly available and proposed datasets. Notice the richness and diversity in the proposed UTRSet-Real and UTRSet-Synth datasets, as compared to existing datasets

ability to compare the performance of different models accurately and to continue advancing research in the field, which our work aims to tackle with the introduction of two new datasets. The following are the datasets used in this paper (also summarized in Fig. 4):

IIITH. This dataset was introduced by [30] in 2017, and it contains only the validation data set of 1610 line images in nearly uniform colour and font. No training data set has been provided. We corrected the ground-truth annotations for this dataset, as we found several mistakes, as highlighted in Fig. 5.

UPTI. Unlike the other two, this is a synthetic data set introduced in 2013 by [51]. It consists of a total of 10,063 samples, out of which 2,012 samples are in the validation set. This data set is also uniform in colour and font and has a vocabulary of 12,054 words.

Sample Image	Original GT	Corrected GT
	ایر گر سو سین رہ کر توا ٴات اج س و کا طقسلق ۃ سذ زتہ س٭ے تطی ٴپ پلبی کھر کزویوش	ایک گروہ کہتے لگا کہ خدا نے ان بتوں کی شکل و صورت سے ملتے جلتے چندمردوں
	:سالہ کہتی لگی : عہدابنا ہتلی حچھت٭ پا ہصل ضیداللد٭ صول٭ ٖللّٰٓی ٴاے ہخ	سالہ کہتی لگی: "ماراپنا مثل حچنک پا محمد شہداللہ رسول اللہ ٣ٴے محمد
	ہ ہیسندا بہ ۔ملر ۔بلر و مر او روٴہ ہا اگر مر ے۔	بجرت مدینہ سے پہلی بیودی أپس میں ان نشانیوں کا تذکرہ کیا کارٹ
	:الناس قدب احکاٴء دینا ، فبیا شروٴض ادۃ وغر بھا"، فرج حذہ	،الناس فی احکام دینھم فی شرق الأرض ع غیربھا۔ فی ھذہ العصور
	٣٩ دقم	٢٤/٣/٢٠٠٢ ٤٤۔١٨۔م رقم
	دیگل بالی: واکو ٴنعظ الصفد۔ و کجادہ اٴ، پا ٴتی شہ۔	زیک بالحکمۃ والفوعظۃ الخسنۃ وَجادلھم بالتی ھی

Fig. 5. Mistakes in IIITH dataset annotations and our corrections.

Proposed UTRSet-Real A comprehensive real-world annotated dataset curated by us, containing a few easy and mostly hard images (as illustrated in Fig. 1). To create this dataset, we collected 130 books and old documents, scanned over 500 pages, and manually annotated the scanned images with line-wise bounding boxes and corresponding ground truth labels. After cropping the lines and performing data cleaning, we obtained a final dataset of 11,161 lines, with 2,096 lines in the validation set and the remaining in the training set. This dataset stands out for its diversity, with various fonts, text sizes, colours, orientations, lighting conditions, noises, styles, and backgrounds represented in the samples, making it well-suited for real-world Urdu text recognition.

Proposed UTRSet-Synth To complement the real-world data in UTRSet-Real for training purposes, we also present UTRSet-Synth, a high-quality synthetic dataset of 20,000 lines with over 28,000 total unique words, closely resembling real-world representations of Urdu text. This dataset is generated using a custom-designed synthetic data generation module that allows for precise control over variations in font, text size, colour, resolution, orientation, noise, style, background etc. The module addresses the challenge of standardizing fonts by collecting and incorporating over 130 diverse fonts of Urdu after making corrections to their rendering schemes. It also addresses the limitation of current datasets, which have very few instances of Arabic words and numerals, Urdu digits etc., by incorporating such samples in sufficient numbers. Additionally, it generates text samples by randomly selecting words from a vocabulary of 100,000 words. The generated UTRSet-Synth has 28,187 unique words with an average word length of 7 characters. The data generation module has been made publicly available on the project page to facilitate further research.

Proposed UrduDoc. In addition to the recognition datasets discussed above, we also present UrduDoc, a benchmark dataset for Urdu text line detection in scanned documents. To the best of our knowledge, this is the first dataset of its kind [4,15]. It was created as a byproduct of the UTRSet-Real dataset generation process, in which the pages were initially scanned and then annotated with horizontal bounding boxes in COCO format [41] to crop the text lines. Comprising of 478 diverse images collected from various sources such as books, documents, manuscripts, and newspapers, it is split into 358 pages for training

Table 1. Experimental results on UrduDoc

Methods	Precision	Recall	Hmean
EAST [75]	70.43	72.56	71.48
PSENet [67]	78.32	77.91	78.11
DRRG [72]	83.05	84.72	83.87
ContourNet [69]	85.36	88.68	86.99

Fig. 6. Sample images from the UrduDoc Dataset: An annotated real-world benchmark for Urdu text line detection in scanned documents

and 120 for validation. The images include a wide range of styles, scales, and lighting conditions, making them a valuable resource for the research community. The UrduDoc dataset will serve as a valuable resource for the research community, advancing research in Urdu document analysis. We also provide benchmark results of a few SOTA text detection models on this dataset using precision, recall, and h-mean, as shown in Table 1. The results demonstrate that the ContourNet model [69] outperforms the other models in terms of h-mean. It is worth noting that as text detection was not the primary focus of our research but rather a secondary contribution, a thorough examination of text detection has not been conducted. This aspect can be considered a future work for researchers interested in advancing the field of Urdu document analysis. We will make this dataset publicly available to the research community for non-commercial, academic, and research purposes associated with Urdu document analysis, subject to request and upon execution of a no-cost license agreement (Fig. 6).

5   Experiments and Results

Experimental Setup. In order to ensure a fair comparison among the existing models in the field, we have established a consistent experimental setup for evaluating the performance of all the models and report all the results in Table 3. Specifically, we have fixed the choice of training to the UTRSet-Real training set, the validation set to be the validation sets of the datasets outlined in Fig. 4. Further, to compare the different available training datasets, we train our proposed

Table 2. The results of the ablation study for UTRNet provide a comprehensive examination of the impact of individual feature extraction and sequence modelling stages and augmentation strategies on the overall performance of the model. By evaluating each stage one-by-one while keeping the remaining stages constant, the results highlight the key factors driving the accuracy of the model, offering valuable insight into optimizing the performance of UTRNet.

Model/Strategy	UTRSet-Real	IIITH	UPTI
Various Multiscale Feature Extraction Backbones			
UNet [50]	90.87	86.35	95.08
AttnNet [46]	91.95	86.45	95.26
ResidualUNet [73]	91.90	87.16	95.23
InceptionUNet [48]	92.10	87.37	95.61
UNetPlusPlus [76]	92.53	87.36	95.84
HRNet [66]	**92.97**	**88.01**	**95.97**
Various Sequential Modelling Backbones			
LSTM	91.20	87.21	94.41
GRU	91.48	87.04	94.58
MDLSTM	91.51	87.38	94.67
BiLSTM	91.53	87.67	94.69
DBiLSTM	**92.97**	**88.01**	**95.97**
Various Strategies to Improve Generalization			
None	91.07	86.80	95.07
Augmentation	92.35	87.53	95.48
Augmentation + Temporal Dropout	**92.97**	**88.01**	**95.97**

UTRNet models on each of them and present the results in Table 4. We have used the AdaDelta optimizer [71], with a decay rate of 0.95, to train our UTRNet models on an NVIDIA A100 40GB GPU. The batch size and learning rate used were 32 and 1.0, respectively. Gradient clipping was employed at a magnitude of 5, and all parameters were initialized using He's method [66]. To improve the robustness of the model, we employed a variety of data augmentation techniques during the training process, such as random resizing, stretching/compressing, rotation, and translation, various types of noise, random border crop, contrast stretching, and various image processing techniques, to simulate different types of imaging conditions and improve the model's ability to generalize to real-world scenarios. The UTRNet-Large model achieved convergence in approximately 7 h. We utilize the standard character-wise accuracy metric for comparison, which uses the edit distance between the predicted output (Pred) and the ground truth (GT):

$$\text{Accuracy} = \frac{\sum (\text{length}(GT) - \text{EditDistance}(Pred, GT))}{\sum (\text{length}(GT))}$$

Table 3. Performance comparison of SOTA OCR models

Models	UTRSet-Real	IIITH	UPTI
A. Baseline OCR models (Hybrid CNN-RNN)			
R2AM [39]	84.12	81.39	92.07
CRNN [58]	83.11	81.45	91.49
GRCNN [65]	84.21	81.09	92.28
Rosetta [11]	84.08	81.94	92.15
RARE [59]	85.63	83.59	92.74
STAR-Net [42]	87.05	84.27	93.59
TRBA [8]	88.92	85.61	94.16
B. Baseline OCR models (Transformer-based)			
Parseq [10]	26.13	25.60	26.41
ViTSTR [7]	34.86	32.63	35.78
TrOCR [40]	38.43	36.10	37.61
ABINet [19]	41.17	40.20	38.96
CDistNet [74]	33.72	34.96	32.48
VisionLAN [68]	28.40	27.82	29.07
C. SOTA Urdu OCR models			
5LayerCNN-DBiLSTM [20]	82.92	81.15	90.67
VGG-BiLSTM [30]	83.11	81.45	91.49
VGG-MDLSTM [44,45]	83.30	81.72	91.17
VGG-LSTM-Attn [62]	84.16	82.21	91.88
VGG-DBiLSTM-Attn [12]	84.58	82.72	92.01
ResNet-BiLSTM [31]	86.96	84.18	93.61
DenseNet-GRU-Attn [6]	91.10	85.32	94.63
UTRNet-Small	90.87	86.35	95.08
UTRNet	**92.97**	**88.01**	**95.97**

5.1   Results and Analysis

In order to evaluate the effectiveness of our proposed architecture, we conducted a series of experiments and compared our results with state-of-the-art (SOTA) models for Urdu OCR, as well as a few for Arabic, including both printed and handwritten ones (Table 3C). Additionally, we evaluated our model against the current SOTA baseline OCR models, primarily developed for Latin-based languages (Tables 3A and 3B). Our proposed model achieves superior performance, surpassing all of the SOTA OCR models in terms of character-wise accuracy on all three validation datasets, achieving a recognition accuracy of 92.97% on the UTRSet-Real validation set. It is worth noting that while Table 3A presents a comparison of our proposed model against hybrid CNN-RNN models, Table 3B presents a comparison against recent transformer-based models. The results clearly show that transformer-based models perform poorly in comparison to our proposed model and even the SOTA CNN-RNN models for Latin OCR. This can be attributed to the fact that these models, which are designed to be trained on massive datasets when applied to the case of Urdu script recognition, overfit the small-size training data and struggle to generalize, thus resulting in poor validation accuracy.

In our analysis of our proposed UTRSet-Real and UTRSet-Synth datasets against the existing UPTI dataset, which is currently the only available training dataset for this purpose, we found that our proposed datasets effectively improve the performance of the UTRNet model. When trained on the UPTI dataset, both UTRNet-Small and UTRNet achieve high accuracy on the UPTI validation set but perform poorly on the UTRSet-Real and IIITH validation sets. This suggests that the UPTI dataset is not representative of real-world scenarios and does not adequately capture the complexity and diversity of printed Urdu text. Our proposed datasets, on the other hand, are specifically designed to address these issues and provide a more comprehensive and realistic representation of the task at hand, as both of them perform significantly well on all datasets, with UTRSet-Real being the best. Furthermore, the results show that combining all three training datasets can further improve the performance, especially on the IIITH and UPTI validation sets.

One of the key insights from our results is the significant difference in accuracy when comparing our proposed UTRNet model with the current state-of-the-art (SOTA) model for printed Urdu OCR [30], as presented in Fig. 2. This highlights the complexity of recognizing the intricate features of Urdu characters and the efficacy of our proposed UTRNet model in addressing these challenges. Our results align with our hypothesis that high-resolution multi-scale feature maps are essential for capturing the nuanced details required for accurate Urdu OCR.

Table 4. Performance after training with different datasets. See that training with UPTI dataset leads to poor performance on real-world validation datasets. (Here, "Mix-All" means a mixture of UPTI, UTRSet-Real & UTRSet-Synth)

Model	Training Data	UTRSet-Real	IIITH	UPTI
UTRNet-Small	UPTI	54.84	72.82	98.63
UTRNet-Small	UTRSet-Real	90.87	86.35	95.08
UTRNet-Small	UTRSet-Synth	75.14	85.47	92.09
UTRNet-Small	Mix-All	91.71	90.04	98.72
UTRNet	UPTI	64.73	76.15	**99.17**
UTRNet	UTRSet-Real	92.97	88.01	95.97
UTRNet	UTRSet-Synth	80.26	87.85	93.49
UTRNet	Mix-All	**93.39**	**90.91**	98.36

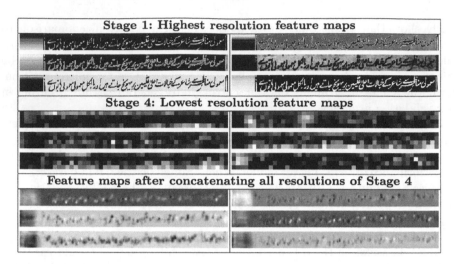

Fig. 7. Visualization of feature maps learnt by various layers of the UTRNet-Large. The small features associated with various characters are lost at low resolution (middle row) but are present in the last row. This illustrates the effectiveness of our proposed high-resolution multi-scale feature extraction technique, leading to improved performance in the printed Urdu text recognition.

To further support this claim, we also present a visualization of feature maps generated from our CNN (as depicted in Fig. 7), which clearly demonstrates the ability of UTRNet to effectively extract and preserve the high-resolution details of the input image. Additionally, we provide a qualitative analysis of the results by comparing our model with the SOTA [30] in Fig. 9.

We also conducted a series of ablation studies to investigate the impact of various components of our proposed UTRNet model on performance. The results of this study, as shown in Table 2, indicate that each component of our model makes a significant contribution to the overall performance. Specifically, since we observed that incorporating a multi-scale high-resolution feature extraction significantly improves the result for Urdu OCR, we tried various other multi-scale CNNs as a part of our ablation study. We also found that the use of generalisation techniques, such as temporal dropout and augmentation, further improved the robustness of our model, making it able to effectively handle a wide range of challenges which are commonly encountered in real-world Urdu OCR scenarios.

6   Web Tool for End-to-End Urdu OCR

We have developed an online website for Urdu OCR that integrates ContourNet [69] model, trained on the Urdu-Doc dataset with our proposed UTRNet model. This integration allows for end-to-end text recognition in real-world documents, making the website a valuable tool for easy and efficient digitization of a large corpus of available Urdu literature (Fig. 8).

Fig. 8. Web tool developed by us for end-to-end Urdu OCR

Sample Image	SOTA Prediction	UTRNet Prediction
ـقـلـربـركـسـاں.كـ.روشتـئـ.مـيـب	عاریرکساب کی رولئنی میپ	قل بر گساں کی رولئننی میں
پہلی بات یہ تو یہ ہے کرشیک خوش فکرکیتیش پیش اک غزخ لکلراورفرز گوہاشاکر	پہلی بات تو یہ ہے کہ یکیت کیت ایک خوش کہ اور فزگرہاع ا	پہلی بات تو یہ ہے کہ ہیھکیر بحثیت ایک خوش فکر اور نفز گوھاعر
حلات بتاری کی تھی کہ وہ جوکچھ کہ ربائھا۔ اس کی نہ اخلاطۃ غلط تھا۔ بیوی تھی اور نہ چچہ	حالت باری تں کہ بچ کھہ ۱۱کھ۔ بی ی ہ بھی ے	حالت بتار یں تھی کہ وہ جوکچھ کہ ربائھا غلط تھا۔ اس کی نہ تو بیوی تھی اور نہ چہ۔
سے دلداری کا رجحان اذرشاہءاند رندی دبیا کی سے دوری کا اظہار	ا لداداری یگ ای ۱۱ای ی ی وا ی ن ک یو ای م ل لم	سے دلداری کا رجحان ادرھاعناند رندی دبیا کی سے دوری کا اظہار
۔اکلمہ پڑھ ربا یے	کلیہ ترھ د با یے؛1؛	"اکلمہ پڑھ ریا یے۔

Fig. 9. The figure illustrates a qualitative analysis of our proposed UTRNet-Small and the SOTA method [30] on the UTRSet-Real dataset. To facilitate a fair comparison, the errors in the transcriptions are highlighted in red. The results showcase the superior accuracy of UTRNet in capturing fine-grained details and accurately transcribing Urdu text in real-world documents.

7   Conclusion

In this work, we have addressed the limitations of previous works in Urdu OCR, which struggle to generalize to the intricacies of the Urdu script and the lack of large annotated real-world data. We have presented a novel approach through the introduction of a high-resolution, multi-scale semantic feature extraction-based model which outperforms previous SOTA models for Urdu OCR, as well as Latin OCR, by a significant margin. We have also introduced three comprehensive datasets: UTRSet-Real, UTRSet-Synth, and UrduDoc, which are significant contributions towards advancing research in printed Urdu text recognition. Additionally, the corrections made to the ground truth of the existing IIITH dataset have made it a more reliable resource for future research. Furthermore, we've also developed a web based tool for end-to-end Urdu OCR which we hope will help in digitizing the large corpus of available Urdu literature. Despite the promising results of our proposed approach, there remains scope for further optimization and advancements in the field of Urdu OCR. A crucial area of focus is harnessing the power of transformer-based models along with large amounts of synthetic data by enhancing the robustness and realism of synthetic data and potentially achieving even greater performance gains. Our work has laid the foundation for continued progress in this field, and we hope it will inspire new and innovative approaches for printed Urdu text recognition.

Acknowledgement. We would like to express our gratitude to the Rekhta Foundation and Arjumand Ara for providing us with scanned images, as well as Noor Fatima and Mohammad Usman for their valuable annotations of the UTRSet-Real dataset. Furthermore, we acknowledge the support of a grant from IRD, IIT Delhi, and MEITY, Government of India, through the NLTM-Bhashini project.

References

1. Ahmed, S.B., Naz, S., Swati, S., Razzak, M.I.: Handwritten Urdu character recognition using 1-dimensional blstm classifier (2017). https://doi.org/10.48550/ARXIV.1705.05455
2. Akram, M.U., Hussain, S.: Word segmentation for Urdu ocr system (2010)
3. Alghazo, J.M., Latif, G., Alzubaidi, L., Elhassan, A.: Multi-language handwritten digits recognition based on novel structural features. J. Imaging Sci. Technol. **63**, 1–10 (2019)
4. Ali, A., Pickering, M.: Urdu-text: a dataset and benchmark for Urdu text detection and recognition in natural scenes. In: 2019 International Conference on Document Analysis and Recognition (ICDAR), pp. 323–328 (2019). https://doi.org/10.1109/ICDAR.2019.00059
5. Althobaiti, H., Lu, C.: A survey on Arabic optical character recognition and an isolated handwritten Arabic character recognition algorithm using encoded freeman chain code. In: 2017 51st Annual Conference on Information Sciences and Systems (CISS), pp. 1–6 (2017). https://doi.org/10.1109/CISS.2017.7926062
6. Anjum, T., Khan, N.: An attention based method for offline handwritten Urdu text recognition. In: 2020 17th International Conference on Frontiers in Handwriting Recognition (ICFHR), pp. 169–174 (2020). https://doi.org/10.1109/ICFHR2020.2020.00040
7. Atienza, R.: Vision transformer for fast and efficient scene text recognition (2021). https://doi.org/10.48550/ARXIV.2105.08582. https://arxiv.org/abs/2105.08582
8. Baek, J., et al.: What is wrong with scene text recognition model comparisons? dataset and model analysis (2019). https://doi.org/10.48550/ARXIV.1904.01906
9. Bahdanau, D., Cho, K., Bengio, Y.: Neural machine translation by jointly learning to align and translate (2014). https://doi.org/10.48550/ARXIV.1409.0473
10. Bautista, D., Atienza, R.: Scene text recognition with permuted autoregressive sequence models (2022). https://doi.org/10.48550/ARXIV.2207.06966. https://arxiv.org/abs/2207.06966
11. Borisyuk, F., Gordo, A., Sivakumar, V.: Rosetta: large scale system for text detection and recognition in images. In: Proceedings of the 24th ACM SIGKDD International Conference on Knowledge Discovery & Data Mining. ACM (2018). https://doi.org/10.1145/3219819.3219861

12. Butt, H., Raza, M.R., Ramzan, M., Ali, M.J., Haris, M.: Attention-based cnn-rnn Arabic text recognition from natural scene images. Forecasting **3**, 520–540 (2021). https://doi.org/10.3390/forecast3030033

13. Byeon, W., Liwicki, M., Breuel, T.M.: Texture classification using 2D LSTM networks. In: 2014 22nd International Conference on Pattern Recognition, pp. 1144–1149 (2014). https://doi.org/10.1109/ICPR.2014.206

14. Chammas, E., Mokbel, C.: Fine-tuning handwriting recognition systems with temporal dropout (2021). ArXiv abs/2102.00511 https://arxiv.org/abs/2102.00511

15. Chandio, A.A., Asikuzzaman, M., Pickering, M., Leghari, M.: Cursive-text: a comprehensive dataset for end-to-end Urdu text recognition in natural scene images. Data Brief **31**, 105749 (2020). https://doi.org/10.1016/j.dib.2020.105749

16. Cho, K., van Merrienboer, B., Bahdanau, D., Bengio, Y.: On the properties of neural machine translation: encoder-decoder approaches (2014). https://doi.org/10.48550/ARXIV.1409.1259

17. Choudhary, P., Nain, N.: A four-tier annotated Urdu handwritten text image dataset for multidisciplinary research on Urdu script. ACM Trans. Asian Low Res. Lang. Inf. Process. **15**(4), 1–23 (2016). https://doi.org/10.1145/2857053

18. Djaghbellou, S., Bouziane, A., Attia, A., Akhtar, Z.: A survey on Arabic handwritten script recognition systems. Int. J. Artif. Intell. Mach. Learn. **11**, 1–17 (2021). https://doi.org/10.4018/IJAIML.20210701.oa9

19. Fang, S., Xie, H., Wang, Y., Mao, Z., Zhang, Y.: Read like humans: autonomous, bidirectional and iterative language modeling for scene text recognition (2021). https://doi.org/10.48550/ARXIV.2103.06495. https://arxiv.org/abs/2103.06495

20. Fasha, M., Hammo, B.H., Obeid, N., Widian, J.: A hybrid deep learning model for Arabic text recognition (2020). ArXiv abs/2009.01987 https://arxiv.org/abs/2009.01987

21. Girshick, R., Donahue, J., Darrell, T., Malik, J.: Rich feature hierarchies for accurate object detection and semantic segmentation (2013). https://doi.org/10.48550/ARXIV.1311.2524

22. Graves, A., Fernández, S., Gomez, F., Schmidhuber, J.: Connectionist temporal classification: labelling unsegmented sequence data with recurrent neural networks. In: Proceedings of the 23rd International Conference on Machine Learning, ICML 2006, p. 369–376 (2006). https://doi.org/10.1145/1143844.1143891

23. Graves, A., Schmidhuber, J.: Offline Arabic handwriting recognition with multidimensional recurrent neural networks, pp. 545–552 (2008)

24. Han, K., et al.: A survey on vision transformer. IEEE Trans. Pattern Anal. Mach. Intell. **45**(1), 87–110 (2023). https://doi.org/10.1109/TPAMI.2022.3152247

25. He, K., Zhang, X., Ren, S., Sun, J.: Deep residual learning for image recognition (2015). https://doi.org/10.48550/ARXIV.1512.03385

26. Huang, G., Liu, Z., van der Maaten, L., Weinberger, K.Q.: Densely connected convolutional networks (2016). https://doi.org/10.48550/ARXIV.1608.06993

27. Husnain, M., Saad Missen, M.M., Mumtaz, S., Coustaty, M., Luqman, M., Ogier, J.M.: Urdu handwritten text recognition: a survey. IET Image Process. **14**(11), 2291–2300 (2020). https://doi.org/10.1049/iet-ipr.2019.0401

28. Hussain, S.: A survey of ocr in Arabic language: applications, techniques, and challenges. Appl. Sci. **13**, 27 (2023). https://doi.org/10.3390/app13074584

29. Jain, M., Mathew, M., Jawahar, C.V.: Unconstrained scene text and video text recognition for Arabic script. In: 2017 1st International Workshop on Arabic Script Analysis and Recognition (ASAR), pp. 26–30 (2017)

30. Jain, M., Mathew, M., Jawahar, C.: Unconstrained ocr for Urdu using deep cnn-rnn hybrid networks. In: 2017 4th IAPR Asian Conference on Pattern Recognition (ACPR), pp. 747–752. IEEE (2017)
31. Kashif, M.: Urdu handwritten text recognition using resnet18 (2021). https://doi.org/10.48550/ARXIV.2103.05105
32. Kassem, A.M., et al.: Ocformer: a transformer-based model for Arabic handwritten text recognition. In: 2021 International Mobile, Intelligent, and Ubiquitous Computing Conference (MIUCC), pp. 182–186 (2021)
33. Khan, K., Ullah, R., Ahmad, N., Naveed, K.: Urdu character recognition using principal component analysis. Int. J. Comput. Appl. **60**, 1–4 (2012). https://doi.org/10.5120/9733-2082
34. Khan, N.H., Adnan, A.: Urdu optical character recognition systems: present contributions and future directions. IEEE Access **6**, 46019–46046 (2018). https://doi.org/10.1109/ACCESS.2018.2865532
35. Khan, N.H., Adnan, A., Basar, S.: An analysis of off-line and on-line approaches in Urdu character recognition. In: 2016 15th International Conference on Artificial Intelligence, Knowledge Engineering and Data Bases (AIKED 2016) (2016)
36. Ko, D., Lee, C., Han, D., Ohk, H., Kang, K., Han, S.: Approach for machine-printed Arabic character recognition: the-state-of-the-art deep-learning method. Electron. Imaging **2018**, 176-1–176-8 (2018)
37. Kolesnikov, A., et al.: An image is worth 16×16 words: transformers for image recognition at scale (2021)
38. Lecun, Y., Bottou, L., Bengio, Y., Haffner, P.: Gradient-based learning applied to document recognition. Proc. IEEE **86**(11), 2278–2324 (1998). https://doi.org/10.1109/5.726791
39. Lee, C.Y., Osindero, S.: Recursive recurrent nets with attention modeling for ocr in the wild (2016). https://doi.org/10.48550/ARXIV.1603.03101. https://arxiv.org/abs/1603.03101
40. Li, M., et al.: Trocr: transformer-based optical character recognition with pre-trained models (2021). https://doi.org/10.48550/ARXIV.2109.10282. https://arxiv.org/abs/2109.10282
41. Lin, T.-Y., et al.: Microsoft COCO: common objects in context. In: Fleet, D., Pajdla, T., Schiele, B., Tuytelaars, T. (eds.) ECCV 2014. LNCS, vol. 8693, pp. 740–755. Springer, Cham (2014). https://doi.org/10.1007/978-3-319-10602-1_48
42. Liu, W., Chen, C., Wong, K.Y., Su, Z., Han, J.: Star-net: A spatial attention residue network for scene text recognition, pp. 43.1–43.13 (2016). https://doi.org/10.5244/C.30.43
43. Mushtaq, F., Misgar, M.M., Kumar, M., Khurana, S.S.: UrduDeepNet: offline handwritten Urdu character recognition using deep neural network. Neural Comput. Appl. **33**(22), 15229–15252 (2021)
44. Naz, S., Ahmed, S., Ahmad, R., Razzak, M.: Zoning features and 2dlstm for Urdu text-line recognition. Procedia Comput. Sci. **96**, 16–22 (2016). https://doi.org/10.1016/j.procs.2016.08.084
45. Naz, S., et al.: Urdu nastaliq recognition using convolutional-recursive deep learning. Neurocomputing **243**, 80–87 (2017). https://doi.org/10.1016/j.neucom.2017.02.081. https://www.sciencedirect.com/science/article/pii/S0925231217304654
46. Oktay, O., et al.: Attention u-net: learning where to look for the pancreas (2018). https://doi.org/10.48550/ARXIV.1804.03999
47. Pal, U., Sarkar, A.: Recognition of printed Urdu script. In: Seventh International Conference on Document Analysis and Recognition, 2003, Proceedings, pp. 1183–1187 (2003). https://doi.org/10.1109/ICDAR.2003.1227844

48. Punn, N.S., Agarwal, S.: Inception u-net architecture for semantic segmentation to identify nuclei in microscopy cell images. ACM Trans. Multimedia Comput. Commun. Appl. **16**(1), 1–15 (2020). https://doi.org/10.1145/3376922

49. Rashid, S.F., Schambach, M.P., Rottland, J., Nüll, S.: Low resolution Arabic recognition with multidimensional recurrent neural networks (2013). https://doi.org/10.1145/2505377.2505385

50. Ronneberger, O., Fischer, P., Brox, T.: U-net: convolutional networks for biomedical image segmentation. In: Navab, N., Hornegger, J., Wells, W.M., Frangi, A.F. (eds.) MICCAI 2015. LNCS, vol. 9351, pp. 234–241. Springer, Cham (2015). https://doi.org/10.1007/978-3-319-24574-4_28

51. Sabbour, N., Shafait, F.: A segmentation free approach to Arabic and Urdu ocr. In: Proceedings of SPIE - The International Society for Optical Engineering, vol. 8658 (2013). https://doi.org/10.1117/12.2003731

52. Sagheer, M.W., He, C.L., Nobile, N., Suen, C.Y.: A new large Urdu database for off-line handwriting recognition. In: Foggia, P., Sansone, C., Vento, M. (eds.) ICIAP 2009. LNCS, vol. 5716, pp. 538–546. Springer, Heidelberg (2009). https://doi.org/10.1007/978-3-642-04146-4_58

53. Sardar, S., Wahab, A.: Optical character recognition system for Urdu. In: 2010 International Conference on Information and Emerging Technologies, pp. 1–5 (2010)

54. Schuster, M., Paliwal, K.K.: Bidirectional recurrent neural networks. IEEE Trans. Signal Process. **45**(11), 2673–2681 (1997)

55. Semary, N., Rashad, M.: Isolated printed Arabic character recognition using knn and random forest tree classifiers, vol. 488, p. 11 (2014)

56. Shahin, A.: Printed Arabic text recognition using linear and nonlinear regression. Int. J. Adv. Comput. Sci. Appl. **8** (2017). https://doi.org/10.14569/IJACSA.2017.080129

57. Shaiq, M.D., Cheema, M.D.A., Kamal, A.: Transformer based Urdu handwritten text optical character reader (2022). https://doi.org/10.48550/ARXIV.2206.04575. https://arxiv.org/abs/2206.04575

58. Shi, B., Bai, X., Yao, C.: An end-to-end trainable neural network for image-based sequence recognition and its application to scene text recognition (2015). https://doi.org/10.48550/ARXIV.1507.05717

59. Shi, B., Wang, X., Lyu, P., Yao, C., Bai, X.: Robust scene text recognition with automatic rectification (2016). https://doi.org/10.48550/ARXIV.1603.03915. https://arxiv.org/abs/1603.03915

60. Simonyan, K., Zisserman, A.: Very deep convolutional networks for large-scale image recognition (2014). https://doi.org/10.48550/ARXIV.1409.1556

61. Singh, A., Bacchuwar, K., Bhasin, A.: A survey of OCR applications. Int. J. Mach. Learn. Comput. (IJMLC) **2**, 314 (2012). https://doi.org/10.7763/IJMLC.2012.V2.137

62. Sobhi, M., Hifny, Y., Elkaffas, S.M.: Arabic optical character recognition using attention based encoder-decoder architecture. In: 2020 2nd International Conference on Artificial Intelligence, Robotics and Control, AIRC 2020, pp. 1–5. Association for Computing Machinery, New York (2021). https://doi.org/10.1145/3448326.3448327

63. Sutskever, I., Vinyals, O., Le, Q.V.: Sequence to sequence learning with neural networks (2014). https://doi.org/10.48550/ARXIV.1409.3215. https://arxiv.org/abs/1409.3215

64. Tabassam, N., Naqvi, S., Rehman, H., Anoshia, F.: Optical character recognition system for Urdu (Naskh font) using pattern matching technique. Int. J. Image Process. **3**, 92 (2009)
65. Wang, J., Hu, X.: Gated recurrent convolution neural network for ocr. In: Proceedings of the 31st International Conference on Neural Information Processing Systems, NIPS 2017, pp. 334–343. Curran Associates Inc., Red Hook (2017)
66. Wang, J., et al.: Deep high-resolution representation learning for visual recognition (2019)
67. Wang, W., et al.: Shape robust text detection with progressive scale expansion network (2019). https://doi.org/10.48550/ARXIV.1903.12473
68. Wang, Y., Xie, H., Fang, S., Wang, J., Zhu, S., Zhang, Y.: From two to one: a new scene text recognizer with visual language modeling network (2021). https://doi.org/10.48550/ARXIV.2108.09661. https://arxiv.org/abs/2108.09661
69. Wang, Y., Xie, H., Zha, Z., Xing, M., Fu, Z., Zhang, Y.: Contournet: taking a further step toward accurate arbitrary-shaped scene text detection (2020). https://doi.org/10.48550/ARXIV.2004.04940
70. Yuan, L., et al.: Tokens-to-token vit: training vision transformers from scratch on imagenet (2021). https://doi.org/10.48550/ARXIV.2101.11986. https://arxiv.org/abs/2101.11986
71. Zeiler, M.D.: Adadelta: an adaptive learning rate method (2012). https://doi.org/10.48550/ARXIV.1212.5701
72. Zhang, S.X., et al.: Deep relational reasoning graph network for arbitrary shape text detection (2020). https://doi.org/10.48550/ARXIV.2003.07493
73. Zhang, Z., Liu, Q., Wang, Y.: Road extraction by deep residual u-net. IEEE Geosci. Remote Sens. Lett. **15**(5), 749–753 (2018). https://doi.org/10.1109/lgrs.2018.2802944
74. Zheng, T., Chen, Z., Fang, S., Xie, H., Jiang, Y.G.: Cdistnet: perceiving multi-domain character distance for robust text recognition (2021). https://doi.org/10.48550/ARXIV.2111.11011. https://arxiv.org/abs/2111.11011
75. Zhou, X., et al.: East: an efficient and accurate scene text detector (2017). https://doi.org/10.48550/ARXIV.1704.03155
76. Zhou, Z., Siddiquee, M.M.R., Tajbakhsh, N., Liang, J.: Unet++: a nested u-net architecture for medical image segmentation (2018). https://doi.org/10.48550/ARXIV.1807.10165
77. Zoizou, A., Zarghili, A., Chaker, I.: A new hybrid method for Arabic multi-font text segmentation, and a reference corpus construction. J. King Saud Univ. Comput. Inf. Sci. **32**, 576–582 (2020)

Layout Analysis of Historical Document Images Using a Light Fully Convolutional Network

Najoua Rahal$^{(\boxtimes)}$ [ID], Lars Vögtlin [ID], and Rolf Ingold [ID]

Document Image and Voice Analysis Group (DIVA), University of Fribourg,
1700 Fribourg, Switzerland
najoua.rahal@unifr.ch

Abstract. In the last few years, many deep neural network architectures, especially Fully Convolutional Networks (FCN), have been proposed in the literature to perform semantic segmentation. These architectures contain many parameters and layers to obtain good results. However, for Historical document images, we show in this paper that there is no need to use so many trainable parameters. An architecture with much fewer parameters can perform better while being lighter for training than the most popular variants of FCN. To have a fair and complete comparison, qualitative and quantitative evaluations are carried out on various datasets using standard pixel-level metrics.

Keywords: Historical document images · Layout analysis · Neural network · Deep learning

1 Introduction

Layout analysis is a crucial process and remains an active research area for historical document image analysis and understanding. The interest in this process has been boosted because of its strong impact on multiple tasks, e.g., text recognition [16], and document categorization [15]. Layout analysis can be very challenging because of the unconstrained handwritten documents with complex and highly variable layouts, degradation, multiple writing styles, scripts, fonts, etc. Given the increasing trends towards deep learning, the latter was successfully incorporated in many research fields of pattern recognition and computer vision [9,10]. Recently, many deep learning-based techniques, especially deep Fully Convolutional Networks (FCN) [11], have been successfully investigated to segment historical document images into semantic regions. Semantic segmentation aims at classifying pixels among given class [21].

This paper addresses two sub-processes of layout analysis: page segmentation and text line detection, as semantic segmentation tasks, in historical document

G. A. Fink et al. (Eds.): ICDAR 2023, LNCS 14191, pp. 325–341, 2023.
https://doi.org/10.1007/978-3-031-41734-4_20

images. We propose a variant of deep Fully Convolutional Networks (FCN) based on dilated convolutions. Our model has been trained on different databases and obtained promising results.

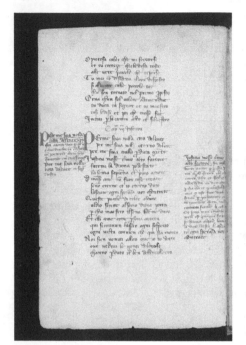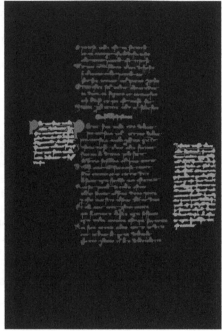

Fig. 1. Page segmentation sample. The colors: black, green, red, and blue represent background, comments, decorations, and main text body, respectively. (Color figure online)

Page Segmentation. Page segmentation (Fig. 1) is an essential step in document image analysis. Its goal is to locate and identify the regions of interest for further processing. The regions can be represented in different ways, either in vectorial form (rectangular bounding boxes, bounded polygons, segments, etc.) or in raster form (pixel-based, with the original or a lower resolution). In our case, we use semantic pixel labelling, which assigns a class to each pixel. The regions are represented by polygons that are adjusted to the text outline.

Text Line Detection. Text line detection (Fig. 2) is considered a special case of page segmentation, where the focus is to detect all the text lines on the page. According to the literature, text lines can be represented either as *vector data* (e.g., bounding box [13]) or *raster data* (e.g., set of pixels [23], baseline [5], or X-height [23] (Fig. 3). In our work, we use a pixel-based binary representation of a strip covering the region delimited by the so-called X-height of each text line. The main contributions of this paper are summarized as follows:

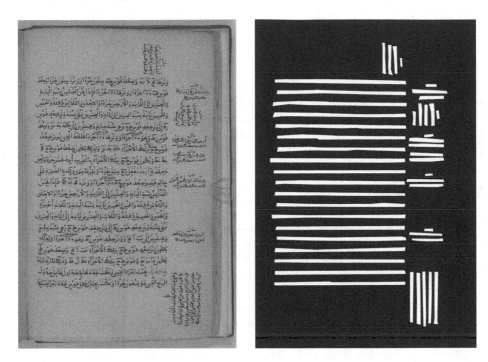

Fig. 2. Text line detection sample at X-height level.

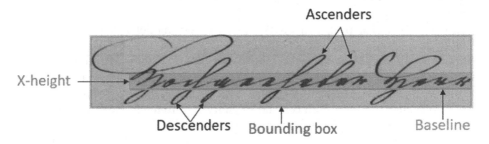

Fig. 3. Illustration of text line representations.

1. We propose L-U-Net, a new variant of U-shape Fully Convolutional Network for semantic pixel labeling related tasks on historical documents.
2. We compare our model with two recent FCN architectures: Doc-UFCN [2], and Adaptive U-Net [12], which are dedicated to text line detection with a reduced number of parameters. We show that training a small and simple architecture with fewer parameters than the existing architectures outperforms the state-of-art performance on various datasets, indicating the effectiveness of our model.

3. It is worth noting that only our work uses the entire images (no patches) as input to train a model for the page segmentation labeling task, which considers the page's overall structure.

The rest of the paper is organized as follows: Sect. 2 presents an overview of some existing related work. Our model and its implementation are described in Sect. 3. In Sect. 4, we present the data used, the experiments carried out to evaluate the performance, and the obtained results. Finally, our conclusion and future work are presented in Sect. 5.

2   Related Work

This section reviews the literature tackling page segmentation and text line detection on historical document images.

Boillet et al. [2] presented the Doc-UFCN model, inspired by the dhSegment model [14], for text line detection. The difference between DocUFCN and dhSegment lies in the encoder used. The encoder of dhSegment was the ResNet-50 [7] architecture, pre-trained on natural scene images. The encoder of Doc-UFCN was smaller than dhSegment's and was fully trained on historical document images. The authors proved that pre-training an FCN model on a few data improved the line detection without needing a huge amount of data to have good results. The proposed model obtained F1-Measure of 91.75%, 88.62%, 79.45%, 78.30%, 85.44% for Balsac [22], Horae [1], READ-BAD [5], and DIVA-HisDB [19], respectively.

Mechi et al. [12] proposed an Adaptive U-Net model for text line detection in handwritten document images, a variant of the U-Net model [18]. Unlike the original U-Net model, which used 64, 128, 256, 512, and 1024 filters at each block in the encoder, the Adaptive U-Net model proposed using 32, 64, 128, 256, and 512 filters. This lead to a reduction of the number of parameters and therefore reduces the memory requirements and the time processing. The model was evaluated on three datasets for X-height-based pixel-wise classifications of text line detection: cBAD [4], DIVA-HisDB [19], and the private ANT[1] datasets. The adaptive U-Net model achieved a performance with an F1-score of 79.00%, 76.00%, 65.00%, 83.00%, 76.00%, and 85.00% on the cBAD, DIVA-HisDB(CB55), DIVA-HisDB(CSG18), DIVA-HisDB(CSG863), ANT(Arabic), and ANT(Latin) datasets, respectively.

Grüning et al. [6] introduced ARU-Net, an extended variant of the U-Net model, for text line detection in historical documents. Two stages were presented in this method. The first stage focused on pixel labeling to one of the three classes: baseline, separator, or other. The second stage performed a bottom-up clustering to build baselines. For evaluation, the proposed method has achieved an F-score of 92.20% and 98.99% for cBAD [4], and DIVA-HisDB [19], respectively.

---

[1] http://www.archives.nat.tn/.

Renton et al. [17] proposed a variant of the U-Net model [18] by using dilated convolutions instead of standard ones for handwritten text line detection in historical document images. The model was applied on an X-height labeling. The experiments of the proposed method were running on the cBAD dataset [4] and achieved an F1-score of 78.30%.

Xu et al. [25] presented a multi-task layout analysis using an FCN, where both page segmentation, text line detection, and baseline detection were performed simultaneously. The proposed model was trained and tested with small patches, not the entire document images. On the DIVA-HisDB database [19], it achieved F1-Measure of 97.52%, 99.07%, and 97.36% for page segmentation, text line detection, and baseline detection, respectively.

A Fully Convolutional Network, proposed by Xu et al. [24], was used for page segmentation for historical handwritten documents on the DIVA-HisDB database [19]. The FCN, based on the VGG 16-layer network [20], was first trained to predict the class pixels as background, main text body, comment, or decoration. However, several modifications are applied to fit the use case of page segmentation, such as low-level processing in an earlier stage of the network and additional convolutional layers before the last stages. Then, heuristic post-processing was adopted to refine the coarse segmentation results by reducing noises and correcting misclassified pixels by analyzing connected components. Finally, the overlap areas were identified by analyzing their sizes and surroundings. The proposed method yielded an F1-Measure of 99.62%.

3 L-U-Net Architecture

We propose L-U-Net architecture, inspired by the general U-Net architecture and, more precisely, by Doc-UFCN [2] for dilated convolutions. The most important novelty is the reduced number of filters, especially for the encoder blocs operating at the lowest resolutions. Probably inspired by classical FCNs without shortcuts, all existing architectures deal with a number of filters that typically are doubled at each block level. They typically use up to 256, 512, and more filters. We argue that, at least for document analysis, there is no justification for using such a progressive series of filter numbers in the contracting part of the network; we believe that the same number of filters used at all levels can capture the characteristics that need to be sent to the decoder or expanding part.

As shown in Fig. 4, L-U-Net architecture is composed of a contraction cascade of a so-called contracting path followed by a symmetrical expansion sequence composed of a so-called Expanding path and a final convolutional layer. The L-U-Net model is publicly available[2].

3.1 Contracting Path

The contracting path consists of four dilated blocks, each composed of four convolutions with 16 filters and dilation 1, 1, 2, and 2, respectively. Each convolution

---

[2] https://github.com/DIVA-DIA/Layout-Analysis-using-a-Light-CNN.

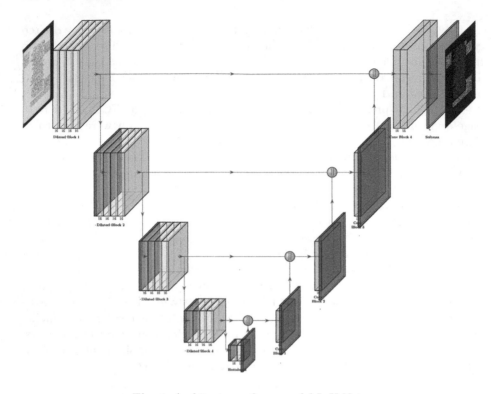

Fig. 4. Architecture of our model L-U-Net.

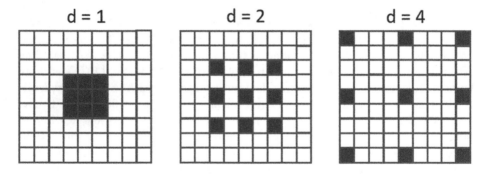

Fig. 5. Dilated convolution representation.

has a 3 × 3 kernel, a stride of one, and adapted padding to maintain the same shape throughout the block. Using dilated convolutions allows the receptive field to be larger and the model to have more contextual information. A visual representation of dilated convolution is given in Fig. 5. All the convolutions of the blocks are followed by a Batch Normalization (BN) layer and a ReLU activation. Each block is followed by a max-pooling layer. The last max-pooling is followed by the bottleneck.

3.2   Expanding Path

The expanding path comprises four convolutional blocks, each consisting of a simple convolution followed by a transposed convolution for up-sampling. Each convolution has a 3×3 kernel, a stride, and a padding of one. Each transposed convolution has a 2×2 kernel and a stride of two. Unlike the Doc-UFCN, the transposed convolutions do not contain BN layers nor ReLU activation.

3.3   Last Convolutional Layer

The output layer returns feature maps representing the number of classes involved in the experiment to get predictions.

4   Experiments and Results

Our experiments focus on finding the best neural network, which presents a constructive compromise between performance and computational cost. We want to show that there is no need to use parameter-intensive architectures to obtain good performances.

4.1   Datasets

To investigate the performance of our proposed L-U-Net model, our experiments are carried out on two datasets: DIVA-HisDB [19], and RASM [3]. DIVA-HisDB datasets are used for page segmentation and text line detection, while the RASM dataset is used for text line detection.

DIVA-HisDB. DIVA-HisDB [19] is a historic manuscript dataset consisting of 120 historical handwritten document images collected from three subsets of medieval manuscripts: CSG18, CSG863, and CB55, with challenging layouts, diverse scripts, and degradations. Each subset includes 20 images for training, 10 for validation, and 10 for testing. The ground truth for page segmentation contains four-pixel classes: background, comments, decorations, and main text. It is worth noting that a pixel can belong to more than one class, known as a multi-label pixel. For example, one pixel can belong to the main text and the decoration class. Like Xu et al. [24], multi-label pixels keep only one of the classes.

RASM. RASM [3], collected from the Qatar digital library, consists of historical Arabic scientific manuscripts. It has been used in ICFHR 2018 Competition on Recognition of Historical Arabic Scientific Manuscripts. The dataset used in our experiments contains 68 document images for training and 17 document images for testing. Among 68 training images, we select 18 images for validation.

4.2 Experimental Protocol

Input Image Size. Our input images and their corresponding masks have been resized into images of size 1152×1728 pixels to keep the original image ratio. This allows a medium resolution considered appropriate to capture the layout structure of entire pages with sufficient precision and reduce the computing time without losing too much information.

Enhancement of DIVA-HisDB Dataset Masks. After resizing, we need to enhance the appearance of DIVA-HisDB dataset mask images. For this purpose, opening and closing morphological operations are investigated to remove noise pixels. As shown in Fig. 6b, the inverted input image is firstly split into three channels Red (R) (Fig. 6c), Green (G)(Fig. 6d), and Blue (B)(Fig. 6e). Afterward, multiplications between channels are implemented, as we can see in Fig. 6f, Fig. 6g, and Fig. 6h, respectively. Then, Opening (O) and closing (C) operations are applied to the three multiplications (Fig. 6i, Fig. 6j, Fig. 6k). Finally, we combine the three multiplications to obtain the cleaned mask, as shown in (Fig. 6l).

Dilation. Using dilated convolutions instead of standard ones brings many advantages: It results in having more context of prediction and provides better performance. The receptive fields are adjusted readily without decreasing the resolution or increasing the number of parameters. In our work, we test four configurations on the CB55 dataset, presented in Table 1.

Dropout. We did three experiments with dropout to see its impact on performance. The first (D1) applies a dropout with $d = 0.4$ after the last dilated block and the bottleneck. The second one (D2) applies the same dropout after each dilated block. The last one (D3) applies the same dropout after every convolution. As shown in Table 2, the application of dropout layers has not positively affected the performances.

Training Settings. To evaluate the performance of our work, a set of experiments has been conducted on several different datasets. The experiments were done with Pytorch. All experiments were handled in the Colab Pro+ offered by Google cloud[3]. In combination with the cross-entropy loss, ADAM optimizer [8] is used for training with an epsilon of $1e^{-5}$ and a learning rate equal to $1e^{-3}$. The cross-entropy loss was used without weights, as preliminary experiments showed no significant impact. In addition, the training is done with a batch size of five during a maximum of 300 epochs for page segmentation and 200 epochs for text line detection. We keep the best model based on the maximum Intersection-over-Union (IOU) value on the validation set.

---

[3] https://colab.research.google.com/signup.

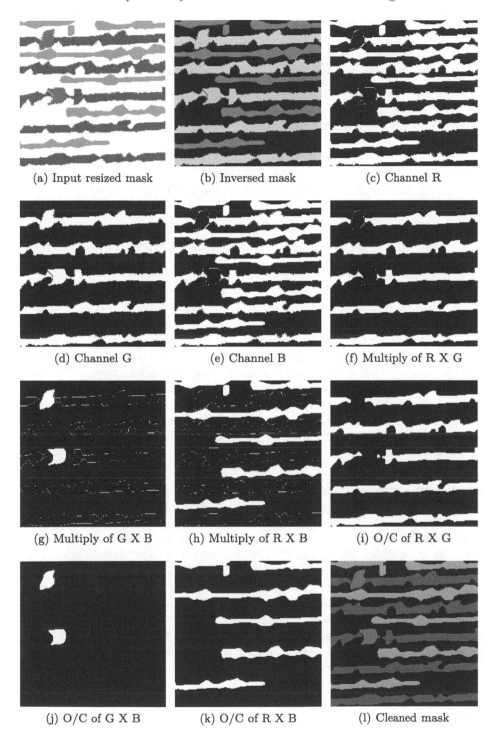

(a) Input resized mask   (b) Inversed mask   (c) Channel R

(d) Channel G   (e) Channel B   (f) Multiply of R X G

(g) Multiply of G X B   (h) Multiply of R X B   (i) O/C of R X G

(j) O/C of G X B   (k) O/C of R X B   (l) Cleaned mask

Fig. 6. Preprocessing steps for enhancement of DIVA-HisDB dataset masks.

Table 1. Impact of the dilation rates.

Dilation	IOU(%)	F1-Measure(%)	Recall(%)	Precision(%)
[1,2,4,8,16]	86.03	91.86	90.81	93.00
[1,1,2,4,8]	70.81	72.83	73.19	72.47
[1,1,2,2,4]	71.11	72.99	73.16	72.99
[1,1,2,2]	**88.31**	**93.35**	**92.77**	**93.95**

Table 2. Impact of the dropout.

Version	IOU(%)	F1-Measure(%)	Recall(%)	Precision(%)
∅	**88.31**	**93.35**	**92.77**	**93.95**
D1	87.48	92.79	93.45	92.14
D2	64.70	69.33	68.61	70.37
D3	42.50	50.14	53.54	62.41

4.3   Results

To show the effectiveness of our model, we implement two recent FCN architectures: Doc-UFCN [2], and Adaptive U-Net [12] dedicated to text line detection, for comparative studies. The reason to choose these two architectures is that they bring major advantages of outperforming the state-of-art results while having a reduced number of parameters, memory requirements, numerical complexity, and training time. To evaluate the performance of our experiments, we computed the following pixel-level metrics: IoU, F1-Measure, Recall, and Precision. We note that the best values are quoted in bold.

Page Segmentation. The page segmentation task refers to segmenting each document into semantic regions by assigning each pixel to the background, comment, main text, or decoration class. In Table 3, we compare the results of our model with those obtained by Doc-UFCN [2] and Adaptive U-Net [12] validated on the CB55, CSG18, and CSG863 datasets. For the CSG863 dataset, the obtained accuracies are less than those of the CB55 and CSG18 datasets. This is due to the class decoration, which does not often occur in the CSG863 dataset compared to the CB55 and CSG18 datasets. Doc-UFCN and Adaptive U-Net report very poor accuracies in this dataset since their IoU and F1-Measure are never higher than 67.00% and 70.00%, respectively. This explains the sensitivity of both architectures to the lack of data; they fail to detect the class decoration in the CSG863 dataset. Contrariwise, our system keeps certain stability of the results obtained for all three datasets. The qualitative results of the proposed and the existing architectures are shown in Fig. 7, where one can notice that our architecture can achieve more accurate segmentation results in the three datasets compared to both existing architectures. However, the L-U-Net sometimes predicts text pixels that belong to the background. Also, some non-textual content is miss-classified as text, representing different types of noise, e.g., back-to-front interference, bleed-through, and ink stains.

Text Line Detection. Text line detection is the main step for text recognition applications and, thus, is of great interest in historical document analysis. It refers to segmenting the document into two categories by classifying each pixel as background or text. In our work, we use the X-height label as text line representation. The quantitative results of the three FCN architectures are reported in Table 3, where we can see a slight difference between them, which does not exceed 0.33% and 0.31% for IoU and F1-Measure, respectively. We observe that the L-U-Net obtains the second-best performance and is slightly behind Adaptive-Unet by 0.02% IOU and 0.01% F1-Measure on the RASM dataset. However, our model performs best for CB55, CSG18, and CSG863 datasets. Figure 8 illustrates the predictions of the best result obtained by the three FCN architectures. We can see that most of the L-U-Net errors are due to weak vertical text line detection. This is due to a small number of training documents with a vertical text line orientation. The other errors are due to the over-segmentation problem (partially detected lines) and the under-segmentation problem (connected close lines). However, the quantitative and qualitative results demonstrate that our proposed architecture is less affected by errors than the existing architectures. We notice that wrongly connected lines crop up in the Adaptive-Unet and Doc-UFCN results more than those of L-U-Net. Most importantly, our architecture proves a better capacity to distinguish close lines than Adaptive-Unet and Doc-UFCN. From the results obtained in the CSG863 dataset, we observe that L-U-Net and Doc-UFCN are less affected by the text line directions, and they help detect the vertical lines where Adaptive-Unet fails.

4.4   Computational Cost

Table 5 shows that the run times and the number of parameters are not sufficiently congruent. Usually, the training time is impacted by the number of trainable parameters. However, we observe that the Doc-UFCN architecture has fewer parameters than the Adaptive-Unet architecture but consumes more training time. This can be justified by the setting of the dilated convolution without growth in the number of parameters. Most importantly, our L-U-Net architecture, having the lowest number of parameters, has the lowest computation time and is way lighter than the Adaptive-Unet and the Doc-UFCN.

4.5   Observations and Recommendations

The following observations and recommendations are derived based on the achieved results to ensure a deductive compromise between obtained results and computation time (Table 4).

- In most cases, we observe an improvement of the L-U-Net results compared to those of the Doc-UFCN and the Adaptive-Unet;

Table 3. Comparison between FCN models for page segmentation.

Datasets	Metrics	L-U-Net	Doc-UFCN	Adaptive-Unet
CB55	IOU(%)	**88.31**	84.08	86.77
	F1-Measure(%)	**93.35**	90.32	92.43
	Recall(%)	**92.77**	89.08	92.19
	Precision(%)	**93.95**	91.95	92.72
CSG18	IOU(%)	**90.67**	84.02	88.91
	F1-Measure(%)	**95.02**	91.07	94.00
	Recall(%)	**95.58**	88.77	94.91
	Precision(%)	**94.47**	93.58	93.14
CSG863	IOU(%)	**83.98**	62.06	67.07
	F1-Measure(%)	**90.64**	67.54	70.70
	Recall(%)	**89.77**	67.87	71.24
	Precision(%)	**91.62**	67.54	70.19

Table 4. Comparison between FCN models for text line detection.

Datasets	Metrics	L-U-Net	Doc-UFCN	Adaptive-Unet
RASM	IOU(%)	87.46	87.42	**87.48**
	F1-Measure(%)	93.06	93.03	**93.07**
	Recall(%)	91.90	**92.97**	92.22
	Precision(%)	**94.32**	93.10	93.97
CB55	IOU(%)	**89.53**	89.19	89.30
	F1-Measure(%)	**94.35**	94.15	94.22
	Recall(%)	93.03	92.73	**93.06**
	Precision(%)	**95.82**	95.75	95.50
CSG18	IOU(%)	**86.56**	86.06	86.46
	F1-Measure(%)	**92.46**	92.15	92.41
	Recall(%)	90.59	90.11	**91.18**
	Precision(%)	**94.59**	94.48	93.74
CSG863	IOU(%)	**89.86**	89.77	89.54
	F1-Measure(%)	**94.55**	94.50	94.36
	Recall(%)	93.55	**93.72**	93.10
	Precision(%)	95.65	95.34	**95.77**

Table 5. Comparison of the three FCN architectures: number of parameters and training time in terms of minutes measured on the CB55 dataset for text line detection.

	L-U-Net	Doc-UFCN	Adaptive-Unet
Nb parameters	65,634	4,096,322	7,760,130
Training time	50m 16s	119m 46s	74m 29s
Memory (MB)	973.76	3681.56	1552.85

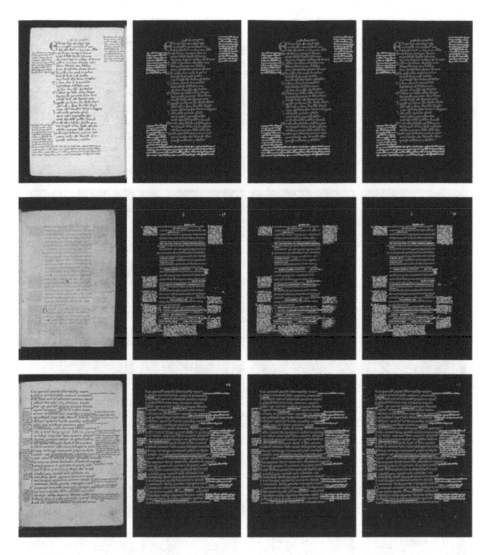

Fig. 7. Page segmentation results on the CB55, CSG18, and CSG863 datasets from top to bottom respectively. The columns from left to right are input and segmentation results of the L-U-Net, Doc-UFCN, and Adaptive-Unet models, respectively.

- The impact of a small model is higher as the amount of data available for training is low;
- The model having a low number of parameters decreases the training time and improves the model's performance;
- When the computation time is taken into consideration, the L-U-Net model has lower complexity compared to the existing models and with comparable performance;

Fig. 8. Text line detection results on the RASM, CB55, CSG18, and CSG863 datasets from top to bottom respectively. The columns left to right are input masks and segmentation results of the L-U-Net, Doc-UFCN, and Adaptive-Unet models, respectively.

- The same architecture with the same parameters can be efficiently trained to perform different segmentation tasks.
- The results obtained from the qualitative and quantitative evaluations confirm our hypothesis that the small network, having less number of parameters, is well-suited for the semantic segmentation of historical document images. It achieves competitive results without being time-consuming or losing information;
- Evaluating and comparing the three neural network models based on the F1-score and IoU would lead to choosing L-U-Net as the best model, having the best trade-off between the best performance and the undermost computational time.

5 Conclusion

In this paper, we have introduced a novel network architecture named L-U-Net characterized by a drastically reduced number of trainable parameters. We have compared its performance with two existing architectures presented in the literature as state-of-the-art solutions for the semantic labeling of historical documents. Two different labeling tasks and four datasets have been considered for the assessment.

As a general observation, we can see that the convergence of the new network is faster, and the obtained results are at least as good as those of the other networks. In several cases, we observe even better results. The quantitative evaluation on standard metrics shows that the improvement with L-U-Net is statistically significant, with a difference of IoU of about 10.93% for page segmentation and 0.24% for text line detection. More interestingly, the qualitative evaluation reveals that the type of errors is even more meaningful. We have observed that L-U-Net results produce much fewer classification errors in the case of page segmentation and fewer line merges (under-segmentation) for text line detection.

Therefore, we argue that the community of historical document analysis should eliminate the heavy multi-purpose network architectures and design new networks that are more focused on their specific tasks. By doing so, the community would also reduce the needed resources and, thus, at least moderately contribute to environmental sustainability.

In future works, we intend to apply our model to other tasks and challenging datasets.

References

1. Boillet, M., Bonhomme, M.L., Stutzmann, D., Kermorvant, C.: Horae: an annotated dataset of books of hours. In: Proceedings of the 5th International Workshop on Historical Document Imaging and Processing, pp. 7–12 (2019)
2. Boillet, M., Kermorvant, C., Paquet, T.: Multiple document datasets pre-training improves text line detection with deep neural networks. In: 2020 25th International Conference on Pattern Recognition (ICPR), pp. 2134–2141. IEEE (2021)

3. Clausner, C., Antonacopoulos, A., Mcgregor, N., Wilson-Nunn, D.: Icfhr 2018 competition on recognition of historical arabic scientific manuscripts-rasm2018. In: 2018 16th International Conference on Frontiers in Handwriting Recognition (ICFHR), pp. 471–476. IEEE (2018)

4. Diem, M., Kleber, F., Fiel, S., Grüning, T., Gatos, B.: cbad: Icdar 2017 competition on baseline detection. In: 2017 14th IAPR International Conference on Document Analysis and Recognition (ICDAR), vol. 1, pp. 1355–1360. IEEE (2017)

5. Grüning, T., Labahn, R., Diem, M., Kleber, F., Fiel, S.: Read-bad: a new dataset and evaluation scheme for baseline detection in archival documents. In: 2018 13th IAPR International Workshop on Document Analysis Systems (DAS), pp. 351–356. IEEE (2018)

6. Grüning, T., Leifert, G., Strauß, T., Michael, J., Labahn, R.: A two-stage method for text line detection in historical documents. Int. J. Doc. Anal. Recogn. (IJDAR) **22**(3), 285–302 (2019)

7. He, K., Zhang, X., Ren, S., Sun, J.: Deep residual learning for image recognition. In: Proceedings of the IEEE Conference on Computer Vision and Pattern Recognition, pp. 770–778 (2016)

8. Kingma, D.P., Ba, J.: Adam: a method for stochastic optimization (2014). arXiv preprint arXiv:1412.6980

9. LeCun, Y., Bengio, Y., Hinton, G.: Deep learning. Nature **521**(7553), 436–444 (2015)

10. Levi, G., Hassner, T.: Age and gender classification using convolutional neural networks. In: Proceedings of the IEEE Conference on Computer Vision and Pattern Recognition Workshops, pp. 34–42 (2015)

11. Long, J., Shelhamer, E., Darrell, T.: Fully convolutional networks for semantic segmentation. In: Proceedings of the IEEE Conference on Computer Vision and Pattern Recognition, pp. 3431–3440 (2015)

12. Mechi, O., Mehri, M., Ingold, R., Amara, N.E.B.: Text line segmentation in historical document images using an adaptive u-net architecture. In: 2019 International Conference on Document Analysis and Recognition (ICDAR), pp. 369–374. IEEE (2019)

13. Moysset, B., Kermorvant, C., Wolf, C., Louradour, J.: Paragraph text segmentation into lines with recurrent neural networks. In: 2015 13th International Conference on Document Analysis and Recognition (ICDAR), pp. 456–460. IEEE (2015)

14. Oliveira, S.A., Seguin, B., Kaplan, F.: dhsegment: a generic deep-learning approach for document segmentation. In: 2018 16th International Conference on Frontiers in Handwriting Recognition (ICFHR), pp. 7–12. IEEE (2018)

15. Paquet, T., Heutte, L., Koch, G., Chatelain, C.: A categorization system for handwritten documents. Int. J. Doc. Anal. Recogn. (IJDAR) **15**(4), 315–330 (2012)

16. Parvez, M.T., Mahmoud, S.A.: Offline Arabic handwritten text recognition: a survey. ACM Comput. Surv. (CSUR) **45**(2), 1–35 (2013)

17. Renton, G., Soullard, Y., Chatelain, C., Adam, S., Kermorvant, C., Paquet, T.: Fully convolutional network with dilated convolutions for handwritten text line segmentation. Int. J. Doc. Anal. Recogn. (IJDAR) **21**(3), 177–186 (2018)

18. Ronneberger, O., Fischer, P., Brox, T.: U-Net: convolutional networks for biomedical image segmentation. In: Navab, N., Hornegger, J., Wells, W.M., Frangi, A.F. (eds.) MICCAI 2015. LNCS, vol. 9351, pp. 234–241. Springer, Cham (2015). https://doi.org/10.1007/978-3-319-24574-4_28

19. Simistira, F., Seuret, M., Eichenberger, N., Garz, A., Liwicki, M., Ingold, R.: Diva-hisdb: a precisely annotated large dataset of challenging medieval manuscripts. In: 2016 15th International Conference on Frontiers in Handwriting Recognition (ICFHR), pp. 471–476. IEEE (2016)
20. Simonyan, K., Zisserman, A.: Very deep convolutional networks for large-scale image recognition (2014). arXiv preprint arXiv:1409.1556
21. Soullard, Y., Tranouez, P., Chatelain, C., Nicolas, S., Paquet, T.: Multi-scale gated fully convolutional densenets for semantic labeling of historical newspaper images. Pattern Recogn. Lett. **131**, 435–441 (2020)
22. Vézina, H., Bournival, J.: An overview of the balsac population database. Past developments, current state and future prospects. Historical Life Course Studies (2020)
23. Vo, Q.N., Lee, G.: Dense prediction for text line segmentation in handwritten document images. In: 2016 IEEE International Conference on Image Processing (ICIP), pp. 3264–3268. IEEE (2016)
24. Xu, Y., He, W., Yin, F., Liu, C.L.: Page segmentation for historical handwritten documents using fully convolutional networks. In: 2017 14th IAPR International Conference on Document Analysis and Recognition (ICDAR), vol. 1, pp. 541–546. IEEE (2017)
25. Xu, Y., Yin, F., Zhang, Z., Liu, C.L., et al.: Multi-task layout analysis for historical handwritten documents using fully convolutional networks. In: IJCAI, pp. 1057–1063 (2018)

Combining OCR Models for Reading Early Modern Books

Mathias Seuret[1]([✉]) [ID], Janne van der Loop[2] [ID], Nikolaus Weichselbaumer[2] [ID], Martin Mayr[1] [ID], Janina Molnar[2] [ID], Tatjana Hass[2] [ID], and Vincent Christlein[1] [ID]

[1] Friedrich-Alexander-Universität Erlangen-Nürnberg, Erlangen, Germany
mathias.seuret@fau.de
[2] University of Mainz, Mainz, Germany
{jannevanderloop,weichsel}@uni-mainz.de

Abstract. In this paper, we investigate the usage of fine-grained font recognition on OCR for books printed from the 15th to the 18th century. We used a newly created dataset for OCR of early printed books for which fonts are labeled with bounding boxes. We know not only the font group used for each character, but the locations of font changes as well. In books of this period, we frequently find font group changes mid-line or even mid-word that indicate changes in language. We consider 8 different font groups present in our corpus and investigate 13 different subsets: the whole dataset and text lines with a single font, multiple fonts, Roman fonts, Gothic fonts, and each of the considered fonts, respectively. We show that OCR performance is strongly impacted by font style and that selecting fine-tuned models with font group recognition has a very positive impact on the results. Moreover, we developed a system using local font group recognition in order to combine the output of multiple font recognition models, and show that while slower, this approach performs better not only on text lines composed of multiple fonts but on the ones containing a single font only as well.

Keywords: OCR · Historical documents · Early modern prints

1 Introduction

Libraries and archives are publishing tremendous amounts of high-quality scans of documents of all types. This leads to the need for efficient methods for dealing with such an amount of data – in many cases, having the documents' content as a text file instead of an image file can be of great help for scholars, as it allows them to look for specific information easily.

In this work, we focus on Optical Character Recognition (OCR) early modern prints – documents printed between the 15th and the 18th centuries – currently

---

M. Seuret and J. van der Loop—Contributed equally to this research.

Supported by the Deutsche Forschungsgemeinschaft (DFG) - Project number 460605811.

G. A. Fink et al. (Eds.): ICDAR 2023, LNCS 14191, pp. 342–357, 2023.
https://doi.org/10.1007/978-3-031-41734-4_21

held by German libraries. Early print frequently used a large range of font styles, much more than today, where most text, such as in this paper, is displayed or printed in some variety of Antiqua. Our investigations are based on the prior assumption that the appearance, or shapes, of fonts, has an impact on the performance of OCR models. In order to evaluate the impact of fonts on OCR and whether they can be used for improving OCR results, we couple fine-grained recognition and OCR. We do not try to differentiate individual fonts – like a particular size of Unger Fraktur – since this information would be too granular and of limited use for OCR. It would also be very challenging to label at a large scale, especially considering that early printers typically had their own, handcrafted fonts. Instead, we differentiate eight font groups – Antiqua, Bastarda, Fraktur, Gotico-Antiqua, Italic, Rotunda, Schwabacher and Textura. These font groups contain large numbers of fonts that share certain stylistic features but may have different sizes, proportions or other individual design peculiarities.

The two main focuses of this paper are the impact of font groups on OCR accuracy and the combination of font group classification and OCR methods. In particular, we (1) evaluate the impact of selecting models fine-tuned on specific font-groups on a line basis, (2) evaluate the impact of splitting text lines according to a fine-grained font-group classifier, and (3) develop a novel combined OCR model denoted as COCR,[1] which outperforms the other systems in case of multiple fonts appearing in one line.[2]

2   State of the Art

Optical Character Recognition. The field of Optical Character Recognition (OCR) and Handwritten Text Recognition (HTR) has seen remarkable progress as a result of the Connectionist Temporal Classification (CTC) decoding, introduced by Graves et al. [11]. Subsequent research, such as Puigcerver et al. [25], combined the CTC loss function with the hybrid architecture of Convolutional Recurrent Neural Networks (CRNNs). Bluche et al. [2] further enhanced this architecture by incorporating gated mechanisms within the convolutional layers to emphasize relevant text features. Michael et al. [22] followed an alternative direction by employing sequence-to-sequence (S2S) models for HTR. These approaches consist of an encoder-decoder structure and often utilize a Convolutional Neural Network (CNN) with a 1-D bidirectional Long Short-Term Memory (LSTM) network as the encoder and a vanilla LSTM network as the decoder, producing one character at a time. The focus of their work lies in the comparison of the attention mechanism linking the encoder and the decoder. Yousef et al. [36] and Coquenet et al. [5–7] achieved state-of-the-art results in HTR on benchmark, such as IAM database [21] and RIMES [12], by using recurrence-free approaches. Kang et al. [16] improved sequential processing of text lines by substituting all LSTM layers with Transformer layers [31]. Wick et al. [33]

---

[1] https://github.com/seuretm/combined-ocr.
[2] On an intermediate version of EMoFoG (Early Modern Font Groups), DOI 10.5281/zenodo.7880739, to appear.

extended this concept by fusing the outputs of a forward-reading transformer decoder with a backward-reading decoder. Diaz et al. [8] examined various architectures with their proposed image chunking process, achieving the best results with a Transformer-CTC-based model in conjunction with a Transformer language model. Li et al. [19] accomplished impressive results on the IAM database through the use of 558 million parameters S2S-Transformer-based model without convolutional layers, as well as through massive pre-training utilizing 684 million text lines extracted from PDF documents and 17 million handwritten text lines. In their study of MaskOCR, Lyu et al. [20] demonstrated the benefits of self-supervised pre-training for OCR on Chinese documents and scene text recognition. For pre-training, the model reconstructed masked-out patches of the text line images using a thin decoder. For the downstream task, the reconstruction decoder was discarded, and the remaining parts of the model underwent fine-tuning. To improve the integration of CTC-predictions and decoder outputs, Wick et al. [34] calculated a CTC-prefix-score to enhance the sequence decoding performance of both the transformer decoder and the language model. An ensemble of N models is used by Reul et al. [26] in a voting strategy. They split their training data into N subsets. Each of these subsets is used as validation data for a single model trained on the remaining $N - 1$ other subsets. Inference is done by averaging the results of all models of the ensemble. Wick et al. [32] refine this approach – instead of training the models independently, they use a masking approach allowing to combine their losses and thus train them together.

Font Type Recognition. This task has been around for a very long time, with, for example, some publications on this topic in 1991 at the first ICDAR conference [9]. Each font is typically considered as being one class. We however are dividing the fonts into different font groups. In [23], the authors produce descriptors based on local binary patterns [24] and compare them with the nearest neighbor algorithm to classify 100 fonts and sizes for Arabic text. In [4], the authors use a ResNet [13] and maximum suppression to identify 1,116 different fonts in born-digital images with high accuracy. A larger dataset with over 3300 fonts is used in [35]. As many of the fonts are visually practically indistinguishable, the authors create groups of fonts by clustering them and evaluate several CNN architectures for either font or cluster classification. In [27], fonts from historical books are classified, but the proposed system is restricted to outputting only one class per page. This is done by average voting on patches, using a DenseNet [15]. In [29], the authors use a ResNet pre-trained on synthetic data both for script and font classification. They do so by combining the scores on randomly selected patches that have been filtered to reject the ones containing only background.

Novelty of this Work. It differs from previous publications in two main points. First, while we also investigate the use of ensembles of OCR models, we combine their outputs using a locally weighted average using font classification, which has not been done in other works. Second, we push the boundary of the font group

Table 1. Base model architecture

Operation	Kernel size	Outputs	Stride	Padding	Comment
Convolution	3×2	8	2×1	–	then ReLU
Convolution	6×4	32	1×1	3×1	then ReLU
Max pooling	4×2	–	4×2	–	–
Convolution	3×3	64	1×1	1×1	then ReLU
Max pooling	1×2	–	1×2	–	then ReLU
Mean (vertical axis)	–	–	–	–	–
Linear	–	128	–	–	–
Bi-directional LSTM [14]	–	2×128	–	–	3 layers, tanh
Linear	–	153	–	–	152 chars + blank

classification granularity as we aim to find where font groups change across a single text line. Another difference, although of lower importance, is that we consider the font group classification as an auxiliary task for OCR, and thus use OCR evaluation as a proxy for investigating the usefulness of font group classification.

3   Methods

3.1   One Model to OCR Them All: Baseline

This baseline approach consists of training a single OCR model on the whole dataset, regardless of any font information. We based our architecture, detailed in Table 1, on the CTC-based model by Wick et al. [34]. However, we employed a finer scale in the first convolutional layer of the feature extractor, a CNN [10], than their original model and halved the kernel and stride sizes. We further decreased the number of neurons in the recurrent layers to 128, for two reasons: First, we make the assumption that OCR does not require an Recurrent Neural Network (RNN) as powerful as HTR, and, second, one of the systems which makes use of multiple instances of the baseline is memory-greedy.

3.2   Font-Group-Specific OCR Models

We produce font-group-specific OCR models by fine-tuning the baseline. Training them from scratch, instead of using the baseline as starting point, led to high Character Error Rate (CER). Note that it is required to know the font group of text lines in order to apply the corresponding model.

3.3   Select One Model per Line: SelOCR

Having font-group-specific models is nice, but of little use unless we know which font group is present in the data to process. For this reason, we developed the

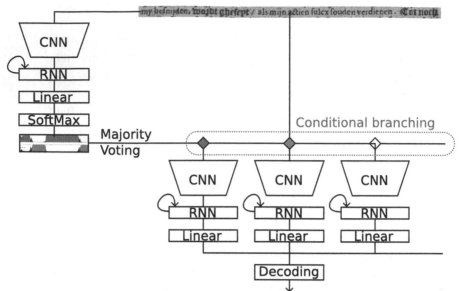

Fig. 1. Pipeline of the SelOCR system.

Selective OCR approach: a network first recognizes the font group of the input text line, and then the corresponding OCR model is applied. This process is illustrated in Fig. 1.

3.4 Splitting Lines in Heterogeneous Parts: SplitOCR

As font groups can change within a text line (even in the middle of a word), we can try to split text lines into homogeneous sequences of a single font group, and process each of them with an adequate OCR model.

We first use a classifier to estimate the font group of every pixel column of the text line to process. Then, any homogeneous segment of a width smaller than the height of the text line is split into two parts, and each gets merged with the adjacent segment. This leads to heterogeneous segments where majority voting is applied to retain only a single class. As the last post-processing step of the classification result, adjacent segments with the same class are merged.

Each segment is then processed with the OCR model corresponding to the majority class of the segments' pixel columns.

3.5 Combining OCR Models: COCR

RNN benefit much from processing longer sequences instead of short, independent inputs. Thus, splitting text lines, as it is done with the SplitOCR approach,

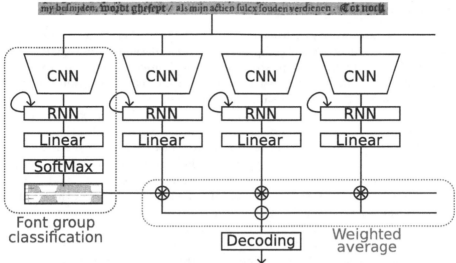

Fig. 2. Pipeline of the COCR system.

might be sub-optimal, as OCR models might not see the whole input and benefit from advantages of processing a sequence, such as language models implicitly learned by RNN.

For this reason, we developed one more approach based on the combination of the outputs of OCR models using font group classification. Font group classification is done with a model having a similar structure as the OCR models, but with only 32 neurons in a single recurrent layer, as many outputs as there are font groups, and a softmax activation function after the output layer. Thus, the classifier and the OCR models all have the same output sequence length, and can easily be merged.

We first apply the classifier and all OCR models[3]. Then, we use the classification scores as weights to make a weighted sum of the OCR outputs, before the decoding step. This is illustrated in Fig. 2.

As this method is differentiable, we can train everything together. However, we started with already-trained networks: the baseline for the different OCR networks, and first trained the classifier on its own. As the OCR system gets to see the whole input and must provide outputs that can be combined, we initialize all of them with the weights of the baseline, not with the font-group-specific models.

---

[3] We found out that applying only OCR models for font groups which have, somewhere in the batch, a score higher than 0.1 improved the speed by roughly 40% with no more than differences of 0.01% of CER. However, these values are highly data-dependent.

3.6   Training Procedure

The text lines are resized to a height of 32 pixels, and their aspect ratio is preserved. This height is a good compromise between data size (and OCR speed) and readability. The font group label arrays, which are attributed to every pixel column of the text lines, are scaled accordingly using the nearest value, as interpolating class numbers would be quite meaningless.

The networks are trained with a batch size of 32, for a maximum of 1000 epochs, and we use early stopping with a patience period of 20 epochs. We selected Adam [17] as the optimizer, mainly for convergence speed reasons, and the CTC [11]. Every 5 epochs without CER improvement, the learning rate is halved. For the baseline, we start with a learning rate of 0.001, and random initial weights. The other models are initialized with the weights of the baseline, and the optimizer is reloaded as well in order to start the training from the point which had the lowest validation CER for the baseline.

The two classifiers, i.e., the one used in the SelOCR and SplitOCR system, and the one used in the COCR system, are trained slightly differently. In the case of the pixel column classifier, we use a batch size of 1, as this seemed to converge faster and better. Regarding the classifier used in the COCR system, experiments have shown that if it is not trained before starting to train the system as a whole, results are not competitive.[4] So, in order to train it, as it outputs a sequence shorter than the labels of the pixel columns, we downscaled the pixel-wise ground truth without interpolation.

Offline data augmentation of three different types (warping, morphological operations, and Gaussian filtering) is applied. As these augmentations are rather fast, applying them at runtime still significantly slowed down the training process. This is why we decided to store the augmented data instead.

Warping is done with ocrodeg; [3] as parameter we set sigma to 5, and maxdelta to 4. These values were selected by trying different combinations and retaining one which provided realistic results on some randomly selected images. As morphological operations, we selected grayscale erosion and dilation with square elements of size 2×2. For the Gaussian filtering, we have two methods as well: Gaussian blur, with its variance randomly selected in $[1, 1.25]$, and unsharp filter using the same parameter range. The verisimilitude of the produced data could likely be enhanced furthermore by adapting all these parameters to the resolution of the images.

To produce offline augmented images, we selected a random non-empty set of these augmentations, with no more than one augmentation of each kind (i.e., erosion, or dilation, or none of these, but not both erosion and dilation). For each original training text line, we created two additional ones. An example of our augmentations is given in Fig. 3.

---

[4] No class labels are shown to the classifier when the COCR system is trained so that the classifier has no restriction in how it merges the OCR outputs.

(a) Original

(b) Warping

(c) Erosion

(d) Dilation

(e) Gaussian

(f) Unsharp

Fig. 3. Illustration of the offline augmentations which we applied.

Moreover, at run time, we use Torchvision [30] to apply further augmentations: shearing (maximum angle of 5°), and modifications of contrast and luminosity (maximum 20%).

3.7 Evaluation

Our main metric is the Character Error Rate (CER). The CER between transcription and the corresponding ground truth, is obtained by dividing the Levenshtein distance [18] between these two strings by the length of the ground truth. In order to avoid averaging together CER obtained on short and long text lines, we consider our test set as a whole, and compute the CER as the sum of the Levenshtein distances for all text lines, by the sum of the length of all ground truth strings. We used the `editdistance` library[5], version 0.6.2. We encode and decode strings using Python iterators, for which reason some special characters are composed of several codepoints.

4   Data

Our data consists of 2,506 pages, taken from 849 books printed between the 15th and 18th centuries. On each of those pages, we have manually drawn bounding boxes around text lines or parts of text lines. We labeled the (only) font group

---

[5] https://pypi.org/project/editdistance/.

Table 2. Amount of bounding boxes and characters in our dataset.

Font group	Training	Validation	Test	Training	Validation	Test
Antiqua	13725	310	592	400685	6166	13361
Bastarda	13528	99	301	582765	5468	12088
Fraktur	10360	163	250	413798	6651	12052
Gotico-Antiqua	8200	143	373	345171	6118	12653
Italic	11315	186	443	410085	6604	14645
Rotunda	7261	132	229	301323	7842	14180
Schwabacher	15808	98	432	684879	5302	16182
Textura	7724	123	265	338632	5714	12875
Total	87929	1254	2885	3477458	49865	108036

of each of these boxes – thus, if a text line contains more than one font group, then it is split into several bounding boxes. We transcribed the content of these bounding boxes. In total, we annotated over 90,000 bounding boxes and transcribed over 3.6 million characters. Our data also contains bounding boxes for Greek, Hebrew, and Manuscripts, however as they are a negligible fraction of our data, we omitted them in this study. More detailed statistics are given in Table 2.

Moreover, we considered two super-groups of fonts: Gothic and Roman. Indeed, as it can happen that Gothic font groups are erroneously considered as *Fraktur* and other ones as *Latin* (e.g., in [1,28], or the Fraktur and non-Fraktur pre-trained models for the German language of Tesseract[6]), we decided to investigate these two subsets of our data. The first one, Gothic, is composed of Bastarda, Fraktur, Rotunda, Schwabacher, and Textura. The second one, the Roman font group, is composed of Italic and Antiqua. We decide not to use the name *Latin*, as some clearly Gothic font groups, such as Textura, are often used for texts in Latin (as a language). We did not include Gotico-Antiqua, as it is by definition halfway between Roman and Gothic font groups.

Our data is split at the book level so that no data from the same book can be present in both training and test sets. To produce the test set, we tried a large number of random combinations for which all font groups have at least 10'000 characters, and kept the one minimizing the variance of the number of characters per font group. Moreover, we manually swapped one Gotico-Antiqua book with another one, as it contained special characters not present in any other book – it would make little sense to test the recognition of characters not present in the training data. The validation set was obtained in the same way but contains a smaller amount of data.

---

[6] https://github.com/tesseract-ocr/tessdata_best.

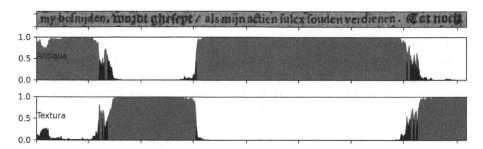

Fig. 4. Classification results of pixel columns. The two plots correspond to the classification scores for Antiqua and Textura. Colors indicate the pixel column with the highest score. Results for other font groups are not shown, as they were all extremely close to zero.

5 Experimental Results

We evaluate the different approaches following three scenarios. The first one is simply using all of our test data. This evaluates how well the different approaches perform if we do not know in advance which font groups are in the data; this is the most realistic use case. The second one is using only text lines containing multiple font groups, as it allows us to see how the different models deal with font changes. The last scenario uses the ground truth to separate the font groups. While the least realistic (it is unlikely an expert would look at each text line to label its font group before passing it through an OCR pipeline), but offers insight into the effect of fine-tuning OCR models for specific font groups. We will first briefly present classification results, and expand more on OCR results afterward.

5.1 Classification

While there is no doubt about which fonts are present in a text line or which font is used for every character, labels for pixel columns are to be considered approximate near class boundaries. Indeed, the column labels were obtained through manually drawn bounding boxes, thus two experts labeling the same file – or the same one labeling it twice – would produce slightly different bounding boxes, especially at font group boundaries in white spaces between words. For this reason, we consider classification accuracy as secondary, as the main task of the classifier is to be included in OCR systems.

An example of pixel column classification is shown in Fig. 4. In this text line, we can see that the two sequences of Antiqua and Textura are mostly correctly identified. We can, however, notice some fluctuations near class boundaries; this is what we smoothen in the post-processing described in Sect. 3.4.

The classifier used in the COCR system produces an output sequence shorter than the text line, thus we downscale without interpolation the pixels' labels in order to have training data usable by this classifier.

5.2   OCR

The results are presented in Table 3. You will notice that the CER for the Gotico-Antiqua font group is significantly higher than for other font groups. As mentioned in Sect. 4, for this font group, we manually selected a document, as the automatic random selection led to having some characters present only in the test data. We checked the validation CER and saw that it matches the ones of other font groups. Moreover, a test done on another document which is not included in our dataset reached a CER only slightly higher than for the other font groups, so we explain the much higher CER of Gotico-Antiqua by the difficulty presented by the selected document. However, for scientific integrity reasons, we did not swap this document with another one and present the results as they are.

First, let us have a look at the baseline and the models with a similar architecture, i.e., the ones which are below it in Table 3. While not being the best, the baseline is performing relatively well on the whole dataset (first column) compared to the other models. However, for all considered subsets of our data, there is at least one other comparable model with a lower CER. Models fine-tuned on specific font groups always perform better than the baseline for the corresponding font group; this is not a surprise but is nevertheless worth being mentioned.

We can also see that results for the fine-tuned models get the best results for their architecture six times out of eight, with the exception of the ones for Schwabacher and Rotunda. Moreover, only one of the complex models performs best on one single font (Italic).

Out of curiosity, we also trained an extra model for Gotico-Antiqua, which we named Gotico-Antiqua[+]. As this font group is by definition halfway between Gothic and Roman font groups, we added training data from the visually most similar font groups, namely Antiqua and Rotunda. While this improves its CER on these two other groups, it also has a negative impact on Gotico-Antiqua.

Now, let us focus on the three other systems, which are based on font group identification. The SplitOCR system has the highest CER of all methods evaluated in this paper. This is due to the fact that when text lines are split into smaller parts – more specifically, we think that shorter text lines do not provide enough contextual information for the RNN layers to perform well. Moreover, another drawback of this method is that text lines have to be processed one by one, and multiple calls to OCR models, one per segment of the text line, have to be made.

In order to investigate this, we computed the baseline CER of all test lines of our dataset, grouped the lines in bins based on their number of characters, and computed the average CER of each bin. The plot we produced is shown in Fig. 5. We can make two main observations. First, shorter text lines tend to have a higher CER, which supports our previous assumption. Second, the variance of the results is higher for shorter text lines. This is not caused by a lower amount of short lines (there are almost 500 of length 1 to 10, and 300 of length 71 to 81), but rather by the higher CER introduced by errors in shorter lines.

Table 3. CER (%) for the different OCR systems, on the test set and its subsets. Systems named as font groups are the ones fine-tuned for this font group. The *All* and *Mult.* sets are the whole test set and its text lines which contain multiple font groups, respectively. The following subsets are text lines with only the mentioned font group. The horizontal and vertical separators help distinguish better results obtained with simple or more complex network architectures, and on subsets with multiple or single font groups, respectively.

System	All	Mult.	Got.	Rom.	Ant.	Bas.	Fra.	G.-A.	Ita.	Rot.	Schw.	Tex.
COCR	**1.81**	**1.95**	**1.17**	**1.89**	1.76	1.36	1.05	4.44	**2.00**	0.99	1.06	1.47
SelOCR	1.82	2.61	**1.17**	1.92	1.76	1.32	1.01	4.35	2.06	1.01	1.07	1.47
SplitOCR	2.19	4.83	1.48	2.19	2.11	2.02	1.59	4.65	2.26	1.07	1.22	1.64
Baseline	1.92	2.59	1.21	2.12	1.98	1.44	1.03	4.57	2.24	1.04	1.08	1.54
Gothic	1.95	2.61	**1.17**	2.35	2.30	**1.26**	**0.98**	4.53	2.40	1.04	1.11	1.49
Roman	2.02	2.40	1.32	2.00	1.87	1.45	1.32	5.06	2.12	1.10	1.23	1.58
Antiqua	1.98	2.53	1.30	1.94	**1.75**	1.44	1.20	4.93	2.11	1.16	1.17	1.57
Bastarda	1.97	2.67	1.24	2.17	2.09	1.31	1.05	4.67	2.25	1.14	1.11	1.63
Fraktur	1.95	2.67	1.21	2.10	2.05	1.39	**0.98**	4.77	2.15	1.11	**1.04**	1.59
Gotico-Antiqua	1.90	2.53	1.23	2.16	2.14	1.38	1.09	**4.30**	2.18	1.06	1.13	1.54
Gotico-Antiqua+	1.91	2.56	1.23	2.07	1.82	1.42	1.10	4.50	2.29	**0.96**	1.15	1.60
Italic	1.90	2.38	1.22	1.93	1.78	1.43	**0.98**	4.81	2.06	1.01	1.13	1.57
Rotunda	1.95	2.53	1.22	2.16	2.12	1.40	1.00	4.69	2.19	1.01	1.12	1.60
Schwabacher	1.97	2.59	1.19	2.24	2.16	1.34	1.03	4.87	2.31	**0.96**	1.06	1.61
Textura	1.94	2.53	1.20	2.10	1.85	1.42	1.10	4.81	2.32	0.99	1.07	**1.45**
Best Fine-Tuned	1.90	2.40	**1.17**	1.93	**1.75**	**1.26**	**0.98**	**4.30**	2.06	**0.96**	**1.04**	**1.45**

The SelOCR system works significantly better than the baseline, with a CER 9.4% lower. It uses the OCR models fine-tuned for the different font groups, and thus differences in CER are entirely due to the results of the classifier which selects which OCR model to apply. At the cost of first classifying text lines, it beats the baseline, as could be expected, in all cases, except for the subset of text lines containing multiple font groups. Also, at least in our implementation, text lines have to be processed one by one, as a condition is used for applying the right OCR model.

The COCR system is slightly better than the SelOCR on the whole dataset. We, however, think that such a small difference is not significant; with a little different test set, the tendency could reverse. For this reason, we would not consider one being better than the other on the whole set. However, if we look at the second column, i.e., text lines containing multiple fonts, we can see that the COCR system does perform significantly better, with a CER a quarter smaller.

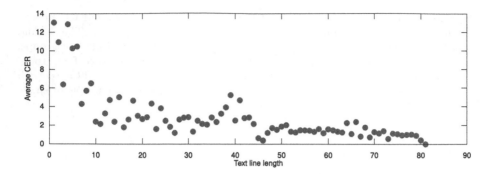

Fig. 5. Mean CER for every text line length in our dataset.

6   Discussion and Conclusion

In the previous sections, we presented and investigated various OCR methods well suited for early modern printed books. We showed that font groups have a significant impact on OCR results, and proved the advantage of fine-tuning models on a granularity finer than Roman and Gothic font groups – this highlights the need to identify these font groups.

Also, we developed and presented a novel method, the COCR system, in which performances are only slightly impacted by the presence of multiple font groups in a text line. However, it is computationally more expensive than the SelOCR, which applies only one OCR model after identifying the main font of a text line. Considering they reach almost identical CER on our data, we would recommend using this approach only for documents that have frequent font group changes inside of text lines.

Now that we showed the potential of combining font group identification and OCR, our future work will consist in investigating this topic further. We are especially interested in knowing if some parts of the OCR models of the COCR system, mentioned in Table 1, can be merged in order to save memory, and process data faster – and be environmentally more friendly by saving energy. Moreover, we would like to examine in detail the outputs of the classifier used in the COCR system, as it surprisingly seems not to perform well for classification once the system has been fine-tuned for OCR.

References

1. Bjerring-Hansen, J., Kristensen-McLachlan, R.D., Diderichsen, P., Hansen, D.H.: Mending Fractured Texts. A Heuristic Procedure for Correcting OCR Data. CEUR-WS (2022)
2. Bluche, T., Messina, R.: Gated convolutional recurrent neural networks for multilingual handwriting recognition. In: 14th IAPR International Conference on Document Analysis and Recognition (ICDAR), vol. 01, pp. 646–651 (2017). https://doi.org/10.1109/ICDAR.2017.111

3. Breul, T.: ocrodeg: document image degradation - github.com (2020). https://github.com/NVlabs/ocrodeg. Accessed 10 Feb 2023
4. Chen, J., Mu, S., Xu, S., Ding, Y.: HENet: forcing a network to think more for font recognition. In: 3rd International Conference on Advanced Information Science and System (AISS), pp. 1–5 (2021)
5. Coquenet, D., Chatelain, C., Paquet, T.: SPAN: a simple predict & align network for handwritten paragraph recognition. In: Lladós, J., Lopresti, D., Uchida, S. (eds.) ICDAR 2021. LNCS, vol. 12823, pp. 70–84. Springer, Cham (2021). https://doi.org/10.1007/978-3-030-86334-0_5
6. Coquenet, D., Chatelain, C., Paquet, T.: End-to-end handwritten paragraph text recognition using a vertical attention network. IEEE Trans. Pattern Anal. Mach. Intell. **45**, 508–524 (2022). https://doi.org/10.1109/TPAMI.2022.3144899
7. Coquenet, D., Chatelain, C., Paquet, T.: Recurrence-free unconstrained handwritten text recognition using gated fully convolutional network. In: 17th International Conference on Frontiers in Handwriting Recognition (ICFHR), pp. 19–24 (2020). https://doi.org/10.1109/ICFHR2020.2020.00015
8. Diaz, D.H., Qin, S., Ingle, R., Fujii, Y., Bissacco, A.: Rethinking text line recognition models (2021). https://doi.org/10.48550/ARXIV.2104.07787. https://arxiv.org/abs/2104.07787
9. Fossey, R., Baird, H.: A 100 font classifier. In: 1st IAPR International Conference on Document Analysis and Recognition (ICDAR) (1991)
10. Fukushima, K.: Neural network model for a mechanism of pattern recognition unaffected by shift in position - neocognitron. IEICE Tech. Rep. A **62**(10), 658–665 (1979)
11. Graves, A., Fernández, S., Gomez, F., Schmidhuber, J.: Connectionist temporal classification: labelling unsegmented sequence data with recurrent neural networks. In: 23rd International Conference on Machine Learning, pp. 369–376 (2006)
12. Grosicki, E., El-Abed, H.: ICDAR 2011 - French handwriting recognition competition. In: 11th IAPR International Conference on Document Analysis and Recognition (ICDAR), pp. 1459–1463 (2011). https://doi.org/10.1109/ICDAR.2011.290
13. He, K., Zhang, X., Ren, S., Sun, J.: Deep residual learning for image recognition. In: IEEE Conference on Computer Vision and Pattern Recognition, pp. 770–778 (2016)
14. Hochreiter, S., Schmidhuber, J.: Long short-term memory. Neural Comput. **9**(8), 1735–1780 (1997)
15. Huang, G., Liu, Z., Van Der Maaten, L., Weinberger, K.Q.: Densely connected convolutional networks. In: IEEE Conference on Computer Vision and Pattern Recognition, pp. 4700–4708 (2017)
16. Kang, L., Riba, P., Rusiñol, M., Fornés, A., Villegas, M.: Pay attention to what you read: non-recurrent handwritten text-line recognition. Pattern Recogn. **129**, 108766 (2022). https://doi.org/10.1016/j.patcog.2022.108766. https://www.sciencedirect.com/science/article/pii/S0031320322002473
17. Kingma, D.P., Ba, J.: Adam: a method for stochastic optimization. arXiv preprint arXiv:1412.6980 (2014)
18. Levenshtein, V.I., et al.: Binary codes capable of correcting deletions, insertions, and reversals. Sov. Phys. Dokl. **10**, 707–710 (1966)
19. Li, M., et al.: TrOCR: transformer-based optical character recognition with pre-trained models (2021). https://doi.org/10.48550/ARXIV.2109.10282. https://arxiv.org/abs/2109.10282

20. Lyu, P., et al.: MaskOCR: text recognition with masked encoder-decoder pretraining (2022). https://doi.org/10.48550/ARXIV.2206.00311. https://arxiv.org/abs/2206.00311
21. Marti, U.V., Bunke, H.: The IAM-database: an English sentence database for offline handwriting recognition. Int. J. Doc. Anal. Recogn. **5**(1), 39–46 (2002). https://doi.org/10.1007/s100320200071
22. Michael, J., Labahn, R., Grüning, T., Zöllner, J.: Evaluating sequence-to-sequence models for handwritten text recognition. In: 15th IAPR International Conference on Document Analysis and Recognition (ICDAR), pp. 1286–1293 (2019). https://doi.org/10.1109/ICDAR.2019.00208
23. Nicolaou, A., Slimane, F., Maergner, V., Liwicki, M.: Local binary patterns for Arabic optical font recognition. In: 11th IAPR International Workshop on Document Analysis Systems (DAS), pp. 76–80. IEEE (2014)
24. Ojala, T., Pietikainen, M., Maenpaa, T.: Multiresolution gray-scale and rotation invariant texture classification with local binary patterns. IEEE Trans. Pattern Anal. Mach. Intell. **24**(7), 971–987 (2002)
25. Puigcerver, J.: Are multidimensional recurrent layers really necessary for handwritten text recognition? In: 14th IAPR International Conference on Document Analysis and Recognition (ICDAR), vol. 01, pp. 67–72 (2017). https://doi.org/10.1109/ICDAR.2017.20
26. Reul, C., Springmann, U., Wick, C., Puppe, F.: Improving OCR accuracy on early printed books by utilizing cross fold training and voting. In: 13th IAPR International Workshop on Document Analysis Systems (DAS), pp. 423–428 (2018). https://doi.org/10.1109/DAS.2018.30
27. Seuret, M., Limbach, S., Weichselbaumer, N., Maier, A., Christlein, V.: Dataset of pages from early printed books with multiple font groups. In: 15th International Workshop on Historical Document Imaging and Processing (HIP), pp. 1–6 (2019)
28. Springmann, U., Reul, C., Dipper, S., Baiter, J.: Ground truth for training OCR engines on historical documents in German Fraktur and Early Modern Latin. arXiv preprint arXiv:1809.05501 (2018)
29. Tensmeyer, C., Saunders, D., Martinez, T.: Convolutional neural networks for font classification. In: 14th IAPR International Conference on Document Analysis and Recognition (ICDAR), vol. 1, pp. 985–990. IEEE (2017)
30. TorchVision maintainers and contributors: TorchVision: PyTorch's Computer Vision library (2016). https://github.com/pytorch/vision
31. Vaswani, A., et al.: Attention is all you need. In: Guyon, I., (eds.) Advances in Neural Information Processing Systems, vol. 30. Curran Associates, Inc. (2017). https://proceedings.neurips.cc/paper/2017/file/3f5ee243547dee91fbd053c1c4a845aa-Paper.pdf
32. Wick, C., Reul, C.: One-model ensemble-learning for text recognition of historical printings. In: Lladós, J., Lopresti, D., Uchida, S. (eds.) ICDAR 2021. LNCS, vol. 12821, pp. 385–399. Springer, Cham (2021). https://doi.org/10.1007/978-3-030-86549-8_25
33. Wick, C., Zöllner, J., Grüning, T.: Transformer for handwritten text recognition using bidirectional post-decoding. In: Lladós, J., Lopresti, D., Uchida, S. (eds.) ICDAR 2021. LNCS, vol. 12823, pp. 112–126. Springer, Cham (2021). https://doi.org/10.1007/978-3-030-86334-0_8
34. Wick, C., Zöllner, J., Grüning, T.: Rescoring sequence-to-sequence models for text line recognition with CTC-prefixes. In: Uchida, S., Barney, E., Eglin, V. (eds.) 15th IAPR International Workshop on Document Analysis Systems (DAS), pp. 260–274. Springer, Cham (2022). https://doi.org/10.1007/978-3-031-06555-2_18

35. Yang, J., Kim, H., Kwak, H., Kim, I.: HanFont: large-scale adaptive hangul font recognizer using CNN and font clustering. Int. J. Doc. Anal. Recogn. (IJDAR) **22**, 407–416 (2019)

36. Yousef, M., Hussain, K.F., Mohammed, U.S.: Accurate, data-efficient, unconstrained text recognition with convolutional neural networks. Pattern Recogn. **108**, 107482 (2020). https://doi.org/10.1016/j.patcog.2020.107482

Detecting Text on Historical Maps by Selecting Best Candidates of Deep Neural Networks Output

Gerasimos Matidis[1,2]([envelope]), Basilis Gatos[1], Anastasios L. Kesidis[2],
and Panagiotis Kaddas[1,3]

[1] Computational Intelligence Laboratory, Institute of Informatics and Telecommunications,
NCSR "Demokritos",, 15310 Athens, Greece
{gmatidis,bgat,pkaddas}@iit.demokritos.gr
[2] Department of Surveying and Geoinformatics Engineering, University of West Attica,
12243 Athens, Greece
akesidis@uniwa.gr
[3] Department of Informatics and Telecommunications, University of Athens,
15784 Athens, Greece

Abstract. The final and perhaps the most crucial step in Object Detection is the selection of the best candidates out of all the proposed regions a framework outputs. Typically, Non-Maximum Suppression approaches (NMS) are employed to tackle this problem. The standard NMS relies exclusively on the confidence scores, as it selects the bounding box with the highest score within a cluster of boxes determined by a relatively high Intersection over Union (IoU) between each other, and then suppresses the remaining ones. On the other hand, algorithms like Confluence determine clusters of bounding boxes according to the proximity between them and select as best the box that is closer to the other ones within each cluster. In this work, we combine these methods by creating clusters of high confidence scores according to their IoU and then we calculate the sums of the Manhattan distances between the vertices of each box and all the others, in order to finally select the one with the minimum overall distance. Our results are compared with the standard NMS and the Locality-Aware NMS (LANMS), an algorithm that is widely used in Object Detection and merges the boxes row by row. The research field that this work explores is the text detection on historical maps and the proposed approach results to average precision that is 2.14–2.94% higher for evaluation IoU in range 0.50 to 0.95 with step 0.05 than the two other methods.

Keywords: Text Detection · Historical Maps · Non-Maximum Suppression

1 Introduction

Historical maps are an important and unique source of information for studying geographical transformations over years. In this paper, the focus is on the text detection task of the digital historical map processing workflow. This task is very important and

G. A. Fink et al. (Eds.): ICDAR 2023, LNCS 14191, pp. 358–367, 2023.
https://doi.org/10.1007/978-3-031-41734-4_22

crucial, since it provides the input for the recognition process that follows. At the same time, this task is extremely challenging due to the complex nature of the historical maps. As it can be observed in the example of Fig. 1, text in the historical maps can be of any size, any orientation, may be curved, with variable spacing and usually overlaps with other graphical map elements.

Current approaches for text detection in historical maps include the use of color and spatial image and text attributes [1, 2]. In [3], text in maps is identified based on the geometry of individual connected components without considering most of the aforementioned text detection challenges. Approach [4] uses 2-D Discrete Cosine Transformation coefficients and Support Vector Machines to classify the pixels of lines and characters on raster maps. Recently, Convolutional Neural Network (CNN) architectures have been proved efficient for text detection in historical maps. In [5] and [6], Deep CNNs are introduced for end-to-end text reading of historical maps. A text detection network predicts word bounding boxes at arbitrary orientations and scales. Several text detection neural network models are evaluated in [7] and [8]. The pixel-wise positions of text regions are detected in [9] by employing a CNN-based architecture.

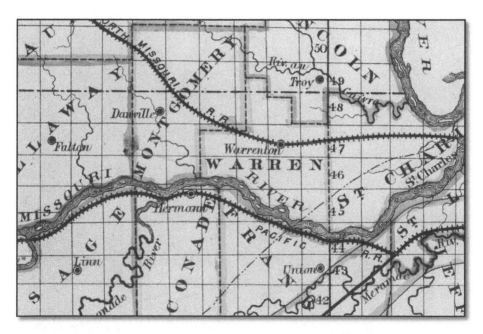

Fig. 1. Part of a historical map from the dataset provided in [5].

Following the recent promising approaches based on CNN architectures, in this work we first apply the deep neural network of [5] and focus on the final and perhaps the most crucial step for text detection, which is the selection of the best candidates out of all the proposed regions coming as output of the CNN framework. As it is demonstrated in Fig. 2, the network output usually corresponds to several overlapping blocks around the text area (Fig. 2b), while the desired final output is just one block around the text area (Fig. 2a). This is a common problem for object detection applications and several Non-Maximum Suppression (NMS) approaches has been proposed for solving it. The standard NMS relies exclusively on the confidence scores, as it selects the bounding box with the highest score within a cluster of boxes determined by a relatively high Intersection over Union (IoU) between them, and then suppresses the remaining ones. The Locality-Aware NMS (LANMS) [10], also used in [5], is based on merging the boxes row by row. On the other hand, algorithms like Confluence [11] determine clusters of bounding boxes according to the proximity between them and they select as best the box that is closer to the other ones within each cluster. However, it can be observed that in NMS-based approaches the most confident box does not always correspond to the best solution (Fig. 2b). Indeed, there may exist other candidate boxes, with slightly lower confidence score than the selected one, which provide more accurate predictions of the desired bounding box. In the proposed work, we try to find the best solution that can be applied to the difficult case of historical maps and combine the standard NMS with the Confluence approach. Initially, we create clusters of high confidence scores according to

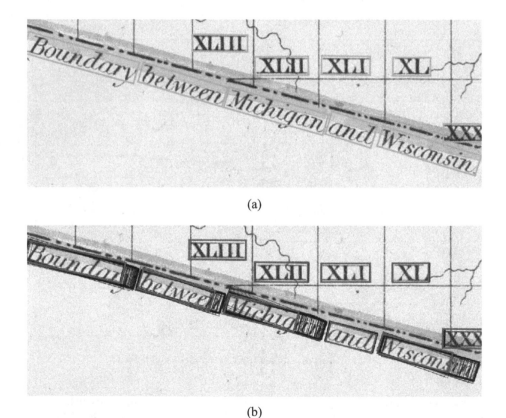

(a)

(b)

Fig. 2. (a) The ground truth bounding boxes (orange), (b) The total high-confident boxes predicted by the network (black) and the ones with the maximum confidence score for each word (green). As it can be observed, the most confident bounding boxes do not always correspond to the best solution.

their IoU. Then, we calculate the sums of the Manhattan distances between the vertices of each box and all the others in the cluster, in order to finally select the one with the minimum overall distance. To achieve this, we also generalize the Confluence algorithm in order to process blocks of any orientation. As it is demonstrated in the experimental results, the proposed method gives an average boost of 2.14–2.94%, concerning the average precision metric for IoU in range 0.50 to 0.95 with step 0.05, when compared with standard NMS and LANMS.

The rest of the paper is organized as follows: Section 2 introduces the proposed text detection method, Section 3 demonstrates the experimental results and Sect. 4 presents the conclusion of this work.

2 Methodology

Recent Object Detection networks usually provide a very large number of bounding box predictions, each one assigned with a confidence score. The number ranges from a few thousand, in cases of region proposal-based [12] or grid-based [13] detectors, to millions, in cases of dense [10] detectors (one prediction per pixel).

As mentioned in the previous section, the standard NMS selects repetitively the bounding box with the maximum confidence score and suppresses the ones that significantly overlap with it. However, it appears that in practice this is not always the best strategy. In particular, in many cases the most confident box misses a considerable part of the text, while other ones capture it more precisely (Fig. 2b). A second issue is that many of the boxes, which are to be suppressed and they have a slightly lower confidence score than the most confident box, correspond to better predictions of the ground truth bounding box. This situation occurs especially when dealing with dense-detection networks. Indeed, in our experiments we reported various cases, where the difference in confidence between the most confident box and some of the suppressed boxes is less than 10^{-4}, while the later ones corresponded to better prediction accuracy.

Considering the above, we repetitively define clusters of boxes, which consist of the most confident and the ones that significantly overlap with it, such as the standard NMS. Then, by calculating the sums of the Manhattan distances between the vertices of each one and the other boxes in the cluster, the one with the minimum overall distance is selected, similarly to [11]. However, we extend [11] in order to include rotated boxes. The Manhattan distances are then calculated for all the four vertices of every pair of boxes, instead of two diagonal ones. Figure 3 depicts the process of the calculation of the distance between two boxes.

In this work, the MapTD network [5], which produces dense predictions, is used for detecting text on the maps. A threshold of 0.95 is applied in order to eliminate the bounding boxes with low confidence scores. Let B denote the list of the remaining candidate boxes. The main steps of the proposed algorithm are as follows:

Step 1: Sort the list B in a descending order with respect to the confidence scores of the candidate bounding boxes.

Step 2: Initiate an empty list F to store the final boxes.

Step 3: Select the first (most confident) box in B and calculate the IoU between this box and every other box in the list.

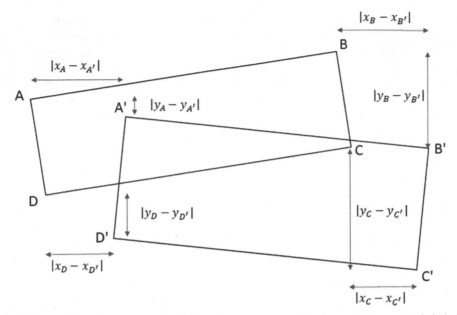

Fig. 3. Example of the distance calculation between two bounding boxes ABCD and A′B′C′D′. The distance is the sum of the Manhattan distance between every pair of similar vertices.

Step 4: Define a cluster by creating a list C including the most confident box and all the boxes that have a minimum overlap with it, which is determined by a predefined IoU threshold. Remove all the boxes that are stored in list C from the main list B.

Step 5: Calculate the distances between each box and every other box in list C. Specifically, t he distance between two bounding boxes b_i with vertices ABCD and b_j with vertices A′B′C′D′ as shown in Fig. 3, is given by calculating the sum of the Manhattan distances between the four pairs of the corresponding vertices. Following the approach described in Sect. 3.2 of [11], the coordinates are normalized as follows:

$$x_i = \frac{x_i - \min(X)}{\max(X) - \min(X)}, \quad y_i = \frac{y_i - \min(Y)}{\max(Y) - \min(Y)}, \quad \forall i \in \{A, B, C, D, A\prime, B\prime, C\prime, D\prime\}$$

$$(1)$$

where

$$X = \{x_A, x_B, x_C, x_D, x_{A\prime}, x_{B\prime}, x_{C\prime}, x_D\}$$

and

$$Y = \{y_A, y_B, y_C, y_D, y_{A\prime}, y_{B\prime}, y_{C\prime}, y_{D\prime}\}$$

The distance between two bounding boxes is calculated as

$$D_{b_i,b_j} = |x_A - x_{A\prime}| + |y_A - y_{A\prime}| + |x_B - x_{B\prime}| + |y_B - y_{B\prime}| +$$
$$+ |x_C - x_{C\prime}| + |y_C - y_{C\prime}| + |x_D - x_{D\prime}| + |y_D - y_{D\prime}|$$

$$(2)$$

The overall distance D_b for every box in cluster C, is the sum of all the distances between bounding box b and all the other boxes as follows:

$$D_b = \sum_{i=1}^{N_C} D_{bb_i}, \forall b \in C, b \neq i \tag{3}$$

where N_C denotes the number of boxes in C.

Step 6: Weight the total distance D_b as:

$$D_b = D_b(1 - S_b + \varepsilon) \tag{4}$$

where $S_b \in [0, 1]$ is its confidence score and ε is a positive number near zero, which ensures that the total distance does not vanish when $S_b \to 1$. In our experiments, ε is set to 0.1.

Step 7: Select the box with the minimum D_b and store it in the list F.

Step 8: Repeat steps 3 to 7 until list B becomes empty.

3 Experimental Results

In order to evaluate the proposed method, we trained MapTD models using the same dataset and training details as [5]. The dataset consists of 31 historical maps of the USA from the period 1866–1927 and it is publicly available[1]. We also follow the same 10-fold cross-validation of [5]. In particular, in every fold we use 27 maps for training and 3 maps for testing. One map is held out for validation across all folds. For the training we use Minibatch Stochastic Gradient Descent with minibatch size $16 \times 512 \times 512$ and Adam optimizer with $\beta_1 = 0.9$, $\beta_2 = 0.999$, $\varepsilon = 10^{-8}$. The learning rate is $\alpha = 10^{-4}$ for the first 2^{17} training steps and $\alpha = 10^{-5}$ for the rest of a total of 2^{20} training steps. For evaluation on each of the 3 test images of each fold, we take predictions on overlapping tiles of size 4096×4096, , with stride equal to 2048. The predicted bounding boxes of the network are filtered with a confidence score threshold of 0.95, in order to keep the most confident ones. All the three NMS methods use the same IoU threshold, equal to 0.1, in order to ensure that even slightly overlapping bounding boxes will belong to the same cluster [5].

Table 1 represents in details the average results across all folds for IoU values in range 0.50 to 0.95 with step 0.05. As it can be observed, the proposed algorithm outperforms the standard NMS and LANMS in average by 2.14% and 2.94%, respectively.

The differences between the proposed and the other two methods increase as the evaluation threshold becomes higher and reach a maximum when the threshold is 0.8. More specifically, the differences are as follows:

- Proposed – standard NMS  **0.50**: 1.41%, **0.55**: 1.83%, **0.60**: 2.62%, **0.65**: 3.18%, **0.70**: 4.17%, **0.75**: 5.17%, **0.80**: 5.17%, **0.85**: 4.33%, **0.90**: 1.55%, **0.95**: 0.03%
- Proposed – LANMS  **0.50**: 1.79%, **0.55**: 2.14%, **0.60**: 2.72%, **0.65**: 2.84%, **0.70**: 3.39%, **0.75**: 3.33%, **0.80**: 3.38%, **0.85**: 1.77%, **0.90**: 0.09%, **0.95**: -0.01%

---

[1] Https://weinman.cs.grinnell.edu/research/maps.shtml#data

Table 1. Overall Evaluation (Confidence threshold = 0.95, NMS threshold = 0.1)

Evaluation IoU threshold	Method	Predicted boxes	Ground Truth boxes	Correctly predicted boxes	Average Precision
0.50	standard NMS	32583	33315	29954	85.99%
	LANMS	32436		29955	85.61%
	Proposed	32583		30081	**87.40%**
0.55	standard NMS	32583	33315	29567	84.06%
	LANMS	32436		29625	83.75%
	Proposed	32583		29734	**85.89%**
0.60	standard NMS	32583	33315	29045	81.44%
	LANMS	32436		29163	81.33%
	Proposed	32583		29310	**84.06%**
0.65	standard NMS	32583	33315	28275	77.72%
	LANMS	32436		28493	78.07%
	Proposed	32583		28587	**80.90%**
0.70	standard NMS	32583	33315	27012	71.67%
	LANMS	32436		27289	72.45%
	Proposed	32583		27453	**75.83%**
0.75	standard NMS	32583	33315	24836	61.79%
	LANMS	32436		25306	63.62%
	Proposed	32583		25423	**66.96%**
0.80	standard NMS	32583	33315	21285	46.83%
	LANMS	32436		21752	48.62%
	Proposed	32583		21924	**52.00%**
0.85	standard NMS	32583	33315	15221	25.14%
	LANMS	32436		15917	27.70%
	Proposed	32583		15996	**29.47%**

(*continued*)

Table 1. (*continued*)

Evaluation IoU threshold	Method	Predicted boxes	Ground Truth boxes	Correctly predicted boxes	Average Precision
0.90	standard NMS	32583	33315	6880	5.45%
	LANMS	32436		7562	6.92%
	Proposed	32583		7573	**7.00%**
0.95	standard NMS	32583	33315	737	0.07%
	LANMS	32436		892	**0.11%**
	Proposed	32583		868	0.10%
Average results	standard NMS				54.02%
	LANMS				54.82%
	Proposed				**56.96%**

Figure 4 depicts two representative examples of how the proposed method results in more accurate bounding boxes. In analogy to the differences on the average precision, the bounding boxes of the proposed algorithm are better than these of LANMS, the boxes of which are better than standard NMS.

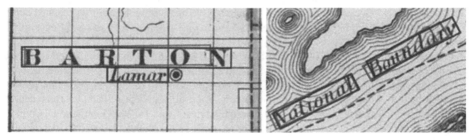

Fig. 4. Two samples with predicted bounding boxes for the three methods: standard NMS (green boxes), LANMS (red boxes) and proposed method (blue boxes). (Color figure online)

4 Conclusion

In this paper, a method is proposed that increases the accuracy of text detection on historical maps. It focuses on a particular post-processing step of the text detection pipeline by selecting the best solution among a large set of candidate bounding boxes, which are predicted by a deep CNN. To this direction, two existing Non-Maximum

Suppression methods are combined, namely, the standard NMS and the Confluence. The proposed method tackles a problem of NMS-based approaches, where the most confident box does not always correspond to the best solution, since there may exist other candidate boxes, with slightly lower confidence score than the selected one, which provide more accurate predictions of the desired bounding box. Instead of eliminating the bounding boxes that significantly overlap with the most confident, the method creates a cluster with all of them and selects as best the box that is closer to all others.

The experimental results show that the proposed method outperforms the standard NMS and LANMS on average precision metric and results to more accurate bounding boxes, a step that is very important, since it provides the input for the recognition process that follows text detection.

Acknowledgments. This research has been partially co-financed by the European Union and Greek national funds through the Operational Program Competitiveness, Entrepreneurship and Innovation, under the call "RESEARCH-CREATE-INNOVATE", project Culdile (Cultural Dimensions of Deep Learning, project code: T1EΔK-03785) and the Operational Program Attica 2014–2020, under the call "RESEARCH AND INNOVATION PARTNERSHIPS IN THE REGION OF ATTICA", project reBook (Digital platform for re-publishing Historical Greek Books, project code: ATTP4-0331172).

References

1. Chiang, Y., Leyk, S., Knoblock, C.: A survey of digital map processing techniques. ACM Comput. Surv. **47**(1) (2014)
2. Velázquez, A., Levachkine, S.: Text/graphics separation and recognition in raster-scanned color cartographic maps. In: Lladós, J., Kwon, Y.-B. (eds.) GREC 2003. LNCS, vol. 3088, pp. 63–74. Springer, Heidelberg (2004). https://doi.org/10.1007/978-3-540-25977-0_6
3. Pouderoux, J., Gonzato, J., Pereira A., Guitton, P.: Toponym recognition in scanned color topographic maps. In: 9th International Conference on Document Analysis and Recognition (ICDAR 2007), pp. 531–535. Curitiba, Brazil (2007)
4. Chiang, Y.-Y., Knoblock, C.A.: Classification of line and character pixels on raster maps using discrete cosine transformation coefficients and support vector machine. In: 18th International Conference on Pattern Recognition (ICPR 2006), pp. 1034–1037. Hong Kong, China (2006)
5. Weinman, J., Chen, Z., Gafford, B., Gifford, N., Lamsal, A., Niehus-Staab, L.: Deep neural networks for text detection and recognition in historical maps. In: 2019 International Conference on Document Analysis and Recognition (ICDAR), pp. 902–909. Sydney, NSW, Australia (2019)
6. Schlegel, I.: Automated extraction of labels from large-scale historical maps. AGILE: GIScience Series, vol. 2, pp. 1–14 (2021)
7. Lenc, L., Martínek, J., Baloun, J., Prantl, M., Král, P.: Historical map toponym extraction for efficient information retrieval. In: Uchida, S., Barney, E., Eglin, V. (eds.) Document Analysis Systems, DAS 2022, LNCS, vol. 13237, pp. 171–183. Springer, Cham (2022). https://doi.org/10.1007/978-3-031-06555-2_12
8. Philipp, J.N., Bryan, M.: Evaluation of CNN architectures for text detection in historical maps. In: Digital Access to Textual Cultural Heritage (DATeCH 2019) (2019)
9. Can, Y.S., Kabadayi, M.E.: Text detection and recognition by using cnns in the austro-hungarian historical military mapping survey. In: The 6th International Workshop on Historical Document Imaging and Processing (HIP 2021), pp. 25–30. Lausanne Switzerland (2021)

10. Zhou, X., et al.: EAST: an efficient and accurate scene text detector. In: The IEEE/CVF Computer Vision and Pattern Recognition Conference (CVPR) 2017, pp. 2642–2651. Honolulu, Hawaii (2017)
11. Andrew, S., Gregory, F., Paul, F.: Confluence: a robust non-IoU alternative to non-maxima suppression in object detection. IEEE Trans. Pattern Anal. Mach. Intell. (2021)
12. Girshick, R., Donahue, J., Darrell, T., Malik, J.: Rich feature hierarchies for accurate object detection and semantic segmentation. In: 2014 IEEE Conference on Computer Vision and Pattern Recognition (CVPR), pp. 580–587. IEEE (2014)
13. Redmon, J., Divvala, S., Girshick, R., Farhadi, A.: You only look once: unified, real-time object detection. In: Proceedings of the IEEE conference on computer vision and pattern recognition (CVPR), pp. 779–788 (2016)

Posters: Graphics

Aligning Benchmark Datasets for Table Structure Recognition

Brandon Smock🄳, Rohith Pesala(✉)🄳, and Robin Abraham🄳

Microsoft, Redmond, WA, USA
{brsmock,ropesala,robin.abraham}@microsoft.com

Abstract. Benchmark datasets for table structure recognition (TSR) must be carefully processed to ensure they are annotated consistently. However, even if a dataset's annotations are self-consistent, there may be significant inconsistency across datasets, which can harm the performance of models trained and evaluated on them. In this work, we show that *aligning* these benchmarks—removing both errors and inconsistency between them—improves model performance significantly. We demonstrate this through a data-centric approach where we adopt one model architecture, the Table Transformer (TATR), that we hold fixed throughout. Baseline exact match accuracy for TATR evaluated on the ICDAR-2013 benchmark is 65% when trained on PubTables-1M, 42% when trained on FinTabNet, and 69% combined. After reducing annotation mistakes and inter-dataset inconsistency, performance of TATR evaluated on ICDAR-2013 increases substantially to 75% when trained on PubTables-1M, 65% when trained on FinTabNet, and 81% combined. We show through ablations over the modification steps that canonicalization of the table annotations has a significantly positive effect on performance, while other choices balance necessary trade-offs that arise when deciding a benchmark dataset's final composition. Overall we believe our work has significant implications for benchmark design for TSR and potentially other tasks as well. Dataset processing and training code will be released at https://github.com/microsoft/table-transformer.

1 Introduction

Table extraction (TE) is a long-standing problem in document intelligence. Over the last decade, steady progress has been made formalizing TE as a machine learning (ML) task. This includes the development of task-specific metrics for evaluating table structure recognition (TSR) models [5,20,26] as well as the increasing variety of datasets and benchmarks [3,21,25,26]. These developments have enabled significant advances in deep learning (DL) modeling for TE [12,13, 16,18,21,25].

In general, benchmarks play a significant role in shaping the direction of ML research [10,17]. Recently it has been shown that benchmark datasets for TSR contain a significant number of errors and inconsistencies [21]. It is well-documented that errors in a benchmark have negative consequences for both

© The Author(s), under exclusive license to Springer Nature Switzerland AG 2023
G. A. Fink et al. (Eds.): ICDAR 2023, LNCS 14191, pp. 371–386, 2023.
https://doi.org/10.1007/978-3-031-41734-4_23

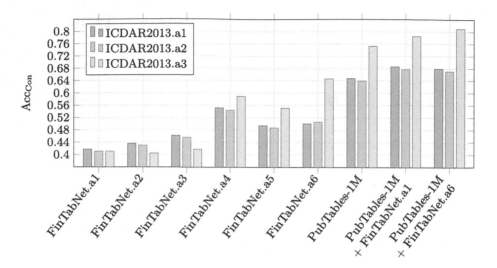

Fig. 1. The improvement in performance (in terms of exact match accuracy) of models trained on FinTabNet and PubTables-1M, and evaluated on ICDAR-2013, as we remove annotation mistakes and inconsistencies between the three datasets.

learning and evaluation [4,6,27]. Errors in the test set for the ImageNet benchmark [14], for instance, have led to top-1 accuracy saturating around 91% [24]. Errors can also lead to false conclusions during model selection, particularly when the training data is drawn from the same noisy distribution as the test set [14].

Compared to annotation mistakes, inconsistencies in a dataset can be more subtle because they happen across a collection of samples rather than in isolated examples—but no less harmful. Even if a single dataset is self-consistent there may be inconsistencies in labeling across different datasets for the same task. We consider datasets for the same task that are annotated inconsistently with respect to each other to be *misaligned*. Misalignment can be considered an additional source of labeling noise. This noise may go unnoticed when a dataset is studied in isolation, but it can have significant effects on model performance when datasets are combined.

In this work we study the effect that errors and misalignment between benchmark datasets have on model performance for TSR. We select two large-scale crowd-sourced datasets for training—FinTabNet and PubTables-1M—and one small expert-labeled dataset for evaluation—the ICDAR-2013 benchmark. For our models we choose a single fixed architecture, the recently proposed Table Transformer (TATR) [21]. This can be seen as a data-centric [22] approach to ML, where we hold the modeling approach fixed and instead seek to improve performance through data improvements. Among our main contributions:

– We remove both annotation mistakes and inconsistencies between FinTabNet and ICDAR-2013, aligning these datasets with PubTables-1M and producing

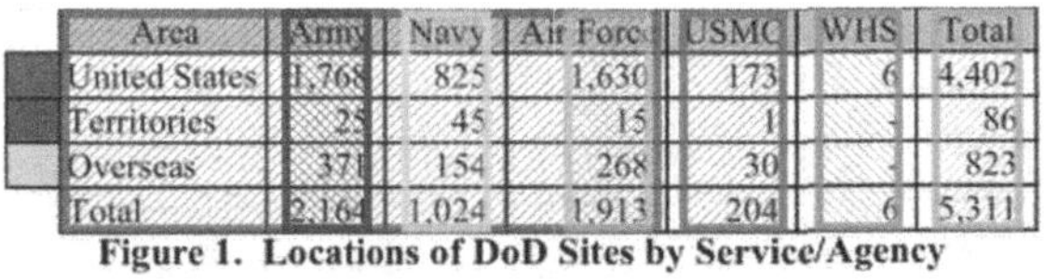

Area	Army	Navy	Air Force	USMC	WHS	Total
United States	1,768	825	1,630	173	6	4,402
Territories	25	45	15	1	-	86
Overseas	371	154	268	30	-	823
Total	2,164	1,024	1,913	204	6	5,311

Figure 1. Locations of DoD Sites by Service/Agency

(a) Original annotation

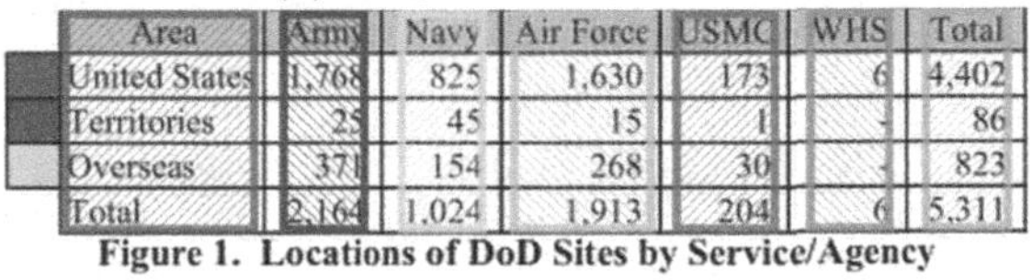

Area	Army	Navy	Air Force	USMC	WHS	Total
United States	1,768	825	1,630	173	6	4,402
Territories	25	45	15	1	-	86
Overseas	371	154	268	30	-	823
Total	2,164	1,024	1,913	204	6	5,311

Figure 1. Locations of DoD Sites by Service/Agency

(b) Corrected annotation

Fig. 2. An example table from us-gov-dataset/us-012-str.xml in the ICDAR-2013 dataset. On the top are the column bounding boxes created when using the dataset's original annotations. On the bottom are the column bounding boxes following correction of the cells' column index labels. Corresponding columns before and after correction are shaded with the same color.

improved versions of two standard benchmark datasets for TSR (which we refer to as FinTabNet.c and ICDAR-2013.c, respectively).
- We show that removing inconsistencies between benchmark datasets for TSR indeed has a substantial positive impact on model performance, improving baseline models trained separately on PubTables-1M and FinTabNet from 65% to 78% and 42% to 65%, respectively, when evaluated on ICDAR-2013 (see Fig. 1).
- We perform a sequence of ablations over the steps of the correction procedure, which shows that canonicalization has a clear positive effect on model performance, while other factors create trade-offs when deciding the final composition of a benchmark.
- We train a single model on both PubTables-1M and the aligned version of FinTabNet (FinTabNet.c), establishing a new baseline DAR of 0.965 and exact match table recognition accuracy of 81% on the corrected ICDAR-2013 benchmark (ICDAR-2013.c).
- We plan to release all of the dataset processing and training code at https://github.com/microsoft/table-transformer.

2 Related Work

The first standard multi-domain benchmark dataset for table structure recognition was the ICDAR-2013 dataset, introduced at the 2013 ICDAR Table Competition [5]. In total it contains 158 tables[1] in the official competition dataset and

[1] The dataset is originally documented as having 156 tables but we found 2 cases of tables that need to be split into multiple tables to be annotated consistently.

	Years Ended	
	December 28, 2013	December 29, 2012
	(dollars in thousands)	
Operating activities..	$ 591,281	$ 553,607
Investing activities...	(597,393)	(27,866)
Financing activities...	93,757	(517,777)
Effect of changes in foreign currency exchange rates on cash........................	(14,578)	(513)
Change in cash and cash equivalents..	73,067	7,451
Cash and cash equivalents at beginning of year...............................	42,796	35,345
Cash and cash equivalents at end of year....................................	$ 115,863	$ 42,796

Fig. 3. An example table from HBI/2013/page_39.pdf in the FinTabNet dataset with cell bounding boxes (shaded in orange) that are inconsistently annotated. Some cells in the first column include the dot leaders in the bounding box while others do not. (Color figure online)

at least 100 tables[2] in a practice dataset. The competition also established the first widely-used metric for evaluation, the directed adjacency relations (DAR) metric, though alternative metrics such as TEDS [26] and GriTS [20] have been proposed more recently. At the time of the competition, the best reported approach achieved a DAR of 0.946. More recent work from Tensmeyer et al. [23] reports a DAR of 0.953 on the full competition dataset.

The difficulty of the ICDAR-2013 benchmark and historical lack of training data for the TSR task has led to a variety of approaches and evaluation procedures that differ from those established in the original competition. DeepDeSRT [18] split the competition dataset into a training set and a test set of just 34 samples. Additionally, the authors processed the annotations to add row and column bounding boxes and then framed the table structure recognition problem as the task of detecting these boxes. This was a significant step at the time but because spanning cells are ignored it is only a partial solution to the full recognition problem. Still, many works have followed this approach [7]. Hashmi et al. [8] recently reported an F-score of 0.9546 on the row and column detection task, compared to the original score of 0.9144 achieved by DeepDeSRT.

While ICDAR-2013 has been a standard benchmark of progress over the last decade, little attention has been given to complete solutions to the TSR task evaluated on the full ICDAR-2013 dataset. Of these, none reports exact match recognition accuracy. Further, we are not aware of any work that points out label noise in the ICDAR-2013 dataset and attempts to correct it. Therefore it is an open question to what extent performance via model improvement has saturated on this benchmark in its current form.

To address the need for training data, several large-scale crowd-sourced datasets [3,11,21,25,26] have been released for training table structure recognition models. However, the quality of these datasets is lower overall than that of expert-labeled datasets like ICDAR-2013. Of particular concern is the potential for ambiguity in table structures [1,9,15,19]. A specific form of label error called oversegmentation [21] is widely prevalent in benchmark TSR datasets, which ultimately leads to annotation inconsistency. However, models developed

---

[2] This is the number of tables in the practice set we were able to find online.

and evaluated on these datasets typically treat them as if they are annotated unambiguously, with only one possible correct interpretation.

To address this, Smock et al. [21] proposed a canonicalization algorithm, which can automatically correct and make table annotations more consistent. However, this was applied to only a single source of data. The impact of errors and inconsistency across multiple benchmark datasets remains an open question. In fact, the full extent of this issue can be masked by using metrics that measure the average correctness of cells [20]. But for table extraction systems in real-world settings, individual errors may not be tolerable, and therefore the accuracy both of cells and of entire tables is important to consider.

3   Data

As our baseline training datasets we adopt PubTables-1M [21] and FinTabNet [25]. PubTables-1M contains nearly one million tables from scientific articles, while FinTabNet contains close to 113k tables from financial reports. As our baseline evaluation dataset we adopt the ICDAR-2013 [5] dataset. The ICDAR-2013 dataset contains tables from multiple document domains manually annotated by experts. While its size limits its usefulness for training, the quality of ICDAR-2013 and its diverse set of table appearances and layouts make it useful for model benchmarking. For the ICDAR-2013 dataset, we use the competition dataset as the test set, use the practice dataset as a validation set, and consider each table "region" annotated in the dataset to be a table, yielding a total of 256 tables to start.

3.1   Missing Annotations

While the annotations in FinTabNet and ICDAR-2013 are sufficient to train and evaluate models for table structure recognition, several kinds of labels that could be of additional use are not explicitly annotated. Both datasets annotate bounding boxes for each cell as the smallest rectangle fully-enclosing all of the text of the cell. However, neither dataset annotates bounding boxes for rows, columns, or blank cells. Similarly, neither dataset explicitly annotates which cells belong to the row and column headers.

For each of these datasets, one of the first steps we take before correcting their labels is to automatically add labels that are missing. This can be viewed as making explicit the labels that are *implicit* given the rest of the label information present. We cover these steps in more detail later in this section.

3.2   Label Errors and Inconsistencies

Next we investigate both of these datasets to look for possible annotation mistakes and inconsistencies. For FinTabNet, due to the impracticality of verifying each annotation manually, we initially sample the dataset to identify types of

Table 1. The high-level data processing steps and ablations for the FinTabNet and ICDAR-2013 datasets. Data cleaning steps are grouped into stages. Each stage represents an addition to the steps of the full processing algorithm. For each stage, a dataset ablation is created with the steps up to and including that stage. But the stages themselves do not occur in the same exact order in the full processing pipeline. For instance, when stage a5 is added to the processing it actually occurs at the beginning of stage a4 to annotate more column headers prior to canonicalization.

Dataset	Stage	Description
FinTabNet		
—	Unprocessed	The original data, which does not have header or row/column bounding box annotations
a1	Completion	Baseline FinTabNet dataset; create bounding boxes for all rows and columns; remove tables for which these boxes cannot be defined
a2	Cell box adjustment	Iteratively refine the row and column boxes to tightly surround their coinciding text; create *grid cells* at the intersection of each row and column; remove any table for which 50% of the area of a word overlaps with multiple cells in the table.
a3	Consistency adjustments	Make the sizing and spanning of columns and rows more consistent; adjust cell bounding boxes to always ignore dot leaders (remove any tables for which this was unsuccessful); remove empty rows and empty columns from the annotations; merge any adjacent header rows whose cells all span the same columns; remove tables with columns containing just cent and dollar signs (we found these difficult to automatically correct).
a4	Canonicalization	Use cell structure to infer the column header and projected row headers for each table; remove tables with only two columns as the column header of these is structurally ambiguous; canonicalize each annotation.
a5	Additional column header inference	Infer the column header for many tables with only two columns by using the cell text in the first row to determine if the first row is a column header.
a6	Quality control	Remove tables with erroneous annotations, including: tables where words in the table coincide with either zero or more than one cell bounding box, tables with a projected row header at the top or bottom (indicating a caption or footer is mistakenly included in the table), and tables that appear to have only a header and no body
ICDAR-2013		
—	Unprocessed	The original data, which does not have header or row/column bounding box annotations
a1	Completion	Baseline ICDAR-2013 dataset; create bounding boxes for all rows and columns; remove tables for which these boxes cannot be defined or whose annotations cannot be processed due to errors
a2	Manual correction	Manually fix annotation errors (18 tables fixed in total)
a3	Consistency adjustments and canonicalization	Apply the same automated steps applied to FinTabNet.a2 through FinTabNet.a4; after processing, manually inspect each table and make any additional corrections as needed.

errors common to many instances. For ICDAR-2013, we manually inspect all 256 table annotations.

For both datasets, we note that the previous action of defining and adding bounding boxes for rows and columns was crucial to catching inconsistencies, as

Table 2. Diversity and complexity of table instances in the baseline and ablation datasets.

Dataset	# Tables[a]	# Unique Topologies	Avg. Tables/Topology	Avg. Rows/Table	Avg. Cols./Table	Avg. Spanning Cells/Table
FinTabNet	**112,875**	9,627	11.72	11.94	4.36	1.01
FinTabNet.a1	112,474	9,387	11.98	11.92	4.35	1.00
FinTabNet.a2	109,367	8,789	12.44	11.87	4.33	0.98
FinTabNet.a3	103,737	7,647	13.57	11.81	4.28	0.93
FinTabNet.a4	89,825	18,480	**4.86**	**11.97**	**4.61**	**2.79**
FinTabNet.a5	98,019	**18,752**	5.23	11.78	4.39	2.57
FinTabNet.a6	97,475	18,702	5.21	11.81	4.39	2.58
ICDAR-2013	256	181	1.41	**15.88**	**5.57**	1.36
ICDAR-2013.a1	247	175	1.41	15.61	5.39	1.34
ICDAR-2013.a2	**258**	181	1.43	15.55	5.45	1.42
ICDAR-2013.a3	**258**	**184**	**1.40**	15.52	5.45	**2.08**

[a] The number of tables in the dataset that we were able to successfully read and process.

it exposed a number of errors when we visualized these boxes on top of the tables. For ICDAR-2013 we noticed the following errors during manual inspection: 14 tables with at least one cell with incorrect column or row indices (see Fig. 2), 4 tables with at least one cell with an incorrect bounding box, 1 table with a cell with incorrect text content, and 2 tables that needed to be split into more than one table.

For FinTabNet, we noticed a few errors that appear to be common in crowd-sourced table annotations in the financial domain. For example, we noticed many of these tables use *dot leaders* for spacing and alignment purposes. However, we noticed a significant amount of inconsistency in whether dot leaders are included or excluded from a cell's bounding box (see Fig. 3). We also noticed that it is common in these tables for a logical column that contains a monetary symbol, such as a \$, to be split into two columns, possibly for numerical alignment purposes (for an example, see NEE/2006/page_49.pdf). Tables annotated this way are done so for presentation purposes but are inconsistent with their logical interpretation, which should group the dollar sign and numerical value into the same cell. Finally we noticed annotations with fully empty rows. Empty rows can sometimes be used in a table's markup to create more visual spacing between rows. But there is no universal convention for how much spacing corresponds to an empty row. Furthermore, empty rows used only for spacing serve no logical purpose. Therefore, empty rows in these tables should be considered labeling errors and should be removed for consistency. For both datasets we noticed oversegmentation of header cells, as is common in crowd-sourced markup tables.

3.3   Dataset Corrections and Alignment

Mistakes noticed during manual inspection of ICDAR-2013 are corrected directly. For both datasets, nearly all of the other corrections and alignments are done by a series of automated processing steps. We list these steps at a high level in Table 1. For each dataset the processing steps are grouped into a series of macro-steps, which helps us to study some of their effects in detail using ablation experiments in Sect. 4.

Table 3. Comparison of the original baseline model for PubTables-1M and the one we study in the current work, which uses additional image cropping during training and a slightly longer training schedule.

Training Data	Baseline Version	Epoch[a]	Test Images	$GriTS_{Con}$	$GriTS_{Loc}$	$GriTS_{Top}$	Acc_{Con}
PubTables-1M	Original	23.72	Padded	0.9846	0.9781	0.9845	0.8138
	Current	29	Tight	**0.9855**	**0.9797**	**0.9851**	**0.8326**

[a] In the current work an epoch is standardized across datasets to equal 720,000 training samples.

As part of the processing we adopt the canonicalization procedure [21] used to create PubTables-1M. This helps to align all three datasets and minimize the amount of inconsistencies between them. We make small improvements to the canonicalization procedure that help to generalize it to tables in domains other than scientific articles. These include small improvements to the step that determines the column header and portions of the row header. Tables with inconsistencies that are determined to not be easily or reliably corrected using the previous steps are filtered out.

When creating a benchmark dataset there are many objectives to consider such as maximizing the number of samples, the diversity of the samples, the accuracy and consistency of the labels, the richness of the labels, and the alignment between the labels and the desired use of the learned model for a task. When cleaning a pre-existing dataset these goals potentially compete and trade-offs must be made. There is also the added constraint in this case that whatever steps were used to create the pre-existing dataset prior to cleaning are unavailable to us and unchangeable. This prevents us from optimizing the entire set of processing steps holistically. Overall we balance the competing objectives under this constraint by correcting and adding to the labels as much as possible when we believe this can be done reliably and filtering out potentially low-quality samples in cases where we do not believe we can reliably amend them. We document the effects that the processing steps have on the size, diversity, and complexity of the samples in Table 2.

4   Experiments

For our experiments, we adopt a data-centric approach using the Table Transformer (TATR) [21]. TATR frames TSR as object detection using six classes and is implemented with DETR [2]. We hold the TATR model architecture fixed and make changes only to the data. TATR is free of any TSR-specific engineering or inductive biases, which importantly forces the model to learn to solve the TSR task from its training data alone.

We make a few slight changes to the original approach used to train TATR on PubTables-1M in order to establish a stronger baseline and standardize the approach across datasets. First we re-define an epoch as 720,000 training samples, which corresponds to 23.72 epochs in the original paper, and extend the training to 30 epochs given the new definition. Second, we increase the amount of cropping

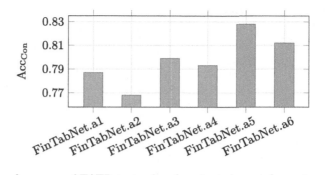

Fig. 4. The performance of TATR trained and evaluated on each version of FinTabNet. As can be seen, performance of FinTabNet evaluated on itself generally improves as a result of the modification steps. Performance is highest for FinTabNet.a5 rather than for the final version of FinTabNet. However, FinTabNet.a6 ends up being better aligned with other benchmark datasets, highlighting the potential trade-offs that must be considering when designing a benchmark dataset.

augmentation during training. All of the table images in the PubTables-1M dataset contain additional padding around the table from the page that each table is cropped from. This extra padding enable models to be trained with cropping augmentation without removing parts of the image belonging to the table. However, the original paper evaluated models on the padded images rather than tightly-cropped table images. In this paper, we instead evaluate all models on more tightly-cropped table images, leaving only 2 pixels of padding around the table boundary in each table image's original size. The use of tight cropping better reflects how TSR models are expected to be used in real-world settings when paired with an accurate table detector (Fig. 4).

For evaluation, we use the recently proposed grid table similarity (GriTS) [20] metrics as well as table content exact match accuracy (Acc_{Con}). GriTS compares predicted tables and ground truth directly in matrix form and can be interpreted as an F-score over the correctness of predicted cells. Exact match accuracy considers the percentage of tables for which all cells, including blank cells, are matched exactly. The TATR model requires words and their bounding boxes to be extracted separately and uses maximum overlap between words and predicted cells to slot the words into their cells. For evaluation, we assume that the true bounding boxes for words are given to the model and used to determine the final output.

In Table 3 we evaluate the performance of the modified training procedure to the original procedure for PubTables-1M. We use a validation set to select the best model, which for the new training procedure selects the model after 29 epochs. Using the modified training procedure improves performance over the original, which establishes new baseline performance metrics.

In the rest of our experiments we train nine TATR models in total using nine different training datasets. We train one model for each modified version of the FinTabNet dataset, a baseline model on PubTables-1M, and two additional

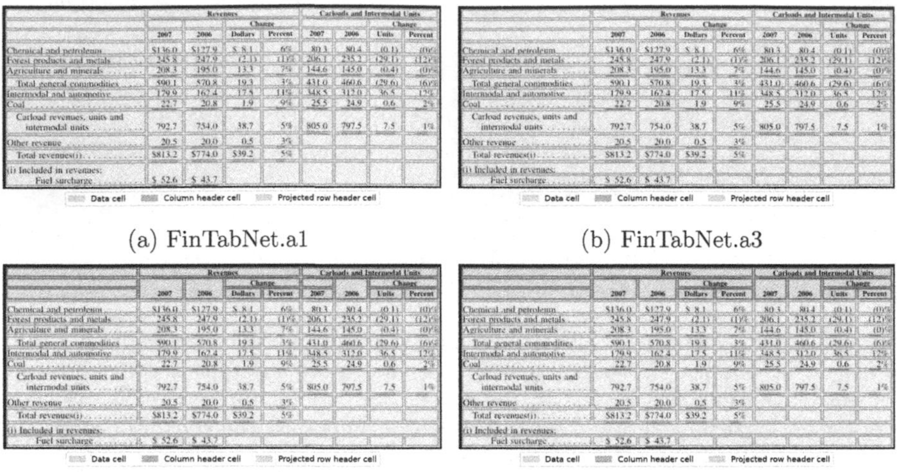

(a) FinTabNet.a1

(b) FinTabNet.a3

(c) FinTabNet.a4

(d) FinTabNet.a5

Fig. 5. Models trained on different versions of FinTabNet produce different table structure output for the same table from FinTabNet's test set. The differences include: how the last two rows are parsed, where the boundary between the first two columns occurs, and how the header is parsed into cells. (Column and row header annotations are also incorporated into the training starting with FinTabNet.a4).

models: one model for PubTables-1M combined with FinTabNet.a1 and one for PubTables-1M combined with FinTabNet.a6. Each model is trained for 30 epochs, where an epoch is defined as 720,000 samples. We evaluate each model's checkpoint after every epoch on a validation set from the same distribution as its training data as well as a separate validation set from ICDAR-2013. We average the values of GriTS$_{Con}$ for the two validation sets and select the saved checkpoint yielding the highest score.

4.1 FinTabNet Self-evaluation

In Table 4, we present the results of each FinTabNet model evaluated on its own test set. As can be seen, the complete set of dataset processing steps leads to an increase in Acc$_{Con}$ from 0.787 to 0.812. In Fig. 5, we illustrate on a sample from the FinTabNet test set how the consistency of the output improves from TATR trained on FinTabNet.a1 to TATR trained on FinTabNet.a5. Note that models trained on FinTabNet.a4 onward also learn header information in addition to structure information.

Models trained on FinTabNet.a5 and FinTabNet.a6 both have higher Acc$_{Con}$ than models trained on FinTabNet.a1 through FinTabNet.a3, despite the tables in a5 and a6 being more complex (and thus more difficult) overall than a1-a3 according to Table 2. This strongly indicates that the results are driven primarily by increasing the consistency and cleanliness of the data, and not by reduc-

Table 4. Object detection and TSR performance metrics for models trained and evaluated on each FinTabNet ablation's training and test sets. For model selection, we train each model for 30 epochs, evaluate the model on its validation set after every epoch, and choose the model with the highest Acc_{Con} on the validation set.

Training Data	Epoch	AP	AP_{50}	AP_{75}	AR	$GriTS_{Con}$	$GriTS_{Loc}$	$GriTS_{Top}$	Acc_{Con}
FinTabNet.a1	27	0.867	0.972	0.932	0.910	0.9796	0.9701	0.9874	0.787
FinTabNet.a2	22	**0.876**	0.974	0.941	0.916	0.9800	0.9709	0.9874	0.768
FinTabNet.a3	14	0.871	0.975	0.942	0.910	0.9845	0.9764	0.9893	0.799
FinTabNet.a4	30	0.874	0.974	0.946	0.919	0.9841	0.9772	0.9884	0.793
FinTabNet.a5	24	0.872	**0.977**	**0.950**	0.917	**0.9861**	0.9795	**0.9897**	**0.828**
FinTabNet.a6	26	0.875	0.976	0.948	**0.921**	0.9854	**0.9796**	0.9891	0.812

ing their inherent complexity. Something else to note is that while $GriTS_{Loc}$, $GriTS_{Con}$, and Acc_{Con} all increase as a result of the improvements to the data, $GriTS_{Top}$ is little changed. $GriTS_{Top}$ measures only how well the model infers the table's cell layout alone, without considering how well the model locates the cells or extracts their text content. This is strong evidence that the improvement in performance of the models trained on the modified data is driven primarily by there being more consistency in the annotations rather than the examples becoming less challenging. However, this evidence is even stronger in the next section, when we evaluate the FinTabNet models on the modified ICDAR-2013 datasets instead.

4.2   ICDAR-2013 Evaluation

In the next set of results, we evaluate all nine models on all three versions of the ICDAR-2013 dataset. These results are intended to show the combined effects that improvements to both the training data and evaluation data have on measured performance. Detailed results are given in Table 5 and a visualization of the results for just Acc_{Con} is given in Fig. 1 (Fig. 6).

Note that all of the improvements to the ICDAR-2013 annotations are verified manually and no tables are removed from the dataset. So observed improvements to performance from ICDAR-2013.a1 to ICDAR-2013.a3 can only be due to cleaner and more consistent evaluation data. This improvement can be observed clearly for each trained model from FinTabNet.a4 onward. Improvements to ICDAR-2013 alone are responsible for measured performance for the model trained on PubTables-1M and FinTabNet.a6 combined increasing significantly, from 68% to 81% Acc_{Con}.

Evaluating improvements to FinTabNet on ICDAR-2013.a3, we also observe a substantial increase in performance, from 41.1% to 64.6% Acc_{Con}. This indicates that not only do improvements to FinTabNet improve its own self-consistency, but also significantly improve its performance on challenging real-world test data. The improvement in performance on test data from both FinTabNet and ICDAR-

Sample Group	Some Year 1 Head Start Participation	No Year 1 Head Start Participation	Total
All Randomly Assigned (N=4,667):			
3-Year-Old Cohort			
Head Start Group	85.1%	14.9%	100%
Control Group	17.3%	82.7%	100%
4-Year-Old Cohort			
Head Start Group	79.8%	20.2%	100%
Control Group	13.9%	86.1%	100%

Data cell   Column header cell   Projected row header cell

(a) FinTabNet.a1

Sample Group	Some Year 1 Head Start Participation	No Year 1 Head Start Participation	Total
All Randomly Assigned (N=4,667):			
3-Year-Old Cohort			
Head Start Group	85.1%	14.9%	100%
Control Group	17.3%	82.7%	100%
4-Year-Old Cohort			
Head Start Group	79.8%	20.2%	100%
Control Group	13.9%	86.1%	100%

Data cell   Column header cell   Projected row header cell

(b) PubTables-1M

Sample Group	Some Year 1 Head Start Participation	No Year 1 Head Start Participation	Total
All Randomly Assigned (N=4,667):			
3-Year-Old Cohort			
Head Start Group	85.1%	14.9%	100%
Control Group	17.3%	82.7%	100%
4-Year-Old Cohort			
Head Start Group	79.8%	20.2%	100%
Control Group	13.9%	86.1%	100%

Data cell   Column header cell   Projected row header cell

(c) FinTabNet.a3

Sample Group	Some Year 1 Head Start Participation	No Year 1 Head Start Participation	Total
All Randomly Assigned (N=4,667):			
3-Year-Old Cohort			
Head Start Group	85.1%	14.9%	100%
Control Group	17.3%	82.7%	100%
4-Year-Old Cohort			
Head Start Group	79.8%	20.2%	100%
Control Group	13.9%	86.1%	100%

Data cell   Column header cell   Projected row header cell

(d) FinTabNet.a6

Sample Group	Some Year 1 Head Start Participation	No Year 1 Head Start Participation	Total
All Randomly Assigned (N=4,667):			
3-Year-Old Cohort			
Head Start Group	85.1%	14.9%	100%
Control Group	17.3%	82.7%	100%
4-Year-Old Cohort			
Head Start Group	79.8%	20.2%	100%
Control Group	13.9%	86.1%	100%

Data cell   Column header cell   Projected row header cell

(e) FinTabNet.a1 + PubTables-1M

Sample Group	Some Year 1 Head Start Participation	No Year 1 Head Start Participation	Total
All Randomly Assigned (N=4,667):			
3-Year-Old Cohort			
Head Start Group	85.1%	14.9%	100%
Control Group	17.3%	82.7%	100%
4-Year-Old Cohort			
Head Start Group	79.8%	20.2%	100%
Control Group	13.9%	86.1%	100%

Data cell   Column header cell   Projected row header cell

(f) FinTabNet.a6 + PubTables-1M

Fig. 6. Table structure output for a table from the ICDAR-2013 test set for models trained on different datasets. This example highlights how both the variety of the examples in the training sets and the consistency with which the examples are annotated affect the result.

2013 clearly indicates that there is a significant increase in the consistency of the data across both benchmark datasets.

These results also hold for FinTabNet combined with PubTables-1M. The final model trained with PubTables-1M and FinTabNet.a6 combined outperforms all other baselines, achieving a final Acc_{Con} of 81%. Few prior works report this metric on ICDAR-2013 or any benchmark dataset for TSR. As we discuss previously, this is likely due to the fact that measured accuracy depends not only on model performance but the quality of the ground truth annotation. These results simultaneously establish new performance baselines and indicate that the evaluation data is clean enough for this metric to be a useful measure of model performance.

While TATR trained on PubTables-1M and FinTabNet.a6 jointly performs best overall, in Fig. 5 we highlight an interesting example from the ICDAR-2013 test set where TATR trained on FinTabNet.a6 alone performs better. For this test case, TATR trained on the baseline FinTabNet.a1 does not fully recognize the table's structure, while TATR trained on the FinTabNet.a6 dataset does. Surprisingly, adding PubTables-1M to the training makes recognition for this example worse. We believe cases like this suggest the potential for future work to

Table 5. Object detection and TSR performance metrics for models trained on each FinTabNet ablation's training set and evaluated on each ICDAR-2013 ablation.

Training	Ep.	Test Data	AP	AP_{50}	AP_{75}	DAR_C	$GriTS_C$	$GriTS_L$	$GriTS_T$	Acc_C
FinTabNet.a1	28	IC13.a1	0.670	0.859	0.722	0.8922	0.9390	0.9148	0.9503	0.417
	28	IC13.a2	0.670	0.859	0.723	0.9010	0.9384	0.9145	0.9507	0.411
	20	IC13.a3	0.445	0.573	0.471	0.8987	0.9174	0.8884	0.9320	0.411
FinTabNet.a2	27	IC13.a1	0.716	0.856	0.778	0.9107	0.9457	0.9143	0.9536	0.436
	27	IC13.a2	0.714	0.856	0.778	0.9049	0.9422	0.9105	0.9512	0.430
	29	IC13.a3	0.477	0.576	0.516	0.8862	0.9196	0.8874	0.9336	0.405
FinTabNet.a3	25	IC13.a1	0.710	0.851	0.765	0.9130	0.9462	0.9181	0.9546	0.462
	25	IC13.a2	0.710	0.850	0.764	0.9091	0.9443	0.9166	0.9538	0.456
	28	IC13.a3	0.470	0.571	0.505	0.8889	0.9229	0.8930	0.9346	0.418
FinTabNet.a4	30	IC13.a1	0.763	0.935	0.817	0.9134	0.9427	0.9170	0.9516	0.551
	30	IC13.a2	0.763	0.935	0.818	0.9088	0.9409	0.9155	0.9503	0.544
	25	IC13.a3	0.765	0.944	0.832	0.9287	0.9608	0.9409	0.9703	0.589
FinTabNet.a5	30	IC13.a1	0.774	0.939	0.838	0.9119	0.9412	0.9075	0.9456	0.494
	30	IC13.a2	0.774	0.940	0.840	0.9098	0.9400	0.9057	0.9443	0.487
	16	IC13.a3	0.773	0.956	0.854	0.9198	0.9548	0.9290	0.9624	0.551
FinTabNet.a6	23	IC13.a1	0.760	0.927	0.830	0.9057	0.9369	0.9117	0.9497	0.500
	21	IC13.a2	0.757	0.926	0.818	0.9049	0.9374	0.9106	0.9491	0.506
	20	IC13.a3	0.757	0.941	0.840	0.9336	0.9625	0.9431	0.9702	0.646
PubTables-1M	29	IC13.a1	0.873	0.972	0.943	0.9440	0.9590	0.9462	0.9623	0.647
	29	IC13.a2	0.873	0.972	0.941	0.9392	0.9570	0.9448	0.9608	0.639
	30	IC13.a3	0.828	0.973	0.934	0.9570	0.9756	0.9700	0.9786	0.753
PubTables-1M +	25	IC13.a1	0.872	0.970	0.940	0.9543	0.9678	0.9564	0.9720	0.686
FinTabNet.a1	25	IC13.a2	0.871	0.970	0.939	0.9501	0.9655	0.9543	0.9704	0.677
	28	IC13.a3	0.820	0.949	0.911	0.9630	0.9787	0.9702	0.9829	0.785
PubTables-1M +	28	IC13.a1	0.881	0.977	0.953	0.9687	0.9678	0.9569	0.9705	0.679
FinTabNet.a6	28	IC13.a2	0.880	0.975	0.951	0.9605	0.9669	0.9566	0.9702	0.671
	29	IC13.a3	0.826	0.974	0.934	0.9648	0.9811	0.9750	0.9842	0.810

explore additional ways to leverage PubTables-1M and FinTabNet.a6 to improve joint performance even further (Figs. 7 and 8).

5 Limitations

While we demonstrated clearly that removing annotation inconsistencies between TSR datasets improves model performance, we did not directly measure the accuracy of the automated alignment procedure itself. Instead, we minimized annotation mistakes by introducing quality control checks and filtering out any tables in FinTabNet whose annotations failed these tests. But it is possible that the alignment procedure may introduce its own mistakes, some of which may not be caught by the quality control checks. It is also possible that some oversegmented tables could require domain knowledge specific to the table content itself to infer

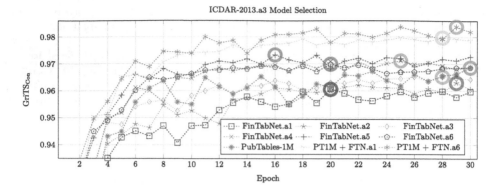

Fig. 7. The performance of each model after every epoch of training evaluated on its validation set. The training epoch with the best performance for each model is circled. PT1M = PubTables-1M, FTN = FinTabNet.

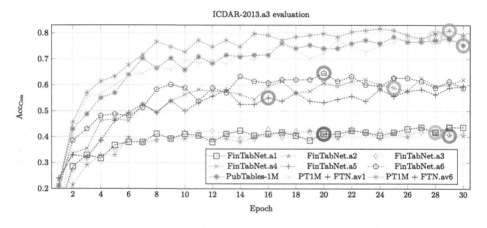

Fig. 8. The performance of each model after every epoch of training evaluated on ICDAR-2013.a3. The training epoch with the best performance for each model on its own validation set is circled. PT1M = PubTables-1M, FTN = FinTabNet.

their canonicalized structure, which could not easily be incorporated into an automated canonicalization algorithm. Finally, the quality control checks themselves may in some cases be too restrictive, filtering out tables whose annotations are actually correct. Therefore it is an open question to what extent model performance could be improved by further improving the alignment between these datasets or by improving the quality control checks.

6   Conclusion

In this work we addressed the problem of *misalignment* in benchmark datasets for table structure recognition. We adopted the Table Transformer (TATR) model and three standard benchmarks for TSR—FinTabNet, PubTables-1M, and ICDAR-2013—and removed significant errors and inconsistencies between them. After data improvements, performance of TATR on ICDAR-2013 increased substantially from 42% to 65% exact match accuracy when trained on FinTabNet and 65% to 75% when trained on PubTables-1M. In addition, we trained TATR on the final FinTabNet and PubTables-1M datasets combined, establishing new improved baselines of 0.965 DAR and 81% exact match accuracy on the ICDAR-2013 benchmark through data improvements alone. Finally, we demonstrated through ablations that canonicalization has a significantly positive effect on the performance improvements across all three datasets.

Acknowledgments. We would like to thank the anonymous reviewers for helpful feedback while preparing this manuscript.

References

1. Broman, K.W., Woo, K.H.: Data organization in spreadsheets. Am. Stat. **72**(1), 2–10 (2018)
2. Carion, N., Massa, F., Synnaeve, G., Usunier, N., Kirillov, A., Zagoruyko, S.: End-to-end object detection with transformers. In: Vedaldi, A., Bischof, H., Brox, T., Frahm, J.-M. (eds.) ECCV 2020. LNCS, vol. 12346, pp. 213–229. Springer, Cham (2020). https://doi.org/10.1007/978-3-030-58452-8_13
3. Chi, Z., Huang, H., Xu, H.D., Yu, H., Yin, W., Mao, X.L.: Complicated table structure recognition. arXiv preprint arXiv:1908.04729 (2019)
4. Frénay, B., Verleysen, M.: Classification in the presence of label noise: a survey. IEEE Trans. Neural Netw. Learn. Syst. **25**(5), 845–869 (2013)
5. Göbel, M., Hassan, T., Oro, E., Orsi, G.: ICDAR 2013 table competition. In: 2013 12th International Conference on Document Analysis and Recognition, pp. 1449–1453. IEEE (2013)
6. Guyon, I., Matić, N., Vapnik, V.: Discovering informative patterns and data cleaning. In: Proceedings of the 3rd International Conference on Knowledge Discovery and Data Mining, pp. 145–156 (1994)
7. Hashmi, K.A., Liwicki, M., Stricker, D., Afzal, M.A., Afzal, M.A., Afzal, M.Z.: Current status and performance analysis of table recognition in document images with deep neural networks. arXiv preprint arXiv:2104.14272 (2021)
8. Hashmi, K.A., Stricker, D., Liwicki, M., Afzal, M.N., Afzal, M.Z.: Guided table structure recognition through anchor optimization. arXiv preprint arXiv:2104.10538 (2021)
9. Hu, J., Kashi, R., Lopresti, D., Nagy, G., Wilfong, G.: Why table ground-truthing is hard. In: Proceedings of Sixth International Conference on Document Analysis and Recognition, pp. 129–133. IEEE (2001)
10. Koch, B., Denton, E., Hanna, A., Foster, J.G.: Reduced, reused and recycled: the life of a dataset in machine learning research. arXiv preprint arXiv:2112.01716 (2021)
11. Li, M., Cui, L., Huang, S., Wei, F., Zhou, M., Li, Z.: TableBank: table benchmark for image-based table detection and recognition. In: Proceedings of The 12th Language Resources and Evaluation Conference, pp. 1918–1925 (2020)

12. Liu, H., Li, X., Liu, B., Jiang, D., Liu, Y., Ren, B.: Neural collaborative graph machines for table structure recognition. In: Proceedings of the IEEE/CVF Conference on Computer Vision and Pattern Recognition (CVPR), June 2022, pp. 4533–4542 (2022)

13. Nassar, A., Livathinos, N., Lysak, M., Staar, P.: TableFormer: table structure understanding with transformers. In: Proceedings of the IEEE/CVF Conference on Computer Vision and Pattern Recognition (CVPR), June 2022, pp. 4614–4623 (2022)

14. Northcutt, C.G., Athalye, A., Mueller, J.: Pervasive label errors in test sets destabilize machine learning benchmarks. arXiv preprint arXiv:2103.14749 (2021)

15. Paramonov, V., Shigarov, A., Vetrova, V.: Table header correction algorithm based on heuristics for improving spreadsheet data extraction. In: Lopata, A., Butkienė, R., Gudonienė, D., Sukackė, V. (eds.) ICIST 2020. CCIS, vol. 1283, pp. 147–158. Springer, Cham (2020). https://doi.org/10.1007/978-3-030-59506-7_13

16. Prasad, D., Gadpal, A., Kapadni, K., Visave, M., Sultanpure, K.: CascadeTabNet: an approach for end to end table detection and structure recognition from image-based documents. In: Proceedings of the IEEE/CVF Conference on Computer Vision and Pattern Recognition Workshops, pp. 572–573 (2020)

17. Raji, I.D., Bender, E.M., Paullada, A., Denton, E., Hanna, A.: AI and the everything in the whole wide world benchmark. arXiv preprint arXiv:2111.15366 (2021)

18. Schreiber, S., Agne, S., Wolf, I., Dengel, A., Ahmed, S.: DeepDeSRT: deep learning for detection and structure recognition of tables in document images. In: 2017 14th IAPR International Conference on Document Analysis and Recognition (ICDAR), vol. 1, pp. 1162–1167. IEEE (2017)

19. Seth, S., Jandhyala, R., Krishnamoorthy, M., Nagy, G.: Analysis and taxonomy of column header categories for web tables. In: Proceedings of the 9th IAPR International Workshop on Document Analysis Systems, pp. 81–88 (2010)

20. Smock, B., Pesala, R., Abraham, R.: GriTS: grid table similarity metric for table structure recognition. arXiv preprint arXiv:2203.12555 (2022)

21. Smock, B., Pesala, R., Abraham, R.: PubTables-1M: towards comprehensive table extraction from unstructured documents. In: Proceedings of the IEEE/CVF Conference on Computer Vision and Pattern Recognition (CVPR), June 2022, pp. 4634–4642 (2022)

22. Strickland, E.: Andrew Ng, AI Minimalist: the machine-learning pioneer says small is the new big. IEEE Spectr. **59**(4), 22–50 (2022)

23. Tensmeyer, C., Morariu, V.I., Price, B., Cohen, S., Martinez, T.: Deep splitting and merging for table structure decomposition. In: 2019 International Conference on Document Analysis and Recognition (ICDAR), pp. 114–121. IEEE (2019)

24. Yu, J., Wang, Z., Vasudevan, V., Yeung, L., Seyedhosseini, M., Wu, Y.: CoCa: contrastive captioners are image-text foundation models. arXiv preprint arXiv:2205.01917 (2022)

25. Zheng, X., Burdick, D., Popa, L., Zhong, X., Wang, N.X.R.: Global table extractor (GTE): a framework for joint table identification and cell structure recognition using visual context. In: Proceedings of the IEEE/CVF Winter Conference on Applications of Computer Vision, pp. 697–706 (2021)

26. Zhong, X., ShafieiBavani, E., Yepes, A.J.: Image-based table recognition: data, model, and evaluation. arXiv preprint arXiv:1911.10683 (2019)

27. Zhu, X., Wu, X.: Class noise vs. attribute noise: a quantitative study. Artif. Intell. Rev. **22**(3), 177–210 (2004)

LineFormer: Line Chart Data Extraction Using Instance Segmentation

Jay Lal[✉][ID], Aditya Mitkari[ID], Mahesh Bhosale[ID], and David Doermann[ID]

Artificial Intelligence Innovation Lab (A2IL), Department of Computer Science and Engineering, University at Buffalo, Buffalo, NY, USA
{jayashok,amitkari,mbhosale,doermann}@buffalo.edu

Abstract. Data extraction from line-chart images is an essential component of the automated document understanding process, as line charts are a ubiquitous data visualization format. However, the amount of visual and structural variations in multi-line graphs makes them particularly challenging for automated parsing. Existing works, however, are not robust to all these variations, either taking an all-chart unified approach or relying on auxiliary information such as legends for line data extraction. In this work, we propose LineFormer, a robust approach to line data extraction using instance segmentation. We achieve state-of-the-art performance on several benchmark synthetic and real chart datasets. Our implementation is available at https://github.com/TheJaeLal/LineFormer.

Keywords: Line Charts · Chart Data Extraction · Chart OCR · Instance Segmentation

1 Introduction

Automated parsing of chart images has been of interest to the research community for some time now. Some of the downstream applications include the generation of new visualizations [20], textual summarization [10], retrieval [26] and visual question answering [19]. Although the related text and image recognition fields have matured, research on understanding figures, such as charts, is still developing. Most charts, such as bar, pie, lines, scatter, etc., typically contain textual elements (title, axis labels, legend labels) and graphical elements (axis, tick marks, graph/plot). Like OCR is often a precursor to any document understanding task, data extraction from these chart images is a critical component of the figure analysis process.

For this study, we have chosen to focus on the analysis of line charts. Despite their widespread use, line charts are among the most difficult to parse accurately (see Figs. 4,5). A chart data extraction pipeline typically involves many auxiliary tasks, such as classifying the chart type and detecting text, axes, and legend

---

A. Mitkari and Mahesh Bhosale—Joint Second Authorship.

© The Author(s), under exclusive license to Springer Nature Switzerland AG 2023
G. A. Fink et al. (Eds.): ICDAR 2023, LNCS 14191, pp. 387–400, 2023.
https://doi.org/10.1007/978-3-031-41734-4_24

components before data extraction (see Fig. 1). Here, we concentrate on the final task of data series extraction, that is, given a chart image, we output an ordered set of points x,y (image coordinates), representing the line data points. Our solution can be easily extended to recover the tabular data originally used to generate the chart figure by mapping the extracted series to given axis units.

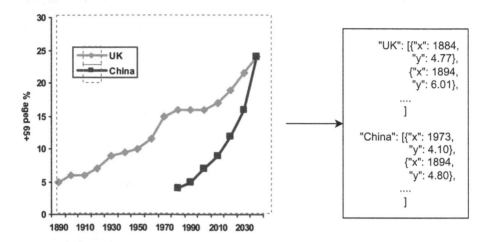

Fig. 1. Data Extraction

Recent works[17,18] show that box and point detectors have achieved reasonable accuracy in extracting data from bar, scatter, and pie charts, as they often have fewer structural or visual variations. However, line charts pose additional challenges because of the following.

– Plethora of visual variations in terms of line styles, markers, colors, thickness, and presence of background grid lines.
– Structural complexities, such as crossings, occlusions, and crowding. (Fig. [2])
– Absence of semantic information, such as legends, to differentiate lines with a similar style or color.

Although some recent works propose specialized solutions for lines, they still fail to address some of the above problems (see Figs. 4, 5). In this work, we propose a robust method for extracting data from line charts by formulating it as an instance segmentation problem. Our main contributions can be summarized as follows.

– We highlight the unique challenges in extracting data from multi-line charts.
– We propose LineFormer, a simple, yet robust approach that addresses the challenges above by reformulating the problem as instance segmentation.
– We demonstrate the effectiveness of our system by evaluating benchmark datasets. We show that it achieves state-of-the-art results, especially on real chart datasets, and significantly improves previous work.

We also open-source our implementation for reproducibility.

2 Related Work

2.1 Chart Analysis

Analyzing data from chart images has been approached from multiple perspectives. Chart-to-text [14] tried to generate natural language summaries of charts using transformer models. Direct Visual Question Answering (VQA) over charts has also been another popular area of research [19]. Many high-level understanding tasks can be performed more accurately if the corresponding tabular data from the chart have been extracted, making the extraction of chart data an essential component for chart analysis.

2.2 Line Data Extraction

Traditional Approaches. Earlier work on data extraction from line charts used image-processing-based techniques and heuristics. These include color-based segmentation [23,27], or tracing using connecting line segments [13] or pixels [22]. Patch-based methods capable of recognizing dashed lines have also been used [4]. These methods work well for simple cases; however, their performance deteriorates as the number of lines increases, adding more visual and structural variations.

Keypoint Based. More recently, deep learning-based methods for extracting chart data have gained popularity. Most of these approaches attempt to detect lines through a set of predicted keypoints and then group those belonging to the same line to differentiate for multi-line graphs. ChartOCR [17] used a CornerNet-based keypoint prediction model and push-pull loss to train keypoint embeddings for grouping. Lenovo[18] used an FPN-based segmentation network to detect line keypoints. Recently, LineEX [25] extended ChartOCR by replacing CornerNet with Vision Transformer and forming lines by matching keypoint image patches with legend patches. A common problem with these keypoint-oriented approaches is that predicted points do not align with the ground-truth line precisely. Furthermore, robust keypoint association requires aggregating contextual information that is difficult to obtain with only low-level features or push-pull-based embeddings. Some segmentation-based tracking approaches have also been proposed, treating the line as a graph with line pixels as nodes and connections as edges [15,28]. Lines are traced by establishing the presence of these edges using low-level features such as color, style, etc., or by matching with the legend to calculate the connection cost. The cost is minimized using Dynamic Programming [28] or Linear Programming [15] to obtain the final line tracking results.

We refer the reader to [6–8] for a more detailed chart comprehension and analysis survey.

2.3 Instance Segmentation

The goal of instance segmentation is to output a pixel-wise mask for each class instance. Earlier approaches, for instance segmentation such as MaskR-

CNN [11], were derived from top-down object detection techniques. However, when segmenting objects with complex shapes and overlapping instances, bottom-up approaches have performed better [9]. Bottom-up approaches, for instance segmentation, extend semantic segmentation networks by using a discriminative loss [9] to cluster pixels belonging to the same instance. More recently, transformer-based architectures have achieved state-of-the-art results on instance segmentation tasks [3,29]. These architectures are typically trained using a set prediction objective and output a fixed set of instance masks by processing object queries.

3   Motivation

3.1   Line Extraction Challenges

As mentioned, line charts, especially multiline graphs, suffer from several structural complexities. We identify the three most common structural patterns: crossings, occlusions, and crowding. Figure 2 shows their examples.

In each of these cases, the visual attributes of the line segments, such as color, style, and markers, are either obscured or blended with adjacent lines. Thus, approaches that primarily rely on low-level image features, such as color, gradients, texture, etc., often fall short. Furthermore, most keypoint-based line extractors [17,25] tend to suffer from two common issues: a.) The inability to predict distinct keypoint for each line at crossings and occlusions. b.) The subsequent keypoint grouping step suffers as it attempts to extract features from an already occluded local image patch. (See Figs. 4,5).

Recent work [15] attempts to address occlusion by modeling an explicit optimization constraint. However, the tracking remains dependent on low-level features and proximity heuristics, which could limit its robustness. It is worth noting that, in such cases, humans can gather contextual information from surrounding areas and fill the gaps.

3.2   Line as Pixel Mask

Due to the shortcomings in the keypoint-based method (Section 3.1), we adopt a different approach inspired by the task of lane detection in autonomous driving systems. In this context, instance segmentation techniques have effectively detected different lanes from road images [24]. Similarly to lanes, we consider every line in the chart as a distinct instance of the class 'line' and predict a binary pixel mask for each such instance. Thus, rather than treating a line as a set of arbitrary keypoints, we treat it as a group of pixels and demonstrate that per-pixel line instance segmentation is a more robust learning strategy than keypoint detection for line charts. Moreover, this approach does not require an explicit keypoint-grouping mechanism since all pixels (or points) segmented in a mask belong to the same line.

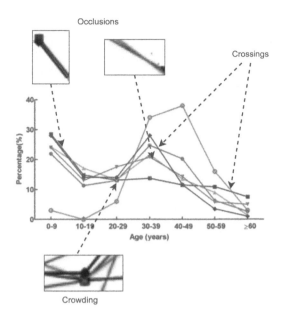

Fig. 2. Line - Structural Complexity Patterns

4   Approach

To address the problem of occlusion and crowding, we adopt an instance segmentation model based on the encoder-decoder transformer [3], which uses a
masked-attention pixel decoder capable of providing the visual context necessary
to detect occluded lines. Our proposed LineFormer system (Fig. 3) is simple and
consists of two modules, Mask Prediction and Data Series Extraction.

4.1   Line Mask Prediction

We adopt a state-of-the-art universal segmentation architecture to predict line
instances from charts [3]. The input line chart image I_{H*W} is processed by a
transformer model, an encoder-decoder network. The encoder extracts multi-
level features f from the input image I, followed by a pixel decoder that upsamples the extracted features f and outputs a 2D per pixel embedding P_{C*H*W}.
A transformer decoder attends to the intermediate layer outputs of the pixel
decoder and predicts a mask embedding vector M_{C*N}, where N is equal to the
number of line queries. Finally, the dot product of P and M gives an independent
per-pixel mask E_{H*W*N}.

4.2   Data Series Extraction

Once the predicted line masks E for an input image are obtained, we find the
start and end x-values of the line $x_{range} = [x_{start}, x_{end}]$. Within x_{range}, we sample foreground points (x_i, y_i) at a regular interval δ_x. The smaller the interval,

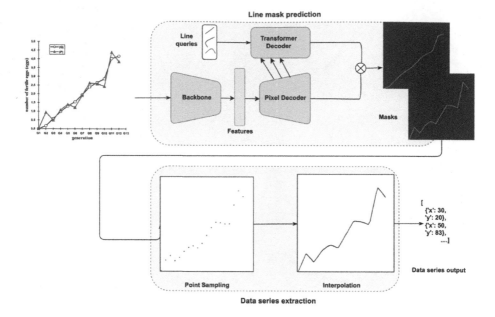

Fig. 3. LineFormer System

δ_x, the more precisely the line can be reconstructed. We used linear interpolation on the initial sampled points to address any gaps or breaks in the predicted line masks. The extracted (x,y) points can be scaled to obtain the corresponding data values if the information on the ground truth axis is available.

5   Implementation Details

We adopt the architecture and hyperparameter settings from [3]. For the backbone, we use SwinTransformer-tiny [16] to balance accuracy and inference speed. To ensure a manageable number of line predictions, we set the number of line queries to 100. As the transformer decoder is trained end-to-end on a set prediction objective, the loss is a linear combination of classification loss and mask prediction loss, with weights 1 and 5, respectively. The latter is the sum of dice loss [21] and cross-entropy loss. All experiments are performed using the MMDetection [2] framework based on PyTorch.

5.1   Generating Ground Truth Line Masks

Training LineFormer requires instance masks as labels. Thus, for a chart image, we generate a separate Boolean ground-truth mask for each line instance. As mentioned earlier, these line masks may have overlapping pixels that reflect the natural structure in the corresponding chart images. Since all datasets for line data extraction provide annotations in the form of x,y points, we need to convert these points into pixel line masks. For this, we use the Bresenham algorithm [1] from computer graphics to generate continuous line masks similar to how the lines would have been rendered in the corresponding plot image. By doing this, we can maintain a 1-1 pixel correspondence between the input image and the output pixel, making the prediction task easier and stabilizing the training. We fix the thickness of lines in ground-truth masks to three pixels, which is empirically found to work best for all the datasets.

6   Experiments

6.1   Baselines

To test our hypothesis, we performed multiple experiments on several real and synthetic chart datasets. We compare LineFormer with keypoint-based approaches as baselines, including line-specific LineEX [25] and unified approaches ChartOCR [17] and Lenovo [18]. We take their reported results for the latter, as they report on the same CHART-Info challenge[7] metrics that we use. For ChartOCR [17] and LineEX [25], we use their publicly available pre-trained models to evaluate on the same competition metrics. We cannot compare with the Linear Programming method [15] due to the inaccessibility of its implementation and because its reported results are based on a different metric.

6.2   Datasets

Training data-extraction models require a large amount of data. Furthermore, no single dataset has a sufficient number of annotated samples in terms of quantity and diversity to train these models. Thus, following suit with recent works [15, 18, 25], we combine multiple datasets for model training.

AdobeSynth19 : This dataset was released from the ICDAR-19 Competition on Harvesting Raw Tables from Infographics 2019 (CHART-Info-19)[5]. It contains samples of different chart types and has nearly 38,000 synthetic line images generated using matplotlib and annotations for data extraction and other auxiliary tasks. We use the competition-provided train test split.

UB-PMC22 : This is the extended version of the UB-PMC-2020 dataset published by CHART-Info-2020 [8]. It consists of real chart samples obtained from PubMed Central that were manually annotated for several tasks, including data extraction. Again, we use the provided train test split, which contains about 1500 line chart samples with data extraction annotation in the training set and 158 in the testing set. It is one of the most varied and challenging real-world chart datasets publicly available for data extraction.

LineEX : This is another synthetic dataset released with LineEX[25] work, which contains more variations in structural characteristics - line shape, crossings, etc., and visual characteristics - line style and color. It was generated using matplotlib and is the largest synthetic dataset available for line chart analysis. We only use a subset, 40K samples from their train set and 10K samples from the test set for evaluation.

7   Evaluation

The particular choice of metric to evaluate line data extraction performance varies amongst existing works, yet they all share a common theme. The idea is to calculate the difference between the y values for the predicted and ground-truth line series and to aggregate that difference to generate a match score. ChartOCR [17] averages these normalized y value differences across all ground truth x,y points, while Figureseer [28] and Linear Programming [15] binarize the point-wise difference by taking an error threshold of 2% to consider a point as correctly predicted. Here, we use the former as it is more precise in capturing the more minor deviations in line point prediction. Furthermore, the CHART-Info challenge formalized the same, proposing the Visual Element Detection Score and the Data Extraction Score, also referred to as task-6a and task-6b metrics [7].

7.1   Pairwise Line Similarity

Formally, to compare a predicted and ground truth data series, a similarity score is calculated by aggregating the absolute difference between the predicted value $\bar{y}_i$ and the ground truth value y_i, for each ground truth data point (x_i, y_i). The predicted value $\bar{y}_i$ is linearly interpolated if absent.

7.2   Factoring Multi-line Predictions

Most charts contain multiple lines. Thus, a complete evaluation requires a mapping between ground-truth and predicted lines. The assignment is done by computing the pairwise similarity scores (as shown in 7.1) between each predicted and ground truth line and then performing a bipartite assignment that maximizes the average pairwise score. This is similar to the mean average precision (mAP) calculation used to evaluate Object Detection performance.

Let N_p and N_g be the number of lines in the predicted and ground-truth data series.

We define K as the number of columns in the pairwise similarity matrix S_{ij} and the bipartite assignment matrix X_{ij}, where i ranges $[1, N_p]$ and j ranges $[1, N_g]$

$$K = \begin{cases} N_g & \text{for Visual Element Detection Score(task 6a)} \\ max(N_g, N_p) & \text{for Data Extraction Score (task 6b)} \end{cases}$$

$$S_{ij} = \begin{cases} 0 & \text{if } j > N_g \\ Similarity(P_i, G_j) & \text{otherwise} \end{cases}$$

$$X_{ij} = \begin{cases} 1 & \text{if } P_i \text{ matched with } G_j \\ 0 & \text{otherwise} \end{cases}$$

The final score is calculated by optimizing the average pairwise similarity score under the one-to-one bipartite assignment constraint.

$$score = \frac{1}{K} * \max_X \sum_i \sum_j S_{ij} X_{ij} \quad s.t \quad \sum_i X_j = 1 \quad and \quad \sum_j X_i = 1$$

The difference between task-6a and task-6b scores is that the former only measures line extraction recall and hence doesn't consider the false positives (extraneous line predictions). In task-6b, however, a score of '0' is assigned for each predicted data series that hasn't been matched with a corresponding ground truth, along with an increase in value of denominator K for normalizing the score.

8 Results

8.1 Quantitative Analysis

The performance of various systems for line data extraction on visual element detection and data extraction is shown in Table 1. It can be observed that the UB-PMC data set proves to be the most challenging of all, as it is diverse and composed of real charts from scientific journals. LineFormer demonstrates state-of-the-art results on most real and synthetic datasets. Furthermore, our results are consistent across the two evaluation metrics: Visual Element Detection (task-6a) and Data Extraction (task-6b). This stability is not reflected in existing work, as they exhibit substantial drops in performance. This discrepancy can be attributed to the fact that Task-6b penalizes false positives, indicating that LineFormer has a considerably higher precision rate. A key highlight of Line-Former is that it is robust across datasets and shows significant improvement on UB-PMC real-world chart data.

8.2 Qualitative Analysis

The qualitative comparison of LineFormer with existing models is carried out by examining line predictions on selected samples from the dataset. This analysis focuses on scenarios that involve crossings, occlusions, and crowding to assess the ability of models to handle such complex situations. The results are presented in Figs. [4,5], which clearly illustrate the difficulties faced by existing approaches as the number of lines and the complexity of the chart increases. LineFormer performs significantly better, keeping track of the line even in occlusions and crowding.

Table 1. LineFormer Quantitative Evaluation

Dataset	AdobeSynth19		UB-PMC22		LineEX	
Work \ Metric	Visual Element Detection[1]	Data Extraction[2]	Visual Element Detection	Data Extraction	Visual Element Detection	Data Extraction
ChartOCR [17]	84.67	55	83.89	72.9	86.47	78.25
Lenovo[3] [18]	**99.29**	**98.81**	84.03	67.01	–	–
LineEX [25]	82.52	81.97	50.23[4]	47.03	71.13	71.08
Lineformer (Ours)	97.51	97.02	**93.1**	**88.25**	**99.20**	**97.57**

[1] task-6a from CHART-Info challenge [7]
[2] task-6b data score from CHART-Info challenge
[3] Implementation not public, hence only reported numbers on AdobeSynth and UB PMC
[4] Ignoring samples without legend, as LineEX doesn't support them

8.3   Ablation Study

Table 2. Performance of different backbones on UB-PMC data

Backbone	PMC - 6a	PMC - 6b
Swin-T [16]	93.1	88.25
ResNet50 [12]	93.17	87.21
Swin-S [16]	**93.19**	88.51
ResNet101 [12]	93.1	**90.08**

We conducted an ablation study to understand the impact of different backbones on line extraction performance. For simplicity, we stick to the UB-PMC dataset, which has the most diverse set of real-world line charts. Table 2 shows the results of the study. We observe that performance remains mostly stable across most backbones, with a general trend towards an increase in 6b scores with an increase in the number of parameters. Furthermore, it should be noted that even the smallest backbone, SwinTransformer-Tiny [16] (Swin-T), surpasses the current keypoint-based state of the art (see Table 1) on the UB-PMC dataset by a large margin.

The ablation study demonstrates that the choice of backbone has a relatively minor impact on the overall performance of line extraction. This suggests that LineFormer is robust to different backbone architectures, and the main advantage is derived from the overall framework and the instance segmentation approach.

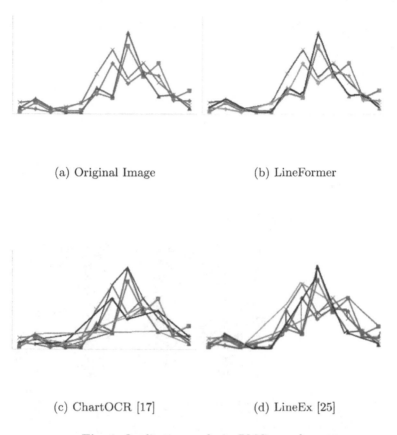

(a) Original Image   (b) LineFormer

(c) ChartOCR [17]   (d) LineEx [25]

Fig. 4. Qualitative analysis: PMC samples

9   Discussion

After extracting the line data series, each line must be associated with its corresponding label in the legend. To accomplish this task, a Siamese network can match the line embeddings with those of the patches in the provided legend.

LineFormer is designed to allow multiple labels for the same pixels to address overlapping lines. This enables LineFormer to predict different masks for lines that have overlapping sections. However, occasionally this increases the likelihood that LineFormer produces duplicate masks for the same line. Fig. [6] shows a plot with four lines where LineFormer has predicted 5, a duplicate mask for line 1.

This issue can be mitigated by selecting the top-n predictions based on legend information and using Intersection Over Union (IoU) based suppression methods in its absence.

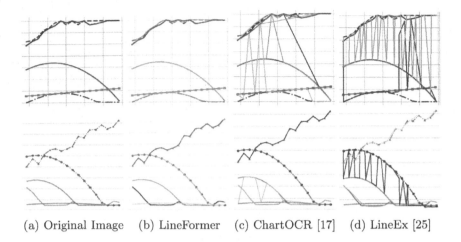

(a) Original Image   (b) LineFormer   (c) ChartOCR [17]   (d) LineEx [25]

Fig. 5. Qualitative analysis: LineEx samples

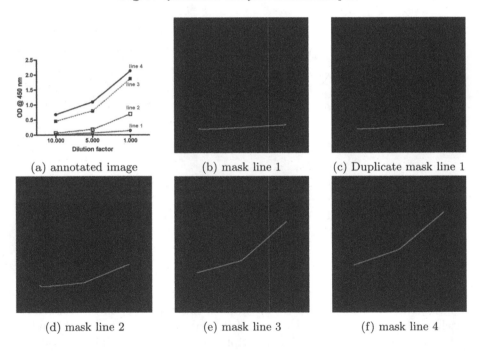

(a) annotated image   (b) mask line 1   (c) Duplicate mask line 1

(d) mask line 2   (e) mask line 3   (f) mask line 4

Fig. 6. Illustration of an occasional issue - repeated line mask for line 1

10   Conclusion

In this work, we have addressed the unique challenges that line charts pose in data extraction, which arise due to their structural complexities. Our proposed approach using Instance Segmentation has effectively addressed these challenges.

Experimental results demonstrate that our method, LineFormer, achieves top-performing results across synthetic and real chart datasets, indicating its potential for real-world applications.

References

1. Bresenham, J.E.: Algorithm for computer control of a digital plotter. IBM Syst. J. **4**(1), 25–30 (1965)
2. Chen, K., et al.: MMDetection: Open mmlab detection toolbox and benchmark. arXiv preprint arXiv:1906.07155 (2019)
3. Cheng, B., Misra, I., Schwing, A.G., Kirillov, A., Girdhar, R.: Masked-attention mask transformer for universal image segmentation. In: Proceedings of the IEEE/CVF Conference on Computer Vision and Pattern Recognition, pp. 1290–1299 (2022)
4. Chester, D., Elzer, S.: Getting computers to see information graphics so users do not have to. In: Hacid, M.-S., Murray, N.V., Raś, Z.W., Tsumoto, S. (eds.) ISMIS 2005. LNCS (LNAI), vol. 3488, pp. 660–668. Springer, Heidelberg (2005). https://doi.org/10.1007/11425274_68
5. Davila, K., et al.: ICDAR 2019 Competition on harvesting raw tables from infographics (CHART-infographics). In: 2019 International Conference on Document Analysis and Recognition (ICDAR), pp. 1594–1599 (Sep 2019), ISSN: 2379–2140
6. Davila, K., Setlur, S., Doermann, D., Kota, B.U., Govindaraju, V.: Chart mining: a survey of methods for automated chart analysis. IEEE Trans. Pattern Anal. Mach. Intell. **43**(11), 3799–3819 (2021)
7. Davila, K., Tensmeyer, C., Shekhar, S., Singh, H., Setlur, S., Govindaraju, V.: ICPR 2020 - competition on harvesting raw tables from infographics. In: Del Bimbo, A., et al. (eds.) ICPR 2021. LNCS, vol. 12668, pp. 361–380. Springer, Cham (2021). https://doi.org/10.1007/978-3-030-68793-9_27
8. Davila, K., Xu, F., Ahmed, S., Mendoza, D.A., Setlur, S., Govindaraju, V.: ICPR 2022: challenge on harvesting raw tables from infographics (CHART-infographics). In: 2022 26th International Conference on Pattern Recognition (ICPR), pp. 4995–5001 (Aug 2022), ISSN: 2831–7475
9. De Brabandere, B., Neven, D., Van Gool, L.: Semantic instance segmentation with a discriminative loss function. arXiv preprint arXiv:1708.02551 (2017)
10. Demir, S., Carberry, S., McCoy, K.F.: Summarizing information graphics textually. Comput. Linguist. **38**(3), 527–574 (2012)
11. He, K., Gkioxari, G., Dollár, P., Girshick, R.: Mask r-cnn. In: Proceedings of the IEEE International Conference on Computer Vision, pp. 2961–2969 (2017)
12. He, K., Zhang, X., Ren, S., Sun, J.: Deep residual learning for image recognition. In: Proceedings of the IEEE Conference on Computer Vision and Pattern Recognition, pp. 770–778 (2016)
13. Hoque, E., Kavehzadeh, P., Masry, A.: Chart question answering: State of the art and future directions. In: Computer Graphics Forum, vol. 41, pp. 555–572. Wiley Online Library (2022)
14. Kanthara, S., et al.: Chart-to-text: A large-scale benchmark for chart summarization. arXiv preprint arXiv:2203.06486 (2022)
15. Kato, H., Nakazawa, M., Yang, H.K., Chen, M., Stenger, B.: Parsing Line Chart Images Using Linear Programming, pp. 2109–2118 (2022)

16. Liu, Z., et al.: Swin transformer: hierarchical vision transformer using shifted windows. In: Proceedings of the IEEE/CVF International Conference on Computer Vision, pp. 10012–10022 (2021)
17. Luo, J., Li, Z., Wang, J., Lin, C.Y.: ChartOCR: Data Extraction From Charts Images via a Deep Hybrid Framework, pp. 1917–1925 (2021)
18. Ma, W., et al.: Towards an efficient framework for data extraction from chart images. In: Lladós, J., Lopresti, D., Uchida, S. (eds.) ICDAR 2021. LNCS, vol. 12821, pp. 583–597. Springer, Cham (2021). https://doi.org/10.1007/978-3-030-86549-8_37
19. Masry, A., Long, D.X., Tan, J.Q., Joty, S., Hoque, E.: ChartQA: A Benchmark for Question Answering about Charts with Visual and Logical Reasoning, arXiv:2203.10244 (Mar 2022) [cs]
20. Mei, H., Ma, Y., Wei, Y., Chen, W.: The design space of construction tools for information visualization: a survey. J. Vis. Lang. Comput. **44**, 120–132 (2018)
21. Milletari, F., Navab, N., Ahmadi, S.A.: V-net: fully convolutional neural networks for volumetric medical image segmentation. In: 2016 Fourth International Conference on 3D Vision (3DV), pp. 565–571. Ieee (2016)
22. Molla, M.K.I., Talukder, K.H., Hossain, M.A.: Line chart recognition and data extraction technique. In: Liu, J., Cheung, Y., Yin, H. (eds.) IDEAL 2003. LNCS, vol. 2690, pp. 865–870. Springer, Heidelberg (2003). https://doi.org/10.1007/978-3-540-45080-1_120
23. Nair, R.R., Sankaran, N., Nwogu, I., Govindaraju, V.: Automated analysis of line plots in documents. In: 2015 13th International Conference on Document Analysis and Recognition (ICDAR), pp. 796–800. IEEE (2015)
24. Neven, D., De Brabandere, B., Georgoulis, S., Proesmans, M., Van Gool, L.: Towards end-to-end lane detection: an instance segmentation approach. In: 2018 IEEE intelligent Vehicles Symposium (IV), pp. 286–291. IEEE (2018)
25. Shivasankaran, V.P., Hassan, M.Y., Singh, M.: LineEX: Data Extraction From Scientific Line Charts, pp. 6213–6221 (2023)
26. Ray Choudhury, S., Giles, C.L.: An architecture for information extraction from figures in digital libraries. In: Proceedings of the 24th International Conference on World Wide Web, pp. 667–672 (2015)
27. Ray Choudhury, S., Wang, S., Giles, C.L.: Curve separation for line graphs in scholarly documents. In: Proceedings of the 16th ACM/IEEE-CS on Joint Conference on Digital Libraries, pp. 277–278 (2016)
28. Siegel, N., Horvitz, Z., Levin, R., Divvala, S., Farhadi, A.: FigureSeer: parsing result-figures in research papers. In: Leibe, B., Matas, J., Sebe, N., Welling, M. (eds.) ECCV 2016. LNCS, vol. 9911, pp. 664–680. Springer, Cham (2016). https://doi.org/10.1007/978-3-319-46478-7_41
29. Wang, Y., et al.: End-to-end video instance segmentation with transformers. In: Proceedings of the IEEE/CVF Conference on Computer Vision and Pattern Recognition, pp. 8741–8750 (2021)

Line-of-Sight with Graph Attention Parser (LGAP) for Math Formulas

Ayush Kumar Shah$^{(\boxtimes)}$ and Richard Zanibbi

Document and Pattern Recognition Lab, Rochester Institute of Technology,
Rochester, NY 14623, USA
{as1211,rxzvcs}@rit.edu

Abstract. Recently there have been notable advancements in encoder-decoder models for parsing the visual appearance of mathematical formulas. These approaches transform input formula images or handwritten stroke sequences into output strings (e.g., LaTeX) representing recognized symbols and their spatial arrangement on writing lines (i.e., a Symbol Layout Tree (SLT)). These sequential encoder-decoder models produce state-of-the-art results but suffer from a lack of interpretability: there is no direct mapping between image regions or handwritten strokes and detected symbols and relationships. In this paper, we present the Line-of-sight with Graph Attention Parser (LGAP), a visual parsing model that treats recognizing formula appearance as a graph search problem. LGAP produces an output SLT from a Maximum Spanning Tree (MST) over input primitives (e.g., connected components in images, or handwritten strokes). LGAP improves the earlier QD-GGA MST-based parser by representing punctuation relationships more consistently in ground truth, using additional context from line-of-sight graph neighbors in visual features, and pooling convolutional features using spatial pyramidal pooling rather than single-region average pooling. These changes improve accuracy while preserving the interpretibility of MST-based visual parsing.

Keywords: math formula recognition · attention masks · spatial pyramidal pooling · line-of-sight neighbors · graph search

1 Introduction

Mathematical notations are widely used in technical documents to represent complex relationships and concepts concisely. However, most retrieval systems only accept text queries, and cannot handle structures like mathematical formulas. To aid in the development of search engines that *can* process queries containing mathematical notation [3, 8, 22, 25, 42, 45], recognition of mathematical expressions is needed for indexing formulas in documents, and to support handwritten input of formulas in queries [21, 23, 28, 50]. These documents include both born-digital PDF documents (e.g., where symbols in a formula may be available, but not formula locations [5, 29]), and scanned documents, for which OCR

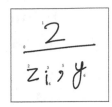

(a) Online handwritten formula

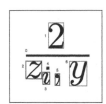
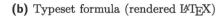
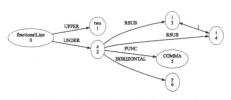

(b) Typeset formula (rendered LaTeX)

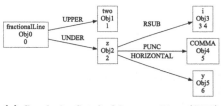

(c) Symbols: Symbol Layout Tree (SLT)

(d) Strokes/CCs: Labeled graph

Fig. 1. Formula structure at the level of symbols (c) and input strokes and connected components (d) for a formula both online handwritten (a) and rendered using LaTeX (b). Numeric identifiers for strokes (a) and their corresponding connected components (b) are shown in blue. Note that the 'i' comprised of strokes/CCs {3,4} is shown in one node in (c), but two nodes in (d). (Color figure online)

results for formulas can be unreliable. We focus here on the task of parsing the visual structure of formulas from images and handwritten strokes after formulas have been detected/isolated, and parsing isolated formulas from connected components in binary images in particular (see Fig. 1).

Parsing mathematical formulas belongs to a broader class of graphical structure recognition tasks that includes visual scene understanding, road type classification, molecular analysis, and others. The Line-of-Sight with Graph Attention Parser (LGAP) presented in this paper is a refinement of the earlier Query-Driven Global Graph Attention formula parser (QD-GGA [20]). Both systems produce *Symbol Layout Trees* (SLTs [43], see Fig. 1c) representing the hierarchical arrangement of symbols on writing lines, and the nesting of writing lines around symbols (e.g., for superscripts and subscripts).

In recent years, there have been several approaches to math formula recognition, including pixel-based encoder-decoder methods that produce state-of-the-art results, but are slow and lack interpretability; syntax-based methods that use grammar-based rules, which are less robust and computationally expensive; and graph search methods such as QD-GGA that are fast and interpretable, but have so far been unable to match the accuracy of encoder-decoder models, in part due to the limited use of context and attention.

In this paper, we describe the LGAP system, which improves the visual features and use of context in QD-GGA. This is achieved by incorporating line-of-sight (LOS) neighbors to capture additional local context [13,14], reducing the loss of spatial information through pyramidal pooling [11], and modifying the representation of punctuation relationships to avoid requiring edges that are missing in many LOS graphs. Our main research questions are:

1. Can recognition accuracy of the parser be increased by modifying the punctuation representation in ground truth?
2. Will increasing use of context in image features by adding LOS graph neighbors for primitives (nodes) and primitive pairs (edges) increase accuracy?
3. Will spatial pyramidal pooling of (SPP) convolutional features improve accuracy over the single window average pooling used in QD-GGA?

2   Related Work

For parsing of mathematical expressions, syntactic (grammar-based) approaches, encoder-decoder approaches converting pixel or strokes to strings, and graph search (tree-based) approaches are commonly used. Chan et al. [6], Zanibbi et al. [43], Zhelezniakov et al. [49], and Sakshi et al. [27] have surveyed a wide range of math recognition systems over the past few decades. The recognition of math expressions dates back to 1967 when Anderson introduced a syntax-directed approach for recognition of two-dimensional handwritten formulas including arithmetic expressions and matrices, using a 2D top-down parsing algorithm employing an attribute grammar [4]. Many systems followed the syntactic approach using a context-free grammar (CFG) of some form, including Stochastic CFGs [2] and a parser to output the parse tree and a LaTeX string [2,5,24,33]. However, there are challenges with syntactic approaches, in that the formula symbols and structures often need to be redefined [41] and it is difficult to design a universal grammar for all notation variants [48], making the syntactic approaches less robust.

In this section, we discuss the contributions and limitations of encoder-decoder models, graph-search based models, and models using graph attention and context, and briefly describe the similarities and differences in our models.

Encoder-decoder models are deep neural network architectures in which an encoder produces a feature embedding for input data in a lower-dimensional space, while the decoder converts the embedded input representation back into the original input, or into another representation. Generally, a Convolution Neural Network (CNN) is used for encoding pixel-level image features, and Recurrent Neural Networks (RNNs) are used as decoders to produce a string representation of a formula (e.g., in LaTeX) with some form of attention mechanism. Encoder-decoder models are used in other sequence prediction problems such as image caption generation [39], speech recognition, and scene text recognition [40], and have produced state-of-the-art results for these problems as well as math formula recognition [32,36,46,47]. The end-to-end Track, Attend and Parse (TAP) system [46] is a popular example. TAP obtained state-of-the-art results for many years, mostly because of the intelligent use of ensemble models that combine online and offline features, as well as different types of attention mechanisms.

Wu at al. [36] aim to improve generalization and interpretability in sequence-based encoder-decoder models by utilizing a tree-based decoder with attention, removing the need for the spatial relationships to be in strict order. A few other variations include a counting aware network [18] using symbol counting as a global symbol-level position information to improve attention, and a stroke constrained network [35] which use strokes as input primitives for their encoder-decoder, to improve the alignment between strokes and symbols.

A limitation of encoder-decoder models is the interpretibility of their results. Error diagnosis in these models is challenging due to the lack of a direct correspondence between the input (image regions or strokes) and the symbols and relationships produced in an output LATEX string or SLT: the exact correspondence between input primitives (pixels, CCs, or strokes) and the output is unknown. Hence, there is no way to identify errors at the input primitive level. They also tend to be computationally expensive: the encoder-decoder parsers use an RNN decoder with attention-based mechanisms, which requires sequential processing and greater memory usage due to maintaining hidden states across the entire input sequence.

Graph-Based Methods and MST-Based Parsing. Representing mathematical formulas as trees is more natural than images or one-dimensional strings. For example, LATEX expresses formulas as a type of SLT with font annotations. Mathematical expression recognition can alternatively be posed as filtering a graph to produce a maximum score or minimum cost spanning tree (MST) representing symbols and their associated spatial relationships. Eto et al. [10] were the first to take this approach, creating a graph with symbol nodes containing alternative labels, and candidate spatial relationships on edges with associated costs. Formula structure was obtained from extracting a minimum spanning tree amongst the relationship edges, along with minimizing a second measure of cost for the global formula structure.

For graph-based parsing, an input graph defined by a complete graph connecting all primitives is natural. However, this leads to lots of computation, and makes statistical learning tasks challenging due to input spaces with high variance in features: this motivates reducing variance through strategic pruning of input graph edges. Line-of-sight (LOS) graphs select edges where strokes in a handwritten formula or connected components in an image are mutually visible [14,19]. Systems such as Hu et al. [13], LPGA [19], and QD-GGA [20] use this LOS input graph constraint, and select the final interpretation as a *directed* Maximum Spanning Tree (MST). QD-GGA [20] extracts formula structure using an MST over detected symbols, extending previous approaches [19,30] by adding a multi-task learning (MTL) framework, and graph-based attention used to define visible primitives in images used to generate visual features for classification. QD-GGA uses Edmond's arborescence algorithm [9] to obtain a directed Maximum Spanning Tree (MST) between detected symbols, maximizing the sum of spatial relationship classification probabilities obtained from an end-to-end CNN network with attention. Using directed MSTs allows many invalid interpretations to be pruned, as the output graph is a rooted directed tree (as SLTs are).

Graph-based parsing is more natural for mathematical expressions. Unlike encoder-decoder models, the mapping between the input and the output can be obtained from labels assigned in the output to the nodes (strokes or connected components) and edges provided in the input [44]. Error metrics at the symbol and primitive levels for segmentation, symbol classification, and relation classification can be computed directly from the labeled output graphs. Also, like encoder-decoder models, these techniques do not require expression grammars, requiring only a vocabulary of symbol and relationship types. However, these models, although fast with easier interpretability, have not been able to match the accuracy of encoder-decoder models. Possible reasons may include a lack of global context, attention, and spatial information in the current models.

The use of graph neural networks with attention has been seeing more use in math parsing to capture context between primitives using graph edges directly. For instance, Peng et al. [26] use a gated graph neural network (GGNN) in the encoder stage as a message passing model to encode CNN features with visual relationships in the LOS graph. Wu et al. [37] use GNN-GNN encoder-decoder (modified Graph Attention Network [34] encoder, modified Graph Convolution Network [16] decoder) to utilize graph context. Tang et al. [32] aims to learn structural relationships by aggregating node and edge features using a Graph Attention Network to produce SLTs by simultaneous node and edge classification, instead of the sequence representation used for encoder-decoder models. They produce the output SLT by filtering 'Background' nodes and 'No-Relation' edges. Note the contrast with the use of Edmonds' algorithm in QD-GGA and LGAP, where all primitives are assumed to belong to a valid symbol, and an MST algorithm selects directed edges between symbols obtained *after* merging primitives predicted to belong to the same symbol to produce an SLT.

This Paper. LGAP is an MST-based parser for math formulas. In LGAP, we focus specially on improving QD-GGA features by intelligent use of context and improved attention mechanisms. In our approach, we employ LOS not only to construct the input graph, but also to incorporate LOS neighbors of primitives as supplementary local context for visual features. Additionally, we adopt spatial pyramidal pooling [11] rather than the isolated average pooling used in QD-GGA to mitigate a loss of spatial information. Likewise, we deal with issues in the LOS graph representation for punctuation. Punctuation was being represented by special 'PUNC' edges between a symbol and its adjacent '.' or ',' when present, but often a symbol blocks the line-of-sight between the parent symbol at left and a punctuation symbol at right.

3   Recognition Task: Punctuation Representation and Metrics for Symbols, Relationships, and Formulas

In this paper we explore improvements in MST-based parsing, focusing on typeset formulas for our initial investigation. Typeset formulas are more visually consistent than handwritten formulas, with lower variance in visual features, and we chose to examine whether our modifications are effective in this simpler setting first. We will consider handwritten formulas at a later time.

In this Section, we define our formula recognition task in terms of inputs (isolated formula CCs), outputs (SLTs), and the evaluation metrics used to quantify performance. In particular, we have modified our task by changing the representation of spatial relationships for punctuation symbols: appearance-wise the position of punctuation relative to its associated symbol at left is more similar to a subscript than horizontal adjacency (e.g., A_1 vs. AB). The use of LOS graphs in LGAP poses some additional challenges, which we address below.

Dataset. We use InftyMCCDB-2[1], a modified version of InftyCDB-2 [31]. The dataset contains binary images for isolated formulas from scanned articles, with formulas containing matrices and grids removed. The training set has 12,551 formulas, and the test set contains 6830 formulas. The dataset includes 213 symbol classes, which may be reduced to 207 by merging visually similar classes (e.g., *ldots* and *cdots*, *minus* and *fractional line*), and 9 relationship classes: *Horizontal*, *Rsub* (right subscript), *Rsup* (right superscript), *Lsub* (left subscript), *Lsup* (left superscript), *Upper*, *Under*, *PUNC* (for punctuation), and *NoRelation* (see Figs. 1d and 2). The training and test sets have approximately the same distribution of symbol classes and relation classes.

Evalation Metrics. We report expression recognition rates for (1) *Structure:* unlabeled SLTs with correct nodes (CC groups for symbols) and edges (for relationships), and (2) labeled SLTs where symbol classes and relationship types must also be correct. For a finer-grained analysis, we also report detection and classification F-scores for symbols and relationships using the LgEval library originally developed for the CROHME handwritten formula recognition competitions [23]. F-scores are the harmonic mean of recall and precision for the detection of target symbols and relationships ($2RP/(R+P)$). We report both: (1) *detection f-scores*: quantify properly detected/segmented CC groups for symbols, and the presence of an edge (relationship) between two properly segmented symbols, and (2) detection + classification f-scores: here a symbol or relationship is correct if has been detected correctly *and* assigned the correct symbol or relationship label.

Modified Punctuation Relationship in Ground Truth SLTs. Mahdavi et al. [19] added a new spatial relationship called *PUNC* in the InftyMCCDB-2 dataset for distinguishing horizontal relationships with punctuation symbols from other horizontal relationships to make their visual features more consistent and thereby identifiable. However, this modification can lead to missing ground truth punctuation edges in the input LOS graph, often when a symbol has a subscript and is followed by a punctuation symbol on its writing line. This happens since there is good chance that the line of sight between the parent symbol and the punctuation is blocked by one or more primitives between them, usually a large subscript. For the example in Fig. 2, the ground truth edge between the symbols 'z' and 'COMMA' is blocked by the primitives corresponding to 'i'. Here

---

[1] https://www.cs.rit.edu/~dprl/data/InftyMCCDB-2.zip.

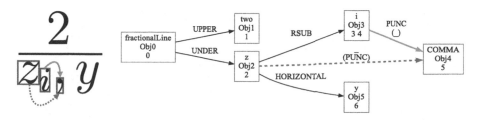

Fig. 2. Modified punctuation (*PUNC*) ground truth representation. The old *PUNC* edge is shown using red dashed arrows, and its corresponding new edge is shown with solid orange arrows. The original *PUNC* edge between nodes 'z' and 'COMMA' is missing in the LOS graph, as can be seen in Fig. 1a and 1b (Color figure online).

the *PUNC* relationship will be missed, due to the missing edge in the input LOS graph representation. Additional heuristics in LPGA [19] modify the LOS graph to add missing punctuation based on angles and relative sizes of symbols, which are not very robust.

To address this issue, we modified the assigned parent node for *PUNC* edges as shown in Fig. 2. Instead of the original parent, which may be far left on the writing line and/or blocked in the LOS graph, we assign the PUNC parent node to the symbol immediately at left of the punctuation symbol. This modification ensures all *PUNC* edges are included in input LOS graphs. This also improves training/learning, as relationships between *PUNC* parent and child nodes using more consistent visual features (see Sect. 5).

4   Line-of-Sight with Graph Attention Parser (LGAP)

In this section we describe the MST-based LGAP parsing model inputs and outputs, its multi-task end-to-end CNN classification model, and finally the steps to select symbols and extract an MST to output a symbol layout tree. LGAP was implemented in PyTorch with Python NetworkX library for representing graphs and running Edmonds' algorithm to obtain MSTs/SLTs.[2]

4.1   Inputs and Outputs

Query and LOS Context Attention Masks. For formula images, we extract CC contours by first smoothing them and then sampling contours to produce trace points. For handwritten formula images, trace points are already available. These trace points are interpolated, normalized, and scaled to a height of 64 pixels while preserving the aspect ratio: this results in formula images with a fixed height but varying width. The full normalized formula image is fed into a CNN encoder backbone to compute the global visual features, and points on sampled contours are used to construct the LOS input graph (described below).

---

[2] LGAP open source implementation: https://gitlab.com/dprl/qdgga-parser.

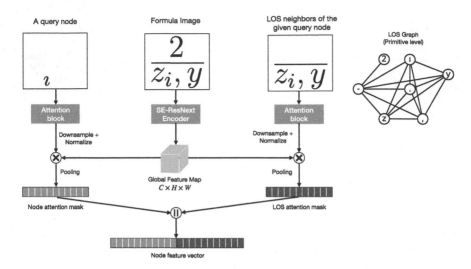

Fig. 3. Binary attention masks in LGAP. The input primitive query mask (represented here by the base of the letter 'i') and its corresponding LOS masks are used to generate the attention masks by performing element-wise multiplication with the global feature map. The two attention masks are concatenated to get the node feature vector that is utilized for symbol classification. Note that the same process is applied to the primitive pair binary masks and LOS mask to generate the primitive pair feature vector for classifying directed edges.

Input regions are processed to produce binary attention masks created for input *queries* corresponding to individual primitives (nodes) or primitive pairs sharing an edge in the LOS graph. Binary masks for directed edges between primitive pairs are concatenated with the binary mask for the parent primitive, to increase identifiability and thus classification accuracy [20]. Query binary masks filter the CNN global feature map, providing focus on image regions pertinent for three classification tasks: (1) masking individual CC/strokes for symbol classification (for nodes), and masking CC/stroke pairs for (2) segmentation and (3) spatial relationship classification (for directed edges).

In LGAP, we introduce an additional binary *LOS mask*, which consists of a binary sum (OR) of the query mask with the binary mask of *all* neighboring primitives in the LOS graph (see Fig. 3). These additional LOS binary masks serves to provide visual spatial context information from the neighbors in the LOS graph. As for primitive and primitive pair binary masks, LOS masks are downsampled and weight the global feature map produced by the encoder in order to focus on the relevant parts of the input image for the three classification tasks (symbols, segments, relationships).

Primitive LOS Graph. Our parser uses as input either a set of online handwritten strokes, or a set of connected components (CCs) from a typeset formula image (see Fig. 1a, 1b). A handwritten stroke is represented by a list of 2D (x,y) coordinates obtained from sampled positions of a pen, trackpad, or other touch device. When parsing online handwritten formulas, our input is a graph over strokes images: strokes are individually rendered using their stroke coordinates. To parse typeset formulas, we construct a graph over connected components. For both handwritten and typeset formulas, we assume that *primitives* (stroke images or CCs) belong to exactly one symbol.

Initially, the input is a complete graph with an edge between every pair of primitives (CCs or strokes). As seen in Fig. 4, we then select edges in the Line of Sight (LOS) graph [7] as suggested by Hu et al. [14]. In the LOS graph, edges exist only for primitive pairs where an unobstructed line can be drawn from the center of one primitive to a point on the convex hull of the contour for the other [19]. This reduces the space of output graphs and number of classifications needed, with few deletions of pertinent edges [14]. In Fig. 4, all edges between the '2' and other CCs other than the fraction line are pruned.

Output: Symbol Layout Tree. Formula appearance is commonly represented as a tree of symbols on writing lines known as a Symbol Layout Tree (SLT, see Fig. 1c [43]). SLTs have popular encodings such as LaTeX and Presentation MathML that include additional formatting information such as fonts, font styles (e.g., italic), and spacing. Because they describe spatial arrangements of complete symbols, SLTs are identical for formulas in handwritten strokes and typeset images. However, the number and arrangement of strokes or CCs may differ for a single formula. For the example in Fig. 1, the primitive label graphs for the image and handwritten version are identical so that CCs and strokes have an exact correspondence. This is not always the case, for example when an 'x' drawn with two handwritten strokes appears as a single CC in a typeset image.

In LGAP we produce an SLT over symbols from our directed LOS input graph over strokes or CCs. Figure 4 shows how a complete graph over primitives is constrained and annotated with classification probabilities in stages, producing an SLT as output. [44]. An example of a labeled LOS graph representing an SLT is shown in Fig. 1d for both handwritten strokes and image CCs. There are bidirectional edges between strokes or CCs belonging to the same symbol, with the edge label matching the symbol class. Spatial relationships are represented by directed labeled with their spatial relationship type. Relation types include *Horizontal*, *Rsub* (right subscript), *Rsup* (right superscript), *Upper*, *Under*, and *PUNC* (for punctuation).

4.2 Multi-task CNN for Classifying Primitives and Primitive Pairs

Image features for primitives and edges in the LOS graph are passed through a CNN architecture to obtain class probability distributions for three tasks: symbol detection (segmentation), symbol classification, and spatial relationship classification. LGAP's CNN architecture is based on QD-GGA [20], an end-to-end trainable multi-task learning model. CNN features are used to estimate probability distributions for each task. These distributions are stored in the cells of three adjacency matrices defined over the input LOS graph. Symbol classification probabilities for primitives are represented along the LOS adjacency matrix

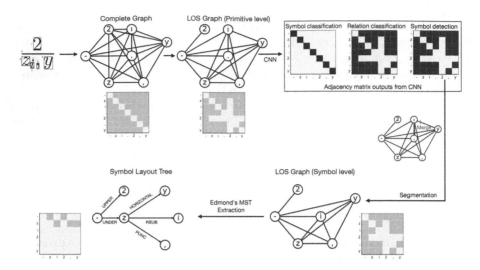

Fig. 4. LGAP formula parsing example. A complete graph over input primitives (here, CCs) is pruned, sub-images corresponding to CCs and CC LOS edges are given symbol, merge/split, and spatial relationship class distributions. Based on merge/split probabilities primitives are merged into symbols (here, for the 'i'), and finally an SLT is produced from remaining spatial relationship edge by extracting a **directed MST** (arrows omitted for legibility). Symbol and relationship class probabilities are averaged when merging primitives into symbols.

diagonal, and merge/split and spatial relationship probabilities are represented for directed LOS edges in off-diagonal entries, as seen at top-right in Fig. 4.

The CNN contains an SE-ResNext backbone [12,38] used to compute a global feature map from the input formula image along with attention modules to produce attention relevance maps from the binary masks described in the previous Section. The SE-ResNext backbone and the attention modules are trained concurrently using a combined cross entropy loss function for all three classification tasks. Attention modules are used to produce 2D relevance maps from the 2D binary masks for nodes, edges and LOS neighbors separately by convolving them through 3 convolutional blocks, trained for each task, where each block has 3 kernels of size 7×7, 5×5, and 5×5 [20].

Modification: For this work, we modified the QD-GGA SE-ResNext encoder by reducing the number of output channels from 512 to 256, to reduce the number of parameters requiring training. This results in a 256-dimensional feature vector for symbol classification, and two 512-dimensional feature vectors for segmentation and relation classification, respectively.

Spatial Pyramidal Pooling and 1-D Context Module. Average pooling uses a single average activation value to represent the convolutional response in a region [1]. Without the use of windowing within an input image, a large amount of important spatial information is lost during average pooling. Jose et al. [15] use a pyramidal pooling method, which integrates spatial information into the feature vectors, producing more compact and location invariant feature vectors. He et al. [11] introduced the spatial pyramid pooling (SPP) strategy for deep CNNs by adopting the widely used spatial pyramid matching (SPM) strategy [17]. SPP captures spatial information by pooling features within equal-sized regions of the feature map for increasing numbers of sub-regions, forming a pyramid of overlapping sub-regions (e.g., whole image, left-right, top-down, 3 horizontal regions, etc.).

In LGAP and QD-GGA, weighted feature maps are pooled to produce feature vectors. A '1D context' module then performs a 1-by-3 convolution along the sequence of query feature vectors used as classification input [20]. The convolution aggregates features from the previous $(i-1)^{th}$ and next $(i+1)^{th}$ query in input order for the ith query. In the input, queries are spatially sorted top-down, left-right by the top-left coordinate of a query's bounding box. Feature vectors are passed to one of three fully-connected output layers for three classification tasks: segmentation and relationships (edges), and symbol classification (nodes).

Modification: For LGAP, Spatial Pyramidal Pooling is used to capture spatial information across multiple horizontal and vertical regions, providing more spatial information and lower variance in features. We use 5 levels with 11 regions in pooling outputs: this includes 1 full feature map, 2 vertical bins, 3 vertical bins, 2 horizontal bins, and 3 horizontal bins. To reduce the growth in parameters due to increasing pooling regions from 1 to 11, we also reduce the number of output channels in the SE-ResNext encoder from 512 to 64 (a factor of 8). The resulting node and edge feature vectors have lengths of 704 (i.e., 64×11) and 1408 (i.e., 128×11), respectively. With the new LOS attention masks, the feature lengths are 1408 (i.e., 704×2) and 2816 (i.e., 1408×2) respectively. The edge features have an additional factor of 2 due to the concatenation of parent primitive attention masks, as mentioned earlier. We also examine removing the 1-by-3 convolutional context module, since its notion of neighbor is not based directly on spatial proximity as discussed earlier (see Sect. 5).

Multi-task Cross Entropy Loss Function and Training. We use the same cross entropy loss (CE) with a softmax activation used in QD-GGA [20], which combines the segmentation, symbol classification and relation classification losses as shown in Eq. 1. In the equation, δ denotes the total loss, N is the set of primitive nodes (strokes or CCs), E is the set of LOS edges in the input graph, D is the set of segmentation ground truth edge labels, R is the set of relationship ground truth edge labels, and S is the set of primitive ground truth symbol labels.

$$\delta(N, E) = \sum_{e=1}^{|E|} (CE(e, D) + CE(e, R)) + \sum_{n=1}^{|N|} CE(n, S) \tag{1}$$

For backpropagation, LGAP uses an Adam optimizer with learning rate 0.0005 and momentum 0.9, with a weight decay (L2 norm) of 0.1 for regularization.

Table 1. Effect of modifying *PUNC* relationship representation for InftyMCCDB-2. F1 and expression rate metrics are defined in Sect. 3

	Symbol (F1)		Relation (F1)		Expression Rate	
	Detect	+Class	Detect	+Class	Structure	+Class
Original relationships	**98.23**	95.21	94.63	94.28	89.21	81.77
Modified PUNC	98.22	**95.23**	**94.93**	**94.56**	**90.45**	**83.00**

4.3   Parsing: Transforming LOS Graphs into SLTs

After LOS nodes and edges have class distributions assigned to them using the multi-task CNN model, we select all LOS segmentation edges with a higher probability for 'merge' than 'split,' and then merge all connected components over primitives in the resulting segmentation graph into symbols. For the example in Fig. 4, primitives corresponding to the symbol 'i' are merged into a single symbol node, along with their corresponding relationship edges.

For merged primitives, we average the symbol and merged edge relationship class probability distributions. We then apply Edmond's arborescence algorithm [9] to obtain a *directed* Maximum Spanning Tree, which forms the Symbol Layout Tree (SLT). The MST maximizes the sum of relationship probabilities in selected directed edges, where the maximum probability relationship is used for each edge. However, this selection may result in duplicate edges of same relationship type between one parent symbol and two or more child symbols. To address this we apply an additional constraint, where a parent symbol may not have more than one edge of each spatial relationship type associated with it (e.g., to prevent having two horizontal relationships from one symbol to two symbols at right). We replace edges that duplicate a relationship by removing the lower-probability edge, and replacing the lower probability edge's label with its next-highest relationship class probability, and then rebuild the MST, repeating until this unique relationship constraint is satisfied.

5   Experiments

In this Section we report some experiments testing the effects of our modifications to the representation of punctuation relationships in ground truth, and changes in the CNN model and visual feature representation. Experiments were run on two servers and two desktop machines, all with hard drives (HDD):

1. 2 x GTX 1080 Ti GPUs (12 GB), 8-core i7-9700KF (3.6 GHz), 32 GB RAM
2. 2 x GTX 1080 Ti GPUs (12 GB), 12-core i7-8700K (3.7 GHz), 32 GB RAM
3. 4 x RTX 2080 Ti (12 GB), 32-core Xeon E5-2667 v4 (3.2 GHz), 512 GB RAM
4. A40 (48 GB), 64-core Xeon Gold 6326 (2.9 GHz), 256 GB RAM

5.1   Effect of Modifying PUNC Relationship in Ground Truth

We saw a 1.23% improvement in formula recognition rate (Structure + Classification) as seen in Table 1 after applying the modification to ground truth described

Table 2. Effect of LGAP Spatial Pyramidal Pooling (SPP) and 1D context module. Feature vector sizes given as *(node-size, edge-size)*. Modified PUNC representation used

Feature Vectors			Symbol (F1)		Relation (F1)		Expression Rate	
POOL	SIZES	1D-CONTEXT	Detect	+Class	Detect	+Class	Structure	+Class
Avg	64,128	True	96.99	90.24	91.18	90.62	85.58	66.41
SPP-Avg	704,1408	True	**98.30**	**95.25**	**94.50**	**94.09**	**88.64**	**80.78**
Avg	64,128	False	89.84	34.68	67.68	65.23	64.32	17.57
SPP-Avg	704,1408	False	95.64	87.55	86.16	84.49	77.76	68.80

Table 3. Effect of Adding LOS Neighborhood Masks to LGAP SPP-Avg Model. Original ground truth used (without PUNC modification)

Feature Vectors			Symbol (F1)		Relation (F1)		Expression Rate	
LOS	SIZES	1D-CONTEXT	Detect	+Class	Detect	+Class	Structure	+Class
False	704,1408	True	97.96	94.48	94.36	93.89	88.70	78.90
True	1408,2816	True	98.32	**95.66**	**94.85**	94.35	89.27	**83.27**
False	256,512	True	98.23	95.21	94.63	94.28	89.21	81.77
True	512,1024	True	**98.39**	95.49	**94.85**	**94.46**	**89.36**	81.89
False	256,512	False	95.16	83.78	86.48	85.22	79.72	65.26
True	512,1024	False	95.40	85.97	86.27	85.07	80.23	70.09

in Sect. 3. Outputs were produced using a reduced encoder architecture with 256 channels and a context module as described in Sect. 4.2. Observing errors for the two conditions using the `confHist` tool in LgEval showed the new representation correctly identified more punctuation relationships that were previously missing or incorrectly labeled.

5.2   Improving Visual Features: SPP and Increased LOS Context

We next check whether our proposed changes in visual features described in Sect. 4 improve accuracy.

Spatial Pyramidal Pooling. Table 2 shows that when the 1D context module is used, the formula recognition rate improves 14.37% after replacing average pooling by spatial pyramidal pooling (first two rows in Table 2). This difference in expression rate increases dramatically to 51.23% when the context module is removed (last two rows in Table 2). These results suggest that spatial pyramidal pooling greatly improves visual features, allowing us to obtain recognition rates close to the QD-GGA model with 256 encoder channels using only 64 encoder channels. However, removing the 1-D context module reduces the expression rate using the new SPP features by 13.17%, and the original single-region average pooling features reduces dramatically (48.83%). This demonstrates that the new SPP features are beneficial, and the importance of context.

Adding LOS Context through Neighborhood Embeddings. Next we evaluate the impact of including additional LOS neighbors in selecting image regions for queries in the attention modules, as outlined in Sect. 4.1. We hypothesized

Table 4. Benchmarking MST-based Parsing Models on InftyMCCDB-2

MST Model	Symbols		Relationships		Formulas	
	Detect.	+Class	Detect.	+Class	Structure	+Class
LGAP (this paper)	98.32	95.66	94.85	94.35	89.27	83.27
QD-GGA	98.50	94.54	94.43	93.96	87.77	76.72
LPGA$_{RF}$	99.34	98.51	97.83	97.56	**93.81**	90.06
LPGA$_{CNN}$	**99.35**	**98.95**	**97.97**	**97.74**	93.37	**90.89**

that incorporating the context from LOS neighbors would reduce ambiguity for visually similar symbols/relationships. Experiments were performed using both the 64-channel encoder output with spatial pyramidal pooling (using 11 bins, resulting a feature vector of lengths 1408 and 2816 (when LOS context is used), as well as the original 256-channel output with average pooling and the original representation. We also investigated the effect of using LOS masks when the context module was removed. The results in Table 3 show improvements in recognition rates at the symbol, relation, and formula levels when LOS neighborhood masks are used. Further, the expression rate accuracy is increased 2.49% over the best SPP model in Table 2.

Error Analysis. An error analysis using the `confHist` tool in LgEval on our best performing model (LOS context, spatial pyramidal pooling, and 1D context; second row in Table 3) reveals the majority of symbol classification errors occur between visually similar symbols, such as $(i, j), (m, n), (l, 1), (\alpha, a)$, and (Left-Paranthesis, Vertical), in decreasing order of frequency. This highlights needed improvements in local visual features for symbols. For relationships, the most frequent errors are missing relationship edges in wide formulas containing a large number of symbols. This type of error can be attributed to the preprocessing step used for inputs: with the height of all formulas fixed at 64 pixels and the aspect ratio preserved, image resolution is noticeably reduced for wide formulas. This leads to features extracted from low-resolution input images being used to locate spatial relationships between connected components for very wide formulas.

We also note that expression rates are influenced by small changes in symbol and relationship recognition accuracy, which may amplify expression rates differences between conditions. For example, if a formula has just one symbol or relationship wrong, it is not counted as correct in the expression rate.

5.3   Benchmarking MST-Based Parsers

As seen in the previous experiments, the LGAP model that obtained the highest expression rate used a combination of Spatial Pyramidal Pooling, line-of-sight attention masks, and a modified punctuation representation in ground truth. We next benchmark this best LGAP model against previous MST-based visual parsers applied to InftyMCCDB-2. Results are presented in Table 4. Performance for LGAP relative to the QD-GGA model it extends is better in every metric, aside from a very small decrease in symbol detection F1 (−0.18%). The expression rate has increased 6.55% over QD-GGA. Unfortunately, performance is weaker than the LPGA models, with an expression rate that is 7.62% lower.

Despite LGAP's slightly weaker performance than LPGA [19], the LGAP offers substantial benefits in speed and efficiency. The encoder and attention modules are trained end-to-end on a joint loss for multiple tasks in a single feed-forward pass, making the training and execution process much faster than LPGA. Running on a desktop system with two GTX 1080 Ti GPUs (12 GB), an 8-core i7-9700KF processor (3.6 GHz) and 32 GB RAM, LGAP requires 25 min per epoch to train on 12,551 training images and 11 min, 12 s to process the 6,830 test formula images (98.4 ms per formula).

Opportunities for further improvements include improving context usage through graph neural networks, as well as more sophisticated graph-based attention models to replace the current 1D context module from QD-GGA.

6   Conclusion

We have introduced the Line-of-sight with Graph Attention Parser (LGAP) that enhances the visual feature representations employed in the MST-based QD-GGA parser through thoughtful use of Line-of-sight neighborhoods and spatial pyramidal pooling, and modified the ground truth representation of spatial relationship edges connecting punctuation with parent symbols in Symbol Layout Trees (SLTs). These modifications have added contextual information, prevented the loss of spatial information than using single region average pooling in QD-GGA produced, and avoided pruning valid punctuation relationships.

In the future, we aim to improve the use of context. Currently context is introduced using a sequential (1D) module that aggregates the two immediate neighbors in the input order. This sometimes misses neighbor relationships and introduces spurious ones because a spatial sorting order rather than actual proximity defines neighbors. We expect that using Graph Neural Networks (GNNs) will avoid these problems, and incorporate actual proximity in aggregation and use the underlying graph structure directly. Additionally, we plan to replace LGAP's attention maps using more sophisticated methods than simple convolutional blocks. We will also assess our model's performance on online handwritten data, such as CROHME. LGAP is available as open source (see Sect. 4).

Acknowledgements. We thank everyone who contributed to the DPRL formula extraction pipeline: Lei Hu, Michael Condon, Kenny Davila, Leilei Sun, Mahshad Mahdavi, Abhisek Dey, and Matt Langsenkamp. This material is based upon work supported by the Alfred P. Sloan Foundation under Grant No. G-2017-9827 and the National Science Foundation (USA) under Grant Nos. IIS-1717997 (MathSeer project) and 2019897 (MMLI project).

References

1. Akhtar, N., Ragavendran, U.: Interpretation of intelligence in CNN-pooling processes: a methodological survey. Neural Comput. Appl. **32**(3), 879–898 (2019). https://doi.org/10.1007/s00521-019-04296-5
2. Alvaro, F., S'nchez, J., Benedi, J.: Recognition of printed mathematical expressions using two-dimensional stochastic context-free grammars. In: 2011 International Conference on Document Analysis and Recognition, pp. 1225–1229, September 2011. https://doi.org/10.1109/ICDAR.2011.247. ISSN: 2379-2140
3. Amador, B., Langsenkamp, M., Dey, A., Shah, A.K., Zanibbi, R.: Searching the ACL anthology with math formulas and text. In: Proceedings ACM SIGIR (2023, to appear)
4. Anderson, R.H.: Syntax-directed recognition of hand-printed two-dimensional mathematics. In: Symposium on Interactive Systems for Experimental Applied Mathematics: Proceedings of the Association for Computing Machinery Inc., Symposium, pp. 436–459. Association for Computing Machinery, New York, NY, USA, August 1967. https://doi.org/10.1145/2402536.2402585
5. Baker, J.B., Sexton, A.P., Sorge, V.: A linear grammar approach to mathematical formula recognition from PDF. In: Carette, J., Dixon, L., Coen, C.S., Watt, S.M. (eds.) CICM 2009. LNCS (LNAI), vol. 5625, pp. 201–216. Springer, Heidelberg (2009). https://doi.org/10.1007/978-3-642-02614-0_19
6. Chan, K.F., Yeung, D.Y.: Mathematical expression recognition: a survey. Int. J. Doc. Anal. Recogn. **3**(1), 3–15 (2000). https://doi.org/10.1007/PL00013549
7. de Berg, M., Cheong, O., van Kreveld, M., Overmars, M.: Computational geometry. In: de Berg, M., Cheong, O., van Kreveld, M., Overmars, M. (eds.) Computational Geometry: Algorithms and Applications, pp. 1–17. Springer, Heidelberg (2008). https://doi.org/10.1007/978-3-540-77974-2_1
8. Diaz, Y., Nishizawa, G., Mansouri, B., Davila, K., Zanibbi, R.: The MathDeck formula editor: interactive formula entry combining latex, structure editing, and search. In: CHI Extended Abstracts, pp. 192:1–192:5. ACM (2021)
9. Edmonds, J.: Optimum branchings. J. Res. Nat. Bureau Stan. Sect. B Math. Math. Phys. **71B**(4), 233 (1967). https://doi.org/10.6028/jres.071B.032. https://nvlpubs.nist.gov/nistpubs/jres/71B/jresv71Bn4p233_A1b.pdf
10. Eto, Y., Suzuki, M.: Mathematical formula recognition using virtual link network. In: Proceedings of Sixth International Conference on Document Analysis and Recognition, pp. 762–767, September 2001. https://doi.org/10.1109/ICDAR.2001.953891
11. He, K., Zhang, X., Ren, S., Sun, J.: Spatial pyramid pooling in deep convolutional networks for visual recognition. In: Fleet, D., Pajdla, T., Schiele, B., Tuytelaars, T. (eds.) ECCV 2014. LNCS, vol. 8691, pp. 346–361. Springer, Cham (2014). https://doi.org/10.1007/978-3-319-10578-9_23
12. Hu, J., Shen, L., Sun, G.: Squeeze-and-excitation networks. In: 2018 IEEE/CVF Conference on Computer Vision and Pattern Recognition, pp. 7132–7141, June 2018. https://doi.org/10.1109/CVPR.2018.00745

13. Hu, L., Zanibbi, R.: MST-based visual parsing of online handwritten mathematical expressions. In: 2016 15th International Conference on Frontiers in Handwriting Recognition (ICFHR), pp. 337–342, October 2016. https://doi.org/10.1109/ICFHR.2016.0070. ISSN: 2167-6445

14. Hu, L., Zanibbi, R.: Line-of-sight stroke graphs and Parzen shape context features for handwritten math formula representation and symbol segmentation. In: 2016 15th International Conference on Frontiers in Handwriting Recognition (ICFHR), pp. 180–186, October 2016. https://doi.org/10.1109/ICFHR.2016.0044. ISSN: 2167-6445

15. Jose, A., Lopez, R.D., Heisterklaus, I., Wien, M.: Pyramid pooling of convolutional feature maps for image retrieval. In: 2018 25th IEEE International Conference on Image Processing (ICIP), pp. 480–484, October 2018. https://doi.org/10.1109/ICIP.2018.8451361. ISSN: 2381-8549

16. Kipf, T.N., Welling, M.: Semi-supervised classification with graph convolutional networks. In: International Conference on Learning Representations, p. 14 (2017). https://openreview.net/forum?id=SJU4ayYgl

17. Lazebnik, S., Schmid, C., Ponce, J.: Beyond bags of features: spatial pyramid matching for recognizing natural scene categories. In: 2006 IEEE Computer Society Conference on Computer Vision and Pattern Recognition (CVPR 2006), vol. 2, pp. 2169–2178, June 2006. https://doi.org/10.1109/CVPR.2006.68. ISSN: 1063-6919

18. Li, B., et al.: When counting meets HMER: counting-aware network for handwritten mathematical expression recognition. In: Avidan, S., Brostow, G., Cissé, M., Farinella, G.M., Hassner, T. (eds.) Computer Vision – ECCV 2022. LNCS, pp. 197–214. Springer, Cham (2022). https://doi.org/10.1007/978-3-031-19815-1_12

19. Mahdavi, M., Condon, M., Davila, K., Zanibbi, R.: LPGA: line-of-sight parsing with graph-based attention for math formula recognition. In: 2019 International Conference on Document Analysis and Recognition (ICDAR), pp. 647–654. IEEE, Sydney, Australia, September 2019. https://doi.org/10.1109/ICDAR.2019.00109. https://ieeexplore.ieee.org/document/8978044/

20. Mahdavi, M., Sun, L., Zanibbi, R.: Visual parsing with query-driven global graph attention (QD-GGA): preliminary results for handwritten math formula recognition. In: 2020 IEEE/CVF Conference on Computer Vision and Pattern Recognition Workshops (CVPRW), pp. 2429–2438. IEEE, Seattle, WA, USA, June 2020. https://doi.org/10.1109/CVPRW50498.2020.00293. https://ieeexplore.ieee.org/document/9150860/

21. Mahdavi, M., Zanibbi, R., Mouchere, H., Viard-Gaudin, C., Garain, U.: ICDAR 2019 CROHME + TFD: competition on recognition of handwritten mathematical expressions and typeset formula detection. In: 2019 International Conference on Document Analysis and Recognition (ICDAR), pp. 1533–1538. IEEE, Sydney, Australia, September 2019. https://doi.org/10.1109/ICDAR.2019.00247. https://ieeexplore.ieee.org/document/8978036/

22. Mansouri, B., Novotný, V., Agarwal, A., Oard, D.W., Zanibbi, R.: Overview of ARQMath-3 (2022): third CLEF lab on answer retrieval for questions on math (working notes version). In: CLEF (Working Notes). CEUR Workshop Proceedings, vol. 3180, pp. 1–27. CEUR-WS.org (2022)

23. Mouchère, H., Zanibbi, R., Garain, U., Viard-Gaudin, C.: Advancing the state of the art for handwritten math recognition: the CROHME competitions, 2011–2014. Int. J. Doc. Anal. Recogn. (IJDAR) 19(2), 173–189 (2016). https://doi.org/10.1007/s10032-016-0263-5

24. Nguyen, C.T., Truong, T.-N., Nguyen, H.T., Nakagawa, M.: Global context for improving recognition of online handwritten mathematical expressions. In: Lladós, J., Lopresti, D., Uchida, S. (eds.) ICDAR 2021. LNCS, vol. 12822, pp. 617–631. Springer, Cham (2021). https://doi.org/10.1007/978-3-030-86331-9_40

25. Nishizawa, G., Liu, J., Diaz, Y., Dmello, A., Zhong, W., Zanibbi, R.: MathSeer: a math-aware search interface with intuitive formula editing, reuse, and lookup. In: Jose, J.M., et al. (eds.) ECIR 2020. LNCS, vol. 12036, pp. 470–475. Springer, Cham (2020). https://doi.org/10.1007/978-3-030-45442-5_60

26. Peng, S., Gao, L., Yuan, K., Tang, Z.: Image to LaTeX with graph neural network for mathematical formula recognition. In: Lladós, J., Lopresti, D., Uchida, S. (eds.) ICDAR 2021. LNCS, vol. 12822, pp. 648–663. Springer, Cham (2021). https://doi.org/10.1007/978-3-030-86331-9_42

27. Sakshi, Kukreja, V.: A retrospective study on handwritten mathematical symbols and expressions: classification and recognition. Eng. Appl. Artif. Intell. **103**, 104292 (2021). https://doi.org/10.1016/j.engappai.2021.104292

28. Sasarak, C., et al.: min: a multimodal web interface for math search. In: Proceedings Human-Centered Information Retrieval (HCIR), Cambridge, MA, USA (2012). https://www.cs.rit.edu/~rlaz/files/HCIRPoster2012.pdf

29. Shah, A.K., Dey, A., Zanibbi, R.: A math formula extraction and evaluation framework for PDF documents. In: Lladós, J., Lopresti, D., Uchida, S. (eds.) ICDAR 2021. LNCS, vol. 12822, pp. 19–34. Springer, Cham (2021). https://doi.org/10.1007/978-3-030-86331-9_2

30. Suzuki, M., Tamari, F., Fukuda, R., Uchida, S., Kanahori, T.: INFTY: an integrated OCR system for mathematical documents. In: Proceedings of the 2003 ACM symposium on Document engineering, DocEng 2003, pp. 95–104. Association for Computing Machinery, New York, NY, USA, November 2003. https://doi.org/10.1145/958220.958239

31. Suzuki, M., Uchida, S., Nomura, A.: A ground-truthed mathematical character and symbol image database. In: ICDAR, pp. 675–679. IEEE Computer Society (2005)

32. Tang, J.M., Wu, J.W., Yin, F., Huang, L.L.: Offline handwritten mathematical expression recognition via graph reasoning network. In: Wallraven, C., Liu, Q., Nagahara, H. (eds.) Pattern Recognition. LNCS, pp. 17–31. Springer, Cham (2022). https://doi.org/10.1007/978-3-031-02375-0_2

33. Toyota, S., Uchida, S., Suzuki, M.: Structural analysis of mathematical formulae with verification based on formula description grammar. In: Bunke, H., Spitz, A.L. (eds.) DAS 2006. LNCS, vol. 3872, pp. 153–163. Springer, Heidelberg (2006). https://doi.org/10.1007/11669487_14

34. Veličković, P., Cucurull, G., Casanova, A., Romero, A., Liò, P., Bengio, Y.: Graph attention networks. In: International Conference on Learning Representations, February 2018

35. Wang, J., Du, J., Zhang, J., Wang, B., Ren, B.: Stroke constrained attention network for online handwritten mathematical expression recognition. Pattern Recogn. **119**, 108047 (2021). https://doi.org/10.1016/j.patcog.2021.108047

36. Wu, C., et al.: TDv2: a novel tree-structured decoder for offline mathematical expression recognition. Proc. AAAI Conf. Artif. Intell. **36**(3), 2694–2702 (2022). https://doi.org/10.1609/aaai.v36i3.20172

37. Wu, J.W., Yin, F., Zhang, Y.M., Zhang, X.Y., Liu, C.L.: Graph-to-graph: towards accurate and interpretable online handwritten mathematical expression recognition. In: Association for the Advancement of Artificial Intelligence, p. 9 (2021). https://www.aaai.org/AAAI21Papers/AAAI-3268.WuJW.pdf

38. Xie, S., Girshick, R., Dollar, P., Tu, Z., He, K.: Aggregated residual transformations for deep neural networks. In: 2017 IEEE Conference on Computer Vision and Pattern Recognition (CVPR), pp. 5987–5995. IEEE, Honolulu, HI, July 2017. https://doi.org/10.1109/CVPR.2017.634

39. Xu, K., et al.: Show, attend and tell: neural image caption generation with visual attention. In: Proceedings of the 32nd International Conference on International Conference on Machine Learning - Volume 37, ICML2015, pp. 2048–2057. JMLR.org, Lille, France, July 2015

40. Yu, D., et al.: Towards accurate scene text recognition with semantic reasoning networks. In: 2020 IEEE/CVF Conference on Computer Vision and Pattern Recognition (CVPR), pp. 12110–12119, June 2020. https://doi.org/10.1109/CVPR42600.2020.01213. ISSN: 2575-7075

41. Zanibbi, R., Blostein, D., Cordy, J.R.: Recognizing mathematical expressions using tree transformation. IEEE Trans. Pattern Anal. Mach. Intell. **24**(11), 1455–1467 (2002). https://doi.org/10.1109/TPAMI.2002.1046157. Conference Name: IEEE Transactions on Pattern Analysis and Machine Intelligence

42. Zanibbi, R., Aizawa, A., Kohlhase, M., Ounis, I., Topic, G., Davila, K.: NTCIR-12 MathIR task overview. In: NTCIR. National Institute of Informatics (NII) (2016)

43. Zanibbi, R., Blostein, D.: Recognition and retrieval of mathematical expressions. Int. J. Doc. Anal. Recogn. (IJDAR) **15**(4), 331–357 (2012). https://doi.org/10.1007/s10032-011-0174-4

44. Zanibbi, R., Mouchère, H., Viard-Gaudin, C.: Evaluating structural pattern recognition for handwritten math via primitive label graphs. In: Document Recognition and Retrieval XX, vol. 8658, p. 865817. International Society for Optics and Photonics, February 2013. https://doi.org/10.1117/12.2008409

45. Zanibbi, R., Yu, L.: Math spotting: retrieving math in technical documents using handwritten query images. In: ICDAR, pp. 446–451. IEEE Computer Society (2011)

46. Zhang, J., Du, J., Dai, L.: Track, attend, and parse (TAP): an end-to-end framework for online handwritten mathematical expression recognition. IEEE Trans. Multimedia **21**(1), 221–233 (2019). https://doi.org/10.1109/TMM.2018.2844689. Conference Name: IEEE Transactions on Multimedia

47. Zhang, J., Du, J., Yang, Y., Song, Y.Z., Dai, L.: SRD: a tree structure based decoder for online handwritten mathematical expression recognition. IEEE Trans. Multimedia **23**, 2471–2480 (2021). https://doi.org/10.1109/TMM.2020.3011316

48. Zhang, X., Gao, L., Yuan, K., Liu, R., Jiang, Z., Tang, Z.: A symbol dominance based formulae recognition approach for PDF documents. In: 2017 14th IAPR International Conference on Document Analysis and Recognition (ICDAR), vol. 01, pp. 1144–1149, November 2017. https://doi.org/10.1109/ICDAR.2017.189. ISSN: 2379-2140

49. Zhelezniakov, D., Zaytsev, V., Radyvonenko, O.: Online handwritten mathematical expression recognition and applications: a survey. IEEE Access **9**, 38352–38373 (2021). https://doi.org/10.1109/ACCESS.2021.3063413

50. Zie, Y., Mouchére, H., et al.: ICDAR CROHME 2023: competition on recognition of handwritten mathematical expressions. In: Proceedings ICDAR (2023) (in this proceedings, to appear)

PyramidTabNet: Transformer-Based Table Recognition in Image-Based Documents

Muhammad Umer[1], Muhammad Ahmed Mohsin[1(✉)], Adnan Ul-Hasan[2], and Faisal Shafait[1,2]

[1] School of Electrical Engineering and Computer Science (SEECS), National University of Sciences and Technology (NUST), Islamabad, Pakistan
{mumer.bee20seecs,mmohsin.bee20seecs,faisal.shafait}@seecs.edu.pk
[2] Deep Learning Laboratory, National Center of Artificial Intelligence (NCAI), Islamabad, Pakistan
adnan.ulhassan@seecs.edu.pk

Abstract. Table detection and structure recognition is an important component of document analysis systems. Deep learning-based transformer models have recently demonstrated significant success in various computer vision and document analysis tasks. In this paper, we introduce PyramidTabNet (PTN), a method that builds upon Convolutionless Pyramid Vision Transformer to detect tables in document images. Furthermore, we present a tabular image generative augmentation technique to effectively train the architecture. The proposed augmentation process consists of three steps, namely, clustering, fusion, and patching, for the generation of new document images containing tables. Our proposed pipeline demonstrates significant performance improvements for table detection on several standard datasets. Additionally, it achieves performance comparable to the state-of-the-art methods for structure recognition tasks.

Keywords: deep learning · image transformer · image processing · data augmentation · table augmentation · table segmentation · table detection · structure recognition

1 Introduction

Table detection and structure recognition are crucial tasks that have numerous applications in fields such as data extraction, document summarization, and information retrieval. Tables play a significant role in presenting data in a structured format, and they are frequently used in a variety of document types, such as research papers, reports, and financial statements. The automatic detection and recognition of tables and their structures is a complex task that requires the integration of several computer vision and pattern recognition techniques.

The most common approach to table detection is to use supervised machine learning techniques [2,12,14,22,24,28,31,33] to learn to identify tables based

G. A. Fink et al. (Eds.): ICDAR 2023, LNCS 14191, pp. 420–437, 2023.
https://doi.org/10.1007/978-3-031-41734-4_26

on features extracted from the input images. These features may include visual features, such as the presence of lines and boxes, as well as layout features, such as the position and size of elements within the document. These supervised machine-learning approaches have shown to be effective at detecting tables, but they often rely on large high-quality datasets and can be sensitive to variations in the input images.

Once a table has been detected, the next step is often to recognize its structure and extract its contents. This can be a complex task, as tables can vary significantly in terms of their layout, formatting, and content. Some approaches to table structure recognition involve analyzing the visual layout of the table, such as the presence and positioning of lines, boxes, and other visual elements. Other approaches may involve analyzing the semantic structure of the table, such as the relationships between different cells and the meaning of their contents.

Detecting and recognizing tables in document images is challenging due to various obstacles. These include inconsistent table structures, poor image quality, complex backgrounds, data imbalance, and insufficient annotated data. Inconsistent table structures result from varying layouts and structures, making it difficult for models to identify tables accurately. Poor image quality, such as blurring, distortion, and low resolution, can also impede recognition as well.

Overall, deep learning techniques present a promising solution for the accurate detection and recognition of tables in document images. However, current techniques are faced with a major challenge, which is the bias arising from image variability. This bias is a result of the complex and diverse nature of document analysis, despite the use of large amounts of training data. As such, there is a need for further research to overcome this challenge and improve the accuracy of document analysis techniques.

In this work, we have developed an end-to-end approach for table recognition in scanned document images that leverages the performance of convolution-less Pyramid Vision Transformer [35]. We also propose and integrate a novel tabular image generative augmentation technique to ensure that different types of table structures are uniformly learned by the model to address the challenges posed by the variability in table appearance and complex backgrounds.

The results of our approach demonstrate that it outperforms many of the recent works in this field, such as HybridTabNet [22], CasTabDetectoRS [14], and Document Image Transformer [18], on a variety of benchmarks. This serves as a measure of the effectiveness of our approach, as well as the value of the data augmentation techniques that we have incorporated into our method.

The rest of the paper is organized as follows: Sect. 2 provides the literature review and current advancements in table recognition using both CNNs and transformers and an overview of novel data augmentation pipelines. Section 3 provides an explanation of the proposed architecture; each component of the model is briefed in-depth. Section 4 provides an overview of the utilized datasets, along with the data augmentation techniques employed. Section 5 provides a comparison of our architecture against the current state-of-the-art along with its analysis. Section 6 concludes this paper along with ideas for future work and further enhancements.

2   Related Work

Table understanding is an important aspect of document image analysis. Deep learning-based approaches have been increasingly exploited to improve the generalization capabilities of table detection systems. This section aims to provide a brief overview of some of these methods.

Among the initial deep learning approaches, Gilani et al. [12] proposed a technique for table detection in which, document images are first subjected to pre-processing before being fed into an RPN. This network is designed to identify regions in the image that are likely to contain tables and later detected using a CNN. Arif and Shafait [2] introduced a method to enhance table detection by utilizing foreground and background features. The technique takes advantage of the fact that most tables contain numeric data, and utilizes color coding to differentiate between textual and numeric information within the table.

Traditional CNNs have a fixed receptive field, making table recognition challenging when tables are present in varying sizes and orientations. Deformable convolution [6], on the other hand, adapts its receptive field to the input, enabling the network to accommodate tables of any layout through customization of the receptive field. Employing a unique combination of Deformable CNN and Faster R-CNN, Siddiqui et al. [31] presented a novel strategy for table detection in documents that leverages the ability to recognize tables with any arrangement.

Qasim et al. [25] proposed using graph neural networks for table recognition tasks, combining the benefits of convolutional neural networks and graph networks for dealing with the input structure. Khan et al. [16] propose a deep learning solution for table structure extraction in document images, using a bidirectional GRU to classify inputs. Khan et al. [17] presented TabAug, a novel table augmentation approach that involves modifying table structure by replicating or deleting rows and columns. TabAug showed improved efficiency compared to conventional methods and greater control over the augmentation process.

Prasad et al. [24] proposed CascadeTabNet, a table detection system built upon Cascade Mask R-CNN HRNet framework and enhanced by transfer learning and image manipulation techniques. Nazir et al. [22] presented HybridTabNet, a pipeline comprising two stages: the first stage extracts features using the ResNeXt-101 network, while the second stage uses a Hybrid Task Cascade (HTC) to localize tables within the document images.

Zheng et al. [36] proposed Global Table Extractor (GTE), a technique that detects tables and recognizes cell structures simultaneously, using any object detection model. GTE-Table, a new training method, is used to improve table detection by incorporating cell placement predictions. Raja et al. [27] presented a novel object-detection-based deep learning model that is designed for efficient optimization and accurately captures the natural alignment of cells within tables. The author proposed a unique rectilinear graph-based formulation to enhance structure recognition to capture long-range inter-table relationships.

Image transformers have garnered a lot of popularity in computer vision and image processing tasks and recently, transformer-based models have been employed for document analysis as well. Smock et al. [32] utilized DEtection

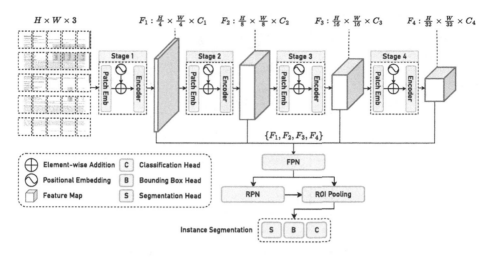

Fig. 1. Model architecture of PyramidTabNet – A convolution-less Pyramid Vision Transformer backbone is attached to a vanilla implementation of the Cascade Mask R-CNN framework to detect the instances and bounding boxes of document tables and their structural components.

TRansformer (DETR) [4] framework for table detection and structure recognition tasks. Document Image Transformer [18] proposed by Xu et al. utilized large-scale unlabeled images for document analysis tasks and achieved state-of-the-art results in document classification as well as table recognition.

Leveraging the success of transformers in the field of document analysis, we integrate a convolution-less transformer backbone with a vanilla Cascade Mask R-CNN framework. The following section provides a comprehensive examination of the end-to-end pipeline for table detection and structure recognition, including a demonstration of how inputs are processed through the architecture.

3   PyramidTabNet: Methodology

3.1   Architecture

Building upon the superiority of Transformer models demonstrated by the Pyramid Vision Transformer (PVT) [34] in dense prediction tasks, PyramidTabNet utilizes the updated PVT v2 architecture [35] with a 3×3 depth-wise convolution in its feed-forward network. The document image is first divided into non-overlapping patches and transformed into a sequence of patch embeddings. These embeddings are then infused with positional information and processed through multiple stages of the Pyramid Vision Transformer to form the backbone of PyramidTabNet. The output of the PVT v2 stage is reshaped to feature maps F_1, F_2, F_3 and F_4 with strides of 4, 8, 16, and 32 respectively. Lastly, the feature pyramid $\{F_1, F_2, F_3, F_4\}$ is forwarded to a vanilla Cascade Mask R-CNN [3] framework to perform instance segmentation as shown in Fig. 1.

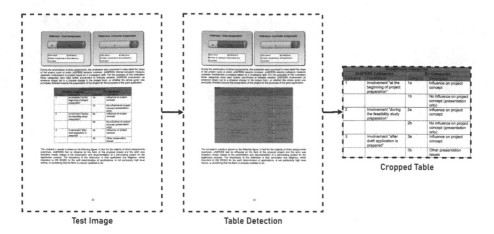

Fig. 2. Table detection pipeline – Instances of tables are detected on an input image and cropped to extract the tabular region.

The proposed end-to-end pipeline can be categorized into two phases: table detection and structure recognition. The details of each stage are discussed in the following sections, along with an exemplary forward pass of the input through the architecture.

Table Detection. In a single feed-forward pass of the input image (a document) to the model, the table detection module detects all instances of tables in the input and performs bounding box regression. The detected bounding box coordinates, in $[x_{min}, y_{min}, x_{max}, y_{max}]$ format, are then used to extract all the tables from the input, and intermediately saved. The bounding boxes are also post-processed to exclude any overlapping detections and undergo sequential expansion to align with the nearest table contour. Figure 2 shows the table detection pipeline on a sample image from ICDAR 2013 [13].

Table Structure Recognition. Detected tables are then passed on to the structure recognition stage, which detects all instances of table columns, table column headers, as well as cells. Overlapping bounding boxes of detected cells are merged into a single cell on the basis of the region area. The cells spanning multiple adjacent cells along its horizontal projection are marked as row identifiers and assigned an identity to capture the row number. sing the structural information of columns and the intersection of the generated cell projections with the detected columns, the row structure is inferred. The overall table structure is further processed by classifying the presence of cells in column and column headers, and the predicted structure is written to an XML file in the same format as in other state-of-the-art methods. Figure 3 shows the table structure recognition pipeline on the extracted table from the detection pipeline.

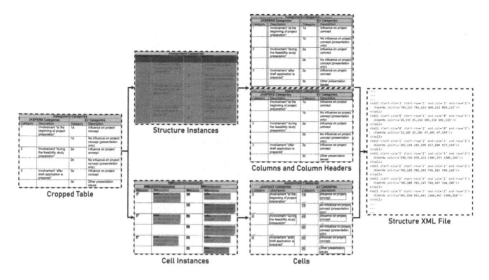

Fig. 3. Table structure recognition pipeline – Instances of cells, columns, and column headers are detected and the row structure is inferred based on the intersection of cell projections with the columns. The overall table structure is inferred based on the positional relation of the cells with the detected columns and generated rows.

3.2 Augmentation Strategy

In this section, we describe the augmentation techniques utilized in our proposed architecture, supplementing the data-hungry nature of transformers.

K-Means Clustering. K-means clustering is an unsupervised learning algorithm that is used to partition n observations into k clusters in which each observation belongs to the cluster with the nearest mean, serving as a prototype of the cluster. In the context of table images, the k-means method can be used to group images in the form of vectors based on visual similarity, thus reducing the overall variation in data that is fused together.

Fusion. Splicing of distinct tables in horizontal and vertical fashion followed by concatenation is collectively termed *fusion* in this paper. Vertical and horizontal lines are detected using probabilistic Hough lines transform [7]. The median horizontal and vertical set of line points in a sorted Hough lines array is selected as the cutoff point to achieve horizontal and vertical splices of a table image, respectively. The resultant images are then fed into an image resizing-concatenation pipeline to generate a new table.

Figure 4 shows an exemplar pipeline of fusion. A batch of images ($n = 2$) is randomly sampled from clusters formed in Sect. 3.2 and the median horizontal contour is selected as the cutoff point after detection of all possible lines, followed by cropping the image to achieve the maximum area. Cropped images are then

Fig. 4. Generation of new table images – Vertical and horizontal contours are detected on two randomly sampled table images from generated clusters. Tables are cropped to the median positional contour and adjacently joined to produce a new table image.

resized along the horizontal axis so that they match in width before they are concatenated to produce a new table image.

Patching. Augmented tables generated as a result of fusion are lastly patched onto existing dataset images. Figure 5 shows an exemplary pipeline of patching. An image is randomly sampled from the training data along with its inverted mask. The center of the largest area in the inverted mask is the point on which a randomly sampled fusion-generated table is pasted on. Collectively, we refer to this process as patching in this paper.

4 Experiments

In this section, we start with an introduction to the datasets utilized to demonstrate the efficacy of our architecture, followed by an introduction to the data augmentation techniques employed in our method. Later, we analyze the results on these datasets and compare them with state-of-the-art methods.

4.1 Datasets

In this section, we will discuss the datasets that are commonly used and publicly available for table detection and table structure recognition.

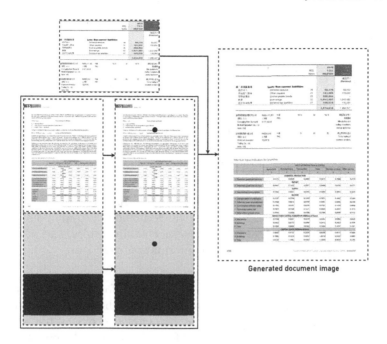

Generated document image

Fig. 5. Patching of table images – A document image along with its semantic information is fed into the pipeline and the best patch point is computed for the fused table to be pasted on in order to generate a new training sample.

ICDAR 2013. The ICDAR 2013 [13] dataset consists of 150 tables, with 75 from EU documents and 75 from US Government documents. The tables are defined by their rectangular coordinates on the page and can span multiple pages. The dataset includes two sub-tasks: identifying the location of tables and determining their structure. In our experiments, we will only utilize the dataset for structure recognition.

ICDAR 2017-POD. ICDAR 2017-POD [11] is a widely used dataset for evaluating various table detection methods. It is significantly larger than the ICDAR 2013 table dataset, containing a total of 2,417 images that include figures, tables, and formulae. This dataset is typically divided into 1,600 images with 731 tabular areas for training and 817 images with 350 tabular regions for testing.

ICDAR 2019 cTDaR. The cTDaR [10] datasets provide separate tracks, both for table detection and table structure recognition. Track A, which targets the task of table detection, is further divided into archival documents and modern documents. In this paper, we focus on modern documents where table annotations are provided for each image. The modern subset consists of 600 training and 240 test images, which contain a broad variety of PDF files. Variability in the images is further enhanced by supplementing English documents with

Chinese documents, both of various formats, including scanned document images, digitally composed documents, etc.

Marmot. The Marmot [8] dataset is a collection of 2,000 PDF pages that includes a net balance of positive and negative samples. It features a diverse range of table types, including ruled and non-ruled, horizontal and vertical, and tables that span multiple columns. The dataset also includes tables that are found inside columns. For evaluation, we utilized the corrected version of this dataset, as in [28], which contains 1,967 images.

UNLV. The UNLV [29] dataset is a widely recognized collection of document images in the field of document analysis, which includes a total of almost 10,000 images. However, only a subset of 427 images contains tables, and in the experiments, only those images with tabular information were used.

TableBank. TableBank [19] proposed an approach for automatically creating a dataset using weak supervision, which generates high-quality labeled data for a diverse range of domains such as business documents, official filings, research papers, and others, making it highly useful for large-scale table recognition tasks. This dataset is comprised of 417,234 labeled tables, along with their corresponding original documents from various domains.

PubLayNet. PubLayNet [37] is a high-quality dataset designed for document layout analysis. The dataset is composed of 335,703 training images, 11,245 validation images, and 11,405 testing images. For table detection, only those images containing at least one table were used, resulting in a total of 86,460 images. The evaluation metrics used in this dataset follow the COCO [20] evaluation protocol, rather than precision, recall, and F1-score, as is used in other table detection datasets.

4.2   Settings

in PyTorch v1.11.0 and were conducted on Google Colaboratory platform with a P100 PCIE GPU of 16 GB GPU memory, Intel® Xeon® CPU @ 2.30 GHz, and 12.72 GB of RAM. The MMDetection toolbox was used to implement the proposed architecture.

Hyperparameters play a crucial role in determining the performance of a deep learning model. They are adjustable settings that are not learned from the data, and must be set before training begins. The choice of hyperparameters can significantly impact the performance of a model, and their optimization is often necessary to achieve the best results. In this paper, we selected hyperparameters based on prior knowledge and empirical studies. This approach has been shown to be effective in selecting reasonable values for the hyperparameters and can save time and resources compared to exhaustive search methods.

The model was optimized using the AdamW algorithm with a batch size of 1 over 180,000 iterations. The learning rate was decayed using a linear decay schedule, with the initial learning rate, betas, epsilon, and weight decay set to 1e-4/1.4, (0.9, 0.999), 1e-8, and 1e-4, respectively. Further studies can be conducted to explore other hyperparameter settings to achieve even better results.

The data augmentation techniques described in Sect. 3.2 were implemented in conjunction with the augmentation policies used to train the DETR [4] architecture. Two auto-augmentation policies were adopted during the training phase. The first policy rescaled the shorter side of each image to a random number in the set {480, 512, 544, 576, 608, 640, 672, 704, 736, 768} while maintaining the aspect ratio. The second policy rescaled the image to a random number in the set {400, 500, 600} before applying an absolute range cropping window of size (384, 600). During the testing phase, the longer side of each image was rescaled to 1024 while maintaining the aspect ratio.

5   Results and Analysis

In this section, we discuss the results of the proposed architecture on the datasets introduced in Sect. 4.1 and compare them with state-of-the-art methods. We also evaluate the efficacy of our augmentation pipeline by training the proposed architecture using different augmentation methodologies.

To assess the effectiveness of the proposed tabular image generative augmentation pipeline, the proposed architecture was trained using three distinct methodologies:

1. **Non-Augmented (NA):** Training images are fed into the transformer without any modifications.
2. **Standard (S):** Standard augmentation techniques such as variations in brightness, exposure, contrast, jitter, etc. combined with strategies employed in DETR [4].
3. **Generative (G) (ours):** Our proposed augmentation pipeline, which consists of sequential clustering, patching, and fusion to generate new images in combination with strategies employed in DETR [4].

5.1   Table Detection

Table 1 presents a summary of the table detection results on various datasets. It does not include table detection performance comparison on PubLayNet and ICDAR 2019 cTDaR as they follow different evaluation criteria and are presented separately. The model is initially trained on a conglomerate of training images of PubLayNet, TableBank, and ICDAR 2019 cTDaR dataset. Additionally, the document images generated by our augmentation technique are also included in the initial training state. Following the strategy of other state-of-the-art methods, we fine-tune these weights on training images of each of the table detection datasets for the computation of respective evaluation metrics.

Table 1. Table detection performance comparison summary – All metrics are computed using models fine-tuned on the training samples of respective datasets – NA: Non-Augmented, S: Standard, G: Generative.

Dataset	Method	Precision	Recall	F1-Score
ICDAR 2017 POD @ IoU = 0.8	CDeC-Net [1]	89.9	96.9	93.4
	DeepTabStR [30]	96.5	97.1	96.8
	YOLOv3 [31]	97.8	97.2	97.5
	HybridTabNet [22]	87.8	**99.3**	93.2
	PyramidTabNet (NA)	95.3	94.7	95.0
	PyramidTabNet (S)	97.8	97.1	97.4
	PyramidTabNet (G)	**99.8**	**99.3**	**99.5**
Marmot @ IoU = 0.5	DeCNT [31]	94.6	84.9	89.5
	CDeC-Net [1]	77.9	94.3	86.1
	HybridTabNet [22]	88.2	91.5	89.8
	CasTabDetectoRS [14]	96.5	**95.2**	95.8
	PyramidTabNet (NA)	92.7	91.1	91.9
	PyramidTabNet (S)	94.6	93.3	93.9
	PyramidTabNet (G)	**97.7**	94.9	**96.3**
UNLV @ IoU = 0.5	DeCNT [31]	91.0	94.6	92.8
	CDeC-Net [1]	91.5	97.0	94.3
	HybridTabNet [22]	**96.2**	96.1	**95.6**
	CasTabDetectoRS [14]	92.8	96.4	94.6
	PyramidTabNet (NA)	89.4	93.2	91.3
	PyramidTabNet (S)	90.7	95.6	93.1
	PyramidTabNet (G)	92.1	**98.2**	95.1
TableBank (LaTeX & Word) @ IoU = 0.5	Li et al. [19]	90.4	95.9	93.1
	CascadeTabNet [24]	95.7	94.4	94.3
	HybridTabNet [22]	95.3	97.6	96.5
	CasTabDetectoRS [14]	98.2	97.4	97.8
	PyramidTabNet (NA)	94.4	94.1	94.2
	PyramidTabNet (S)	96.5	95.6	96.0
	PyramidTabNet (G)	**98.9**	**98.2**	**98.5**

All the metrics in Table 1 are computed at the same IoU threshold for a single multi-cell row (same dataset). In the evaluation of the ICDAR 2017-POD (Page Object Detection) dataset, we achieved an F1-score of 99.5 on the detection of the table class at 0.8 IoU threshold, pushing further the state-of-the-art metrics. It should be noted that the results reported are after the inclusion of post-processing techniques, as also observed in the original competition.

Table 2. Table detection performance comparison on PubLayNet – Evaluation metrics follow the same protocol as in the COCO [20] detection challenge – NA: Non-Augmented, S: Standard, G: Generative.

Method	$AP^{0.5:0.95}$	$AP^{0.75}$	$AP^{0.95}$
CDeC-Net [1]	96.7	-	-
RobusTabNet [21]	97.0	97.8	92.0
DiT-L (Cascade) [18]	97.8	-	-
PyramidTabNet (NA)	94.6	95.4	92.7
PyramidTabNet (S)	96.9	97.6	94.3
PyramidTabNet (G)	**98.1**	**98.8**	**96.4**

On the Marmot dataset, our model achieves the highest precision of 97.7 and F1-score of 96.3 at 0.5 IoU threshold. The direct comparison of our results with CasTabDetectoRS [14] and HybridTabNet [22] on the Marmot dataset proves that we have pronounced the new state-of-the-art.

On the UNLV dataset, we achieved the highest recall of 98.2 at 0.5 IoU threshold, indicating that our method correctly identified the highest number of tables in the dataset. However, a decrease in precision was observed in comparison to the performance on other datasets. We attribute this to the presence of a large proportion of low-quality document images in the UNLV dataset. Our model, which was trained on a diverse set of modern document images, may not have the ability to fine-tune on the UNLV dataset as well as it does on other modern datasets.

On the TableBank dataset, we achieved the highest precision, recall, and F1-score over a 0.5 IoU threshold, achieving the new state-of-the-art. We evaluate the TableBank dataset on both of its parts, LaTeX and Word document image subsets, as we believe it signifies the robustness of the proposed method to different types of modern document images.

The results of table detection on the PubLayNet dataset are shown in Table 2. Utilizing the same evaluation protocol as the COCO detection challenge [20], our method achieved the highest AP of 96.4 at 0.95 IoU threshold and a value of 98.1 for precision averaged over IoUs from 0.5 to 0.95 in steps of 0.05. These results further push the state-of-the-art and demonstrate the fine-grained object detection capabilities of our method.

The results of table detection on the ICDAR 2019 cTDaR dataset are shown in Table 3. As the number of samples in this dataset is relatively small, it aims to evaluate the few-short learning capabilities of models under low-resource scenarios. In Table 3, Our model performs better than the current baselines on all fronts, while observing an increase of 0.9% in weighted F1-score over the recent cascaded DiT-L [18], pushing further the state-of-the-art. It is worth noting that, like DiT, metrics of IoU@{0.9} are significantly performant, indicating that the proposed architecture has better fine-grained object detection capabilities.

Table 3. Table detection performance comparison on ICDAR 2019 cTDaR – F1-scores are computed at different IoU thresholds and the weighted F1-score is used to determine the overall performance – NA: Non-Augmented, S: Standard, G: Generative.

Method	$IoU_{0.6}$	$IoU_{0.7}$	$IoU_{0.8}$	$IoU_{0.9}$	$wF1$
CascadeTabNet [24]	94.3	93.4	92.5	90.1	90.1
HybridTabNet [22]	95.3	94.2	93.3	92.7	92.8
CDeC-Net [1]	95.9	95.6	95.0	91.5	94.3
GTE [36]	96.9	96.9	95.7	91.9	95.1
TableDet [9]	-	-	95.5	89.5	94.0
DiT-L (Cascade) [18]	97.9	97.2	97.0	93.9	96.3
PyramidTabNet (NA)	95.8	95.6	94.4	92.3	94.3
PyramidTabNet (S)	97.0	96.9	96.3	94.1	95.9
PyramidTabNet (G)	**98.7**	**98.7**	**98.0**	**94.5**	**97.2**

5.2   Structure Recognition

To evaluate the efficacy of our model on table structure recognition relative to other state-of-the-art methods, each cell is marked into its corresponding location in the detected table. The location of each cell is represented by [*start row, end row, start column, end column, box coordinates*]. We employ the same IoU threshold as in other methods to compute precision, recall, and F1-score.

The results of structure recognition on the ICDAR 2013 dataset are shown in Table 4. We achieved an F1-score of 93.8, lacking just behind GuiderTSR [15], which proved the efficacy of estimation of viable anchors for the detection of rows and columns over naively applied object detection algorithms.

Table 4. Table structure recognition results on ICDAR 2013 – Performance of fine-tuned models is compared without post-processing techniques as is done in the current state-of-the-art method – NA: Non-Augmented, S: Standard, G: Generative.

Method	Precision	Recall	F1-Score
SPLERGE [33]	**96.9**	90.1	93.4
TableNet [23]	92.2	89.9	91.0
GraphTSR [5]	88.5	86.0	87.2
TabStruct-Net [26]	92.7	91.1	91.9
GuidedTSR [15]	93.6	94.4	**94.2**
PyramidTabNet (NA)	91.1	90.4	90.7
PyramidTabNet (S)	91.4	92.6	92.0
PyramidTabNet (G)	92.3	**95.3**	93.8

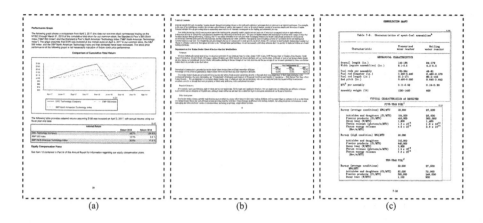

Fig. 6. Table detection examples with (a) incorrect detection, (b) unidentified table, (c) partial detection.

Source	Lead time (years)			
	1	2	3	4
	Absolute value of percentage difference between actual and projected values			
Projections of Education Statistics to 2017	†	0.7	1.1	1.4
Projections of Education Statistics to 2018	0.4	0.7	0.8	1.1
Projections of Education Statistics to 2019	#	0.1	0.2	†
Projections of Education Statistics to 2020	0.2	0.4	†	†
	Mean absolute percentage error			
Example	0.2	0.5	0.7	1.3

(a)

Sample Group	Some Year 1 Head Start Participation	No Year 1 Head Start Participation	Total
All Randomly Assigned (N=4,667):			
3-Year-Old Cohort			
Head Start Group	85.1%	14.9%	100%
Control Group	17.3%	82.7%	100%
4-Year-Old Cohort			
Head Start Group	79.8%	20.2%	100%
Control Group	13.9%	86.1%	100%

(b)

Fig. 7. Structure recognition examples with (a) column error (b) row error.

5.3 Analysis

In this section, we analyze the detection outputs of the proposed architecture and provide potential reasons for the incorrectly detected tables and their structural components. It provides valuable insights into the strengths and limitations of the model and will be useful for guiding future improvements to the model.

The three common types of errors in table detection are depicted in Fig. 6. In Fig. 6a, our model mistakenly identifies the figure legend as a table due to the presence of the x axis label above it. Conversely, in Fig. 6b, the model fails to detect a table that is very small in relation to the overall image size. Figure 6c showcases the instance when the table size encompasses the entire image and our model fails to detect it as a whole, instead recognizing it in parts. This behavior is attributed to the use of patching augmentation techniques, which ensure the presence of a minimum of two tables in a single document image.

The error types in structure recognition are illustrated in Fig. 7. The figure depicts the issues that arise from under-identified rows or over-identified columns. As shown in Fig. 7a, the failure of our model to detect the second column header cell leads to a partially broken structure, demonstrating the critical dependence of our structure recognition pipeline on correctly identifying all column header cells. In Fig. 7b, post-processing techniques result in the merging of two cells from separate rows, leading to an extra row in the end table structure. Despite this, post-processing techniques only counteract the model predictions for a limited number of test images. Thus, it is retained as the final stage of our proposed method after empirical evaluation.

6   Conclusion and Future Work

In this paper, we present PyramidTabNet, an end-to-end approach to table detection and structure recognition in image-based documents based on convolution-less image transformers. To make up for the data-hungry nature of transformers, PyramidTabNet employs a tabular image generative augmentation technique, resulting in an architecture with fine-grained object detection capabilities. Consequently, and through experimental results, we have shown that PyramidTab-Net outperforms several strong baselines in the task of table detection, especially at a high IoU threshold, and achieves competitive and comparable performance on table structure recognition tasks.

For future work, we will study the effects of training PyramidTabNet on even larger datasets to further push the state-of-the-art results on table recognition. We are also exploring the effects of integrating AI image upscaling on detected table images to improve the evaluation metrics on the task of table structure recognition, however, with an added latency overhead.

References

1. Agarwal, M., Mondal, A., Jawahar, C.: CDeC-Net: Composite Deformable Cascade Network for Table Detection in Document Images. In: 2020 25th International Conference on Pattern Recognition (ICPR), pp. 9491–9498. IEEE (2021)
2. Arif, S., Shafait, F.: Table Detection in Document Images using Foreground and Background Features. In: 2018 20th Digital Image Computing: Techniques and Applications (DICTA), pp. 1–8. IEEE (2018)
3. Cai, Z., Vasconcelos, N.: Cascade R-CNN: Delving Into High Quality Object Detection. In: 2018 Conference on Computer Vision and Pattern Recognition (CVPR), pp. 6154–6162. IEEE (2018)
4. Carion, N., Massa, F., Synnaeve, G., Usunier, N., Kirillov, A., Zagoruyko, S.: End-to-End Object Detection With Transformers. In: 2020 16th European Conference on Computer Vision (ECCV), pp. 213–229. Springer (2020)
5. Chi, Z., Huang, H., Xu, H.D., Yu, H., Yin, W., Mao, X.L.: Complicated Table Structure Recognition. arXiv preprint arXiv:1908.04729 (2019)

6. Dai, J., et al.: Deformable Convolutional Networks. In: 2017 16th International Conference on Computer Vision (ICCV), pp. 764–773. IEEE (2017)
7. Duan, D., Xie, M., Mo, Q., Han, Z., Wan, Y.: An Improved Hough Transform for Line Detection. In: 2010 International Conference on Computer Application and System Modeling (ICCASM). vol. 2, pp. 354–357 (2010)
8. Fang, J., Tao, X., Tang, Z., Qiu, R., Liu, Y.: Dataset, Ground-Truth and Performance Metrics for Table Detection Evaluation. In: 2012 10th IAPR International Workshop on Document Analysis Systems (DAS), pp. 445–449 (2012)
9. Fernandes, J., Simsek, M., Kantarci, B., Khan, S.: TableDet: An End-to-End Deep Learning Approach for Table Detection and Table Image Classification in Data Sheet Images. In: Neurocomputing. vol. 468, pp. 317–334. Elsevier (2022)
10. Gao, L., et al.: ICDAR 2019 Competition on Table Detection and Recognition (cTDaR). In: 2019 16th International Conference on Document Analysis and Recognition (ICDAR), pp. 1510–1515 (2019)
11. Gao, L., Yi, X., Jiang, Z., Hao, L., Tang, Z.: ICDAR 2017 Competition on Page Object Detection. In: 2017 14th IAPR International Conference on Document Analysis and Recognition (ICDAR). vol. 1, pp. 1417–1422 (2017)
12. Gilani, A., Qasim, S.R., Malik, I., Shafait, F.: Table Detection Using Deep Learning. In: 2017 14th IAPR International Conference on Document Analysis and Recognition (ICDAR). vol. 1, pp. 771–776. IEEE (2017)
13. Göbel, M., Hassan, T., Oro, E., Orsi, G.: ICDAR 2013 Table Competition. In: 2013 12th International Conference on Document Analysis and Recognition (ICDAR), pp. 1449–1453 (2013)
14. Hashmi, K.A., Pagani, A., Liwicki, M., Stricker, D., Afzal, M.Z.: CasTabDetectoRS: Cascade Network for Table Detection in Document Images With Recursive Feature Pyramid and Switchable Atrous Convolution. In: Journal of Imaging. vol. 7, p. 214. MDPI (2021)
15. Hashmi, K.A., Stricker, D., Liwicki, M., Afzal, M.N., Afzal, M.Z.: Guided Table Structure Recognition Through Anchor Optimization. In: IEEE Access. vol. 9, pp. 113521–113534. IEEE (2021)
16. Khan, S.A., Khalid, S.M.D., Shahzad, M.A., Shafait, F.: Table Structure Extraction with Bi-Directional Gated Recurrent Unit Networks. In: 2019 15th IAPR International Conference on Document Analysis and Recognition (ICDAR), pp. 1366–1371. IEEE (2019)
17. Khan, U., Zahid, S., Ali, M.A., Ul-Hasan, A., Shafait, F.: TabAug: Data Driven Augmentation for Enhanced Table Structure Recognition. In: 2021 16th IAPR International Conference on Document Analysis and Recognition (ICDAR). vol. 2, pp. 585–601. Springer (2021)
18. Li, J., Xu, Y., Lv, T., Cui, L., Zhang, C., Wei, F.: DiT: Self-Supervised Pre-training for Document Image Transformer. In: 2022 30th ACM International Conference on Multimedia (ACM MM), pp. 3530–3539 (2022)
19. Li, M., Cui, L., Huang, S., Wei, F., Zhou, M., Li, Z.: TableBank: Table Benchmark for Image-Based Table Detection and Recognition. In: 2020 12th Language Resources and Evaluation Conference (LREC), pp. 1918–1925 (2020)
20. Lin, T.-Y., et al.: Microsoft COCO: common objects in context. In: Fleet, D., Pajdla, T., Schiele, B., Tuytelaars, T. (eds.) ECCV 2014. LNCS, vol. 8693, pp. 740–755. Springer, Cham (2014). https://doi.org/10.1007/978-3-319-10602-1_48
21. Ma, C., Lin, W., Sun, L., Huo, Q.: Robust Table Detection and Structure Recognition from Heterogeneous Document Images. In: Pattern Recognition. vol. 133, p. 109006. Elsevier (2023)

22. Nazir, D., Hashmi, K.A., Pagani, A., Liwicki, M., Stricker, D., Afzal, M.Z.: Hybridtabnet: Towards Better Table Detection in Scanned Document Images. In: Applied Sciences. vol. 11, p. 8396. MDPI (2021)
23. Paliwal, S.S., Vishwanath, D., Rahul, R., Sharma, M., Vig, L.: TableNet: Deep Learning Model for End-To-End Table Detection and Tabular Data Extraction from Scanned Document Images. In: 2019 15th International Conference on Document Analysis and Recognition (ICDAR), pp. 128–133. IEEE (2019)
24. Prasad, D., Gadpal, A., Kapadni, K., Visave, M., Sultanpure, K.: CascadeTabNet: An Approach for End-to-End Table Detection and Structure Recognition from Image-Based Documents. In: 2020 Conference on Computer Vision and Pattern Recognition Workshops (CVPRW), pp. 572–573 (2020)
25. Qasim, S.R., Mahmood, H., Shafait, F.: Rethinking Table Recognition Using Graph Neural Networks. In: 2019 15th IAPR International Conference on Document Analysis and Recognition (ICDAR), pp. 142–147. IEEE (2019)
26. Raja, S., Mondal, A., Jawahar, C.V.: Table structure recognition using top-down and bottom-up cues. In: Vedaldi, A., Bischof, H., Brox, T., Frahm, J.-M. (eds.) ECCV 2020. LNCS, vol. 12373, pp. 70–86. Springer, Cham (2020). https://doi.org/10.1007/978-3-030-58604-1_5
27. Raja, S., Mondal, A., Jawahar, C.: Visual Understanding of Complex Table Structures from Document Images. In: Proceedings of the IEEE/CVF Winter Conference on Applications of Computer Vision, pp. 2299–2308 (2022)
28. Schreiber, S., Agne, S., Wolf, I., Dengel, A., Ahmed, S.: DeepDeSRT: Deep Learning for Detection and Structure Recognition of Tables in Document Images. In: 2017 14th IAPR International Conference on Document Analysis and Recognition (ICDAR), vol. 1, pp. 1162–1167 (2017)
29. Shahab, A., Shafait, F., Kieninger, T., Dengel, A.: An Open Approach Towards The Benchmarking of Table Structure Recognition Systems. In: Proceedings of the 9th IAPR International Workshop on Document Analysis Systems, pp. 113–120 (2010)
30. Siddiqui, S.A., Fateh, I.A., Rizvi, S.T.R., Dengel, A., Ahmed, S.: DeepTabStR: Deep Learning Based Table Structure Recognition. In: 2019 15th IAPR International Conference on Document Analysis and Recognition (ICDAR), pp. 1403–1409 (2019)
31. Siddiqui, S.A., Malik, M.I., Agne, S., Dengel, A., Ahmed, S.: DeCNT: Deep Deformable CNN for Table Detection. In: IEEE Access. vol. 6, pp. 74151–74161. IEEE (2018)
32. Smock, B., Pesala, R., Abraham, R.: PubTables-1M: Towards Comprehensive Table Extraction from Unstructured Documents. In: 2022 Conference on Computer Vision and Pattern Recognition (CVPR), pp. 4634–4642 (2022)
33. Tensmeyer, C., Morariu, V.I., Price, B., Cohen, S., Martinez, T.: Deep Splitting and Merging for Table Structure Decomposition. In: 2019 15th IAPR International Conference on Document Analysis and Recognition (ICDAR), pp. 114–121. IEEE (2019)
34. Wang, W., et al.: Pyramid Vision Transformer: A Versatile Backbone for Dense Prediction Without Convolutions. In: 2021 17th International Conference on Computer Vision (ICCV), pp. 568–578. IEEE (2021)
35. Wang, W., et al.: PVT v2: improved baselines with pyramid vision transformer. Comput. Visual Media 8, 1–10 (2022). https://doi.org/10.1007/s41095-022-0274-8

36. Zheng, X., Burdick, D., Popa, L., Zhong, X., Wang, N.X.R.: Global Table Extractor (GTE): A Framework for Joint Table Identification and Cell Structure Recognition using Visual Context. In: 2021 Winter Conference on Applications of Computer Vision (WACV), pp. 697–706 (2021)
37. Zhong, X., Tang, J., Yepes, A.J.: PubLayNet: Largest Dataset Ever for Document Layout Analysis. In: 2019 15th IAPR International Conference on Document Analysis and Recognition (ICDAR), pp. 1015–1022. IEEE (2019)

Line Graphics Digitization: A Step Towards Full Automation

Omar Moured[1,2]([✉]), Jiaming Zhang[1], Alina Roitberg[1], Thorsten Schwarz[2], and Rainer Stiefelhagen[1,2]

[1] CV:HCI Lab, Karlsruhe Institute of Technology, Karlsruhe, Germany
{jiaming.zhang,alina.roitberg}@kit.edu
[2] ACCESS@KIT, Karlsruhe Institute of Technology, Karlsruhe, Germany
{omar.moured,thorsten.schwarz,rainer.stiefelhagen}@kit.edu
https://github.com/moured/Document-Graphics-Digitization.git

Abstract. The digitization of documents allows for wider accessibility and reproducibility. While automatic digitization of document layout and text content has been a long-standing focus of research, this problem in regard to graphical elements, such as statistical plots, has been under-explored. In this paper, we introduce the task of fine-grained visual understanding of mathematical graphics and present the *Line Graphics (LG)* dataset, which includes pixel-wise annotations of 5 coarse and 10 fine-grained categories. Our dataset covers 520 images of mathematical graphics collected from 450 documents from different disciplines. Our proposed dataset can support two different computer vision tasks, *i.e.*, *semantic segmentation* and *object detection*. To benchmark our LG dataset, we explore 7 state-of-the-art models. To foster further research on the digitization of statistical graphs, we will make the dataset, code and models publicly available to the community.

Keywords: Graphics Digitization · Line Graphics · Semantic Segmentation · Object Detection

1 Introduction

With the rapid growth of information available online[1], access to knowledge has never been easier. However, as the volume of information continues to grow, there is a need for more efficient ways to extract useful information from documents such as papers and presentation slides. This is particularly important for individuals with special needs, such as visually impaired individuals [23], for whom traditional methods of accessing information may not be feasible (Fig. 1).

During courses, graphs are a vital supplement to lecturers' speech as they effectively summarize complex data or visualize mathematical functions. However, one downside of this medium is the difficulty of automatic information extraction, as graphs contain very fine-grained elements, such as fine lines, small

---

[1] https://www.statista.com/statistics/871513/worldwide-data-created/.

© The Author(s), under exclusive license to Springer Nature Switzerland AG 2023
G. A. Fink et al. (Eds.): ICDAR 2023, LNCS 14191, pp. 438–453, 2023.
https://doi.org/10.1007/978-3-031-41734-4_27

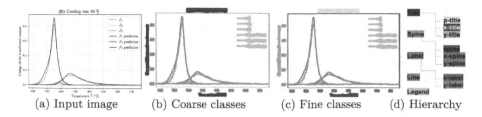

(a) Input image   (b) Coarse classes   (c) Fine classes   (d) Hierarchy

Fig. 1. Hierarchical semantic segmentation on the proposed Line Graphics (LG) benchmark. (a) Each image includes coarse-to-fine two-level semantic segmentation, *i.e.*, (b) coarse 5 classes and (c) 10 fine classes. (d) The coarse-to-fine hierarchy shows the two-level semantic segmentation classes.

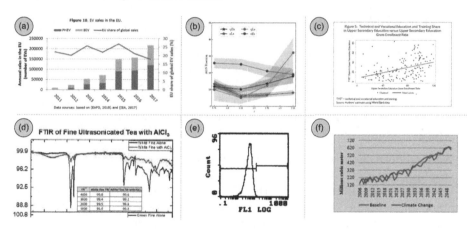

Fig. 2. Diverse mathematical graphics covered in our Line Graphics (LG) dataset, including 100 bar charts (a), 320 line graphics (b, d–f) and 100 scatter plots (c). These samples pose significant challenges for existing document analysis methods.

numbers or axes descriptions, while the traditional document analysis frameworks focus on coarse structures within complete pages [5,9,34] or slides [18,19]. The process of separating distinct regions of a plot and assigning them a semantic meaning at a pixel-level, known as graph segmentation, is an important prerequisite step for graph understanding. One application of using pixel-level data to fully automate the process is to generate an imposed document or 2D refreshable tactile display that can be easily interpreted through touch for people with blindness or visual impairment. Hence, end-to-end full automation of plot digitization could be achieved.

Presumably, due to the lack of annotated datasets for fine-grained analysis of plots, the utilization of modern deep semantic segmentation architectures has been rather overlooked in the context of mathematical graphs.

In this paper, we introduce the task of fine-grained visual understanding of mathematical graphics and present the Line Graphics (LG) dataset, which includes pixel-wise annotations of 10 different categories. Our dataset covers 520 images of mathematical graphics collected from 450 documents from different disciplines, such as physics, economics, and engineering. Figure 2 provides several examples of statistical plots collected in our dataset. By providing pixel-wise and bounding box annotations, we enable our dataset to support two different computer vision tasks: *instance, semantic segmentation* and *object detection.*

To benchmark our LG dataset, we explore 7 state-of-the-art models, including efficiency- and accuracy-driven frameworks, (e.g., MobileNetV3 [20] and Seg-NeXt [16]) with SegNeXt yielding the best results with 67.56% mIoU. Our results show that while we have achieved high overall accuracy in our models, the accuracy varies depending on the type of object. Specifically, we found that spine-related categories and plot title were the hardest to recognize accurately. To foster further research on the digitization of statistical graphs, we will make the dataset, code, and models publicly available to the community.

The key findings and contributions of this paper are can be summarized as:

- We introduce the task of fine-grained visual understanding of mathematical graphics, aimed at reducing manual user input when digitalizing documents.
- We collect and publicly release the Line Graphics (LG) dataset as a benchmark for semantic segmentation and object detection in line graphics. The dataset includes plots from papers and slides from various fields and is annotated with 10 fine-grained classes at both pixel and bounding box levels.
- We perform extensive evaluations on 7 state-of-the-art semantic segmentation models, analyzing the impact of factors such as image resolution and category types on the performance. Our findings demonstrate the feasibility of the proposed task, with the top model achieving a mean Intersection over Union of 67.56%. However, further advancement is needed in certain categories, such as plot title or spines, as well as for low-resolution data.

2   Related Work

2.1   Document Graphics Analysis

Visual document analysis is a well-studied research area, mostly focusing on text [21,27] and layout analysis [4,5,10] of complete pages originating from scientific papers [7,14], presentation slides [19], magazines [10], historical handwritten documents [4,5] or receipts [21]. In comparison, the research of chart analysis is more limited, with an overview of the existing approaches provided by [13]. In particular, several learning-based methods have been used for (1) localizing and extracting charts from pages [9,24,31], or (2) harvesting text and tabular data from charts [8,14,25,28]. Seweryn et al. [30] propose a framework covering chart

Table 1. Overview of the five most related datasets. Our LG dataset for the first time addresses fine-grained semantic segmentation of line graphs.

Dataset	Task	Labels Type	Year
PDFFigures 2.0 [9]	Figure and caption detection	bounding boxes	2016
Poco et al. [29]	Text Pixel Classification and localization	text binary mask and metadata	2017
Dai et al. [11]	Classification and text recognition	bounding boxes	2018
DocFigure [22]	Figure classification	one hot encoded images	2019
ICDAR Charts 2019 [12]	Text content extraction and recognition	bounding boxes	2019
Ours	Semantic segmentation and detection	segmentation masks for 10 classes	2023

classification, detection of essential elements, and generation of textual descriptions of four chart types (lines, dot lines, vertical bar plots and horizontal bar plots). [7] focuses on exploring the semantic content of alt text in accessibility publications, revealing that the quality of author-written alt text is mixed. The authors also provide a dataset of such alt text to aid the development of tools for better authoring and give recommendations for publishers. [35] developed a program for fully automatic conversions of line plots (png files) into numerical data (CSV files) by using several Deep-NNs. [2] introduce a fully automated chart data extraction algorithm for circular-shaped and grid-like chart types. Semantic segmentation of documents [1,3,15,17–19,34] has close ties to computer vision, where segmentation model performance has greatly improved with deep learning advancements [6,16,33,36,37]. Despite this, research in fine-grained semantic segmentation of mathematical graphs lags due to a lack of annotated examples for training. To address this, our dataset seeks to close the gap and provide a public benchmark for data-driven graph segmentation methods.

2.2 Document Graphics Datasets

Table 1 provides an overview of the five published datasets most related to our benchmark. The PDFFigures 2.0 dataset is a random sample of 346 papers from over 200 venues with at least 9 citations collected from Semantic Scholar, covering bounding box annotations for captions, figures, and tables [9]. The dataset of Poco et al. [29] comprises automatically generated and manually annotated charts from Quartz news and academic paper. The data for each image includes the bounding boxes and transcribed content of all text elements. The authors further investigate automatic recovery of visual encodings from chart images using text elements and OCR. They present an end-to-end pipeline that detects text elements, classifies their role and content, and infers encoding specification using a CNN for mark type classification. Dai et al. [11] collect a benchmark that covers bar charts collected from the web as well as synthetic charts randomly generated through a script. The authors present Chart Decoder – a deep learning-based system which automatically extracts textual and numeric information from such charts. DocFigure [22] is a scientific figure classification dataset consisting of 33,000 annotated figures from 28 categories found in scientific articles. The authors also designed a web-based annotation tool to efficiently categorize a

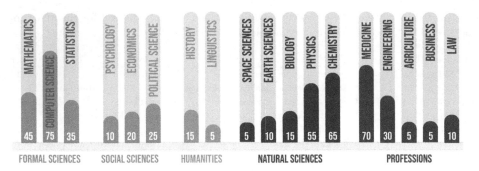

Fig. 3. Statistical distribution of documents in our dataset grouped by different disciplines. Our dataset was collected from 18 distinct disciplines from formal-, social and natural sciences as well as humanities and professions.

large number of figures. The ICDAR 2019 CHART-Infographics competition [12] aimed to explore automatic chart recognition, which was divided into multiple tasks, including image classification, text detection, text role classification, axis analysis and plot element detection. A large synthetic training set was provided and systems were evaluated on synthetic charts and real charts from scientific literature.

In comparison to these datasets, *LG* targets *semantic segmentation* of line graphs with > 500 examples collected from documents of 18 different disciplines, with manual pixel-level annotations at two levels of granularity and 15 labels in total (5 coarse and 10 fine-grained categories). Our LG dataset aims to establish a public benchmark for data-driven graph segmentation methods and will be made publicly accessible upon publication.

3   *LG* Dataset

In this paper, we present the first segmentation dataset to analyze line charts and keep pace with the advancements in the AI community. Our dataset contains 520 mathematical graphics extracted manually from 450 documents. Among these, 7238 human annotated instances. The goal is to facilitate automatic visual understanding of mathematical charts by offering a suitable and challenging benchmark. Next, we provide a comprehensive description of the data collection and annotation process, followed by a thorough analysis of the dataset's features and characteristics.

3.1   Data Collection and Annotation

Classes. To ensure a comprehensive and robust labelling process, we set out to categorize line chart pixels into 5 coarse and 10 fine-grained classes. The primary focus was on creating fine-grained categories that offer a wide range

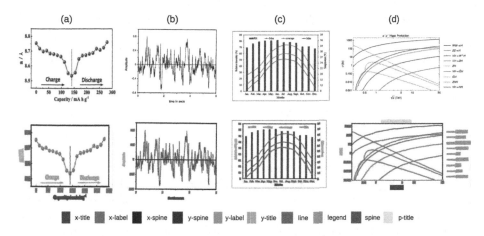

Fig. 4. Example annotations of our Line Graphics (LG) dataset. From top to bottom are the challenging line graphics and the ground truth with fine-grained annotations of 10 classes, which are complemented by 5 coarse categories.

of variations and challenges for further analysis. This was achieved through a thorough review of charts by three annotators with research experience, who identified the most frequent and critical object types encountered in such charts. Based on this review, as well as an inspection of related work, we arrived at 10 relevant categories. Some of these can be further categorized into three coarse categories, namely, *Title* class (e.g. plot title), *Spine* class (e.g. "spine" with no label data), and the *Label* class (e.g. x-axis labels). As detailed in Table 2, in this work, we conduct experiments with the 10 classes, which are *p-title, x-title, y-title, x-spine, y-spine, spine, x-label, y-label, legend and line*.

Collection. In addition to ensuring that the source documents in *LG* dataset are free from intellectual property constraints, we have imposed certain requirements for the documents to adhere to. First, all collected documents should contain at least one complex line chart, regardless of the document type (scanned, digital, slide, etc.) or field. Second, to represent different time periods, both old and new charts were collected. Third, the similarity levels between cropped images were maintained as low as possible to ensure that each image presents a unique and challenging point for analysis. To achieve broad coverage of all fields, documents in *LG* dataset were collected from 5 different disciplines and their top published subcategories as shown in Fig. 3. The collection process involved a manual search using scientific keywords, and carefully inspecting each document downloaded from sources such as arXiv and Google Scholar. This approach helped ensure a consistent and uniform distribution of documents across all categories.

Table 2. The number of semantic instances of each data split.

Split	#images	Title			Spine			Label		Legend	Line
		ptitle	*xtitle*	*ytitle*	*xspine*	*yspine*	*spine*	*xlabel*	*ylabel*	*legend*	*line*
Train	312	39	268	235	245	255	431	1484	1512	382	581
Validation	104	11	92	78	82	82	163	539	484	152	183
Test	104	14	91	70	79	78	145	478	459	161	171

Annotation. Fine-grained pixel-level annotations were provided for the 10 chart classes as depicted in Fig. 4. This level of detail was necessary due to the presence of fine structures in the charts, such as lines. Using bounding boxes alone would not be sufficient as it would result in the incorrect annotation of background pixels as foreground and difficulty in distinguishing between different lines in the plot. Bounding boxes were still provided despite the pixel-level ones, as we believe they may be useful for certain classes such as the text content, except lines, for which a plotting area bounding box category was labelled instead. The annotation process was initiated with the provision of 20 pages of guidelines and 100 sample images to each of the three annotators, resulting in a mean pairwise label agreement of 80%. Further, annotators were given a batch of images to annotate and each annotation was reviewed by the other two annotators. To facilitate instance-level segmentation, we provide annotations for each instance separately in COCO *JSON* format. For example, each line has a separate ID in the line mask.

3.2   Dataset Properties

Fully automating the mathematical graphics digitization process involves retrieving the metadata and converting it into a machine-readable format, such as a spreadsheet. This requires a thorough understanding of fine-grained elements, such as axes to project the lines and obtain pixel (x, y) entries, axes labels to calibrate retrieved line values from the pixels domain, and legend to match and describe the lines respectively. In our review of existing datasets, we found a mix of both private and public datasets focused on graphics digitization. However, despite their similar goal, these datasets often lack diversity in terms of plot variations, richness, and classes.

Split. The LG dataset consists of three subsets - training, validation, and test - each of which is split with a reasonable proportion of the total instances. As depicted in Table 2, some classes exhibit limited numbers, however, this accurately reflects their low-frequency occurrence, such as plot titles that are typically found in figure captions. Despite this, our experiments have demonstrated that the richness of the data was crucial in overcoming this challenge.

Table 3. Variations of line chart categories found in LG dataset.

Class	Variations
Plot	combination chart, multi-axis chart, gridlines
Title, Label	orientation, type (integer, date, etc.), font variations
Line, Legend	style, markers, annotations, position, count
Spine	length, width, count, style
Background	colour (gradient, filled, etc.), grid

Variations. Table 3 below demonstrates the diversity and inclusiveness of our dataset, as it includes a wide range of instances counts, styles, and locations, without any aforementioned limitations. Our dataset includes a comprehensive range of variations for all classes, as summarized in Table 3. We have covered a wide range of plot types, including those that feature multiple chart types like bar, scatter, and line charts as in Fig. 2 (a) (c) and (d), as well as plots with repeated classes like multiple y or x-axes and ticks. The text content in our dataset is annotated with variations in integer, decimal, and DateTime formats, as well as tilt. Furthermore, we have taken into account different markers, patterns, and sizes for line and spine classes, and added the class "other" to represent the annotated plot area explanatory text, focus points, and arrows. The background variations in our dataset include colour (single or multiple), gradient, and RGB images.

Spatial Distribution Visualization. We have additionally analyzed the statistical localization information of all ground-truth instances. As shown in Fig. 5, the frequency of occurrence is visualized in heatmaps. We can see that the Title class has a strong prior position, as titles are typically standalone text at the edges of the chart. Spines on the other side reveal that the majority of charts are box format with two axes (left and bottom edge). Cartesian-type with intersecting x and y axes are observed less frequently. Our heatmap evaluation shows that spines have an average width of 3 pixels, with a minimum and maximum of 1 and 7 pixels, respectively, making them one of the challenging classes to segment. Interestingly, as we see in the legend heat map, they are predominantly positioned at the top and on either side of the plotting area, but they can also appear in other locations. According to statistics, 44% of the legends are located at the top.

4   Experiments

4.1   Implementation Details

We perform experiments utilizing both Jittor and Pytorch. Our implementation is based on the MMsegmentation[2] library and the models were trained on

---

[2] https://github.com/open-mmlab/mmsegmentation.

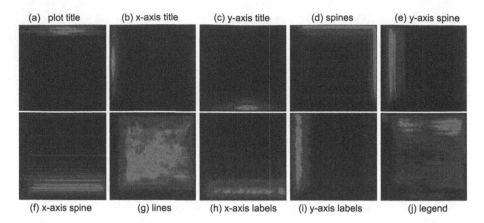

Fig. 5. Location heat maps for the ten different fine-grained labels in our LG dataset. We observe clear category-specific biases, (e.g., plot title mostly appears at the top, while lines are nearly uniformly distributed along the document space), which is expected for real document data.

an A40 GPU with an input resolution of (2048, 1024). Our evaluation metric is Mean Intersection over Union (mIoU). During training, we applied common data augmentation techniques such as random flipping, scaling (ranging from 0.5 to 2), and cropping. The batch size was set to 8 with an initial learning rate of $6e-5$, using a poly-learning rate decay policy. The models were trained for 50K iterations. For testing, we employed a single-scale flip strategy to ensure fairness in comparison. To understand the choice of models, we further analyze the properties of the selected models in conjunction with our proposed line graphic segmentation task.

4.2   Baselines

We consider 7 state-of-the-art semantic segmentation models for this task:

MobileNetV3 [20] is designed for image segmentation in both high and low-resource environments. It incorporates the depthwise separable convolution from MobileNetV1 and the Inverted Residual with Linear Disability from MobileNetV2 to balance accuracy and computational cost. Additionally, the model introduces the V3 lightweight attention mechanism, which enhances its ability to selectively focus on important features. These improvements make MobileNetV3 a good choice for resource-constrained applications of line graphic segmentation.

HRNet [32] leverages multi-scale feature representations to effectively handle high-resolution image understanding tasks such as human pose estimation and semantic segmentation. Throughout the network, HRNet utilizes repeated information exchange across multiple scales and a multi-scale feature output that is fed into a task-specific head. This innovative architecture enables easy feature reuse and efficient computation, making HRNet a top choice for high-resolution graphic segmentation task.

DeepLabv3+ [6] is a CNN semantic segmentation model that utilizes a decoder module to obtain sharper object boundaries and a more fine-grained segmentation, which is crucial for the proposed line graphic segmentation. In addition, an ASPP module captures multi-scale context information, while a lightweight Xception architecture provides efficient computation.

PSPNet [37] proposed the pyramid pooling module, which is able to extract an effective global contextual prior. Extracting pyramidal features since the model can perceive more context information of the input line graphic.

Swin [26] introduced the Transformer model with a shifted window operation and a hierarchical design. The shifted window operation includes non-overlapping local windows and overlapping cross-windows. It provides the locality of convolution to the graphic segmentation task, and on the other hand, it saves computation as compared to the original transformer models.

SegFormer [33] leverages the Transformer architecture and self-attention mechanisms. The model consists of a hierarchical Transformer encoder and a lightweight all-MLP decoder head, making it both effective and computationally efficient. The design enables SegFormer to capture long-range dependencies within an image, *i.e.,* a line graphic in this work.

SegNeXt [16] proposed a new design of convolutional attention by rethinking the self-attention in Transformer models. In this work, the resource-costly self-attention module is replaced by using depth-wise convolution with large sizes. As a result, the multi-scale convolutional attention with a large kernel can encode context information more effectively and efficiently, which is crucial for the line graphic segmentation task.

5   Evaluation

5.1   Quantitative Results

In Table 4, the efficiency-oriented CNN model MobileNetV3 with only 1.14M parameters obtains 56.22% mIoU score on the proposed LG dataset. The high-resolution model HRNet has 57.60% in mIoU and the DeepLabv3+ model has 61.64%, but both have parameter >60M. We found that the PSPNet with pyramid pooling module in the architectural design can achieve better results with 62.04%. In Table 4, the recent Transformer-based models achieve relatively better results than the CNN-based models. For example, the SegFormer model with pyramid architecture and with 81.97M parameters can obtain 65.59% in mIoU with a +3.55% gain compared to PSPNet. The Swin Transformer with hierarchical design and shifted windows has 66.61% in mIoU, but it has the highest number of parameters. However, the state-of-the-art CNN-based SegNeXt utilizes multi-scale convolutional attention to evoke spatial attention, leading to the highest mIoU score of 67.56% in our LG dataset. Furthermore, the SegNeXt achieve 4 top scores on 5 coarse classes, including *Title, Spine, Label* and *Legend*. Besides, it obtains 6 top scores out of 10 fine classes, which are *xtitle, ytitle,*

Table 4. Semantic segmentation results of CNN- and Transformer-based models on the *test* set of LG dataset. **#P**: the number of model parameters in millions; **GFLOPs**: the model complexity calculated in the same image resolution of 512×512; **Per-class IoU** (%): the Intersection over Union (IoU) score for each of coarse and fine classes; **mIoU** (%): the average score across all of 10 fine classes. The best score is highlighted in bold.

Model	Backbone	#P(M)	GFLOPs	Coarse Per-class IoU					mIoU
				Title	*Spine*	*Label*	*Legend*	*Line*	
MobileNetV3 [20]	MobileNetV3-D8	1.14	4.20	45.06	43.68	68.74	60.86	62.12	56.22
HRNet [32]	HRNet-W48	65.86	93.59	52.48	44.4	67.95	53.34	61.91	57.60
DeepLabv3+ [6]	ResNet-50	62.58	79.15	55.41	46.14	74.72	67.07	32.97	61.46
PSPNet [37]	ResNetV1c	144.07	393.90	57.12	43.77	78.52	67.75	62.30	62.04
SegFormer [33]	MiT-B5	81.97	51.90	58.36	54.13	76.79	67.09	69.67	65.59
Swin [26]	Swin-L	233.65	403.78	62.26	52.85	76.57	68.91	**71.01**	66.61
SegNeXt [16]	MSCAN-L	49.00	570.0	**63.79**	**54.61**	**80.29**	**69.09**	65.07	**67.56**

Model	Fine Per-class IoU									
	ptitle	*xtitle*	*ytitle*	*xspine*	*yspine*	*spine*	*xlabel*	*ylabel*	*legend*	*line*
MobileNetV3 [20]	09.03	55.36	70.81	53.47	40.21	37.36	67.83	69.65	60.86	62.12
HRNet [32]	31.30	55.85	70.30	44.68	50.20	38.36	65.42	70.48	53.34	61.91
DeepLabv3+ [6]	30.30	62.21	73.74	49.47	47.57	41.39	73.65	75.79	67.07	32.97
PSPNet [37]	22.92	68.09	80.39	47.09	53.11	31.12	77.28	79.76	67.75	62.30
SegFormer [33]	37.25	61.10	76.75	60.37	55.83	**46.21**	73.35	80.23	67.09	69.67
Swin [26]	**49.21**	59.44	78.15	59.48	55.33	43.74	71.96	**81.54**	68.91	**71.01**
SegNeXt [16]	36.57	**73.95**	**80.85**	**61.77**	**56.20**	45.88	**80.68**	79.91	**69.09**	65.07

Table 5. Ablation on different models, backbone variants, and input sizes

Model	Backbone	Input Size	mIoU
DeepLabv3+ [6]	ResNet-50	512×512	22.88
Swin [26]	Swin-S		42.39
SegFormer [33]	MiT-B0		36.45
SegNext [16]	MSCA-S		**48.84**
DeepLabv3+ [6]	ResNet-101	512×512	29.11
Swin [26]	Swin-T		45.78
SegFormer [33]	MiT-B5		38.32
SegNext [16]	MSCA-L		**48.65**
DeepLabv3+ [6]	ResNet-50	2048×1024	63.69
Swin [26]	Swin-S		65.25
SegFormer [33]	MiT-B0		60.14
SegNext [16]	MSCA-S		**67.08**

xpsine, yspine, xlabel, and *legend.* The results show that a stronger architecture for the semantic segmentation task can achieve better results in the proposed LG benchmark, yielding reliable and accessible mathematical graphics.

5.2 Ablation Study

Apart from the qualitative analysis of state-of-the-art models, we further perform an ablation study on the aforementioned CNN- and Transformer-based models. As shown in Table 5, the ablation study is two-fold. First, to understand the impact of model scales on segmentation performance, different model scales are ablated. For example, the tiny (T) and small (S) versions of Swin Transformer, are evaluated on our LG dataset. Besides, to analyze the effect of the image resolution, two image sizes (*i.e.*, 512×512 and 2048×1024) are involved in the ablation study. According to the results shown in Table 5, we obtain an insight that higher resolution of input images can achieve larger gains than using models with higher complexity. For example, SegFormer-B0 in resolution of 2048×1024

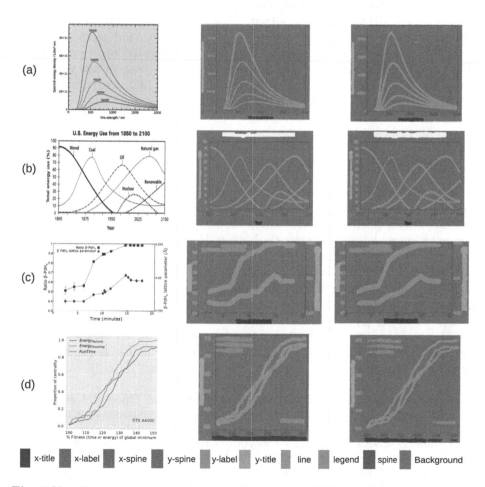

Fig. 6. Visualization of semantic segmentation results of LG dataset. From left to right are the input RGB line graphs, the ground truth labels, and the predicted masks from SegNeXt-L model.

outperforms SegFormer-B5 in 512×512 with a 21.82% gain in mIoU. Regarding the ablation study, it is recommended to use a larger image size as input than to use a model with greater complexity. Another benefit of this setting is to maintain the high efficiency of the trained model, which is more practical and promising on graphic segmentation applications.

5.3 Qualitative Results

To further understand the line graph semantic segmentation task, we visualize some examples in Fig. 6. From left to right are the input RGB line graphs, the ground truth labels, and the segmentation results generated by SegNeXt-L model. Although the backgrounds of these input images have different colors and textures, the model can accurately segment them (in purple). We found that SegNeXt with >67% mIoU can output surprisingly good segmentation results, including precise masks for thin objects, such as *xspine* and *yspine*. Besides, in the bottom row, the intersecting lines can be segmented accurately. Apart from the positive results, the *xspine* and *yspine* in the bottom row cannot be recognized well, which means that there is still room for improvement in the LG benchmark. Nonetheless, the other classes, such as *labels*, *titles* and *legend*, can be segmented correctly.

6 Conclusion

In conclusion, this paper presents the first line plot dataset for multi-task deep learning, providing support for object detection, semantic segmentation, and instance-level segmentation. Our comprehensive evaluations of state-of-the-art segmentation models demonstrate the potential for an end-to-end solution. Moreover, this work has the potential to greatly improve accessibility for visually impaired and blind individuals. The ability to accurately detect and recognize mathematical graphics could lead to more accessible educational materials and support the digitization of mathematical information. We are actively working to expand the scope of the dataset by including more types of mathematical graphics and incorporating instances relationship into the metadata. This will continue to drive advancements in this field and enable further research into the digitization of mathematical graphics.

Acknowledgments. This work was supported in part by the European Union's Horizon 2020 research and innovation program under the Marie Sklodowska-Curie Grant No. 861166, in part by the Ministry of Science, Research and the Arts of Baden-Württemberg (MWK) through the Cooperative Graduate School Accessibility through AI-based Assistive Technology (KATE) under Grant BW6-03, and in part by the Federal Ministry of Education and Research (BMBF) through a fellowship within the IFI programme of the German Academic Exchange Service (DAAD). This work was partially performed on the HoreKa supercomputer funded by the MWK and by the Federal Ministry of Education and Research.

References

1. Amin, A., Shiu, R.: Page segmentation and classification utilizing bottom-up approach. Int. J. Image Graph. **1**(02), 345–361 (2001)
2. Bajić, F., Orel, O., Habijan, M.: A multi-purpose shallow convolutional neural network for chart images. Sensors **22**(20), 7695 (2022)
3. Breuel, T.M.: Robust, simple page segmentation using hybrid convolutional MDL-STM networks. In: 2017 14th IAPR International Conference on Document Analysis and Recognition (ICDAR), vol. 1, pp. 733–740. IEEE (2017)
4. Chen, K., Seuret, M., Hennebert, J., Ingold, R.: Convolutional neural networks for page segmentation of historical document images. In: 2017 14th IAPR International Conference on Document Analysis and Recognition (ICDAR), vol. 1, pp. 965–970. IEEE (2017)
5. Chen, K., Seuret, M., Liwicki, M., Hennebert, J., Ingold, R.: Page segmentation of historical document images with convolutional autoencoders. In: 2015 13th International Conference on Document Analysis and Recognition (ICDAR), pp. 1011–1015. IEEE (2015)
6. Chen, L.-C., Zhu, Y., Papandreou, G., Schroff, F., Adam, H.: Encoder-decoder with Atrous separable convolution for semantic image segmentation. In: Ferrari, V., Hebert, M., Sminchisescu, C., Weiss, Y. (eds.) ECCV 2018. LNCS, vol. 11211, pp. 833–851. Springer, Cham (2018). https://doi.org/10.1007/978-3-030-01234-2_49
7. Chintalapati, S., Bragg, J., Wang, L.L.: A dataset of alt texts from IICI publications: analyses and uses towards producing more descriptive alt texts of data visualizations in scientific papers. arXiv preprint arXiv:2209.13718 (2022)
8. Choi, J., Jung, S., Park, D.G., Choo, J., Elmqvist, N.: Visualizing for the non-visual: enabling the visually impaired to use visualization. In: Computer Graphics Forum, pp. 249–260. Wiley Online Library (2019)
9. Clark, C., Divvala, S.: PDFFigures 2.0: mining figures from research papers. In: Proceedings of the 16th ACM/IEEE-CS on Joint Conference on Digital Libraries, pp. 143–152 (2016)
10. Clausner, C., Antonacopoulos, A., Pletschacher, S.: ICDAR 2017 competition on recognition of documents with complex layouts-RDCL2017. In: 2017 14th IAPR International Conference on Document Analysis and Recognition (ICDAR), vol. 1, pp. 1404–1410. IEEE (2017)
11. Dai, W., Wang, M., Niu, Z., Zhang, J.: Chart decoder: generating textual and numeric information from chart images automatically. J. Vis. Lang. Comput. **48**, 101–109 (2018)
12. Davila, K., et al.: ICDAR 2019 competition on harvesting raw tables from infographics (chart-infographics). In: 2019 International Conference on Document Analysis and Recognition (ICDAR), pp. 1594–1599. IEEE (2019)
13. Davila, K., Setlur, S., Doermann, D., Kota, B.U., Govindaraju, V.: Chart mining: a survey of methods for automated chart analysis. IEEE Trans. Pattern Anal. Mach. Intell. **43**(11), 3799–3819 (2020)
14. Davila, K., Tensmeyer, C., Shekhar, S., Singh, H., Setlur, S., Govindaraju, V.: ICPR 2020 - competition on harvesting raw tables from infographics. In: Del Bimbo, A., et al. (eds.) ICPR 2021. LNCS, vol. 12668, pp. 361–380. Springer, Cham (2021). https://doi.org/10.1007/978-3-030-68793-9_27
15. Drivas, D., Amin, A.: Page segmentation and classification utilising a bottom-up approach. In: Proceedings of 3rd International Conference on Document Analysis and Recognition, vol. 2, pp. 610–614. IEEE (1995)

16. Guo, M.H., Lu, C.Z., Hou, Q., Liu, Z., Cheng, M.M., Hu, S.M.: SegNeXt: rethinking convolutional attention design for semantic segmentation. arXiv preprint arXiv:2209.08575 (2022)
17. Ha, J., Haralick, R.M., Phillips, I.T.: Document page decomposition by the bounding-box project. In: Proceedings of 3rd International Conference on Document Analysis and Recognition, vol. 2, pp. 1119–1122. IEEE (1995)
18. Haurilet, M., Al-Halah, Z., Stiefelhagen, R.: SPaSe-multi-label page segmentation for presentation slides. In: 2019 IEEE Winter Conference on Applications of Computer Vision (WACV), pp. 726–734. IEEE (2019)
19. Haurilet, M., Roitberg, A., Martinez, M., Stiefelhagen, R.: Wise-slide segmentation in the wild. In: 2019 International Conference on Document Analysis and Recognition (ICDAR), pp. 343–348. IEEE (2019)
20. Howard, A., et al.: Searching for MobileNetV3. In: Proceedings of the IEEE/CVF International Conference on Computer Vision, pp. 1314–1324 (2019)
21. Huang, Z., et al.: ICDAR 2019 Competition On Scanned Receipt OCR and information extraction. In: 2019 International Conference on Document Analysis and Recognition (ICDAR), pp. 1516–1520. IEEE (2019)
22. Jobin, K., Mondal, A., Jawahar, C.: DocFigure: a dataset for scientific document figure classification. In: 2019 International Conference on Document Analysis and Recognition Workshops (ICDARW), vol. 1, pp. 74–79. IEEE (2019)
23. Keefer, R., Bourbakis, N.: From image to XML: monitoring a page layout analysis approach for the visually impaired. Int. J. Monit. Surveill. Technol. Res. (IJMSTR) **2**(1), 22–43 (2014)
24. Li, P., Jiang, X., Shatkay, H.: Figure and caption extraction from biomedical documents. Bioinformatics **35**(21), 4381–4388 (2019)
25. Liu, X., Klabjan, D., NBless, P.: Data extraction from charts via single deep neural network. arXiv preprint arXiv:1906.11906 (2019)
26. Liu, Z., et al.: Swin transformer: hierarchical vision transformer using shifted windows. In: ICCV (2021)
27. Long, S., Ruan, J., Zhang, W., He, X., Wu, W., Yao, C.: TextSnake: a flexible representation for detecting text of arbitrary shapes. In: Ferrari, V., Hebert, M., Sminchisescu, C., Weiss, Y. (eds.) ECCV 2018. LNCS, vol. 11206, pp. 19–35. Springer, Cham (2018). https://doi.org/10.1007/978-3-030-01216-8_2
28. Methani, N., Ganguly, P., Khapra, M.M., Kumar, P.: PlotQA: reasoning over scientific plots. In: Proceedings of the IEEE/CVF Winter Conference on Applications of Computer Vision, pp. 1527–1536 (2020)
29. Poco, J., Heer, J.: Reverse-engineering visualizations: recovering visual encodings from chart images. In: Computer Graphics Forum, pp. 353–363. Wiley Online Library (2017)
30. Seweryn, K., Lorenc, K., Wróblewska, A., Sysko-Romańczuk, S.: What will you tell me about the chart? – automated description of charts. In: Mantoro, T., Lee, M., Ayu, M.A., Wong, K.W., Hidayanto, A.N. (eds.) ICONIP 2021. CCIS, vol. 1516, pp. 12–19. Springer, Cham (2021). https://doi.org/10.1007/978-3-030-92307-5_2
31. Siegel, N., Lourie, N., Power, R., Ammar, W.: Extracting scientific figures with distantly supervised neural networks. In: Proceedings of the 18th ACM/IEEE on Joint Conference on Digital Libraries, pp. 223–232 (2018)
32. Wang, J., et al.: Deep high-resolution representation learning for visual recognition. TPAMI **43**, 3349–3364 (2021)
33. Xie, E., Wang, W., Yu, Z., Anandkumar, A., Alvarez, J.M., Luo, P.: SegFormer: simple and efficient design for semantic segmentation with transformers. In: NeurIPS (2021)

34. Yang, X., Yumer, E., Asente, P., Kraley, M., Kifer, D., Lee Giles, C.: Learning to extract semantic structure from documents using multimodal fully convolutional neural networks. In: Proceedings of the IEEE Conference on Computer Vision and Pattern Recognition, pp. 5315–5324 (2017)
35. Yoshitake, M., Kono, T., Kadohira, T.: Program for automatic numerical conversion of a line graph (line plot). J. Comput. Chem. Jpn. **19**(2), 25–35 (2020)
36. Zhang, J., Ma, C., Yang, K., Roitberg, A., Peng, K., Stiefelhagen, R.: Transfer beyond the field of view: dense panoramic semantic segmentation via unsupervised domain adaptation. IEEE Trans. Intell. Transp. Syst. **23**(7), 9478–9491 (2021)
37. Zhao, H., Shi, J., Qi, X., Wang, X., Jia, J.: Pyramid scene parsing network. In: CVPR (2017)

Linear Object Detection in Document Images Using Multiple Object Tracking

Philippe Bernet[ID], Joseph Chazalon[(✉)][ID], Edwin Carlinet[ID],
Alexandre Bourquelot[ID], and Elodie Puybareau[ID]

EPITA Research Lab (LRE), Le Kremlin-Bicêtre, France
{philippe.bernet,joseph.chazalon,edwin.carlinet,
alexandre.bourquelot,elodie.puybareau}@epita.fr

Abstract. Linear objects convey substantial information about document structure, but are challenging to detect accurately because of degradation (curved, erased) or decoration (doubled, dashed). Many approaches can recover some vector representation, but only one closed-source technique introduced in 1994, based on Kalman filters (a particular case of Multiple Object Tracking algorithm), can perform a pixel-accurate instance segmentation of linear objects and enable to selectively remove them from the original image. We aim at re-popularizing this approach and propose: 1. a framework for accurate instance segmentation of linear objects in document images using Multiple Object Tracking (MOT); 2. document image datasets and metrics which enable both vector- and pixel-based evaluation of linear object detection; 3. performance measures of MOT approaches against modern segment detectors; 4. performance measures of various tracking strategies, exhibiting alternatives to the original Kalman filters approach; and 5. an open-source implementation of a detector which can discriminate instances of curved, erased, dashed, intersecting and/or overlapping linear objects.

Keywords: Line segment detection · Benchmark · Open source

1 Introduction

The detection of linear structures is often cast as a boundary detection problem in Computer Vision (CV), focusing on local gradient maxima separating homogeneous regions to reveal edges in natural images. Document images, however, often exhibit very contrasted and thin strokes which carry main information, containing printings and writings, overlaid on some homogeneous "background". It is therefore common to observe a local maximum and a local minimum in a short

Supplementary Information The online version contains supplementary material available at https://doi.org/10.1007/978-3-031-41734-4_28.

range along the direction normal to such stroke. As a result, methods based on region contrast tend to fail on document images, rejecting text and strokes as noise or produce double-detections around linear structures.

In this work, we are interested in enabling a fast and accurate pre-processing which can detect and eventually remove decorated, degraded or overlapping linear objects in document images. Figure 2 illustrates the two major outputs such method needs to produce: a vector output containing the simplified representation of all start and end coordinates of approximated line segments; and a pixel-accurate instance segmentation where each pixel is assigned zero (background), one, or more (intersections) linear object label(s), enabling their removal while preserving the integrity of other shapes (i.e. without creating gaps).

Our work aims at reviving a technique proposed in 1994 [23,27,28] which introduced a way to leverage Kalman filters to detect and segment instances of linear objects in document images, for which no public implementation is available. We propose a re-construction of this method with the following new contributions: 1. we introduce a general framework for accurate instance segmentation of linear objects in document images using Multiple Object Tracking (MOT), a more general approach, which enables to separate the stages of the approach properly and eventually leverage deep image filters, and also to extend the method with new trackers; 2. we introduce datasets and metrics which enable both vector- and pixel-based evaluation of linear object detector on document images: from existing and newly-created contents we propose vector and pixel-accurate ground-truth for linear object and release it publicly; 3. we show the performance of MOT approaches against modern segment detectors on a reproducible benchmark using our public datasets and evaluation code; 4. we compare the performance of various tracking strategies, exhibiting alternatives to the original Kalman filters approach and introducing, in particular, a better parametrization of the Kalman filters; and 5. we release the first open-source implementation of a fast, accurate and extendable detector which can

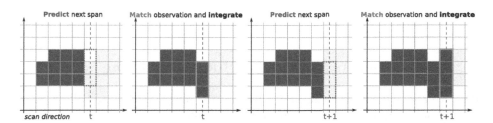

Fig. 1. Tracking process of linear objects. Images are scanned in a single left-to-right (resp. top-to-bottom) pass. At each step t, the tracker predicts the next *span* (red box) of a linear object from past (opaque) observations, then it matches candidate observations, selects the most probable one (green box), and adjust its internal state accordingly. Multiple Object Tracking (MOT) handles intersections, overlaps, gaps, and detects boundaries accurately.

discriminate instances of curved, dashed, intersecting and/or overlapping linear objects.

This study is organized as follows. We review existing relevant methods for linear object detection in Sect. 2, detail the formalization of our proposed framework in Sect. 3, and present the evaluation protocol (including the presentation of datasets and metrics), as well as the results of our benchmarks for vector and pixel-accurate detection, in Sect. 4.

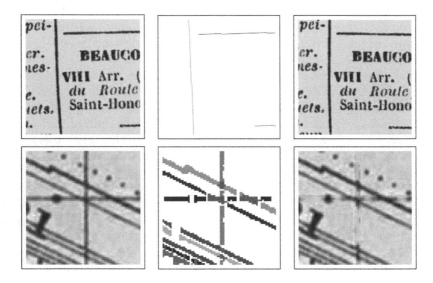

Fig. 2. Two possible applications. *First row* illustrates an orientation detection task, relying on locating endpoints of vertical lines. *Second row* illustrates a multi-stage information extraction task, relying on discriminating between building shapes and georeferencing lines (horizontal and vertical) in map images. *Left*: image inputs, *center*: detections, *right*: rotated or inpainted result.

2   State of the Art in Linear Object Detection

The two complementary goals addressed in this work were traditionally tackled by distinct method categories. Among these categories, some produce vector outputs (with end coordinates, but also sometimes sparse or dense coordinates of points laying on the object), and some others provide some pixel labeling of linear objects. Such pixel labeling is often limited to a *semantic segmentation* [19], i.e. a binary "foreground" vs "background" pixel-wise classification, and, as we will see, only tracking-based approaches can propose an *instance segmentation* [19], i.e. effectively assign each pixel to a set of objects. Due to the lack of datasets for linear object detection in document images, we are particularly interested, in this current work, in techniques which have light requirements on training data, and leave for future research the implementation of a larger benchmark. Linear object detection is very related to the problem of *wireframe parsing*, well studied in CV, for which several datasets were proposed [11,16], but their applicability to document images, which exhibit distinct structural and textural properties, were not specifically studied. We review a selection of approaches which pioneered key ideas in each category of the taxonomy we propose here.

Pixel-Wise Edge Classifiers. The Sobel operator [32] is a good example of an early approach which predicts, considering a limited spatial context, the probability or score that a particular image location belongs to an edge. This large family includes all hourglass-shaped deep *semantic segmentation* networks, notably any approach designed around some U-Net variant [24,31] like the HED [36] and BDCN [15] edge detectors. Such approaches can make pixel-accurate predictions but cannot discriminate object instances. To enable their use for vectorization, Xue et al. [38] proposed to predict an Attraction Field Map which eases the extraction of 1-pixel-thin edges after a post-processing stage. EDTER [29] pushes the *semantic segmentation* approach as far as possible with a transformer encoder, enabling to capture a larger context. While all these techniques can produce visually convincing results, they cannot assign a pixel to more than one object label and require a post-processing to produce some vector output. Furthermore, their training requirements and computational costs forces us to remove them from our comparison, but they could be integrated as an image pre-processing in the global framework we propose.

Hough Transform-Based Detectors. The Hough transform [17] is a traditional technique to detect lines in images. It can be viewed as a kind of meta-template matching technique: line evidence L with (ρ, θ) polar parameters is supported by pixel observations (x_i, y_i) according to the constraint that $(L) : \rho = x_i \cos\theta + y_i \sin\theta$. In practice, such technique relies on a binarization pre-processing to identify "foreground" pixels which can support a set of (ρ, θ) values in polar space. Line detection is performed by identifying maximal clusters in this space, which limits the accuracy the detection to a very coarse vectorization, while relying on a high algorithmic complexity. Several variants [26] were proposed to overcome those limitations: the Randomized Hough Transform [21] and Probabilistic Hough Transform [20] lowered the computational cost thanks to some random sampling of pixels, and the Progressive Probabilistic Hough Transform [13] added the ability to predict start and end line segment coordinates, turning the approach into an effective linear object detector. This method can tolerate gaps and overlaps to some extents, and is highly popular thanks to its implementation in the OpenCV library [3]. However, it is very sensitive to noise, to object length, rather inaccurate with slightly curved objects, and cannot assign pixels to linear object instances.

Region Growing Tracers. The Canny edge detector [5] is the most famous example of this category. It is based on the following steps: a gradient computation on a smoothed image, the detection of local gradient maximums, the elimination of non-maximal edge pixels in the gradient's direction, and the filtering of edge pixels (based on gradient's magnitude) in the edge direction. All region growing approaches improve this early algorithm, probably introduced by Burns et al. [4], which effectively consists in finding salient edge stubs and tracing from there initial position. LSD [14] was the first approach which took linear object detection to the next level in natural images. Fast and accurate, it is based on a sampling of gradient maximums which are connected only if the gradient flow between these candidates is significantly strong compared to a background noise model, according to the *Helmholtz principle* proposed by [12]. EDLine [2] and AG3line [39] improved the routing scheme to connect distant gradient maximums using Least Square fitting and a better sampling, accelerating and enhancing the whole process. CannyLines [25] proposed a parameter-free detection of local gradient maximums, and reintegrated the *Helmholtz principle* in the routing. Ultimately, ELSED [33] further improved this pipeline to propose the fastest detector to date, with a leading performance among learning-less methods. These recent approaches are very fast and accurate, require no training stage, and can handle intersections. However, they do not detect all pixels which belong to a stroke, and have limited tolerance to gaps and overlaps. Finally, these methods require a careful tuning to be used on document images as the integrated gradient computation step tends to be problematic for thin linear objects, and leads to double detections or filters objects (confusing them with noise).

Deep Linear Object Detectors. Region Proposal Networks, and their ability to be trained end-to-end using RoI pooling, where introduced in the Faster R-CNN architecture [30]. This 2-stage architecture was adapted to linear object detection by the L-CNN approach [40]. L-CNN uses a junction heatmap to generate line segment proposals, which are fed to classifier using a Line of Interest pooling, eventually producing vector information containing start and end coordinates. HAWP [38] accelerated the proposal stage by replacing the joint detection stage with the previously-introduced Attraction Field Map [38]. F-Clip [10] proposed a similar, faster approach using a single-stage network which directly predicts center, length and orientation for each detected line segment. These approaches produce very solid vector results, but they are still not capable to assign pixels to a particular object instance. Furthermore, their important computation and training data requirements forces us to remove them from our current study.

Vertex Sequence Generators. Recently, the progress of decoders enabled the direct generation of vector data from an image input, using a sequence generator on top of some feature extractor. Polygon-RNN [1,7] uses a CNN as feature extractor, and an RNN decoder which generates sequences of vertex point coordinates. LETR [37] uses a full encoder/decoder transformer architecture and reaches the best wireframe parsing accuracy to date. However, once again, such architectures are currently limited to vector predictions, and their computation and training data requirements make them unsuitable for our current comparison.

Linear Object Trackers. A neglected direction, with several key advantages and which opens interesting perspectives, was introduced in 1994 in a short series of papers [23,27,28]. This approach leverages the power of Kalman filters [18], which were very successful for sensors denoising, to stabilize the detection of linear objects over the course of two image scans (horizontal and vertical). By tracking individual object candidates, eventually connecting or dropping them, this approach proposed a lightweight solution, with few tunable parameters, which can *segment instances* of linear objects by assigning pixels to all the objects they belong to, but also deal with noise, curved objects, gaps and noise. However, no comparison against other approaches, nor public implementation, were disclosed. This called for a revival and a comparison against the fastest methods to date.

3   MOT Framework for Linear Object Detection

As previously mentioned, we extend the original approach of [23,27,28], which performs 2 scans over an image, plus a post-processing, to detect linear objects. We will describe the horizontal scan only: the vertical one consists in the same process applied on the transposed image. During the horizontal scan, the image is read column by column in a single left-to-right pass; each column being considered as a *1-dimensional scene* containing linear object *spans*, i.e. slices of dark pixels in the direction normal to linear object (Fig. 1). Those spans are tracked scene by scene, and linked together to retrieve objects with pixel accuracy. To ensure accurate linking, and also to tolerate gaps, overlaps and intersections, each coherent sequence of *spans* is modeled as a *Kalman filter* which stores information about past *spans*, and can predict the attributes (position, thickness, luminance...) of the next most probable one. By matching such *observations* with *predictions*, and then correcting the internal parameters of the *filter*, observations are aggregated in a self-correcting model instance for each linear object. Once the horizontal and vertical scans are complete, object deduplication is required to merge double detections close to 45°. Finally, using pixel-accurate information about each object instance, several outputs can be generated, and in particular: a mapping which stores for each pixel the associated object(s), and a simplified vector representation containing first and last span coordinates only.

Framework Overview. We propose to abstract this original approach into a more general Multiple Objects Tracking (MOT) framework. This enables us to explicit each stage of such process, and introduce variants. Linear object detection using 2-pass MOT can be detailed as follows.

Pre-processing. In Sect. 4, we will report results with grayscale images only, as this is the simplest possible input for this framework. However, some preprocessing may be used to enhance linear object detection. In the case of uneven background contrast, we obtained good results using a *black top hat*, and to be able to process very noisy images, or images with rich textures, it may be possible to train a semantic segmentation network which would produce some edge probability map.

Processing. The horizontal and vertical scans, which can be performed in parallel, aim at initializing, updating and returning a set of *trackers*, a generalization of the "filters" specific to Kalman's model. Like in the original approach, each detected instance is tracked by a unique *tracker* instance. *Trackers* are structures which possess an internal State S containing a variable amount of information, according to the variant considered; two key methods "predict" and "integrate" which will be described hereafter; and an internal list of *spans* which compose the linear object. The State and the two methods can be customized to derive alternate tracker implementations (we provide some examples later in this section). For each *scene t* (column or line) read during the scan, the following steps are performed. Steps 1 and 2 can be performed in parallel, as well as steps 4, 5 and 6.

1. Extract the set of observations O_t^j, $j \in [0, nobs_t[$. Observations represent linear object *spans*, and contain information about their *position* in the scene, *luminance* and *thickness*. At this step, some observations may be rejected because there size is over a certain threshold. This threshold is a parameter of the method. The algorithm and the associated illustration in Fig. 3 detail our implementation this step based on our interpretation of the original approach [23, 27, 28].

2. Predict the most probable next observation X_t^i for each tracker i, using its current internal State S_{t-1}. Predictions have the same structure and attributes as observations.

3. Match extracted observations O_t^j with predicted ones X_t^i. Matching is a two-step process performed for each tracker i. First, candidate observations $O_t^{i,j}$ are selected based on a distance threshold, and *slope*, *thickness* and *luminance* compatibility (they must be inferior to 3 times the standard deviation of each parameter, computed over a window of past observations). Second, the closest observation is matched: $\hat{O}_t^i = \arg\min |O_t^{i,j}(\text{position}) - X_t^i(\text{position})|$. The same observation can be matched by multiple trackers when lines are crossing. Unmatched observations $\bar{O}_t$ are kept until step 5.

4. Integrate new observations $\hat{O}_t^i$ into trackers' States S_t^i, considering (in the more general case) current State S_{t-1}^i, scene t, matched observation $\hat{O}_t^i$ and prediction X_t^i. This enables each tracker to adapt its internal model to the particular object being detected.

5. Initialize new trackers from unmatched observations $\bar{O}_t$. New trackers are added to the active pool of trackers to consider at each scene t.

6. Stop trackers of lost objects. When a tracker does not match any observation for too many t, it is removed from the pool of active trackers. The exact threshold depends on the current size of the object plus an absolute thresholds for acceptable gap size. Those two thresholds are parameters of our method.

Post-processing. *Deduplication* is required because 45°-oriented segments may be detected twice. Duplications are removed by comparing and discarding object instances with high overlap. An optional *attribute filtering* may then be performed, that consists in filtering objects according to their length, thickness or angle. This filtering must be performed after the main processing stage to avoid missing intersection and overlaps with other objects, would these be linear or not: this makes possible to handle interactions between handwritten strokes and line segments, for instance. Finally, *outputs* can be generated by decoding the values stored in each *tracker* object.

Observation extraction algorithm

1: **for** $n = 0$ to max_n **do**
2:     **if** $Image_t[n] < l_{mm}$ **then**
3:         $s \leftarrow 0$; **do** { $s \leftarrow s + 1$ } **while** $(Image_t[n + s] < l_{mm})$;
4:         l_{min} $\leftarrow$ min of $Image_t[n \dots n + s]$
5:         l_{max} $\leftarrow$ max of $Image_t[n \dots n + s]$
6:         contrast $\leftarrow l_{max} - l_{min}$
7:         l_{stab} $\leftarrow l_{min} + r * $contrast
8:         $n_i \leftarrow n$; **while** $(Image_t[n_i] > l_{stab})$ { $n_i \leftarrow n_i + 1$ }
9:         $n_f \leftarrow n$; **while** $(Image_t[n_f] < l_{stab})$ { $n_f \leftarrow n_f - 1$ }
10:         obs_{pos} $\leftarrow (n_i + n_f) / 2$
11:         obs_{thick} $\leftarrow n_f - n_i + 1$
12:         obs_{lum} $\leftarrow$ mean of $Image_t[n_i \dots n_f]$

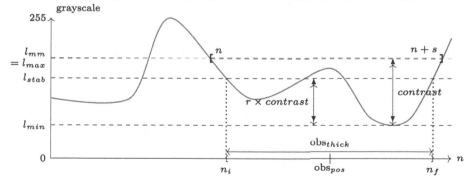

Fig. 3. Algorithm and illustration of the observation extraction process. The red curve represents the luminosity profile of a line (or a column) of pixels. The algorithm looks for a range of pixels $[n \dots n + s]$ whose value do not exceed l_{mm} (L.3) and computes its contrast (L.4–6). Then, this range is refined to extract the largest sub-range whose contrast does not exceed r (a parameter of the method set to 1 in all our application but kept from the original method) times *contrast* (L.8–9) and gives the *observation* with features computed on lines 10–12. (Color figure online)

Tracker Variants. Each tracker variant features a different internal State structures (which needs to be initialized), and specific prediction and integration functions. However, in the context of Linear Object Detection, we keep the original model of the IRISA team [23,27,28]: all observations have luminance, position and thickness attributes, and States have a similar structure, generally adding a *slope* attribute to capture position change. In this work, we consider the following variants. To simplify the following equations, we refer to $t = 0$ as the time of the first observation of a tracker, and consider only one tracker at a time.

Last Observation. A naive baseline approach which uses as prediction the last matched observation $(X_t = \hat{O}_{t-1})$. Updating its States simply consists in storing the last matched observation in place of the previous one.

Simple Moving Average (SMA). Another baseline approach which stores the k last matched observations and extrapolates the prediction based on them. This tracker uses a *slope* attribute. Prediction for the attribute a is the average of the previous observations $X_t(a) = (\sum_{j=t-k}^{t-1} \hat{O}_j)/k$, except for the position, which is computed using the last observed position and the Exponential Moving Average (EMA) of the slope. Integration of new observations consists in adding them to the buffer. We used a buffer of size $k = \min(t, 30)$ in our experiments.

Exponential Moving Average (EMA). A last baseline approach very similar to the previous one, which requires only to store the last matched observation $\hat{O}_{t-1}$ and the last prediction X_{t-1}. The prediction of some attribute a is computed by: $X_t(a) = \alpha * \hat{O}_{t-1}(a) + (1 - \alpha) * X_{t-1}(a)$ where α tunes importance of the new observation. The prediction of the position is computed using the last observed position and the EMA of the slope. We used a value of $\alpha = 2/(\min(t, 16) + 1)$ in our experiments.

Double Exponential [22]. This tracker uses a double exponential smoothing algorithm which is faster than Kalman filters and simpler to implement. $X_t = (2 + \frac{\alpha}{1-\alpha}) \cdot S_{X_{t-1}} - (1 + \frac{\alpha}{1-\alpha}) \cdot S_{X[2]_{t-1}}$ where $S_{X_t} = \alpha \cdot O_t + (1-\alpha) \cdot S_{X_{t-1}}$, $S_{X[2]t} = \alpha \cdot S_{X_t} + (1-\alpha) \cdot S_{X[2]_{t-1}}$ and $\alpha \in [0, 1]$ (set to 0.6 in our experiments) a smoothing parameter.

1€ Filter [6]. This tracker features a more sophisticated approach, also based on an exponential filter, which can deal with uneven signal sampling. It adjusts its low-pass filtering stage according to signal's derivative, and has only two parameters to configure: a minimum cut-off frequency that we set to 1 in our experiments, and a β parameter that we set to 0.007. We refer the reader to the original publication for the details of this approach.

Kalman Filter [18,35], IRISA variant [23,27,28]. This tracker is based on our implementation of the approach proposed by the IRISA team. The hidden State is composed of $n = 4$ attributes $S \in \mathcal{R}^n :$ [pos. slope tick. lum.]$^\mathsf{T}$, and the process is governed by the following equation: $S_t = AS_{t-1} + w_{k-1}$ where $A \in \mathcal{R}^{n,n}$ is the transition model and w_{k-1} some process noise. Process States S can be projected to observation space $O \in \mathcal{R}^m :$ [pos. tick. lum.]$^\mathsf{T}$ according to the measurement equation $O_t = HS_t + v_k$ where H is a simple projection matrix discarding the slope and v_k the measurement noise. By progressively refining the estimate of the covariance matrix Q (resp. R) of w (resp. v), the Kalman filter recursively converges toward a reliable estimate of the hidden State S and its internal error covariance matrix P.

The prediction step consists in (1) projecting the State ahead according to a noise-free model: $S_{t+1}^- = AS_t$; and (2) projecting the error covariance ahead: $P_t^- = AP_{t-1}A^T + Q$ where Q is the process noise covariance matrix. The integration step consists in (1) computing the Kalman gain: $K_t = P_t^- H^T (HP_t^- H^T + R)^{-1}$ where R is the measurement noise covariance matrix; (2) updating estimate with measurement $\hat{O}_t$: $S_t = AS_t^- + K_t(\hat{O}_t - HS_t^-)$ where H relates the State to the measurement; and (3) updating the error covariance: $P_t = (I - K_t H)P_t^-$.

The following initialization choices are made, according to our experiments and the original publications [23, 27, 28]. We initialize each State with the values of the first observation, with a slope of 0. We use $P_0 = I_4$ as initial value for the error covariance matrix, I_n being the identity matrix in $\mathcal{R}^n$. Q, R, H and A assumed to be constant with the following values:

$$A = \begin{bmatrix} 1 & 1 & 0 & 0 \\ 0 & 1 & 0 & 0 \\ 0 & 0 & 1 & 0 \\ 0 & 0 & 0 & 1 \end{bmatrix}, H = \begin{bmatrix} 1 & 0 & 0 & 0 \\ 0 & 0 & 0 & 0 \\ 0 & 0 & 1 & 0 \\ 0 & 0 & 0 & 1 \end{bmatrix}, Q = I_4 \cdot 10^{-5}, \text{ and } R = \begin{bmatrix} 1 & 1 & 4 \end{bmatrix}^T.$$

4   Experiments

We report here the results obtained by comparing the performance of *training-less* approaches on two tasks: a pure *vectorization task*, where only two endpoint coordinates are required for each line segment, and an *instance segmentation task*, where pixel-accurate labeling of each line segment instance is required. This separation is due to the limitations of some approaches which can only generate vector output, as well as the limitations of existing datasets.

While the methods studied are *training-less*, they are not *parameter-free*, and such parameters require to be tuned to achieve the best performance. In order to ensure an unbiased evaluation, we manually tuned (because grid-search and other optimization techniques were not usable here) each approach on the *train set* of each dataset, then evaluate their performance on the *test set*.

4.1   Vectorization Task

Dataset. To our knowledge, no dataset for line segment detection in vector format exists for document images. We introduce here a small new, public dataset which contains endpoints annotations for line segments in 195 images of 19th trade directories. In these documents, line segment detection can be used for image deskewing. The train (resp. test) set is composed of 5 (resp. 190) images, containing on average 4.3 line segments to detect each. Images samples are available in the extra material.

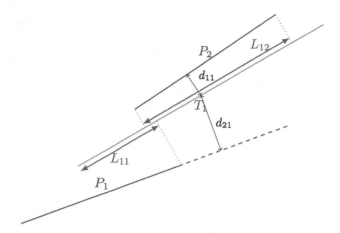

Fig. 4. Matching and scoring process. P_1 and P_2 are associated to T_1 because angles are compatible and overlaps L_{11} and L_{22} are large enough. In this example, *precision* is $\frac{|L_{11}|+|L_{21}|}{|P_1|+|P_2|}$ and is < 1 because P_1 does not fully match the ground-truth. *Recall* is $\frac{|L_{11}\cup L_{21}|}{|T_1|}$ and also is < 1 because T_1 is not fully matched.

Metric. Our evaluation protocol is a slightly modified version of the one proposed by Cho *et al.* [9], and is defined as follows. Let $P = \{P_1, \cdots, P_n\}$ and $T = \{T_1, \cdots, T_m\}$ be respectively the set of *predicted* and *targets* line segment (LS). Let L_{ij} be the projection of P_i over T_j. We note $|X|$, the length of the LS X. A predicted LS P_i matches the *target* T_j if: 1. the prediction overlaps the target over more than 80% ($|L_{ij}|/|P_i| \geq 0.8$), 2. the perpendicular distance d_{ij} between T_j' center and P_i is less than 20 pixels, and 3. the orientations of P_i and P_j differ at most by 5°. For each P_i, we associate the closest LS (in terms of d_{ij}) among the set of *matching* LSs from the targets. We note $A = \{A_{ij}\} \in (P \times T)$ the mapping that contains $A_{ij} = (P_i, T_j)$ whenever P_i matches T_j. Our protocol allows a *prediction* to match a single *target*, but a *target* may be matched by many *predictions*. This allows for *target* fragmentation (1-to-many relation). We then compute the *precision* and *recall* scores as follows:

$$precision = \frac{\sum_{(i,j)\in A} |L_{i,j}|}{\sum_i |P_i|} \qquad recall = \frac{|\bigcup_{(i,j)\in A} L_{i,j}|}{\sum_j |T_j|}$$

Roughly said, *precision* stands for the quality of the *predictions* to match and cover the ground-truth, while *recall* assess the coverage of the *targets* that were matched, as illustrated in Fig. 4.

Our protocol differs from the original one [9] which allows *prediction* to match many *targets* (many-to-many relation). This case can happen when segments are close to each other, and results in a *precision* score that can exceed *1.0*. Our variant ensure precision remains in the $[0,1]$ range. Even with this modification, these metrics still have limitations: duplication of detections and fragmentation

Table 1. Vectorization performance and compute time of various MOT strategies on the *trade directories* dataset. F-score and F-score$_2$ are computed per-page and averaged on the dataset (standard deviation is shown between brackets).

	Time (ms)	F-Score		F-score$_2$	
		Train	Test	Train	Test
Last observation	**616**	**95.2 (±7.5)**	90.0 (±24.1)	**92.7 (±13.0)**	87.2 (±24.7)
SMA	652	**95.2 (±7.5)**	90.0 (±24.0)	**92.7 (±13.0)**	87.4 (±24.7)
EMA	617	92.6 (±8.6)	89.7 (±24.3)	88.3 (±16.9)	86.5 (±24.8)
Double exp. [22]	623	94.6 (±7.2)	87.3 (±25.6)	85.6 (±15.9)	81.7 (±26.4)
One euro [6]	627	**95.2 (±7.5)**	**90.1 (±24.0)**	90.8 (±17.2)	87.2 (±24.7)
Kalman [27]	633	**95.2 (±7.5)**	**90.1 (±24.0)**	**92.7 (±13.0)**	**87.6 (±24.6)**

Table 2. Vectorization performance and compute time of various line segment detection approaches on the *trade directories* dataset. Even on this rather simple dataset, these results show the superiority of MOT-based approaches for line segment detection in document images.

	Time (ms)	F-Score		F-score$_2$	
		Train	Test	Train	Test
MOT (Kalman)	633	**95.2 (±7.5)**	**90.1 (±24.0)**	**92.7 (±13.0)**	**87.6 (±24.6)**
AG3Line [39]	434	66.2 (±23.8)	72.5 (±35.4)	25.9 (±9.9)	24.2 (±13.7)
CannyLines [25]	551	81.2 (±22.7)	84.4 (±24.2)	39.0 (±13.5)	34.2 (±14.0)
EDLines [2]	314	83.2 (±23.6)	87.4 (±24.0)	35.5 (±8.3)	30.5 (±12.3)
ELSED [33]	**264**	91.1 (±11.3)	87.0 (±26.6)	45.3 (±9.6)	35.2 (±13.7)
Hough [13]	419	80.5 (±14.5)	64.8 (±30.0)	23.5 (±9.2)	18.2 (±10.1)
LSD [14]	2338	18.7 (±10.3)	12.5 (±8.5)	1.6 (±1.5)	0.5 (±0.6)
LSD II	2206	76.7 (±28.6)	53.3 (±43.7)	47.6 (±24.7)	20.7 (±17.9)

of targets are not penalized. We thus propose an updated *precision* as:

$$precision_2 = \frac{1}{\sum_i |P_i|} \times \sum_{(i,j) \in A} \frac{|L_{i,j}|}{|A_{(*,j)}|}$$

with $|A_{(*,j)}| = |\{A_{kj} \in A\}|$ being the number of matches with target T_j (number of fragments). From *precision* (resp. *precision$_2$*) and *recall*, the F-score (resp. F-score$_2$) is then computed and used as the final evaluation metric.

Results. We chose to only compare Hough based and Region growing algorithms featuring a rapid, public implementation and requiring no training. In Table 1, we show that the original Kalman strategy performs the best on this dataset. Nevertheless, the other tracking strategies reach almost the same level of performance (the first five are within a 2% range). The differences between the F-score and F-score$_2$ columns are explained by many very short detections that match a ground truth line and are more penalized with F-score$_2$. In Table 2, state-of-the-art detectors are compared: MOT-based (using *Kalman* tracker), Edlines [2],

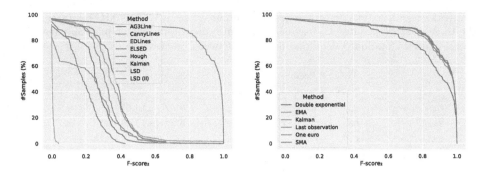

Fig. 5. Distribution of the scores obtained by traditional methods (*left*) and tracking systems (*right*) on the *trade directories* dataset. The distribution $F(\alpha)$ counts the number of pages that gets a F-score$_2$ higher than α.

Hough (from OpenCV) [20], CannyLines [25], LSD [14], LSD II with a filtering on segment lengths, ELSED [33], and AG3Line [39]. All these techniques performs much lower than the MOT-based approach because of thick lines being detected twice (especially with the F-score$_2$ metric where splits are penalized). The two previous tables exhibit a large standard deviation of the performance because of some bad quality pages (noise, page distortions...) where most methods fail to detect lines. This behavior is shown on Fig. 5 where the F-score$_2$ distributions of the dataset samples are compared. It shows some outliers at the beginning where detectors get a null score for some documents.

4.2 Instance Segmentation Task

Dataset. To demonstrate the feasibility of line segment instance segmentation, we adapted the dataset of the ICDAR 2013 music score competition for staff removal [34] to add information about staff line instances. This was, to our knowledge, the only dataset for document images which features some form of line segmentation annotation we could leverage easily. As the original competition dataset contain only binary information, i.e. for a given pixel whether it belong to any staff line, we annotated the 2000 binary images from the competition test set to assign to each pixel some unique identifier according to the particular staff line it belongs to, effectively enabling to evaluate the instance segmentation performance of the MOT-based line segment detectors. However, this enriched dataset contains only binary images with horizontal, non-intersecting staff lines. To enable hyperparameter calibration, we randomly selected 5 images to use as a train set, and kept the 1995 others in the test set.

Metrics. We use the COCO Panoptic Quality (PQ) metric [19] to measure instance segmentation quality. The metric is based pairings between proposed and ground-truth regions which overlap over more than 50%. The Intersection over Union (IoU) is used to score the pairings and enables to compute an instance

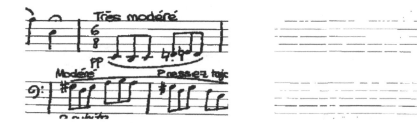

Fig. 6. Excerpt of the original image of staff lines (700 × 400) and its instance segmentation performed by the Kalman Filter predictor in 60 ms.

Table 3. Instance segmentation performance and compute time of MOT-based approaches on the *music sheets* dataset adapted from ICDAR 2013 music score competition [34]. For completeness, we also report the winner of the binary segmentation contest, though such method only performs a semantic segmentation.

	Time	Panoptic Quality		F-Score (ICDAR'13)	
	(ms)	Train	Test	Train	Test
Last observation	323	86.3 (± 5.2)	83.7 (± 11.1)	95.9 (± 2.1)	95.4 (± 2.7)
SMA	323	67.6 (± 17.0)	66.0 (± 17.6)	90.7 (± 6.5)	89.9 (± 7.4)
EMA	322	74.0 (± 14.9)	65.5 (± 18.4)	92.5 (± 4.7)	89.6 (± 7.7)
Double exp. [22]	**320**	55.4 (± 16.2)	51.7 (± 15.8)	87.3 (± 5.0)	83.8 (± 8.6)
One euro [6]	327	**87.2 (± 5.9)**	**85.1 (± 9.6)**	**95.9 (± 2.1)**	**95.7 (± 2.2)**
Kalman [27]	328	85.0 (± 7.1)	80.7 (± 15.6)	95.3 (± 2.5)	94.1 (± 5.6)
LRDE-bin [34]					**97.1**

segmentation quality ("COCO PQ") that we report here. Even if COCO PQ has become the standard metric for instance segmentation evaluation, it has two main limits: 1. it does not measure fragmentation as it considers only 1-to-1 matches; 2. IoU is somehow unstable on thin and long objects. For the sake of completeness, we also report results from the binary segmentation metric used in the original ICDAR 2013 competition [34]. The problem is seen as binary classification of pixels from which an F-Score is computed. This evaluation is not really relevant in our case because it does not include any instance information, but is shows that MOT-based techniques can be used for binary detection (Fig. 6).

Results. Table 3 shows the performance of the various trackers to identify staff lines. All trackers successfully retrieve the staff lines. The score difference is explained by the way the trackers handle line intersections (when a staff line is hidden behind an object). The *one-euro* and *last observation* tracking strategies perform almost equally and are the ones that handle the best such cases.

Qualitative Evaluation on the Municipal Atlases of Paris. Finally, to assess the capabilities of MOT-based line segment detectors, we processed a selection of

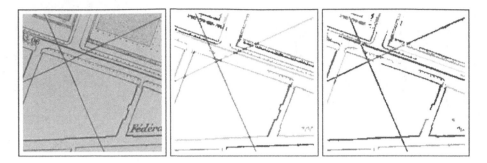

Fig. 7. Instance segmentation on map images. Left: original image, center: Kalman tracker, right: EMA tracker. With the same filtering parameters, the EMA tracker is more sensitive to gaps and overlaps and fragments objects.

images from the dataset of the ICDAR 2021 competition on historical map segmentation [8]. The challenges of this competition relied on an accurate detection of boundaries of building blocks, of map content on the sheet, and of georeferencing lines; all of them being mostly linear objects in the map images. We retained 15 images of average size 5454×3878 (21 Mpix) for our experiment: 3 for the training set, and 12 for the test set. Because no pixel-level ground truth exists for these images, we report qualitative results. The computation times of the trackers are similar: about 5–6s per 21 Mpix maps image. Time variations are due to the number of trackers to update during the process that depends on the observations integrated with the tracking strategy. From a quality standpoint (outputs available in extra material), the Kalman strategy outperforms all the other tracking strategies as shown in Fig. 7. Indeed, these documents contain many overlaps and noise that create discontinuities when extracting observations. Kalman filters enable recovering the lines even if there are hidden behind some object. On the other hand, the other trackers with simpler prediction models (such as the *one-euro* or the *last observation* trackers) integrate wrong observations and lead to line segment fragmentation.

5   Conclusion

Our goal was to implement a line segment detection method that was first proposed in the 1990s [23, 27, 28]. However, we generalized it within the Multiple Object Tracking framework that we proposed. We demonstrated the efficiency of the original proposal, which was based on Kalman filters, and also suggested some competitive alternatives. These approaches are highly robust to noise, overlapping contents, and gaps. They can produce an accurate instance segmentation of linear objects in document images at the pixel level. Results can be reproduced using open code and data available at https://doi.org/10.5281/zenodo.7871318.

Acknowledgements. This work is supported by the French National Research Agency under Grant ANR-18-CE38-0013 (SoDUCo project).

References

1. Acuna, D., Ling, H., Kar, A., Fidler, S.: Efficient interactive annotation of segmentation datasets with polygon-RNN++. In: Proceedings of the IEEE conference on Computer Vision and Pattern Recognition, pp. 859–868 (2018). https://doi.org/10.1109/cvpr.2018.00096. https://github.com/fidler-lab/polyrnn-pp
2. Akinlar, C., Topal, C.: EDLines: a real-time line segment detector with a false detection control. Pattern Recogn. Lett. **32**(13), 1633–1642 (2011). https://doi.org/10.1016/j.patrec.2011.06.001. https://github.com/CihanTopal/ED_Lib
3. Bradski, G.: The OpenCV library. Dr. Dobb's J. Softw. Tools (2000). https://opencv.org/
4. Burns, J.B., Hanson, A.R., Riseman, E.M.: Extracting straight lines. IEEE Trans. Pattern Anal. Mach. Intell. **8**(4), 425–455 (1986). https://doi.org/10.1109/TPAMI.1986.4767808
5. Canny, J.: A computational approach to edge detection. IEEE Trans. Pattern Anal. Mach. Intell. **8**(6), 679–698 (1986). https://doi.org/10.1109/TPAMI.1986.4767851
6. Casiez, G., Roussel, N., Vogel, D.: 1€ filter: a simple speed-based low-pass filter for noisy input in interactive systems. In: Proceedings of the SIGCHI Conference on Human Factors in Computing Systems, pp. 2527–2530 (2012). https://doi.org/10.1145/2207676.2208639
7. Castrejon, L., Kundu, K., Urtasun, R., Fidler, S.: Annotating object instances with a polygon-RNN. In: Proceedings of the IEEE Conference on Computer Vision and Pattern Recognition, pp. 5230–5238 (2017). https://doi.org/10.1109/cvpr.2017.477
8. Chazalon, J., et al.: ICDAR 2021 competition on historical map segmentation. In: Proceedings of the 16th International Conference on Document Analysis and Recognition (ICDAR 2021), Switzerland, Lausanne (2021). https://doi.org/10.1007/978-3-030-86337-1_46
9. Cho, N.G., Yuille, A., Lee, S.W.: A novel Linelet-Based Representation for Line segment detection. IEEE Trans. Pattern Anal. Mach. Intell. **40**(5), 1195–1208 (2018). https://doi.org/10.1109/tpami.2017.2703841. https://github.com/NamgyuCho/Linelet-code-and-YorkUrban-LineSegment-DB
10. Dai, X., Gong, H., Wu, S., Yuan, X., Yi, M.: Fully convolutional line parsing. Neurocomputing **506**, 1–11 (2022). https://doi.org/10.1016/j.neucom.2022.07.026. https://github.com/Delay-Xili/F-Clip
11. Denis, P., Elder, J.H., Estrada, F.J.: Efficient edge-based methods for estimating Manhattan frames in urban imagery. In: Forsyth, D., Torr, P., Zisserman, A. (eds.) ECCV 2008. LNCS, vol. 5303, pp. 197–210. Springer, Heidelberg (2008). https://doi.org/10.1007/978-3-540-88688-4_15
12. Desolneux, A., Moisan, L., Morel, J.M.: From Gestalt Theory to Image Analysis: A Probabilistic Approach, vol. 34. Springer, New York (2007). https://doi.org/10.1007/978-0-387-74378-3
13. Galamhos, C., Matas, J., Kittler, J.: Progressive probabilistic Hough transform for line detection. In: Proceedings. 1999 IEEE Computer Society Conference on Computer Vision and Pattern Recognition. pp. 554–560. IEEE Computer Society, Fort Collins, CO, USA (1999). https://doi.org/10.1109/cvpr.1999.786993. https://github.com/rmenke/ppht

14. Grompone von Gioi, R., Jakubowicz, J., Morel, J.M., Randall, G.: LSD: a fast line segment detector with a false detection control. IEEE Trans. Pattern Anal. Mach. Intell. **32**(4), 722–732 (2010). https://doi.org/10.1109/TPAMI.2008.300

15. He, J., Zhang, S., Yang, M., Shan, Y., Huang, T.: Bi-directional cascade network for perceptual edge detection. In: IEEE/CVF Conference on Computer Vision and Pattern Recognition (CVPR), pp. 3823–3832. IEEE (2019). https://doi.org/10.1109/CVPR.2019.00395

16. Huang, K., Wang, Y., Zhou, Z., Ding, T., Gao, S., Ma, Y.: Learning to parse wireframes in images of man-made environments. In: Proceedings of the IEEE Conference on Computer Vision and Pattern Recognition, pp. 626–635 (2018). https://doi.org/10.1109/cvpr.2018.00072

17. Illingworth, J., Kittler, J.: A survey of the Hough transform. Comput. Vis. Graph. Image Process. **44**(1), 87–116 (1988). https://doi.org/10.1016/0734-189x(88)90071-0

18. Kalman, R.E.: A new approach to linear filtering and prediction problems. J. Basic Eng. **82**(1), 35–45 (1960). https://doi.org/10.1115/1.3662552

19. Kirillov, A., He, K., Girshick, R., Rother, C., Dollár, P.: Panoptic segmentation. In: Proceedings of the IEEE/CVF Conference on Computer Vision and Pattern Recognition, pp. 9404–9413 (2019). https://doi.org/10.1109/cvpr.2019.00963

20. Kiryati, N., Eldar, Y., Bruckstein, A.M.: A probabilistic Hough transform. Pattern Recogn. **24**(4), 303–316 (1991). https://doi.org/10.1016/0031-3203(91)90073-e

21. Kultanen, P., Xu, L., Oja, E.: Randomized Hough transform (RHT). In: Proceedings. 10th International Conference on Pattern Recognition, vol. 1, pp. 631–635. IEEE (1990). https://doi.org/10.1109/ICPR.1990.118177

22. LaViola, J.J.: Double exponential smoothing: an alternative to Kalman filter-based predictive tracking. In: Proceedings of the Workshop on Virtual Environments 2003, pp. 199–206 (2003). https://doi.org/10.1145/769953.769976

23. Leplumey, I., Camillerapp, J., Queguiner, C.: Kalman filter contributions towards document segmentation. In: Proceedings of 3rd International Conference on Document Analysis and Recognition, vol. 2, pp. 765–769. IEEE (1995). https://doi.org/10.1109/icdar.1995.602015

24. Lin, T.Y., Dollár, P., Girshick, R., He, K., Hariharan, B., Belongie, S.: Feature pyramid networks for object detection. In: IEEE Conference on Computer Vision and Pattern Recognition (CVPR), pp. 936–944 (2017). https://doi.org/10.1109/CVPR.2017.106

25. Lu, X., Yao, J., Li, K., Li, L.: CannyLines: a parameter-free line segment detector. In: IEEE International Conference on Image Processing (ICIP), pp. 507–511. IEEE (2015). https://doi.org/10.1109/icip.2015.7350850. https://cvrs.whu.edu.cn/cannylines/

26. Mukhopadhyay, P., Chaudhuri, B.B.: A survey of Hough transform. Pattern Recogn. **48**(3), 993–1010 (2015). https://doi.org/10.1016/j.patcog.2014.08.027

27. Poulain d'Andecy, V., Camillerapp, J., Leplumey, I.: Kalman filtering for segment detection: application to music scores analysis. In: Proceedings of 12th International Conference on Pattern Recognition, vol. 1, pp. 301–305, October 1994. https://doi.org/10.1109/ICPR.1994.576283

28. Poulain d'Andecy, V., Camillerapp, J., Leplumey, I.: Analyse de partitions musicales. Traitement du Signal **12**(6), 653–661 (1995)

29. Pu, M., Huang, Y., Liu, Y., Guan, Q., Ling, H.: EDTER: edge detection with transformer. In: IEEE/CVF Conference on Computer Vision and Pattern Recognition (CVPR), pp. 1392–1402. IEEE (2022). https://doi.org/10.1109/CVPR52688.2022.00146

30. Ren, S., He, K., Girshick, R., Sun, J.: Faster R-CNN: towards real-time object detection with region proposal networks. In: Cortes, C., Lawrence, N., Lee, D., Sugiyama, M., Garnett, R. (eds.) Advances in Neural Information Processing Systems, vol. 28. Curran Associates, Inc. (2015). https://proceedings.neurips.cc/paper/2015/file/14bfa6bb14875e45bba028a21ed38046-Paper.pdf

31. Ronneberger, O., Fischer, P., Brox, T.: U-net: convolutional networks for biomedical image segmentation. In: Navab, N., Hornegger, J., Wells, W.M., Frangi, A.F. (eds.) MICCAI 2015. LNCS, vol. 9351, pp. 234–241. Springer, Cham (2015). https://doi.org/10.1007/978-3-319-24574-4_28

32. Sobel, I., Feldman, G.: An isotropic 3×3 image gradient operator (2015). https://doi.org/10.13140/RG.2.1.1912.4965

33. Suárez, I., Buenaposada, J.M., Baumela, L.: ELSED: enhanced line segment drawing. Pattern Recogn. **127**, 108619 (2022) https://doi.org/10.1016/j.patcog.2022.108619. https://github.com/iago-suarez/ELSED

34. Visani, M., Kieu, V., Fornés, A., Journet, N.: ICDAR 2013 music scores competition: staff removal. In: 2013 12th International Conference on Document Analysis and Recognition, pp. 1407–1411. IEEE (2013). https://doi.org/10.1109/icdar.2013.284

35. Welch, G., Bishop, G.: An introduction to the Kalman filter. Technical report TR 95–041, Department of Computer Science, University of North Carolina at Chapel Hill, Chapel Hill, NC 27599–3175 (2006). https://www.cs.unc.edu/~welch/media/pdf/kalman_intro.pdf

36. Xie, S., Tu, Z.: Holistically-nested edge detection. In: IEEE International Conference on Computer Vision (ICCV), pp. 1395–1403 (2015). https://doi.org/10.1109/ICCV.2015.164

37. Xu, Y., Xu, W., Cheung, D., Tu, Z.: Line segment detection using transformers without edges. In: Proceedings of the IEEE/CVF Conference on Computer Vision and Pattern Recognition, pp. 4257–4266 (2021). https://doi.org/10.1109/cvpr46437.2021.00424. https://github.com/mlpc-ucsd/LETR

38. Xue, N., Bai, S., Wang, F., Xia, G.S., Wu, T., Zhang, L.: Learning attraction field representation for robust line segment detection. In: Proceedings of the IEEE/CVF Conference on Computer Vision and Pattern Recognition, pp. 1595–1603 (2019). https://doi.org/10.1109/cvpr.2019.00169. https://github.com/cherubicXN/afm_cvpr2019

39. Zhang, Y., Wei, D., Li, Y.: AG3line: active grouping and geometry-gradient combined validation for fast line segment extraction. Pattern Recogn. **113**, 107834 (2021). https://doi.org/10.1016/j.patcog.2021.107834. https://github.com/weidong-whu/AG3line

40. Zhou, Y., Qi, H., Ma, Y.: End-to-end wireframe parsing. In: Proceedings of the IEEE/CVF International Conference on Computer Vision, pp. 962–971 (2019). https://doi.org/10.1109/iccv.2019.00105. https://github.com/zhou13/lcnn

TRACE: Table Reconstruction Aligned to Corner and Edges

Youngmin Baek[1,2](✉) , Daehyun Nam[3] , Jaeheung Surh[4] , Seung Shin[3] , and Seonghyeon Kim[5]

[1] NAVER Cloud, Seongnam-si, Gyeonggi-do, Korea
youngmin.baek@navercorp.com
[2] WORKS MOBILE Japan, Shibuya City, Tokyo, Japan
[3] Upstage, Yongin-si, Gyeonggi-do, Korea
{daehyun.nam,seung.shin}@upstage.ai
[4] Bucketplace, Seocho-gu, Seoul, Korea
jh.surh@bucketplace.net
[5] Kakao Brain, Seongnam-si, Gyeonggi-do, Korea
matt.mldev@kakaobrain.com

Abstract. A table is an object that captures structured and informative content within a document, and recognizing a table in an image is challenging due to the complexity and variety of table layouts. Many previous works typically adopt a two-stage approach; (1) Table detection(TD) localizes the table region in an image and (2) Table Structure Recognition(TSR) identifies row- and column-wise adjacency relations between the cells. The use of a two-stage approach often entails the consequences of error propagation between the modules and raises training and inference inefficiency. In this work, we analyze the natural characteristics of a table, where a table is composed of cells and each cell is made up of borders consisting of edges. We propose a novel method to reconstruct the table in a bottom-up manner. Through a simple process, the proposed method separates cell boundaries from low-level features, such as corners and edges, and localizes table positions by combining the cells. A simple design makes the model easier to train and requires less computation than previous two-stage methods. We achieve state-of-the-art performance on the ICDAR2013 table competition benchmark and Wired Table in the Wild(WTW) dataset.

Keywords: Table reconstruction · Table recognition · Split-Merge · separator segmentation

1 Introduction

Tables are structured objects that capture informative contents and are commonly found in various documents, such as financial reports, scientific papers, invoices, application forms, etc. With the growth of automated systems, the need to recognize tables in an image has increased. The table recognition task can be

G. A. Fink et al. (Eds.): ICDAR 2023, LNCS 14191, pp. 472–489, 2023.
https://doi.org/10.1007/978-3-031-41734-4_29

(a) Previous two-stage approach

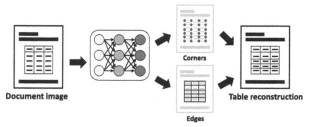

(b) Proposed bottom-up approach

Fig. 1. Comparison of our bottom-up approach with previous two-stage ones.

divided into two sub-tasks: Table Detection (TD) and Table Structure Recognition (TSR). TD is a task to localize tabular objects in an image, and TSR is a job to identify the adjacency relationship between cells (or contents in cells) within the corresponding table area [7]. In the past, the two tasks were conducted using heuristics after extracting the encoded meta-data in the PDF format [32, 38]. The rule-based approach is out of our scope, and this paper focuses on fully image-based table recognition.

Recognizing a table in an image is very challenging because tables have complex layouts and styles. It usually consists of ruling lines, but cells in borderless tables should be semantically separated. Cells and tables have specific considerations as opposed to general objects; there are many empty cells, vast spaces between cells (or contents), extreme aspect ratios, and various sizes.

With the recent advancement of Deep Learning, TD and TSR performances have been significantly improved [8]. Most previous methods take the two-stage approach: performing TD and TSR independently, as shown in Fig. 1(a). This approach inevitably has two weaknesses. The first is the high training and inference costs. Creating a table recognition system requires more effort since TD and TSR models must be trained separately. Moreover, the overall table recognition system should serve two models in a row, which causes inefficiencies within the inference process. The second is a constrained end performance bounded by the TD's results. Although TD and TSR are highly correlated and are complementary tasks, there is no interaction between the models to improve each model's performance. In summary, handling TD and TSR with a single model is highly desirable from a practical point of view. To the best of our knowledge, only two such researches [27, 28] have been studied so far.

Toward an end-to-end approach, we analyzed the essential elements consti-
tuting tabular objects. A table comprises cells; each cell can be represented using
corners and edges. This insight led us to think in the opposite way. Once corners
and edges are found, we can reconstruct cells and tables in a bottom-up manner,
as depicted in Fig. 1(b). Consequently, we propose a novel table reconstruction
method called *TRACE (Table Reconstruction Aligned to Corners and Edges)*.
In TRACE, a single segmentation model predicts low-level features (corners and
edges) rather than the bounding box of cells or contents. After that, simple
post-processing can reconstruct the tabular structure.

Using a single model significantly increases the time efficiency in the training
and inference phases. We made it possible to reconstruct complex tables by
classifying the edges into explicit and implicit lines. Our method shows robust
and stable performance on public datasets, including ICDAR2013 and WTW,
with state-of-the-art performance.

The main contributions of our paper are summarized as following:

- We propose a novel end-to-end table reconstruction method with a single
  model solely from an image.
- In a bottom-up manner, we propose a table reconstruction method starting
  from the basic elements of cells, such as corners and edges.
- The proposed method reconstructs complex tables by classifying edges into
  explicit and implicit lines.
- The proposed method achieves superior performance in both clean document
  datasets (ICDAR13) and natural image benchmarks (WTW).

2   Related Work

There are two input types for table recognition; PDF and image. The PDF file
contains content data, including textual information and coordinates, which are
leveraged by the table analysis methods in the early stage [4]. These rule-based
methods rely on visual clues such as text arrangement, lines, and templates. The
conventional method only works on PDF-type inputs. As technology advanced,
image-input table recognizers have been proposed mainly based on a statistical
machine learning approach as listed in [12]. However, it still requires much human
effort to design handcrafted features and heuristics. In the era of deep learning,
many methods have been studied, showing superior performance compared to
the conventional ones [8]. In this paper, we mainly address deep-learning-based
related works.

Table Detection(TD). TD methods are roughly divided into two categories;
object detection-based, and semantic segmentation-based.

The object detection-based approaches adopt state-of-the-art generic object
detectors to table detection problems. For example, Faster R-CNN was adopted
by [6, 36, 37, 42, 44], and YOLO was used in [11]. Recently, thanks to its instance
segmentation ability, Mask R-CNN-based methods [1, 28] were studied. These

methods invented techniques such as data augmentation, image transformation, and architecture modification to mitigate a discrepancy between the nature of tables and objects in terms of aspect ratio and sparse visual features.

There were several attempts to apply the semantic segmentation method FCN [23] on table localization, such as [9,13,27,47]. However, differing from the object segmentation task, the unclear boundary and sparse visual clues in the table region limit the capabilities of these approaches to find accurate table segments.

Table Structure Recognition(TSR). The primary purpose of TSR is to identify the structural information of cellular objects. Structure information can be (1) row, column coordinates, and IDs, (2) structural description, (3) connections between contents. According to the problem definition, TSR methods can be categorized into three approaches; detection-based, markup generation-based, and graph-based.

Firstly, many studies have utilized a detection-based approach. However, detection targets are different; row/column regions, cell/content bounding boxes, and separators. Researchers in [37,40,41] have proposed to detect row/column regions based on segmentation and off-the-shelf object detectors. Several studies have proposed detectors for detecting cells or their contents [28,30,31,49]. In recent years, a split-merge approach has emerged as a popular technique for TSR, in which the separators between cells are initially detected and then subsequently merged [14,19,25,45,48]. This strategy is particularly effective in representing the complex layout of tables.

Secondly, markup generation-based approaches try to generate LaTex code or HTML tag directly rather than identifying the coordinates of cellular objects. Collecting LaTex and HTML tags of the table is beneficial for synthetic data generation. Therefore, a large size synthetic table datasets by rendering from the tags have been released such as Table 2Latex [3], TableBank [17], and PubTabNet [50] In general, the encoder-decoder architecture converts images into markup tags. TableFormer differs from other generation methods in that it decodes not only structural tags but also cell box coordinates [26]. However, these approaches need a relatively large dataset and are difficult to handle complex tables in natural scenes.

Lastly, graph-based methods that treat words or cell contents as nodes have been proposed. They analyze the connection between cell relationships using a graph neural network [2,18,29,34]. When a node pair is found, it determines whether two nodes are the same row or column and further performs table localization. However, the biggest problem with this method is that it requires a content detection process or additional input from PDF to acquire content. At the same time, it is difficult to deal with empty cells.

End-to-End Table Recognition. A few researchers proposed end-to-end table analyses that include TD and TSR. Most used the two-stage pipeline using two separate models for TD and TSR. DeepDeSRT [37] adopted Faster-RCNN [33]

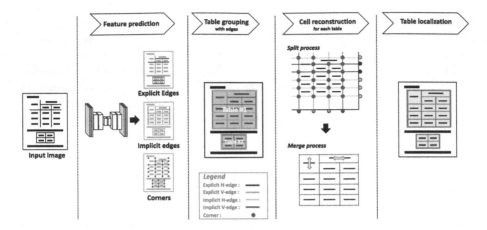

Fig. 2. Schematic illustration of our network architecture. (Color figure online)

as a table detector, and SSD [21] as a cell detector. RobusTabNet [25] used CornerNet [16] for TD, and proposed line prediction model for TSR. Recently, Zheng et al. proposed Global Table Extractor (GTE) [49] that used two separate table and cell detectors based on Faster-RCNN.

We found only a couple of literature using a single model so far. TableNet [27] identifies table and column region. The limitation to separate rows was solved using additional heuristics. CascadeTabNet [28] proposed a novel approach to classify table and cell regions simultaneously even though they used a single model. However, the method still has difficulties handling the blank cell, which needs separated branches for handling bordered and borderless tables after model prediction.

3   Methodology

3.1   Overview

Our method predicts corner and edge segmentation maps that form cells using a deep learning model. The model outputs five segmentation maps; the first channel is used to detect cell corners and the other four channels are used to detect horizontal and vertical edges of a cell box. We define an edge as a separation line between cells, and we predict four types of edges since each edge could represent an explicit or implicit line, either vertically or horizontally. Following the binarization process of the aggregated edge maps, a candidate table region is obtained through connected component labeling. This approach enables the reconstruction of multiple tables with a single inference. In the cell reconstruction step, we simply calculate the position of the separation by projecting horizontal edges to the y-plane and vertical edges to the x-plane for each table candidate region. Here, corners are also used in the search for the separation lines. Then, the spanned cells are merged if there is no edge between the cell bounding boxes. Finally, the table regions are localized through the combination of individual cells. An overview of the pipeline is illustrated in Fig. 2. Note that as shown in the legend, the four types of edges are indicated by different colors; explicit horizontal edges in blue, explicit vertical edges in green, implicit horizontal edges in yellow, and implicit vertical edges in purple.

3.2   Corner and Edges Prediction Model

The required corner and edge information are trained using a CNN-based segmentation model. We adopted ResNet-50 [10] as the feature extraction backbone, and the overall architecture is similar to that of U-Net [35], which aggregates low- and high-level features. The final output has five channels; a corner map, explicit horizontal/vertical edge maps, and implicit horizontal/vertical edge maps.

For ground truth label generation, we need the cell bounding boxes and properties of each edge. For the corner map, we render a fixed sized Gaussian heatmap centered on every corner point of the cell. For the edge map, a line segment is drawn with a fixed thickness on every side of the cell. Here, horizontal- and vertical- edge ground truths are generated on the different channels. Also, a property of the edge indicates whether the line segment is visible or not. If visible, we use the *Explicit Edge* channels, if not, the *Implicit Edge* channels are used.

We use the MSE loss for the objective L, defined as,

$$L = \sum_p \sum_i ||S_i(p) - S_i^*(p)||_2^2, \tag{1}$$

where $S_i(p)$ denotes the ground truth of i-th segmentation map, and $S_i^*(p)$ denote the predicted segmentation map at the pixel p.

Note that the TRACE method has two notable features that set it apart from traditional cell detection methods relying on the off-the-shelf object detectors. Firstly, TRACE detects low-level visual features rather than high-level semantic elements such as cell bounding boxes or content bounding boxes. This approach allows for easier learning due to the distinctiveness of low-level visual cues and enables the effective handling of empty cells, which is challenging for other methods. Secondly, TRACE is capable of identifying both explicit and implicit edges when generating separators. The distinction between visible and non-visible lines helps to find separators from the table image. The table reconstruction is achieved through a series of heuristic techniques in the post-processing after the low-level features have been obtained.

3.3   Data Preparation

As explained in the previous section, we need cell bounding box data with attributes. Unfortunately, public datasets only have two types of bounding box annotation; content bounding box [2, 5, 7] and wired cell bounding box [24]. Some

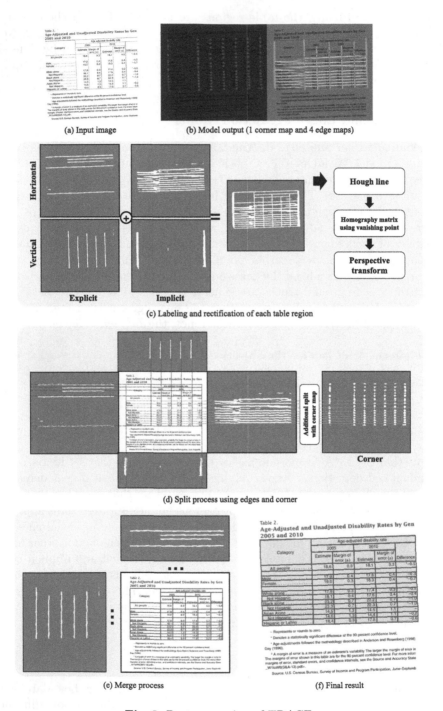

Fig. 3. Post-processing of TRACE.

public datasets for structure recognition do not provide any coordinates for content or cell, but structure markup such as HTML tags or LaTex symbols [17,50]. This lack of available data for our approach motivated us to collect our own dataset.

We collected document images from the web including invoices and commonly used government documents. Also, some came from TableBank [17]. The dataset includes complicated tables with visible and invisible separation lines. The annotators were asked to annotate line segments and its attributes for us to create cell bounding boxes with properties.

It is easy to maintain data consistency of explicit edges since the visual cues are clear. However, data inconsistency issue arise when dealing with implicit edges because it is not clear where to form a line between the cells. To alleviate this issue, the annotators were guided to construct an equidistant bisector line that equally separates the cells.

3.4   Post-processing for Reconstruction

The procedure for table reconstruction by TRACE is illustrated in Fig. 3. Given an input image, the TRACE model predicts corner and edge maps, which are then processed in the following steps.

In the image rectification step, we apply binarization to the inferred segmentation maps, and then use the combination of all binary edge maps to approximate the location of each table. The Hough line transform is used to detect lines in the binary edges. We rectify the table image by making horizontal and vertical lines perpendicular. This rectification is not necessary for most scanned document images.

TRACE employs a split-merge strategy, inspired by SPLERGE [45], to reconstruct tables. The split process is demonstrated in the Fig. 4. It begins by projecting explicit horizontal edge maps onto the y-plane and explicit vertical edge maps onto the x-plane. The midpoint of the projected edge group on the plane is considered to be the location of each separation line.

In the case of implicit horizontal lines, the projection may not clearly separate them. When a row in the table includes empty cells, the implicit horizontal lines may be combined due to the ambiguity of the content borders. To resolve this issue, we use the corner map. We apply the same binarization and projection processes to the corner map. If the projected line group is thick, with more than two peak points of corners, the final split point is calculated based on the peak points of the corner groups, rather than the midpoint of the line group.

After the split process, the cell merge process is initiated. This process utilizes both explicit and implicit edge maps. The basic rule applied in this process is that if two adjacent cells lacked a binary edge map in the midpoint of their separator, the separator is removed, and the cells are merged.

Finally, the quadrilateral of the table location is determined by computing the coordinates of the top-left, top-right, bottom-left, and bottom-right corners of all detected cells. If the table image is transformed due to the rectification

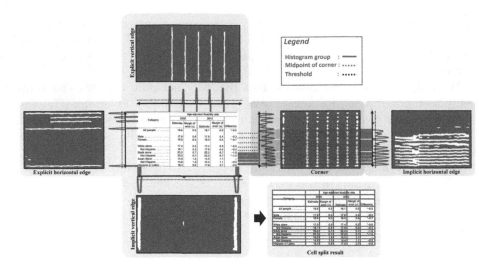

Fig. 4. The split process in TRACE is performed to separate the cells of a table. After the binarization of the edge maps, the mid-point of the grouped binary edge maps is determined as the separator. This process is repeated for both the horizontal and vertical axes. To address the issue of ambiguity in implicit horizontal lines, the peak positions of the projected corner map are also utilized.

process, the cell coordinates and table coordinate are unwarped to the image coordinate system.

It is important to note that while the basic post-processing for TRACE is relatively straightforward, the requirements may vary depending on the dataset. For instance, the WTW dataset does not provide annotations for implicit lines, but some images in the dataset contain inner tables with invisible edges. Thus, heuristics may need to be designed based on the end task. This will be discussed in the experiment section, where examples will be presented.

4   Experiments

We conducted experiments on public benchmarks, including the ICDAR2013 table competition dataset and Wired Table in the Wild (WTW) dataset, to validate the proposed method.

4.1   Dataset and Evaluation

To train TRACE, we need cell bounding box data with visibility flags. However, there is no public dataset that satisfies this condition. We collected document images from the web, and manually annotated tables. Our in-house dataset mainly consists of financial and scientific documents. The number of total images is 9717, which are divided into 7783 training images, 971 validation ones, and

963 testing ones. We will soon make a portion of our in-house dataset available to the public. The dataset will include an adjacency relation-based evaluation metric for assessing performance.

ICDAR2013 Table Competition benchmark [7] is the table dataset that is commonly evaluated. The dataset is composed of 156 tables in PDF format from the EU/US government website. We only tested our method on this dataset without finetuning. We used the official evaluation protocol. For table detection, we calculated the character-level recall, precision and F1-score along with *Purity* and *Completeness*. *Purity* increases if a detected table does not include any character that are not also in the GT region, *Completeness* counts whether a table includes all characters in the GT region. For the table structure recognition evaluation, we used adjacency relations-based recall precision measures.

WTW dataset [24] contains not only document images but also images of scenes from the wild. Therefore, it has a variety of tables in terms of types, layouts, and distortions. There are 10,970 training images and 3,611 testing images. WTW only focuses on wired tables. We follow the evaluation metric by the authors; (1) for cell detection, cell box-level recall/precision with IoU=0.9 are used, (2) for structure recognition, cell adjacency relationship-level recall/precision from the matched tabular cells with IoU=0.6 are used.

Our method is unable to evaluate widely-used table datasets for tag generation, such as SciTsr [2] and TableBank [17] since evaluation metrics cell content with text attributes. This necessitates the use of additional OCR results, which beyond the scope of image-based table reconstruction.

4.2 Implementation Detail

In the training process, we used a ResNet50 backbone pretrained on ImageNet. The longer side of the training images are resized to 1280 while preserving the aspect ratio. The initial learning rate is set to 1e-4, and decayed at every 10k iteration steps. We train the model up to 100k iterations with a batch size of 12. The basic deep learning techniques such as *ADAM* [15] optimizer, *On-line Hard Negative Mining* [39] and data augmentations including color variations, random rotations, and cropping are applied.

In our methodology, the parameters for post-processing were determined empirically. Specifically, binary thresholds were set to 0.5 for explicit edge maps and 0.2 for implicit edge maps. And, if the length of a vertical or horizontal edge was less than 25% of the corresponding table height or width, the edge was deemed not suitable for the split process and was discarded.

4.3 Result on Document Dataset(ICDAR2013)

The table detection results of various methods on ICDAR2013 benchmark are listed in Table 1. TRACE's F1-measure shows competitive performance with GTE [49], and it achieved a higher score in terms of *Purity* and *Completeness*.

Some TD methods are not directly compared here, because 1) CascadeTab-Net [28] used a subset of test images for evaluation, and 2) CDeC-Net [1] and

Table 1. Table Detection Results on ICDAR13 dataset.

Method	Input type	Recall	Precision	F1	Complete	Pure
Nurminen [7]	PDF	90.77	92.1	91.43	114	151
Silva [43]	PDF	98.32	92.92	95.54	149	137
TableNet [27]	Image	95.01	95.47	95.47	–	–
Tran et al. [46]	Image	96.36	95.21	95.78	147	141
TableBank [17]	Image	–	–	96.25	–	–
DeepDeSRT [37]	Image	96.15	97.40	96.77	–	–
GTE [49]	Image	**99.77**	**98.97**	**99.31**	147	146
Ours (TRACE)	Image	98.08	97.67	97.53	**150**	**147**

Table 2. Table Structure Recognition Results on ICDAR13 dataset.

Method	Venue	Recall	Precision	F1	Approach	GT	PDF
Nurminen	ICDAR13	94.09	95.12	94.60	TSR	✓	✓
TabStructNet	ECCV20	89.70	91.50	90.60	TSR	✓	–
SPLERGE	ICDAR19	90.44	91.36	90.89	TSR	✓	–
Split-PDF+Heuristics	ICDAR19	94.64	95.89	95.26	TSR	✓	✓
GTE	WACV21	92.72	94.41	93.50	TD+TSR	–	–
GTE(with GT)	WACV21	95.77	96.76	96.24	TSR	✓	–
LGPMA	ICDAR21	93.00	97.70	95.30	TSR	✓	–
Ours(TRACE)	–	**96.69**	**98.47**	**97.46**	E2E	–	–

other methods [11,36,42] reported their performance using IoU-based metrics. The results of PDF-based methods are only for reference since they detect table regions from PDF-metadata, not from images.

For the table structure recognition task, TRACE achieved the highest score when comparing with previous works as shown in Table 2. The important point we want to emphasize is that TRACE is the only end-to-end approach. TSR-only methods require the cropped table region in the image, but ours does not. Performing both TD and TSR tasks simultaneously is difficult since the inaccurate table detection results in lowering the end performance. For example, when comparing the result with and without table detection in GTE, the performance dropped by 2.7%. Our bottom-up approach proved its robustness both in TD and TSR tasks.

As with the TD results, some of previous TSR methods cannot be listed in the result table directly because the authors randomly chose 34 images for testing in order to overcome the lack of training images (e.g. [27,37])

4.4 Result on Wild Scene Dataset (WTW)

The WTW dataset only contains wired tables, so, TRACE was trained on this dataset without incorporating implicit edge maps. For the experiment on the

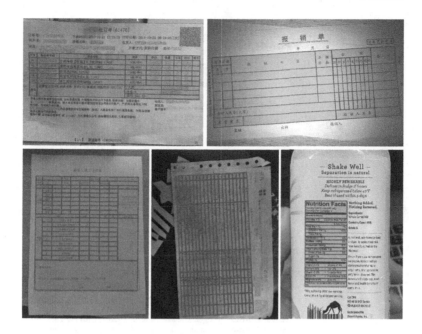

Fig. 5. Table reconstruction results on WTW datasets.

Table 3. Cell-level detection and adjacency relation-level structure recognition results on the WTW dataset.

Method	Physical Coordinates			Adjacency Relation			Approach	GT Table
	R	P	F1	R	P	F1		
Cycle-Centernet [24]	**78.5**	**78.0**	**78.3**	91.5	93.3	92.4	TD+TSR	-
TSRFormer [19]	-	-	-	93.2	93.7	93.4	TSR	✓
NCGM [20]	-	-	-	**94.6**	93.7	94.1	TSR	✓
Ours (TRACE)	63.8	65.7	64.8	93.5	**95.5**	**94.5**	E2E	-

WTW dataset, no additional data was utilized for the training process. The results on WTW are visualized in Fig. 5.

Cell-level evaluation is conducted on WTW instead of table-level detection. The results on WTW is shown in Table 3. TRACE achieved the best performances in the structure recognition task. It is also noted that we compared the TSR methods that require GT table regions. TRACE is the result obtained by performing the task end-to-end without table cropping.

We evaluated our method using official parameters provided by the authors of the WTW dataset. Here, the IoU threshold was set to 0.9 for cell detection

Table 4. Cell detection result with various IoU thresholds for matching cells on WTW dataset.

IoU	R	P	F1	Ref
0.9	63.8	65.7	64.8	Parameter for detection
0.8	89.5	92.1	90.8	–
0.7	93.6	96.3	94.9	–
0.6	94.8	97.5	96.1	Parameter for TSR

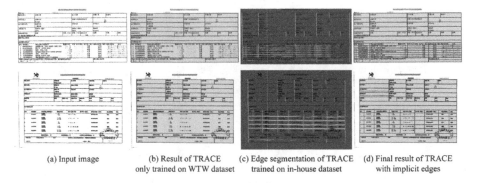

(a) Input image   (b) Result of TRACE   (c) Edge segmentation of TRACE   (d) Final result of TRACE
                     only trained on WTW dataset   trained on in-house dataset   with implicit edges

Fig. 6. Comparison of TRACE results with and without implicit edge. Even complex documents like custom declarations are correctly recognized with TRACE.

evaluation. However, we found false positive cases even for sufficiently reasonable results, which means 0.9 is too strict parameter for matching of small cells. Also, some imprecise annotations were found, which is critical for small cell boxes. We estimate that 0.7 is a more appropriate parameter for IoU threshold based on the quantitative analysis. For reference, further experiments with various IoU were conducted as shown in Table 4.

Figure 6 visualizes the table reconstruction result of TRACE on WTW dataset. Most of the images in WTW contain wired tables. Yet, there are document images including borderless tables in which product items are listed. The user's expectation is to reconstruct cellular objects that are semantically listed up. Here, we additionally investigate table reconstruction results on customs declarations through TRACE trained on our in-house dataset. The final result was qualitatively better than the human annotated ground truth. By showing correct table reconstruction results on these mixed-type tables, TRACE proved its high usability and generalization ability.

4.5   Discussion

Bordered and Borderless Table Comparison. We separate ICDAR2013 benchmark into bordered and borderless tables, and evaluated them separately. We investigated this to check the difference in difficulty although the number

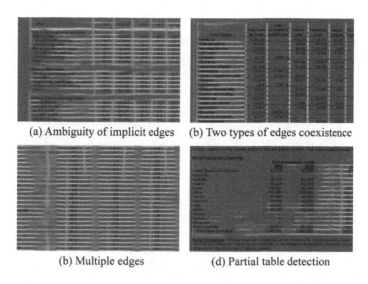

(a) Ambiguity of implicit edges   (b) Two types of edges coexistence

(b) Multiple edges   (d) Partial table detection

Fig. 7. Failure cases caused by implicit edges in the ICDAR2013 dataset.

of files differ: 30 bordered and 31 borderless sheets. In the TSR task, the F1-score on bordered and borderless tables are 99.4% and 93.85%, respectively. Like this, the score on bordered tables are quite high, and the main reason for the performance drop is with the borderless cases. To improve this, implicit edges could be counted as separation line candidates, and additional heuristics such as the existence of content can be applied further.

Borderless Table Failure Cases. The causes of failure on most borderless tables are analyzed in detail, as shown in Fig. 7. They are roughly classified into the following four cases. (a) Two different rows merged due to implicit horizontal edges, (b) Explicit and implicit edges coexist in the same separation line, (c) Multiple vertical edges due to a wide interval between cell contents, (d) Partial table detection due to a large interval. Most of the problems arise from the ambiguity of the separation line position. To mitigate this, we need to utilize global attention-based machine learning techniques like SwinTransformer [22] as a future work.

5   Conclusion

We have proposed a novel end-to-end table reconstruction method, TRACE, which performs both table detection and structure recognition with a single model. This method differs from conventional approaches, which rely on a two-stage process, as it reconstructs cells and tables from fundamental visual elements such as corners and edges, in a bottom-up manner. Our model effectively recognizes tables, even when they are rotated, through the use of simple and effective post-processing techniques. We have achieved state-of-the-art performance on both the clean document dataset (ICDAR2013) and tables in the wild dataset (WTW). In future work, we plan to address weakly-supervised techniques for training the model with more diverse data.

References

1. Agarwal, M., Mondal, A., Jawahar, C.: Cdec-net: Composite deformable cascade network for table detection in document images. In: 2020 25th International Conference on Pattern Recognition (ICPR), pp. 9491–9498. IEEE (2021)
2. Chi, Z., Huang, H., Xu, H.D., Yu, H., Yin, W., Mao, X.L.: Complicated table structure recognition. arXiv preprint arXiv:1908.04729 (2019)
3. Deng, Y., Rosenberg, D., Mann, G.: Challenges in end-to-end neural scientific table recognition. In: 2019 International Conference on Document Analysis and Recognition (ICDAR), pp. 894–901. IEEE (2019)
4. Embley, D.W., Hurst, M., Lopresti, D., Nagy, G.: Table-processing paradigms: a research survey. IJDAR 8(2), 66–86 (2006)
5. Gao, L., et al.: Icdar 2019 competition on table detection and recognition (ctdar). In: 2019 International Conference on Document Analysis and Recognition (ICDAR), pp. 1510–1515. IEEE (2019)
6. Gilani, A., Qasim, S.R., Malik, I., Shafait, F.: Table detection using deep learning. In: 2017 14th IAPR international conference on document analysis and recognition (ICDAR). vol. 1, pp. 771–776. IEEE (2017)
7. Göbel, M., Hassan, T., Oro, E., Orsi, G.: Icdar 2013 table competition. In: 2013 12th International Conference on Document Analysis and Recognition, pp. 1449–1453. IEEE (2013)
8. Hashmi, K.A., Liwicki, M., Stricker, D., Afzal, M.A., Afzal, M.A., Afzal, M.Z.: Current status and performance analysis of table recognition in document images with deep neural networks. IEEE Access 9, 87663–87685 (2021)
9. He, D., Cohen, S., Price, B., Kifer, D., Giles, C.L.: Multi-scale multi-task fcn for semantic page segmentation and table detection. In: 2017 14th IAPR International Conference on Document Analysis and Recognition (ICDAR). vol. 1, pp. 254–261. IEEE (2017)
10. He, K., Zhang, X., Ren, S., Sun, J.: Deep residual learning for image recognition. In: Proceedings of the IEEE Conference On Computer Vision And Pattern Recognition, pp. 770–778 (2016)
11. Huang, Y., et al.: A yolo-based table detection method. In: 2019 International Conference on Document Analysis and Recognition (ICDAR), pp. 813–818. IEEE (2019)
12. Jorge, A.M., Torgo, L., et al.: Design of an end-to-end method to extract information from tables. IJDAR 8(2), 144–171 (2006)
13. Kavasidis, I., et al.: A saliency-based convolutional neural network for table and chart detection in digitized documents. In: Ricci, E., Rota Bulò, S., Snoek, C., Lanz, O., Messelodi, S., Sebe, N. (eds.) ICIAP 2019. LNCS, vol. 11752, pp. 292–302. Springer, Cham (2019). https://doi.org/10.1007/978-3-030-30645-8_27
14. Khan, S.A., Khalid, S.M.D., Shahzad, M.A., Shafait, F.: Table structure extraction with bi-directional gated recurrent unit networks. In: 2019 International Conference on Document Analysis and Recognition (ICDAR), pp. 1366–1371. IEEE (2019)

15. Kingma, D.P., Ba, J.: Adam: A method for stochastic optimization. arXiv preprint arXiv:1412.6980 (2014)

16. Law, H., Deng, J.: Cornernet: Detecting objects as paired keypoints. In: Proceedings of the European conference on computer vision (ECCV), pp. 734–750 (2018)

17. Li, M., Cui, L., Huang, S., Wei, F., Zhou, M., Li, Z.: Tablebank: Table benchmark for image-based table detection and recognition. In: Proceedings of The 12th Language Resources And Evaluation Conference, pp. 1918–1925 (2020)

18. Li, Y., Huang, Z., Yan, J., Zhou, Y., Ye, F., Liu, X.: GFTE: graph-based financial table extraction. In: Del Bimbo, A., et al. (eds.) ICPR 2021. LNCS, vol. 12662, pp. 644–658. Springer, Cham (2021). https://doi.org/10.1007/978-3-030-68790-8_50

19. Lin, W., et al.: Tsrformer: Table structure recognition with transformers. In: Proceedings of the 30th ACM International Conference on Multimedia, pp. 6473–6482 (2022)

20. Liu, H., Li, X., Liu, B., Jiang, D., Liu, Y., Ren, B.: Neural collaborative graph machines for table structure recognition. In: Proceedings of the IEEE/CVF Conference on Computer Vision and Pattern Recognition, pp. 4533–4542 (2022)

21. Liu, W., et al.: SSD: single shot multibox detector. In: Leibe, B., Matas, J., Sebe, N., Welling, M. (eds.) ECCV 2016. LNCS, vol. 9905, pp. 21–37. Springer, Cham (2016). https://doi.org/10.1007/978-3-319-46448-0_2

22. Liu, Z., et al.: Swin transformer: Hierarchical vision transformer using shifted windows. In: Proceedings of the IEEE/CVF International Conference on Computer Vision, pp. 10012–10022 (2021)

23. Long, J., Shelhamer, E., Darrell, T.: Fully convolutional networks for semantic segmentation. In: Proceedings of the IEEE Conference on Computer Vision and Pattern Recognition, pp. 3431–3440 (2015)

24. Long, R., et al.: Parsing table structures in the wild. In: Proceedings of the IEEE/CVF International Conference on Computer Vision, pp. 944–952 (2021)

25. Ma, C., Lin, W., Sun, L., Huo, Q.: Robust table detection and structure recognition from heterogeneous document images. Pattern Recogn. **133**, 109006 (2023)

26. Nassar, A., Livathinos, N., Lysak, M., Staar, P.: Tableformer: Table structure understanding with transformers. In: Proceedings of the IEEE/CVF Conference on Computer Vision and Pattern Recognition, pp. 4614–4623 (2022)

27. Paliwal, S.S., Vishwanath, D., Rahul, R., Sharma, M., Vig, L.: Tablenet: Deep learning model for end-to-end table detection and tabular data extraction from scanned document images. In: 2019 International Conference on Document Analysis and Recognition (ICDAR), pp. 128–133. IEEE (2019)

28. Prasad, D., Gadpal, A., Kapadni, K., Visave, M., Sultanpure, K.: Cascadetabnet: An approach for end to end table detection and structure recognition from image-based documents. In: Proceedings of the IEEE/CVF Conference on Computer Vision and Pattern Recognition Workshops, pp. 572–573 (2020)

29. Qasim, S.R., Mahmood, H., Shafait, F.: Rethinking table recognition using graph neural networks. In: 2019 International Conference on Document Analysis and Recognition (ICDAR), pp. 142–147. IEEE (2019)

30. Qiao, L., et al.: LGPMA: complicated table structure recognition with local and global pyramid mask alignment. In: Lladós, J., Lopresti, D., Uchida, S. (eds.) ICDAR 2021. LNCS, vol. 12821, pp. 99–114. Springer, Cham (2021). https://doi.org/10.1007/978-3-030-86549-8_7

31. Raja, S., Mondal, A., Jawahar, C.V.: Table structure recognition using top-down and bottom-up cues. In: Vedaldi, A., Bischof, H., Brox, T., Frahm, J.-M. (eds.) ECCV 2020. LNCS, vol. 12373, pp. 70–86. Springer, Cham (2020). https://doi.org/10.1007/978-3-030-58604-1_5

32. Rastan, R., Paik, H.Y., Shepherd, J.: Texus: A task-based approach for table extraction and understanding. In: Proceedings of the 2015 ACM Symposium on Document Engineering, pp. 25–34 (2015)
33. Ren, S., He, K., Girshick, R., Sun, J.: Faster r-cnn: Towards real-time object detection with region proposal networks. In: Advances in Neural Information Processing Systems, vol. 28 (2015)
34. Riba, P., Dutta, A., Goldmann, L., Fornés, A., Ramos, O., Lladós, J.: Table detection in invoice documents by graph neural networks. In: 2019 International Conference on Document Analysis and Recognition (ICDAR), pp. 122–127. IEEE (2019)
35. Ronneberger, O., Fischer, P., Brox, T.: U-Net: convolutional networks for biomedical image segmentation. In: Navab, N., Hornegger, J., Wells, W.M., Frangi, A.F. (eds.) MICCAI 2015. LNCS, vol. 9351, pp. 234–241. Springer, Cham (2015). https://doi.org/10.1007/978-3-319-24574-4_28
36. Saha, R., Mondal, A., Jawahar, C.: Graphical object detection in document images. In: 2019 International Conference on Document Analysis and Recognition (ICDAR), pp. 51–58. IEEE (2019)
37. Schreiber, S., Agne, S., Wolf, I., Dengel, A., Ahmed, S.: Deepdesrt: Deep learning for detection and structure recognition of tables in document images. In: 2017 14th IAPR international Conference on Document Analysis and Recognition (ICDAR). vol. 1, pp. 1162–1167. IEEE (2017)
38. Shigarov, A., Mikhailov, A., Altaev, A.: Configurable table structure recognition in untagged pdf documents. In: Proceedings of the 2016 ACM Symposium On Document Engineering, pp. 119–122 (2016)
39. Shrivastava, A., Gupta, A., Girshick, R.: Training region-based object detectors with online hard example mining. In: Proceedings of the IEEE Conference On Computer Vision And Pattern Recognition, pp. 761–769 (2016)
40. Siddiqui, S.A., Fateh, I.A., Rizvi, S.T.R., Dengel, A., Ahmed, S.: Deeptabstr: Deep learning based table structure recognition. In: 2019 International Conference on Document Analysis and Recognition (ICDAR), pp. 1403–1409. IEEE (2019)
41. Siddiqui, S.A., Khan, P.I., Dengel, A., Ahmed, S.: Rethinking semantic segmentation for table structure recognition in documents. In: 2019 International Conference on Document Analysis and Recognition (ICDAR), pp. 1397–1402. IEEE (2019)
42. Siddiqui, S.A., Malik, M.I., Agne, S., Dengel, A., Ahmed, S.: Decnt: deep deformable CNN for table detection. IEEE access 6, 74151–74161 (2018)
43. Silva, A.: Parts that add up to a whole: a framework for the analysis of tables. Edinburgh University, UK (2010)
44. Sun, N., Zhu, Y., Hu, X.: Faster r-cnn based table detection combining corner locating. In: 2019 International Conference on Document Analysis and Recognition (ICDAR), pp. 1314–1319. IEEE (2019)
45. Tensmeyer, C., Morariu, V.I., Price, B., Cohen, S., Martinez, T.: Deep splitting and merging for table structure decomposition. In: 2019 International Conference on Document Analysis and Recognition (ICDAR), pp. 114–121. IEEE (2019)
46. Tran, D.N., Tran, T.A., Oh, A., Kim, S.H., Na, I.S.: Table detection from document image using vertical arrangement of text blocks. Int. J. Contents 11(4), 77–85 (2015)
47. Yang, X., Yumer, E., Asente, P., Kraley, M., Kifer, D., Lee Giles, C.: Learning to extract semantic structure from documents using multimodal fully convolutional neural networks. In: Proceedings of the IEEE Conference on Computer Vision and Pattern Recognition, pp. 5315–5324 (2017)
48. Zhang, Z., Zhang, J., Du, J., Wang, F.: Split, embed and merge: an accurate table structure recognizer. Pattern Recogn. 126, 108565 (2022)

49. Zheng, X., Burdick, D., Popa, L., Zhong, X., Wang, N.X.R.: Global table extractor (gte): A framework for joint table identification and cell structure recognition using visual context. In: Proceedings of the IEEE/CVF Winter Conference On Applications Of Computer Vision, pp. 697–706 (2021)
50. Zhong, X., ShafieiBavani, E., Jimeno Yepes, A.: Image-based table recognition: data, model, and evaluation. In: Vedaldi, A., Bischof, H., Brox, T., Frahm, J.-M. (eds.) ECCV 2020. LNCS, vol. 12366, pp. 564–580. Springer, Cham (2020). https://doi.org/10.1007/978-3-030-58589-1_34

Contour Completion by Transformers and Its Application to Vector Font Data

Yusuke Nagata⬡, Brian Kenji Iwana⁽✉⁾⬡, and Seiichi Uchida⬡

Kyushu University, Fukuoka, Japan
yusuke.nagata@human.ait.kyushu-u.ac.jp, {iwana,uchida}@ait.kyushu-u.ac.jp

Abstract. In documents and graphics, contours are a popular format to describe specific shapes. For example, in the True Type Font (TTF) file format, contours describe vector outlines of typeface shapes. Each contour is often defined as a sequence of points. In this paper, we tackle the contour completion task. In this task, the input is a contour sequence with missing points, and the output is a generated completed contour. This task is more difficult than image completion because, for images, the missing pixels are indicated. Since there is no such indication in the contour completion task, we must solve the problem of missing part detection and completion simultaneously. We propose a Transformer-based method to solve this problem and show the results of the typeface contour completion.

Keywords: Vector font generation · Contour completion · Transformer

1 Introduction

Contours are a popular method of representing vector objects in document and graphics recognition. For example, digital-born fonts are often formatted using the True Type Font (TTF) file format. This format consists of a sequence of contour points for each character in the font set. The contour points, called *control points*, contain a point identifier, contour identifier, two-dimensional coordinates, and an on-curve/off-curve flag. The point identifier identifies the location in the sequence, the contour identifier identifies the particular contour that the point lies on, the coordinates describe the coordinates on a relative scale and the flags determine whether the point is on a curve or an angle.

In this paper, we perform vector completion of vector character contours using TTF data. However, unlike other completion problems, in our problem set, there is no indication of which parts of the contour data are missing. For example, as shown in Fig. 1 (a), in standard image completion problems, the missing pixels are specified with a mask or fixed pixel values (e.g., gray or negative values).

This research was partially supported by MEXT-Japan (Grant No. JP22H00540 and JP23K16949).

G. A. Fink et al. (Eds.): ICDAR 2023, LNCS 14191, pp. 490–504, 2023.
https://doi.org/10.1007/978-3-031-41734-4_30

(a) Image-based completion problem

(b) Indicator-free contour completion problem

Fig. 1. Comparison between image-based completion and vector-based completion. Image-based completion uses a bounding box or mask to indicate the missing pixels. The proposed task does not have any indication of where the missing control points are.

For time series, interpolation and regression are similar; they require knowledge of the location of predicted data. Conversely, we perform indication-free contour completion in that there is no indication, no mask, and no information about the missing coordinates. Therefore, the problem set is more difficult to solve due to needing to identify the missing points and estimate the missing values simultaneously.

There are two reasons for addressing the character contour completion problem. The first is the application value. Consider an application that converts bitmap character information into vector contour information (i.e., raster-vector transformation). In this application, if the character is partially overlapped, hidden by a design element, damaged, or incomplete, the result will be poor. By using automatic contour completion, such cases might be resolved. Also, in font design, automatic completion is possible without all of the contour points. The second is the suitability for machine learning. There are a vast number of character fonts, and TTF is the most common format. It is possible to leverage a large amount of data to train robust models to learn the trends in font and character tendency.

We consider the missing contour completion problem as a problem of converting a missing sequence into a completed sequence. To solve this problem, we propose using an Encoder-Decoder Transformer network [18]. Transformers were originally proposed for the Natural Language Processing (NLP) domain. However, they have been expanded to use for images [4] and sequences [20]. The Transformer is trained to convert an input character contour sequence with missing control points (unknown to the Transformer) into a complete contour sequence.

Transformers have two major advantages for this problem. The first is the ease of handling series of varying lengths. Being able to tackle series of varying lengths is important because the contour sequences are not fixed in length. In our model, neither the input nor the output could have a fixed length. The second advantage is due to self-attention being suitable for our application. Character contours have structures that are both global and local to specific areas. For example, serifs exist in specific parts of characters. If the character is missing a serif, then the contour completion model needs to recognize that the serif is missing and reconstruct the serif based on the other serifs in the character. Thus, because Transformers have global receptive fields, we believe they are suitable for character contour completion. In fact, the font style recognition task has shown that high accuracy can be achieved by using contour data with Transformers [11].

To evaluate the performance of the proposed method, we conducted a contour completion experiment using fonts collected from Google Fonts [1]. We formulated two problem setups: random deletion and burst deletion. In random deletion, randomly selected contour points were deleted. This is evaluated to demonstrate that the proposed method can identify and repair individual points. In burst deletion, large segments of the contour sequences are removed. Burst deletion provides a difficult problem because there is less information for the model to infer the correct contour from.

The main contributions of this paper are as follows.

– We use a Transformer to simultaneously identify missing control points in character contours and estimate the missing values.
– We propose the use of a multifaceted loss that considers the different features in the TTF file format for vector fonts.
– We trained the above model using character data collected from Google Fonts and quantitatively and qualitatively evaluated the extent to which contour completion is possible.

2   Related Work

2.1   Transformers

Transformers [18] are neural networks that use layers of Multi-Head Self-Attention and fully-connected layers and are designed for sequences recognition. They were originally proposed for Natural Language Processing (NLP) [18] such as translation and text classification. They have had many successes in the NLP field [5,9].

However, in recent years, Transformers have been applied to other fields such as image recognition [6] and time series recognition [20]. To do this, instead of word token sequences for NLP, other embedding methods are used to create the input sequences. For example, Vision Transformers (ViT) [4] use sequences of patched embedded in vector space. For time series, many works use a linear embedding of time series elements along with positional embeddings for the inputs to the Transformers [8,21].

2.2   Vector Font Generation

There have been attempts related to font generation using vector formats. Campbell et al. [2] pioneered outline-based font generation using a fixed-dimensional font contour model as a low-dimensional manifold. Lopes et al. [10] proposed a Variational Autoencoder (VAE) based model that extracts font style features from images and uses a Scalable Vector Graphic (SVG) based decoder to convert it to SVG format. In a similar work, Reddy et al. [12] proposed Im2vec which generates a vector from an image. DeepSVG [3] is a font generation model using a hierarchical generative network to create SVG format fonts. DeepVecFont [19] is a network that supports both image and sequential fonts using an encoder-decoder structure. In another work by Reddy et al. [13], they choose not to use SVG and instead learn font representations defined by implicit functions.

Unlike these font generation methods, we are focusing on automatic font repair and completion. Specifically, instead of generating vector fonts from images, we use sequence-to-sequence font generation to restore missing elements of fonts. In addition, vector-based font generation methods typically have indications of where the missing data is or how much is missing. To the best of our knowledge, none of the previous methods simultaneously detect missing points and fill in the values.

3   Missing Contour Completion with Transformers

We propose using a Transformer to automatically complete contour data when given a contour with missing points. Figure 2 shows an overview of the proposed method.

3.1   Contour Data Representation

To represent the contours of fonts as a sequence for the Transformer, we use the True Type Font (TTF) file format of each font. The TTF files contain the vector information of fonts in the form of the contours of each character in that font. Specifically, the outlines of each character are expressed as a variable number of point coordinates. Each point is represented by a 5-dimensional vector f_i^c, which consists of x and y coordinates, a Contour ID, a Point ID, and a curve flag. The x and y coordinates are the relative locations of the control point. The Contour ID $c \in \{-1, 1, \ldots, C\}$ and Point ID $i \in \{-1, 1, \ldots, I^c\}$ are identifiers that label the contour number C and point number I^c, respectively. The IDs, $c = -1$ and $i = -1$, are used as placeholders for the start-of-sequence and end-of-sequence tokens. In the contour data, $f_1^c = f_{I^c}^c$, since the start and end points of each contour indicate exactly the same position. Thus, the character data is represented by a set of $N = \sum_c P^c$ vectors, which are input to the Transformer in the order $\mathbf{f} = f_{-1}^{-1}, f_1^1, \ldots, f_{I^1}^1, f_1^2, \ldots, f_i^c, \ldots, f_{I^C}^C$.

Finally, the curve flag is an indicator if the control point is *on-curve* or *off-curve*. If the control point is on-curve, then B-spline is used to define the curvature. If it is off-curve, then an angle is used. Furthermore, for the purpose of our input data representation, the start-of-sequence and end-of-sequence flags are grouped with the curve flags.

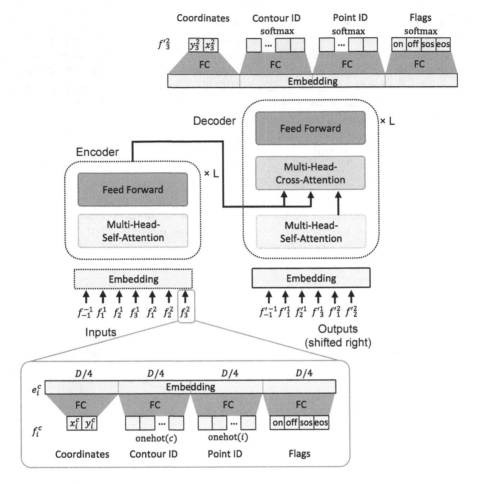

Fig. 2. Overview of the proposed Encoder-Decoder Transformer for contour completion.

3.2   Encoder-Decoder Transformer for Contour Completion

We propose the use of an Encoder-Decoder Transformer to perform contour completion, as shown in Fig. 2. The input to the Transformer is a character contour with missing control points and the output is the completed character contour. This is analogous to a translation task in that a variable length input is translated to a variable length completed contour output. Notably, the input has no indication of the location of missing control points. Therefore, the proposed method simultaneously detects missing control points and infers the values of the missing control points.

The proposed method is different from the original Transformer [18] in three ways:

- We adapt the token embedding of the Transformer for the use of vector contours, i.e. TTF file format.
- We consider the problem as a multitask problem to predict the coordinates, Contour ID, Point ID, and flags, separately.
- We propose a new loss to optimize the Transformer in order to consider each task.

Embedding. To adapt the standard token embedding for contours, each element of $\mathbf{f}$ is embedded into a D-dimensional vector. As shown in Fig. 2, to do this, we split the features of the TTF file format into four groups, the coordinates, the Contour ID, the Point ID, and the flags. Each group is then used with a fully-connected (FC) layer to embed it into a $D/4$-dimensional vector. The resulting four $D/4$-dimensional vectors are concatenated to create a D-dimensional vector embedding e_i^c for each input f_i^c in $\mathbf{f}$. The sequence of embeddings is the input to the Transformer.

Namely, the 2-dimensional coordinate values (x_i^c, y_i^c) are embedded using the FC layer to the first segment of e_i^c. Next, the Contour ID and Point ID are converted to one-hot representations, and then embedded using the FC layer. The final $D/4$ dimensions use a 4-dimensional one-hot input representing an on-curve flag, an off-curve flag, a start-of-sequence (sos) flag, and an end-of-sequence (eos) flag.

Encoder-Decoder Structure. The Transformer is arranged in an Encoder-Decoder structure, similar to how it is used in NLP. The model is divided into an Encoder and a Decoder. The input of the Encoder is the previously mentioned sequence of embeddings. Similar to a standard Transformer, the Encoder is made of L Transformer layers that each contain a Multi-Head Self-Attention layer and a Feed Forward layer. The output of the Encoder is an element-wise embedding that represents each input in a sequence of N number of D-dimensional vectors.

The Decoder is structured similarly to Vaswani et al. [18]. The input of the Decoder is the generated sequence of control points, shifted right. The output of the Decoder is the next predicted control point in the sequence. To connect the Encoder and the Decoder, the output of the Encoder is used as the key K and value V of the Multi-Head Cross-Attention layer. The query Q is the out of the Multi-Head Self-Attention layer of the Decoder. Much like the Encoder, the Decoder also has L number of Transformer layers.

Output. One proposed modification of the Transformer is to use a multitask prediction for the output of the Decoder. Specifically, as shown in the top of Fig. 2, the D-dimensional vector representation of the Decoder is divided into four parts, each part relating to the coordinates, Contour ID, Point ID, and flags, respectively. Then, an FC layer is used with each part to determine the prediction of each part. The FC layer for the coordinates has two nodes, one for x_i^c and one for y_i^c. The FC layers for the Contour ID, Point ID, and flags use softmax to perform the one-hot prediction.

Loss Functions. In order to train the network, we propose a multifaceted loss to consider all of the information in the TTF file format. The loss function consists of four terms: contour loss $\mathcal{L}_{\text{contour}}$, point loss $\mathcal{L}_{\text{point}}$, coordinate loss $\mathcal{L}_{\text{coord}}$, and flag loss $\mathcal{L}_{\text{flag}}$. The contour loss $\mathcal{L}_{\text{contour}}$ ensures that the predicted point is part of the correct contour. It is defined as the cross entropy loss between the predicted contour ID q_c and the true contour ID p_c, or:

$$\mathcal{L}_{\text{contour}} = -\frac{1}{C}\sum_{c=1}^{C} p_c \log(q_c), \tag{1}$$

where C is the number of contours. The point loss $\mathcal{L}_{\text{point}}$ is similar to the contour except that is for the control point ID, or:

$$\mathcal{L}_{\text{point}} = -\frac{1}{I}\sum_{i=1}^{I^c} p_i \log(q_i), \tag{2}$$

where I^c is the number of control points in the contour c, p_i is the true point ID, and q_i is the predicted point ID. This loss makes sure that the model learns the location of the missing control point in the sequence. The coordinate loss is the difference between the predicted coordinates $\hat{y}_i$ and the true coordinates y_i. For this loss, we use L1 loss, or:

$$\mathcal{L}_{\text{coord}} = \sum_{i=1}^{I^c} |y_i - \hat{y}_i|. \tag{3}$$

Finally, the flag loss $\mathcal{L}_{\text{flag}}$ is the loss between the four flags p_u and the predicted four flags q_u, or:

$$\mathcal{L}_{\text{flag}} = -\frac{1}{4}\sum_{u=1}^{4} p_u \log(q_u). \tag{4}$$

The four losses are summed in the total loss $\mathcal{L}$:

$$\mathcal{L} = \mathcal{L}_{\text{contour}} + \mathcal{L}_{\text{point}} + \mathcal{L}_{\text{coord}} + \mathcal{L}_{\text{flag}}. \tag{5}$$

The Transformer is trained using the total loss $\mathcal{L}$.

4   Experiment

4.1   Dataset

Font Database. For the experiments, we used the Google Fonts [1] database because it is one of the most popular font sets for font analysis research [14–16]. It also contains a large number of fonts in TTF file format. They are annotated with five font style categories (*Serif, Sans-Serif, Display, Handwriting,* and *Monospace*). From Google Fonts, we selected the Latin fonts that were used in STEFANN [14]. We then omitted *Monospace* fonts because they are mainly

(a) Completed contour   (b) Random deletion   (c) Burst deletion

Fig. 3. Comparison of the deletion methods used to simulate missing data. The red points are on-curve points, the blue points are off-curve points, and the gray points are deleted control points.

characterized by their kerning rule (i.e., space between adjacent letters); therefore, some *Monospace* fonts have the same style of another font type, such as *Sans-Serif*. We set the upper limit of Contour IDs and Point ID to $C = 4$ and $P = 102$, respectively. Consequently, we used 489 *Serif* fonts, 1,275 *Sans-Serif*, 327 *Display*, and 91 *Handwriting* fonts. The fonts are split into three font-disjoint subsets; that is, the training set with 1,777 fonts, the validation set with 200 fonts, and the test set with 205 fonts.

Simulation of Missing Data. To simulate missing contour data, the control points were removed from the font contours. We modify the fonts in two ways, *random deletion* and *burst deletion*. In random deletion, as shown in Fig. 3 (b), D points are randomly removed from the characters. For burst deletion, intervals or segments of control points are removed. To perform burst deletion, as shown in Fig. 3 (c), a random control point is removed along with the $D - 1$ points surrounding it. In the experiments, we compare different deletion rates. Thus, the number of deleted points D is determined by $D = rN$, where r is the deletion rate and N is the total number of control points in a character.

4.2   Implementation Details

The Transformer used in the experiment is shown in Fig. 2. The Encoder and Decoder use $L = 4$ number of the Transformer layers each. We set Mutli-head Attention heads as $M = 6$ and The internal dimensionality of FC layers set at $D = 256$. To train the network, a batch size of 400 is used with an Adam [7] optimizer with an initial learning rate of 0.0001. Training of the network was stopped by no decrease in validation loss for 30 epochs.

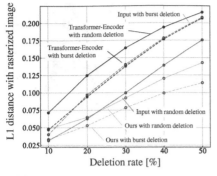 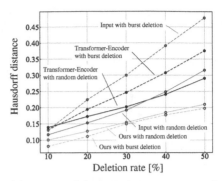

(a) L1 distance between ours and the comparative method

(b) **Hausdorff distance between ours and** the comparative method

Fig. 4. L1 distance with rasterized image and Hausdorff distance

4.3 Comparative Method

To evaluate the proposed method, we use a Transformer-Encoder model. This model is similar to the Encoder used in the proposed method, except the number of layers and Multi-Head Self-Attention layers are increased to be equal to the proposed method. The model receives the same embedding as the proposed method and the output is performed in the same manner as the proposed method. The difference between the two is that the comparative model is like a standard neural network in that the layers are stacked and not parallel like the Encoder-Decoder Transformer. For fairness, the network has the same hyperparameters and is trained with the same optimizer and training scheme as the proposed method.

Note that, unlike the proposed method which can have different length inputs and outputs, the comparative method can only predict same length outputs as the inputs. This means that a placeholder must be used for the missing control points. This is much like traditional completion tasks where the missing information is explicitly indicated. Thus, only the value prediction is required, which reduces the difficulty of the problem.

4.4 Quantitative Evaluation

To compare the proposed method with the comparative method, we evaluate them with different deletion rates with both random deletion and burst deletion. Figure 4 shows the L1 distance and Hausdorff distance [17] between the contour completion methods and the ground truth. To calculate the L1 distance d_{L1}, the contours are rasterized into bitmap images of 250×250 pixels. The bitmaps are then compared using:

$$L1(\mathbf{f}, \hat{\mathbf{f}}) = \sum_{(x,y)} |R(\mathbf{f})_{(x,y)} - R(\hat{\mathbf{f}})_{(x,y)}|, \tag{6}$$

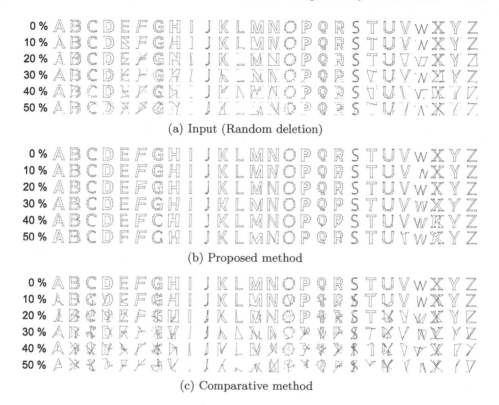

(a) Input (Random deletion)

(b) Proposed method

(c) Comparative method

Fig. 5. The results on random deletion. The columns are characters in the same font style, and the rows are the same missing percentages. The first row is the ground truth complete contour and second to sixth row have a deletion ratio of 10% to 50%, respectively. The colors indicate the Contour ID.

where (x, y) is the coordinate of each pixel, $R(\mathbf{f})$ is the rasterized font, and $R(\hat{\mathbf{f}})$ is the rasterized predicted font. For the Hausdorff distance, the sequences are considered point clouds. The Hausdorff distance can be calculated as the distance between two point clouds with no point-to-point correspondence. It is widely used in 3D point cloud generation. It is defined as:

$$\breve{H}(\mathbf{f}, \hat{\mathbf{f}}) = \max_{f \in \mathbf{f}} \{\min_{f' \in \hat{\mathbf{f}}} \{\|f, f'\|\}\}, \tag{7}$$

$$H(\mathbf{f}, \hat{\mathbf{f}}) = \max\{\breve{H}(\mathbf{f}, \hat{\mathbf{f}}), \breve{H}(\hat{\mathbf{f}}, \mathbf{f})\}, \tag{8}$$

where $\|\cdot\|$ is the L2 distance between the nearest points in the respective sequences, and f is a point in the ground truth $\mathbf{f}$ and f' is a point in the predicted $\hat{\mathbf{f}}$. For both distances, smaller distances mean that the contours are more accurately reproduced.

For the L1 distance and the Hausdorff distance evaluations, the proposed method is better than the comparative method at all deletion rates and deletion

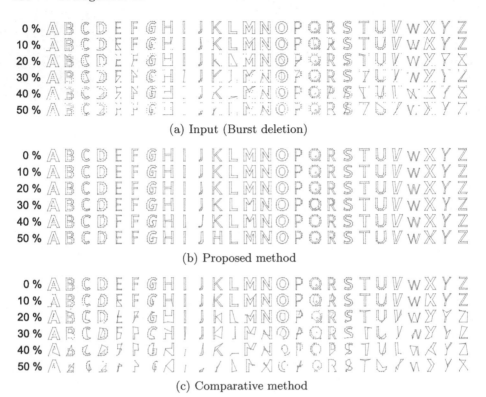

(a) Input (Burst deletion)

(b) Proposed method

(c) Comparative method

Fig. 6. The results from burst deletion. The columns are characters in the same font style, and the rows are the same missing percentages. The first row is the ground truth complete contour and second to sixth row have a deletion ratio of 10% to 50%, respectively. The colors are the different contours of each character.

methods. Despite burst deletion having a very high L1 distance and Hausdorff distance, the proposed method was able to perform the contour completion task very well. However, at 10% and 20%, the L1 distance for the proposed method is larger than the input. This is because, at low deletion rates, most of the control points are intact. That means the rasterized characters are mostly correct, whereas the generated contours might have very slight translations causing the pixel-based distance to be larger than the inputs with incomplete contours.

Also, another interest interesting result is that both Transformers were able to have better results for L1 distance on the input with burst deletion than random deletion despite the burst deletion having higher distances originally. This means that quantitatively, the Transformers were able to recover large continuous segments better than randomly deleted points.

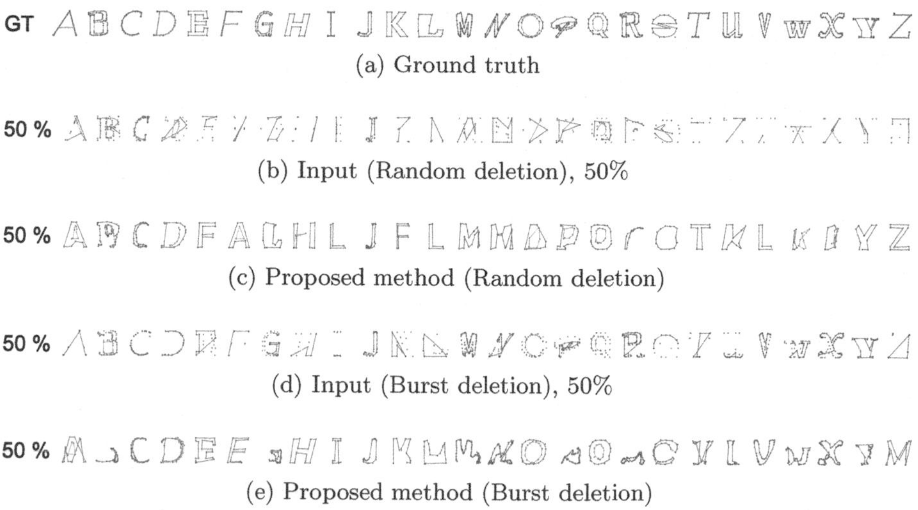

GT $A\mathbb{B}CD\mathbb{E}FG\mathcal{H}IJK\mathbb{L}\mathbb{M}\mathcal{N}O\mathcal{P}QR\mathbb{S}TU\mathbb{V}\mathrm{w}\mathcal{X}\mathrm{Y}Z$

(a) Ground truth

50% $A\mathbb{B}C\mathcal{D}F\mathcal{Y}\mathcal{Z}:I|IJ\mathcal{Z}\setminus\mathcal{R}\mathbb{M}\mathcal{Z}\mathcal{P}Q\mathbb{F}\mathbb{S}\mathbb{T}\mathcal{Z}.\mathcal{Z}.\pi\mathcal{X}\setminus\backslash\mathbb{F}$

(b) Input (Random deletion), 50%

50% $A\mathbb{P}CDFA\mathbb{L}HLJFLM\mathbb{M}\Delta\mathbb{P}O\Gamma\mathbb{C}TKL\mathbb{K}\mathbb{P}YZ$

(c) Proposed method (Random deletion)

50% $\wedge\mathbb{B}CD\mathcal{R}\Gamma\mathbb{S}\mathbb{W}\colon IJ\mathbb{K}\mathbb{L}\mathbb{W}\mathcal{N}O\mathcal{P}\mathbb{Q}\mathbb{P}O\Gamma\mathbb{L}V\mathrm{w}\mathcal{X}\mathrm{Y}\mathcal{Z}$

(d) Input (Burst deletion), 50%

50% $\mathbb{M}\mathbb{L}CD\mathbb{E}F\mathbb{S}HIJK\mathbb{L}M\mathbb{M}\mathbb{K}O\mathbb{A}O\mathbb{A}CV\mathbb{L}V\mathrm{w}\mathcal{X}\mathbb{J}M$

(e) Proposed method (Burst deletion)

Fig. 7. Examples of Poor Reconstructions by the Proposed Method

4.5 Qualitative Evaluation

Figure 5 shows the generated results for random deletion and Fig. 6 shows the generated results for the burst deletion. In both cases, the proposed method outputs contour data closer to the correct contour data. Notably, the proposed method was able to reconstruct many of the characters, despite having only 50% of the original control points.

Supporting Fig. 4, the figures show that the proposed method was more effective on the data with burst deletion than random deletion. It was able to complete the contours more accurately. On the random deletion characters in Fig. 5 (b), there is more overlap between the contours. For example, the "B" and "O" at high deletion rates had more overlapping orange and blue contours. Furthermore, "G," "C," "N," "S," and "J" have more regions where the contours are too thin, when compared to Fig. 6 (b). This indicates that the random deletion removes more important information that cannot be recovered by the proposed method when compared to burst deletion. The proposed method is able to reconstruct lost segments from burst deletion more accurately based on the intact features.

Examples of poor reconstructions are shown in Fig. 7. In these examples, the proposed method was unable to reconstruct the characters accurately. Many of the burst deletion characters only had small regions to base the reconstructions on. Thus, the proposed method was unable to extrapolate the characters.

4.6 Comparison Between Styles

The degree of difficulty varies depending on the font category. For example, Fig. 8 has examples of reconstructed contours from the proposed method when

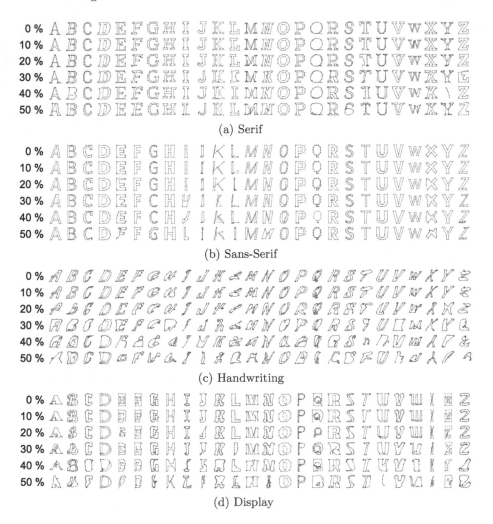

Fig. 8. Examples of generated results using the proposed method on burst deletion separated by font category.

burst deletion is used. Each row is different deletion rates with 10% at the top and 50% on the bottom of each subsection. As seen in Fig. 8 (a) and (b), reconstruction was relatively easy for Serif and Sans-Serif, despite having very little information to reconstruct the character from. This is due to Serif and Sans-Serif fonts having less variation compared to Display and Handwriting fonts. Display fonts in particular have a large variation between fonts, thus the results of 50% deletion rate are poor.

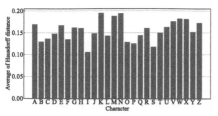

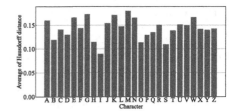

| (a) Hausdorff distance with random deletion | (b) Hausdorff distance with burst deletion |

Fig. 9. Average distance between the proposed method and the ground truth when separated by character class

4.7 Comparison Between Characters

Certain characters tend to be more difficult to reconstruct. A comparison of characters is shown in Fig. 9. In the figure, the Hausdorff distance is calculated for each character with inputs with random deletion and burst deletion. The average from 10–50% was used for the evaluation. It can be seen that there is a difference in accuracy for each character. The characters that obtained high accuracy with the proposed model were "I," "B," "P," and "S." On the contrary, "K," "U," "V," and "W' were the characters that did not obtain high accuracy. In addition, "X" was one of the characters for which the difference between the random deletion and the burst deletion was significant.

5   Conclusion

In this paper, we proposed the use of a Transformer-based model to tackle contour completion for vector characters. The proposed method is indication-free in that it does not require any indication of where, if, or how much the contour is missing. It simultaneously solves missing point identification and value estimation. Specifically, we proposed an Encoder-Decoder Transformer with a multitask output and loss.

To evaluate the proposed method, we constructed two datasets of characters in a variety of fonts. The characters are represented by sequences of control points. Then, in order to test the contour completion ability of the proposed method, we perform corruptions of the vector characters. To do this, in one dataset, we randomly delete points and in the other, we remove a continuous segment from the character. In this paper, we demonstrated that the proposed method can effectively reconstruct the corrupted characters. Future work could include the introduction of an image-based loss function to improve the quality of the output contour data.

References

1. Google fonts. https://fonts.google.com/. Accessed 20 Jan 2023
2. Campbell, N.D., Kautz, J.: Learning a manifold of fonts. ACM Trans. Graph. **33**(4), 1–11 (2014)
3. Carlier, A., Danelljan, M., Alahi, A., Timofte, R.: DeepSVG: a hierarchical generative network for vector graphics animation. In: NeurIPS, vol. 33, pp. 16351–16361 (2020)
4. Dosovitskiy, A., et al.: An image is worth 16 × 16 words: transformers for image recognition at scale. arXiv preprint arXiv:2010.11929 (2020)
5. Kenton, J.D.M.W.C., Toutanova, L.K.: BERT: pre-training of deep bidirectional transformers for language understanding. In: NAACL-HLT, pp. 4171–4186 (2019)
6. Khan, S., Naseer, M., Hayat, M., Zamir, S.W., Khan, F.S., Shah, M.: Transformers in vision: a survey. ACM Comput. Surv. **54**(10s), 1–41 (2022)
7. Kingma, D.P., Ba, J.: Adam: a method for stochastic optimization. In: ICLR (2015)
8. Li, S., et al.: Enhancing the locality and breaking the memory bottleneck of transformer on time series forecasting. In: NeurIPS, vol. 32 (2019)
9. Liu, Y., et al.: Roberta: a robustly optimized BERT pretraining approach. arXiv preprint arXiv:1907.11692 (2019)
10. Lopes, R.G., Ha, D., Eck, D., Shlens, J.: A learned representation for scalable vector graphics. In: ICCV, pp. 7929–7938 (2019)
11. Nagata, Y., Otao, J., Haraguchi, D., Uchida, S.: TrueType transformer: character and font style recognition in outline format. In: Uchida, S., Barney, E., Eglin, V. (eds.) Document Analysis Systems, DAS 2022. LNCS, vol. 13237, pp. 18–32. Springer, Cham (2022). https://doi.org/10.1007/978-3-031-06555-2_2
12. Reddy, P., Gharbi, M., Lukac, M., Mitra, N.J.: Im2Vec: synthesizing vector graphics without vector supervision. In: CVPR, pp. 7342–7351 (2021)
13. Reddy, P., Zhang, Z., Wang, Z., Fisher, M., Jin, H., Mitra, N.: A multi-implicit neural representation for fonts. In: NeurIPS, pp. 12637–12647 (2021)
14. Roy, P., Bhattacharya, S., Ghosh, S., Pal, U.: STEFANN: scene text editor using font adaptive neural network. In: CVPR, pp. 13228–13237 (2020)
15. Srivatsan, N., Barron, J., Klein, D., Berg-Kirkpatrick, T.: A deep factorization of style and structure in fonts. In: EMNLP-IJCNLP, pp. 2195–2205 (2019)
16. Srivatsan, N., Wu, S., Barron, J., Berg-Kirkpatrick, T.: Scalable font reconstruction with dual latent manifolds. In: EMNLP, pp. 3060–3072 (2021)
17. Taha, A.A., Hanbury, A.: An efficient algorithm for calculating the exact Hausdorff distance. IEEE Trans. Pattern Anal. Mach. Intell. **37**(11), 2153–2163 (2015)
18. Vaswani, A., et al.: Attention is all you need. In: NeurIPS, vol. 30, pp. 5998–6008 (2017)
19. Wang, Y., Lian, Z.: DeepVecFont: synthesizing high-quality vector fonts via dual-modality learning. ACM Trans. Graph. **40**(6), 1–15 (2021)
20. Wen, Q., et al.: Transformers in time series: a survey. arXiv preprint arXiv:2202.07125 (2022)
21. Zhou, H., et al.: Informer: beyond efficient transformer for long sequence time-series forecasting. In: AAAI, pp. 11106–11115 (2021)

Towards Making Flowchart Images Machine Interpretable

Shreya Shukla(ID), Prajwal Gatti(ID), Yogesh Kumar(✉)(ID), Vikash Yadav(ID),
and Anand Mishra(✉)(ID)

Vision, Language and Learning Group (VL2G), IIT Jodhpur,
Jodhpur, India
{shukla.12,pgatti,kumar.204,yadav.41,mishra}@iitj.ac.in

Abstract. Computer programming textbooks and software documenta-
tions often contain flowcharts to illustrate the flow of an algorithm or procedure. Modern OCR engines often tag these flowcharts as graphics and
ignore them in further processing. In this paper, we work towards making
flowchart images machine-interpretable by converting them to executable
Python codes. To this end, inspired by the recent success in natural language to code generation literature, we present a novel transformer-based
framework, namely FLoCo-T5. Our model is well-suited for this task,
as it can effectively learn semantics, structure, and patterns of programming languages, which it leverages to generate syntactically correct code.
We also used a task-specific pre-training objective to pre-train FLoCo-
T5 using a large number of logic-preserving augmented code samples.
Further, to perform a rigorous study of this problem, we introduce the
FLoCo dataset that contains 11,884 flowchart images and their corresponding Python codes. Our experiments show promising results, and
FLoCo-T5 clearly outperforms related competitive baselines on code
generation metrics. We make our dataset and implementation publicly
available (https://vl2g.github.io/projects/floco).

Keywords: Flowchart Understanding · Code Generation · Large
Language Models

1 Introduction

Flowcharts are widely used across documents to represent algorithms, processes,
or workflows in a graphical manner and provide a clear and concise understanding of complex processes. They contain short textual commands or conditions
inside various intent-specific shapes, e.g., diamond for decision-making block,
rhomboid for input, and output. These shapes are connected with directed or
undirected arrows to define a sequential flow of information and processing. In
computer programming textbooks and software documentation, flowcharts are
more often used as a program-planning tool to communicate the complex logic of
programs and keep track of the data flow through a process, as shown in Fig. 1.

© The Author(s), under exclusive license to Springer Nature Switzerland AG 2023
G. A. Fink et al. (Eds.): ICDAR 2023, LNCS 14191, pp. 505–521, 2023.
https://doi.org/10.1007/978-3-031-41734-4_31

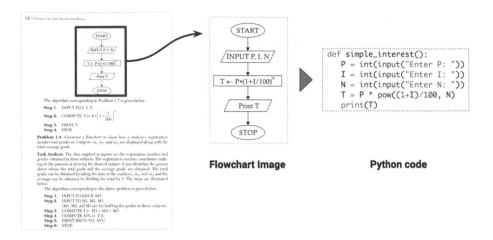

Flowchart Image   **Python code**

Fig. 1. Flow2Code. A scanned document from a programming text book [10] containing a flowchart is shown here. Our aim is to convert flowchart images to executable computer programs. We scope ourselves to cropped flowchart images and Python codes in this work.

These visual depictions help beginners in programming to focus on formulating the logic behind the program while ignoring the intricacies of the syntax of different programming languages. Machine interpretation of these flowcharts followed by automatic code generation would not only help school students and people from non-software engineering backgrounds but also speed up software development. In order to make these flowchart images machine-interpretable, we study the problem of automatically converting flowchart images to a computer program (code) in a high-level language. This problem is referred to as FLOW2CODE [16]. Despite its practical importance and utility, FLOW2CODE has not been rigorously explored in the literature.

There is no existing large-scale dataset in flowchart-to-code literature for performing a rigorous experimental evaluation of FLOW2CODE task. To fill this gap, we introduce the first dataset, namely FLOCO. The FLOCO contains 11,884 flowchart images along with corresponding Python codes. Inspired by the success of transformer-based approaches in natural language and code generation tasks [2, 3, 8, 17, 26], we present FLOCO-T5 – a novel framework to convert flowchart images to the Python code. In our proposed architecture, we first convert the flowchart images into a sequence encoding by automatically detecting different shapes and reading text using off-the-shelf OCR engines [1, 20]. Then, to adapt our transformer model to this novel domain, we pre-train it on the masked token modeling objective using a large number of logic-preserving data-augmented code samples. This pre-training step helps the model to understand the structure and semantics of the programming language as well as the flowchart encoding. Finally, we fine-tune the model on train split as a sequence-to-sequence generation problem where flowchart encoding and expected Python code are used as input and output sequence, respectively. We conducted extensive experiments with the sequence encoding of the flowchart images (as shown in Table 2) and compared the code generation performance of our model against competitive baselines, namely Vanilla Transformer [32], BART [19], PLBART [2] and CodeT5 [33]. Our experiments show that FLoCo-T5 outperforms all other baselines on different code generation metrics, showing the efficacy of the proposed pre-training objective and data augmentation techniques. Qualitative results and further analysis (Figs. 6 and 7) demonstrate that our model effectively learns the structure and pattern of programming languages and the logical data flow and generates syntactically correct code corresponding to the flowchart images.

Contributions: The major contributions of this work are three folds:

1. We study the FLOW2CODE task in a "large-scale setting" and introduce an accompanying dataset – FLoCo containing 11,884 flowchart images and corresponding Python codes. This dataset shall enable future research in this under-explored space (Sect. 3).
2. We propose a novel framework viz. FLoCo-T5 to address the task in hand, which involves generating flowchart encodings, pre-training CodeT5 on the task-specific objective with augmented codes, and finally fine-tuning for the code generation task (Sect. 4).
3. We conducted extensive experiments with various baselines and proposed task-specific code augmentation and pre-training strategy. We achieve BLEU, CodeBLEU, and exact match scores of 67.4, 75.7, and 20.0, respectively. Towards the end, we show that our model can be adopted to hand-drawn flowchart images as well (Sect. 5).

2   Related Work

2.1   Flowchart Understanding

There have been several attempts to build software for flowchart-to-code conversion, such as authors in [12], and [30] introduced interactive user interfaces to convert flowcharts to codes on-the-fly in various programming languages. These rule-based approaches, however, impose restrictions and do not support the conversion for offline flowchart images like ours. In [35], a platform was designed to recognize flowcharts and convert them to ANSI-C code using structure identification. In [16], a method was proposed for handwritten flowcharts, using rule-based techniques for preprocessing and generating pseudo code. In [9], improved results were achieved in flowchart recognition by combining statistical and structural information. In [28], the Faster RCNN object detection system was extended with an arrow keypoint predictor to recognize handwritten flowcharts. In [13], DrawnNet was proposed, a keypoint-based detector for handwritten diagram recognition.

A recent work [31] introduced a novel benchmark and dataset for question-answering over flowcharts. However, their flowchart images are unsuited for programming tasks and can not be used for our problem. The work closest to our setting is [23], which targets the digitization of handwritten flowchart images with Faster RCNN and OCR-techniques, followed by converting them to codes in C programming language using a CNN-LSTM based model. In this task, the authors propose a dataset of 775 handwritten flowchart images in Spanish and English languages. However, this dataset is unsuited for FLOW2CODE as it only consisted of hand-drawn flowchart images, with many samples consisting of only box drawings with no text, and the corresponding C codes were publicly unavailable. In this work, we consider FLOW2CODE as a sequence-to-sequence generation problem and address it using a state-of-the-art transformer-based technique in a data-driven manner. Further, we curate a dataset of 11.8K samples containing both digitized and handwritten flowchart images along with their corresponding Python codes to provide a more suitable benchmark for this task.

2.2   Large-Scale Pre-trained Language Models

The introduction of the transformer [32] architecture has brought a remarkable revolution in natural language processing. Further, to deal with the scarcity of labeled data and build a general-purpose model for a wide range of NLP applications, Radford et al. [25] proposed GPT, which is based on a transformer-decoder and pre-trained with an unlabeled pool of data in a self-supervised fashion. However, it follows a unidirectional autoregressive approach and is not suitable for tasks utilizing information from the entire sequence. Kenton et al. introduced BERT [17], a transformer-encoder-based method trained in a similar self-supervised fashion. BERT [17] follows a bidirectional autoencoder nature and is unsuitable for generation tasks that utilize information from the previously generated tokens in the sequence. To deal with the shortcomings of GPT [25] and BERT [17], Lewis et al. introduced BART [19], a denoising autoencoder that uses a bidirectional encoder and an auto-regressive decoder. These large-scale language models are often fine-tuned with a small set of labeled data for the supervised downstream task. In general, there are other well-explored pre-trained transformer-based methods such as T5 [26], MASS [29], ELECTRA [11], and RoBERTa [21]. In this work, we utilize CodeT5 [33], which adopts the encoder-decoder-based transformer model viz. T5 [26], and is pre-trained on programming language data.

2.3   Language Modeling for Code Generation

A significant amount of effort has been invested in automating software engineering using deep learning. Recent work has focused on transferable representations rather than task-specific ones. Pre-trained NLP models like BERT [17], GPT [25], and BART [19] have demonstrated transferability to programming languages, yielding positive results for a range of code-related tasks.

Feng et al. [14] introduced CodeBERT, which utilized BERT architecture pre-trained on programming language and natural language used in the software development domain, with masked language modeling objective. Guo et al. [15] proposed GraphCodeBERT as an improvement upon CodeBert by leveraging

Table 1. Statistics of the FLOCO dataset.

Property	Value
Total number of samples	11,884
Avg. length of the program (in tokens)	46
Avg. length of the program (in lines)	4.6
Train set size	10,102
Test set size	1,188
Validation set size	594

dataflow in source code through two additional pre-training tasks, predicting code structure edges, and aligning representations between source code and code structure. Ahmad et al. [2] introduced PLBART, a bidirectional and autoregressive transformer pre-trained on unlabeled natural language and programming language data, with denoising autoencoding objective, where the noising strategies employed were token masking, token deletion, and text infilling. Wang et al. [33] proposed CodeT5 by extending T5 [33] to programming languages. Similar to PLBART [2], it is a unified encoder-decoder transformer model, but it has task-specific fine-grain pre-training objectives such as masked span prediction, identifier tagging, masked identifier prediction, and bimodal dual generation objectives. As CodeT5 [33] has the advantage of task-specific pre-training strategies, we adopted it for our main method. We generated sequential encodings from flowchart images to treat FLOW2CODE as a sequence-to-sequence problem. We pre-trained the CodeT5 model with masked token modeling objective on a large number of logic-preserved augmented codes. Finally, we fine-tuned the pre-trained model for code generation.

3 FLOCO: A Novel Dataset for Flowchart Image to Python Code Conversion

We introduce a novel large-scale dataset for Flowchart images to python Code conversion. We refer to this dataset as FLOCO. It contains 11,884 flowchart images and corresponding python codes. A selection of representative examples from the FLOCO dataset is depicted in Fig. 2. We make FLOCO publicly available for download[1].

Flowchart-related research has been under-explored in the literature. However, there exist some related datasets such as (a) OHFCD dataset [5] has 419 handwritten flowcharts; however, it does not contain the corresponding codes as their focus is reading handwritten flowchart images and not code generation, (b) a more recent dataset namely FlowchartQA [31] introduces a synthetic dataset for question answering and reasoning on flowcharts. (c) in [23], authors introduced a collection of 775 handwritten flowchart images and corresponding C

---

[1] https://vl2g.github.io/projects/floco/.

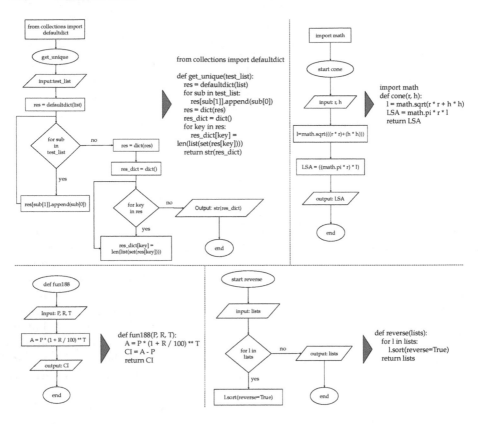

Fig. 2. Samples from the proposed FLoCo dataset. Each flowchart image is associated with the corresponding ground truth code.

programming languages. However, codes for this dataset are not publicly available. Our new dataset, viz. FLoCo has been introduced in this work to fill the research gap in the literature.

The FLoCo dataset contains 11,884 flowchart images and python code pairs. The dataset has been generated by writing a few codes from scratch and gathering and cleaning codes from the MBPP (Mostly Basic Python Programs) [4], and code-to-text dataset of CodeXGleu [22] datasets. The digitized flowchart images corresponding to the codes are generated using the pyflowchart[2] and diagrams[3] libraries. FLoCo is divided into train, test, and validation sets following an 85:10:5 ratio split respectively. Our data comprises a diverse collection of Python programs spanning a spectrum of complexity and uniqueness in their designated tasks. A few examples of these designated tasks include *calculating the* N^{th} *Fibonacci number, determining binomial coefficients, checking if a binary*

---

[2] https://pypi.org/project/pyflowchart/.
[3] https://diagrams.mingrammer.com/.

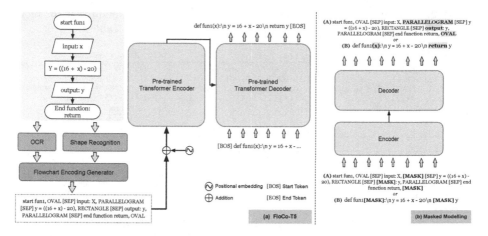

Fig. 3. Overview of the proposed method, viz. FloCo-T5: (a). Given flowchart image is converted into sequence encoding using the off-the-shelf OCR techniques. The encoder of the pre-trained CodeT5 model takes flowchart encoding added with positional encodings as input. The Decoder initially takes the start token as input and has access to encoder output, then autoregressively generates the expected code word by word. **(b).** Shows the token mask modeling. Before fine-tuning, CodeT5 is pre-trained on a token mask modeling task, where some tokens of flowchart encoding are masked and reconstructed by the decoder in an unsupervised learning fashion [**Best viewed in color**].

tree is balanced, and *finding n^{th} Catalan number*. Detailed tatistics related to FLOCO are provided in Table 1.

4 Proposed Approach

The goal of FLOW2CODE is to generate code from a given flowchart image. We approach this task as a sequence-to-sequence generation problem involving two different modalities: image (flowchart) and text (code) and propose a framework, namely FLOCO-T5 (Flowchart-to-Code T5 Model) that involves: (i) reading and converting the flowchart image into a sequence encoding, and then (ii) autoregressively generating code using the flowchart encoding. Figure 3 illustrates the proposed framework. We describe the two steps in the following subsections:

4.1 Flowchart Encoding Generation

In this step, we encode flowchart images into intermediate sequence encodings in the form of text. Given the flowchart image, we first detect and recognize the flowchart blocks, namely process (rectangle), decision (diamond), input/output (rhomboid), and terminal (oval), using the Hough transform-based shape detectors [7]. We further employ an off-the-shelf OCR method viz. easyOCR [1] to recognize the text within the boxes and on arrowheads for digitized flowchart

Table 2. Examples of different flowchart encoding. Refer to the main text for more details. A comparative study of different encodings is provided in Table 4.

Tuple encodings	String encodings	Modified string encodings
[('start fun1', 'OVAL'), ('input: X', 'PARALLELOGRAM'), ('y = ((16 + x) - 20)', 'RECTANGLE'), ('output: y', 'PARALLELOGRAM'), ('end function return', 'OVAL')]	{startfun1,OVAL},{input: X,PARALLELOGRAM},{y = ((16 + x) - 20),RECTANGLE},{output: y,PARALLELOGRAM},{end function return,OVAL}	start fun1, OVAL [SEP] input: X, PARALLELOGRAM [SEP] y = ((16 + x) - 20), RECTANGLE [SEP] output: y, PARALLELOGRAM [SEP] end function return, OVAL

Original Code	Function augmented

import numpy as np
def theta(self, s):
    s = np.where(s < -709, -709, s)
    return 1 / (1 + np.exp((-1) * s))
	import numpy as np
	def he9GxMm5QgFn(self, s):
	    s = np.where(s < -709, -709, s)
	    return 1 / (1 + np.exp((-1) * s))

Variable augmented	Function-variable augmented

import numpy as np
def theta(self, kO9):
    kO9 = np.where(kO9 < -709, -709, kO9)
    return 1 / (1 + np.exp((-1) * kO9))
	import numpy as np
	def he9GxMm5QgFn(self, Lyv):
	    Lyv = np.where(Lyv < -709, -709, Lyv)
	    return 1 / (1 + np.exp((-1) * Lyv))

Fig. 4. Data augmentation: Example of data augmentation used in code samples during pre-training of FLoCo-T5. We propose three logic-preserving augmentations that include changing the function names, variable names, and both. Augmented names are highlighted in red color. (Color figure online)

images. We then match the recognized shapes and text using their respective coordinates, i.e., a text is paired with the name of a block only if the text coordinates lie within the shape coordinates. The final flowchart encoding is a text sequence combining all the recognized text and shapes in the form of a key-value pair in the order in which they appear (from start to end). To this end, we experiment with three different strategies for encoding representation: (i) *tuple encodings*, wherein each step of the flowchart is represented as a tuple of the text and the box shape, each within quotes; (ii) *string encodings*, which eliminates quotes from text and shapes, and make use of braces to separate each step of the flowchart; and the optimized (iii) *modified string encodings*, where we utilize the [SEP] special tokens in the vocabulary of transformers to get rid of any additional braces or quotes and delineate each step of the flowchart. We provide an example for each of these encoding representations in Table 2. We experiment with all three encoding forms and compare their effectiveness on our target task in Table 4.

4.2   Code Generation

Inspired by the recent success of large-scale pre-trained code generation models, we adapt Code-T5 [33] – a transformer-based baseline trained for code generation, to our task. To this end, we initially pre-train it on a large number of logic-preserving augmented codes on the masked modeling objective in a self-supervised setting. The pre-training process adds knowledge of flowchart structure and code semantics to the model. Finally, we fine-tuned the pre-trained Code-T5 on the training set of FloCo. The data augmentation, pre-training, and fine-tuning process are performed as follows:

Data Augmentation: In order to increase the size of the dataset while keeping the logic of codes intact, we explored data augmentation. This has been achieved by changing function names and variable names. We augmented the training subset of the FloCo dataset. Replacing all functions and variables with a specific set of characters would make the dataset biased. Therefore, the function and variable names were constructed randomly using uppercase/lowercase letters, underscore, and/or digits while keeping the naming conventions for the Python programming language in mind. The length of the function names was chosen randomly from the range of 4–13; for variable names, the range was 1–3. Thus, each program was augmented in three different ways: changing the function or variable names or changing both function and variable names together. Figure 4 depicts all the augmentations corresponding to a sample code. After augmentation, the train dataset size increased from $10,102$ to $40,408$. These $30,306$ augmented codes have been utilized at the pre-training stage of our method.

Masked Modeling Objective: Inspired by the success of the Masked Language Modeling (MLM) pre-training objective in BERT [17], we propose an analogous objective specific to our problem. We adopted the pre-trained CodeT5 model and trained it on the augmented codes and flowchart encodings of the train set of FloCo. Tokens in the pre-training dataset are masked randomly at a probability of 0.15, and we aim to optimize the loss associated with the reconstruction of the original sample, as shown below:

$$L_{mml}(E, \bar{E}) = -\sum_{t=1}^{N} log(e_t|e_{0:t-1}, \bar{E}). \tag{1}$$

where $E = < e_1, e_2 \ldots, e_N >$ and $\bar{E} = < f_r(e_1), f_r(e2), \ldots, f_r(e_N)) >$ represent the ground truth and masked encodings/code, respectively. $\bar{E}$ is obtained by applying the function $f_r(e_i)$ to the ground truth encoding, which randomly replaces token e_i with the mask token $[MASK]$ with a probability of 0.15. N, e_0 denotes the length of the flowchart encoding and start token, respectively. Figure 5 shows examples of masked modeling implemented for encoding and a code sample. For the encoding input, if we mask the shape of a block (*PARALLELOGRAM* in the given example), the model must be able to infer the correct shape based on the context and the pattern it has learned during training.

Fine-Tuning: After pre-training FloCo-T5 on augmented data, we further fine-tuned it on the training data of FloCo, for FLOW2CODE task. Figure 3 (a) shows the training pipeline; the given flowchart image is first converted into

Encoder Input	Decoder Output
"start average_tuple,OVAL [SEP] input: nums,[MASK] [SEP] result = [(sum([MASK]) / len(x)) for x in zip(*nums)],RECTANGLE [SEP] output:[MASK],PARALLELOGRAM [SEP] end,None"	"start average_tuple,OVAL [SEP] input: nums,PARALLELOGRAM [SEP] result = [(sum(x) / len(x)) for x in zip(*nums)],RECTANGLE [SEP] output: result,PARALLELOGRAM [SEP] end,None"
| ```
def sector_area(r, a):
    pi = 22 / 7
    if a >= 360:
        [MASK] None
    sectorarea = (pi * r**2) * (a / 360)
    return [MASK]
``` | ```
def sector_area(r, a):
    pi = 22 / 7
    if a >= 360:
        return None
    sectorarea = (pi * r**2) * (a / 360)
    return sectorarea
``` |

Fig. 5. Masked Modelling: Example encoder inputs and decoder outputs during mask token prediction of FLoCo-T5.

sequence encoding by detecting shapes and the text inside the shapes using an off-the-shelf OCR technique. Positional encodings are added to the flowchart encodings before feeding them to the encoder. The decoder has access to the output of the encoder. It starts with a start token, and auto-regressively generates code token-by-token. To this end, during fine-tuning, we employed a language modeling loss expressed as follows:

$$L(X, E) = - \sum_{t=1}^{M} log(x_t | x_{0:t-1}, E). \tag{2}$$

where $X = < x_1, x_2, \ldots, x_M >$ denotes the ground truth code. Additionally, M, and x_0 represent the length of the code and start token, respectively. Note that during both the pre-training and the fine-tuning stages, we include different flowchart box shapes as a special token in the transformer model's vocabulary.

5   Experiments and Results

In this section, we present an extensive experimental analysis on the FLoCo benchmark to verify the efficacy of our proposed model.

5.1   Evaluation Metrics

Following the code generation literature [2], we evaluated the performance of our baselines and proposed model using the following three metrics:

i. **BLEU** [24]: a widely used word-overlap metric for assessing the quality of machine-translated text by comparing the n-grams of the generated code to the reference (ground truth) code and counting the number of matches.
ii. **CodeBLEU** [27]: a specialized metric that evaluates the quality of generated code, taking into account syntactical and logical correctness and the code's structure as reflected in the abstract syntax tree and data flow, in addition to comparing n-grams.
iii. **Exact Match (EM)**: a binary metric that checks if the generated code sequence is exactly the same as the ground-truth code.

Table 3. On the FLOCO test set, we compared FLOCO-T5 to competitive transformer-based baselines and found that our method achieved higher scores for all metrics.

Method	BLEU	CodeBLEU	EM
Vanilla Transformer [32]	10.3	26.8	0.0
BART [19]	31.1	40.7	2.2
PLBART [2]	55.7	63.7	19.4
CodeT5 [33]	63.8	71.8	17.8
FloCo-T5	**67.4**	**75.7**	**20.0**

5.2 Baseline Models

To evaluate the effectiveness of the proposed method, we compared it against the following four competitive baselines:

Vanilla Transformer [32] is the attention-based encoder-decoder architecture upon which the transformer-based pre-trained models are built. By comparing the proposed method with this baseline, we can observe the specific advantages of pre-training.

BART [19] is a pre-trained, bidirectional, autoregressive encoder-decoder architecture that was pre-trained on unlabelled natural language data and optimized using reconstruction loss. The noising techniques used were token masking, token deletion, text infilling, sentence permutation, and document rotation.

PLBART [2] is an extension of BART and was pre-trained on a large-scale dataset containing unlabelled natural language and programming language data. The pre-training objective was denoising autoencoding, and the noising strategies used were token masking, deletion, and infilling.

CodeT5 [33] adopted the T5 (pre-trained on natural language) architecture and was pre-trained on natural language and programming language data. The pre-training objectives were span prediction, identifier tagging, masked identifier prediction, and bimodal dual generation.

By comparing our proposed method with these baselines, we can observe how our method outperforms them and understand how it leverages the pre-training.

5.3 Implementation Details for Reproducibility

FLOCO-T5 is implemented using the Huggingface library [34] and utilizes the implementation of CodeT5 [33], using the 'Salesforce/codet5-base' pre-trained model available on Huggingface. The model contains 222.9 million trainable parameters. It consists of 12 encoder, 12 decoder layers, and 12 attention heads in each layer. The input encodings are truncated or padded to a maximum length of 512 tokens. We optimize training using the Adam [18] optimizer with a learning rate of $1e-5$, a warmup for 2450 steps, and a batch size of 16. We use the

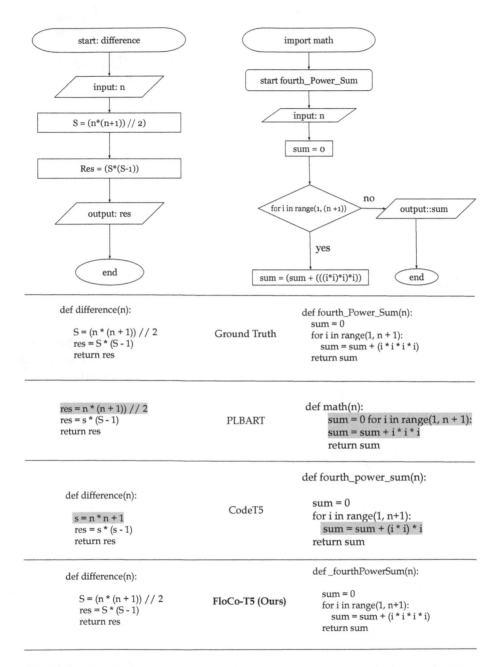

Fig. 6. Qualitative comparison: Results on two flowchart images using PLBART, CodeT5 and our method. Errors are highlighted in dark red color. The codes generated by our method are similar to the ground truth as compared to the PLBART and CodeT5. CodeT5 has fewer errors as compared to PLBART. (Color figure online)

Table 4. Comparision of different flowchart encoding representations on the performance of FLoCo-T5.

Method	BLEU	CodeBLEU	EM
Tuple encodings	16.7	37.7	0.2
String encodings	50.1	63.4	11.1
Modified string encodings (Ours)	**67.4**	**75.7**	**20.0**

same training configuration in both the pre-training and fine-tuning stages. All the baselines were trained on a single NVIDIA A6000 GPU with 48 GB VRAM. The training process requires nearly 12 h to reach convergence. We make our implementation available here: https://vl2g.github.io/projects/floco/.

5.4 Results and Discussions

We evaluate our method on the proposed FLoCo dataset and compare it against competitive baselines, namely vanilla transformer [32], BART [19], PLBART [2], and CodeT5 [33]. Table 3 shows the performance of the implemented baselines and proposed FLoCo-T5 on three evaluation metrics. Vanilla Transformer [32] is trained from scratch in contrast to other baselines, pre-trained on large-scale unlabelled data with different self-supervised pre-training objectives. Hence, Vanilla Transformer lacks the understanding of language and programming semantics and structure, resulting in the lowest performance for all the metrics. BART [19] is pre-trained on natural language text and thus, has a better understanding of the semantics and structure of the sequential data, as natural text also has rules, structure, and other syntactical properties. It results in better performance as compared to the Vanilla Transformer for all of the metrics. PLBART is pre-trained on the natural text and programming language, which means it has a better understanding of code structure and semantics, resulting in better performance compared to BART and Vanilla Transformer on all metrics. CodeT5 [33] is pre-trained with programming-language-specific, fine-grained identifier-aware denoising tasks, which help in exploiting code semantics and structure in a more exquisite way, resulting in significant improvement over other baselines. In the proposed FLoCo-T5, we adopted a pre-trained CodeT5 model, which has task-specific knowledge, and further pre-trained it on augmented training samples for the mask token generation task. As expected, FLoCo-T5 outperforms all baselines for all the metrics used for evaluation, showing the efficacy of the proposed code augmentation and pre-training strategy.

Figure 6 shows the generated codes for two flowchart samples. We compare the ground truth codes with the ones generated from PLBART, CodeT5, and our method. FLoCo-T5 is able to generate codes syntactically correct codes, which are similar to the ground truth codes, while other baselines fall short in generating correct codes. This observation is same across other test samples as numerically summarized by Table 3.

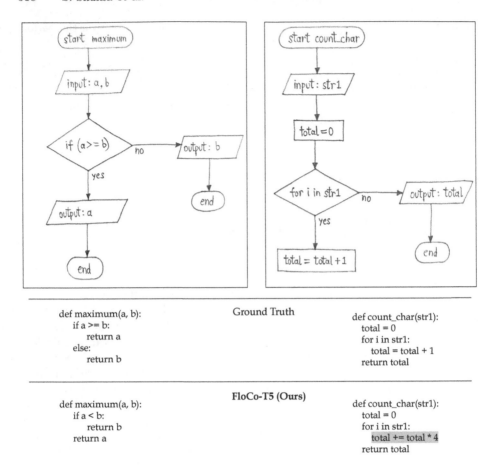

Ground Truth

```
def maximum(a, b):
    if a >= b:
        return a
    else:
        return b
```

```
def count_char(str1):
    total = 0
    for i in str1:
        total = total + 1
    return total
```

FloCo-T5 (Ours)

```
def maximum(a, b):
    if a < b:
        return b
    return a
```

```
def count_char(str1):
    total = 0
    for i in str1:
        total += total * 4
    return total
```

Fig. 7. Results on hand-drawn flowchart images using FloCo-T5. Errors are highlighted in dark red color. We observe an error in the program due to incorrect recognition of the handwritten character '*' by our OCR module. (Color figure online)

We further conducted an experiment with three flowchart image encoding methods, shown in Table 2, and results presented in Table 4. The modified string encoding method utilized a [SEP] token to separate each step of the flowchart, and removed extra braces, enhancing the preservation of the flowchart's structure, and consequently outperforming other encoding methods.

Can the Proposed Approach Work for Hand-Drawn Flowchart Images? We evaluated FLOCO-T5 on hand-drawn flowchart images using 40 samples created by three human annotators. Flowchart block detection and recognition were performed with OpenCV [7]. For handwritten text recognition, we employed CRAFT text detection [6] and TrOCR text recognition [20]. FLOCO-T5 achieved a BLEU score of 21.4% and a CodeBLEU score of 34.6% on

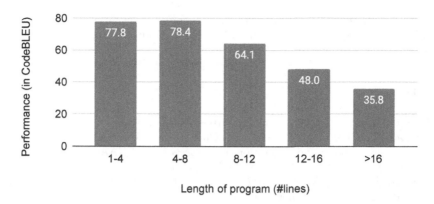

Fig. 8. Performance of FLoCo-T5 across various program lengths.

these hand-drawn flowcharts. Figure 7 displays Python codes generated for two sample hand-drawn flowcharts. These results indicate our approach's suitability for hand-drawn flowcharts, and performance can be significantly enhanced with advances in handwritten text recognition.

Limitations: We observed that code generation performance of our model is higher for shorter programs ($< \approx 12$ lines) but drops for longer programs ($> \approx 15$ lines) due to the dataset's bias towards shorter programs (average length: 4.6 lines) as shown in Fig. 8. To address this issue, we propose increasing the number of training samples for longer programs. Future work will focus on expanding the FLoCo dataset to include longer and more complex code and researching information flow in state and block diagrams.

6   Conclusion

We introduced the FLoCo-T5 framework for generating Python code from flowchart images and presented the FLoCo dataset to benchmark the FLOW2CODE task. FLOW2CODE is modeled as a sequence-to-sequence problem, where flowcharts are first encoded by detecting the shapes of blocks and reading the text within, and further transformed into code using competitive transformer baselines. FLoCo-T5's task-specific pre-training results in significant improvements over related baselines. The recent advancements in Large Language Models (LLMs), such as ChatGPT, have revolutionized the field of code generation, and they can be adapted to solve our task. However, ensuring that these massive models have not seen our test data is not a trivial task. Furthermore, despite these advancements, we firmly believe that our dataset can be used to study open problems such as development of lightweight and interpretable models for generating code from flowchart images. We leave these as future directions to work on.

Acknowledgements. This work was partly supported by MeitY, Govt. of India (project number: S/MeitY/AM/20210114). Yogesh Kumar is supported by a UGC fellowship.

References

1. Easy OCR (2022). https://pypi.org/project/easyocr/1.6.2/
2. Ahmad, W., Chakraborty, S., Ray, B., Chang, K.W.: Unified pre-training for program understanding and generation. In: Proceedings of NAACL-HLT (2021)
3. Akermi, I., Heinecke, J., Herledan, F.: Transformer based natural language generation for question-answering. In: Proceedings of the 13th International Conference on Natural Language Generation (2020)
4. Austin, J., et al.: Program synthesis with large language models. CoRR abs/2108.07732 (2021)
5. Awal, A.M., Feng, G., Mouchère, H., Viard-Gaudin, C.: Handwritten flowchart dataset (OHFCD). Document Recognition and Retrieval XVIII, January 2011, San Fransisco, United States, pp. 7874–78740A (2011). https://doi.org/10.1117/12.876624
6. Baek, Y., Lee, B., Han, D., Yun, S., Lee, H.: Character region awareness for text detection. In: Proceedings of CVPR (2019)
7. Bradski, G.: The OpenCV library. Dr. Dobb's J. Softw. Tools **120**, 122–125 (2000)
8. Brown, T., et al.: Language models are few-shot learners. In: Proceedings of NeurIPS (2020)
9. Carton, C., Lemaitre, A., Coüasnon, B.: Fusion of statistical and structural information for flowchart recognition. In: Proceedings of ICDAR (2013)
10. Chaudhuri, A.: Flowchart and Algorithm Basics: The Art of Programming. Mercury Learning and Information (2020)
11. Clark, K., Luong, M.T., Le, Q.V., Manning, C.D.: ELECTRA: pre-training text encoders as discriminators rather than generators. In: Proceedings of ICLR (2020)
12. Cook, D.: Flowgorithm (2022). https://www.flowgorithm.org/
13. Fang, J., Feng, Z., Cai, B.: DrawnNet: offline hand-drawn diagram recognition based on keypoint prediction of aggregating geometric characteristics. Entropy **24**(3), 425 (2022)
14. Feng, Z., et al.: CodeBERT: a pre-trained model for programming and natural languages. In: Findings of the ACL: EMNLP (2020)
15. Guo, D., et al.: GraphCodeBERT: pre-training code representations with data flow. In: Proceedings of ICLR (2020)
16. Herrera-Camara, J.I., Hammond, T.: Flow2Code: from hand-drawn flowcharts to code execution. In: Proceedings of the Symposium on Sketch-Based Interfaces and Modeling (2017)
17. Kenton, J.D.M.W.C., Toutanova, L.K.: BERT: pre-training of deep bidirectional transformers for language understanding. In: Proceedings of NAACL-HLT (2019)
18. Kingma, D.P., Ba, J.: Adam: a method for stochastic optimization. In: Bengio, Y., LeCun, Y. (eds.) Proceedings of ICLR (2015)
19. Lewis, M., et al.: BART: denoising sequence-to-sequence pre-training for natural language generation, translation, and comprehension. In: Proceedings of ACL (2020)
20. Li, M., et al.: TrOCR: transformer-based optical character recognition with pre-trained models. arXiv preprint arXiv:2109.10282 (2021)

21. Liu, Y., et al.: RoBERTa: a robustly optimized BERT pretraining approach. CoRR abs/1907.11692 (2019)
22. Lu, S., et al.: CodeXGLUE: a machine learning benchmark dataset for code understanding and generation. arXiv (2021)
23. Montellano, C.D.B., Garcia, C.O.F.C., Leija, R.O.C.: Recognition of handwritten flowcharts using convolutional neural networks. Int. J. Comput. Appl. (2022)
24. Papineni, K., Roukos, S., Ward, T., Zhu, W.J.: BLEU: a method for automatic evaluation of machine translation. In: Proceedings of ACL (2002)
25. Radford, A., Narasimhan, K., Salimans, T., Sutskever, I., et al.: Improving language understanding by generative pre-training. OpenAI (2018)
26. Raffel, C., et al.: Exploring the limits of transfer learning with a unified text-to-text transformer. J. Mach. Learn. Res. **21**, 1–67 (2020)
27. Ren, S., et al.: CodeBLEU: a method for automatic evaluation of code synthesis. CoRR abs/2009.10297 (2020)
28. Schäfer, B., Stuckenschmidt, H.: Arrow R-CNN for flowchart recognition. In: Proceedings of ICDAR Workshop (2019)
29. Song, K., Tan, X., Qin, T., Lu, J., Liu, T.Y.: MASS: masked sequence to sequence pre-training for language generation. In: Proceedings of ICML (2019)
30. Supaartagorn, C.: Web application for automatic code generator using a structured flowchart. In: 2017 8th IEEE International Conference on Software Engineering and Service Science (ICSESS) (2017)
31. Tannert, S., Feighelstein, M., Bogojeska, J., Shtok, J., Staar, A.A.P., Karlinsky, A.S.J.K.L.: FlowchartQA: the first large-scale benchmark for reasoning over flowcharts. In: Document Intelligence Workshop @ KDD (2022)
32. Vaswani, A., et al.: Attention is all you need. In: Proceedings of NeurIPS (2017)
33. Wang, Y., Wang, W., Joty, S., Hoi, S.C.: CodeT5: identifier-aware unified pre-trained encoder-decoder models for code understanding and generation. In: Proceedings of EMNLP (2021)
34. Wolf, T., et al.: Transformers: state-of-the-art natural language processing. In: Proceedings of EMNLP: System Demonstrations (2020)
35. Wu, X.H., Qu, M.C., Liu, Z.Q., Li, J.Z., et al.: Research and application of code automatic generation algorithm based on structured flowchart. J. Softw. Eng. Appl. **4**, 534–545 (2011)

Formerge: Recover Spanning Cells in Complex Table Structure Using Transformer Network

Nam Quan Nguyen[1][✉], Anh Duy Le[1], Anh Khoa Lu[1], Xuan Toan Mai[2], and Tuan Anh Tran[2][✉]

[1] Viettel Cyberspace Center, Viettel Group, Vietnam Lot D26 Cau Giay New Urban Area, Yen Hoa Ward, Hanoi, Cau Giay District, Vietnam
ngnamquan@gmail.com, leanhduy497@gmail.com, luanhkhoa123@gmail.com
[2] Faculty of Computer Science and Engineering, Ho Chi Minh City University of Technology (HCMUT), VNU-HCM, Ho Chi Minh City, Vietnam
{mxtoan,trtanh}@hcmut.edu.vn

Abstract. Table structure recognition (TSR) task is indispensable in a robust document analysis system. Recently, the split-and-merge-based approach has attracted many researchers to develop the TSR problem. It is a two-stage method: firstly, split table region into row/column separation and obtain grid cells of the table; then recover spanning cells by merging some grid cells and complete the table structure. Most recent proposals focus on the first stage, with few solutions for the merge task. Therefore, this paper proposes a novel method to recover spanning cells using Transformer networks called Formerge. This model contains a Transformer encoder and two parallel left-right/top-down decoders. With grid structure output from a split branch, Formerge extracts cell features with RoIAlign and passes them into the encoder to enhance features before decoding to detect spanning cells. Our technique outperforms other methods on two benchmark datasets, including SciTSR and ICDAR19-cTDaR modern.

Keywords: Table structure recognition · Split-and-Merge · Spanning cells · Transformer

1 Introduction

Document image analysis is a crucial problem in the digital transformation process. Automatically extracting information from a document is not trivial and requires an enormous effort. Many components in a document need to be well-recognized, and the table is one of them. Table embeds data in a structured way and includes rich information in a brief format. Therefore, table structure recognition (TSR) is an indispensable process in a current robust document analysis system.

G. A. Fink et al. (Eds.): ICDAR 2023, LNCS 14191, pp. 522–534, 2023.
https://doi.org/10.1007/978-3-031-41734-4_32

TSR aims for structural analysis of cell locations [1] in table images. More particularly, TSR will extract all coordinates of grid cell boxes and row/column spanning information. With the diversity and intricacy of table layouts, this task becomes very challenging for many researchers. Implicit separators, ambiguous row/column groups, complicated backgrounds, and spanning cells are all obstacles in the process. A spanning cell is a table cell that occupies at least two columns, or rows [2]. In complex table structures, spanning cells is critical because they contain more valuable semantic information than other grid cells.

So far, many methods have been proposed to solve the TSR problem, but basically, they can be divided into two groups: one-stage and two-stage. The approach with one-stage training is data-hungry and mainly depends on a large amount of data, while the generalization is not commensurate with spent resources. In recent years, two-stage has become a more attractive approach to many researchers. [3,5] proposed that table structure recognition could be divided into two stages: splitting and merging. The split stage predicts the fine grid structure of the table, and the merge stage indicates which grid elements should be merged to recover spanning cells. The final table structure can be obtained by parsing all table cells. [6-9] made impressive improvements from the split-and-merge-based model and proved generalized capabilities in different kinds of tables.

However, most approaches still mainly focus on the split stage. The merge stage was used in SPLERGE [5], SEM [7], and TSRFormer [9] still have some drawbacks: (1) SPLERGE separated two stages in the training scheme and used a vanilla architecture for merge model; (2) SEM used a time-consuming decoding process to predict spanning cell and involved textual features from BERT [11] which needs an additional OCR step to extract text information. (3) TSR-Former used a simple relation network [12] to predict the possibility of merging between adjacent cells, which do not take advantage of global features of grid cells structure and still needed handcraft features.

This paper proposes a novel Transformer-based model called Formerge to recover spanning cells from grid cells. Starting from the grid cell structure from the split model, RoIAlign [13] operation is applied to extract visual features from the output of the backbone at each grid cell. After that, features and positional embedding are passed into a Transformer encoder to enhance global information in cell structure. Finally, two parallel decoders predict the possibility of merging in a horizontal/vertical direction. The single decoder is a linear transformation to predict the probability of cell merging. We have conducted many experiments on two benchmark datasets, including SciTSR [14] and ICDAR19-cTDaR modern [17]. The results demonstrate that Formerge outperforms other merging approaches.

Our major contributions can be summarized as follows:

- Introducing a merger in a two-stage TSR model called Formerge to handle the current problems in spanning cell recovery tasks.
- Realizing extensive ablation studies and analysis on the performances of our proposed method in a two-stage model.

2   Related Work

2.1   Table Structure Recognition

Early studies on TSR methods such as [4,16,21–23] strongly depended on hand-crafted features and heuristics logic. These methods mainly use traditional techniques in computer vision to detect table structures like space analysis, connected component extraction, text block arrangement, and vertical/horizontal alignment. Usually, rule-based methods require deep insight into the dataset to design heuristics logic and adjust parameters manually. These approaches have impressive performance and rational analysis in small amounts of datasets, but their generalization still needs to be improved and appropriate for diverse structure data. In recent years, many deep learning-based approaches have outperformed traditional methods conspicuously in accuracy and scalability. These approaches can be divided into two categories: end-to-end methods and two-stage methods.

One-stage based methods directly detect table components from the image like row, column, and cell boundary and then combine them to form a final table structure. These methods usually leverage models in object detection [28–30,33] or semantic segmentation methods [25,32] to detect cell, row, column. These methods have difficulty handling complicated table structures, such as spanning cells, empty cells, or cells containing multi-line text.

Besides, other approaches also treat TSR as an image-to-text generation problem. They convert each table image into a text sequence describing table structure and cell contents. The text output is formatted in two standard styles, LaTeX symbols [34] and HTML tags [15,27]. These methods are usually challenging to train their model and rely on a large amount of training data, while adaptation ability on other datasets has struggled.

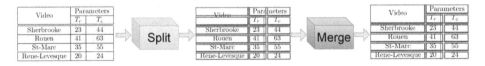

Fig. 1. Overview of the two-stage method. *Split* module predicts grid cell structure and *Merge* module recovers spanning cells.

Two-Stage Based methods have two clear phases, which are recognizing primary grid cells and recovering spanning cells (see Fig. 1). [5] proposed the SPLERGE architecture with a split model and merge model to solve each step in this approach. SPLERGE splits the table using Row Projection Networks and Column Projection Networks with novel projection pooling and then uses grid pooling from table cell regions to predict the probability of up-down/left-right merges. [3,6] proposed RNN-based networks instead of projection networks in SPLERGE. In SEM [7], the performance is significantly improved compared with the original SPLERGE by the additional text information and attention mechanisms. SEM's framework is an end-to-end exploit that helps to overcome the weak point in SPLERGE, which were the isolated train schemes of the two modules. [8,9] proposed a new split-and-merge-based method to deal with the distorted table by incorporating techniques in a split module such as spatial CNN module [24], line regression [35] and a relation network in merge module.

2.2 Spanning Cell Recovery

As only tables with complex structures are considered for spanning cell recovery, there still needs to be a solution to this problem. Among them, SPLERGE [5], SEM [7], relation network [8] and graph network [18] can be mentioned. In SPLERGE, the merge module uses average pooling in grid cells and multiple dilated convolutions. This model is separated from the split module. Posterior methods replace the pooling feature using average pixel with RoIAlign operation and take into account end-to-end manner to train split and merge modules. SEM uses an encoder-decoder with 2D coverage attention mechanism [36] and enhances features with multi-modality in visual and text features. [8,9] predict the merged probability of each adjacent cell by a relation network. [18] leverages graph representations of grid cell structure and apply a Graph Convolution Network (GCN) to detect row/column indices of each cell.

3   Method

The overall TSR pipeline, shown in Fig. 2, includes the CNN backbone, Split, and merge module. The CNN backbone and split module are presented in [8] while the merge module is our proposal, Formerge. The CNN backbone uses Resnet18 [10] as the visual feature encoder with Feature Pyramid Network (FPN) [19] built on top to extract multi-resolution features. Split and merge modules share jointly feature map $P_2 \in R^{\frac{H}{4} \times \frac{W}{4} \times C}$ generated by the CNN backbone.

3.1   Split Model

In splitter, two semantic segmentation branches are built on P_2 to predict row separation mask S_{row} and column separation mask S_{col}. Like the column branch, we take the line prediction branch of row separation as an example. Firstly, P_2 is down-sampled to $P_2' \in R^{\frac{H}{4} \times \frac{W}{32} \times C}$ by 1×2 max-pooling layer. Then, two cascaded spatial CNN networks (SCNN) [24] are applied consecutively to enhance the feature in the W axis. The first SCNN propagates the information from the leftmost sequentially to the rightmost by performing the convolution operator on $\frac{W}{32}$ slices. The second SCNN behaves similarly but in the opposite direction from right to left. Finally, the previous enhanced feature map is up-sampled by a factor of $(4, 32)$ to generate an output feature map $S_{row} \in R^{H \times W \times 1}$.

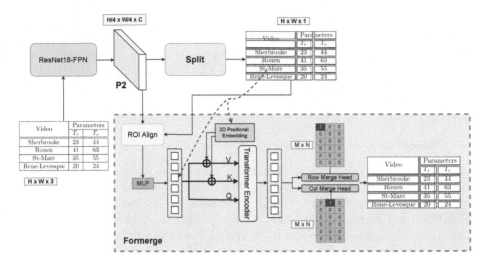

Fig. 2. The architecture of Formerge

3.2   Merge Model

We introduce Formerge as the merged model to recover spanning cells wrongly segmented by a splitter. Similar to modeling the merger in [5], the output of Formerge is two $M \times N$ matrixes: up-down merges D and left-right merges R. Denoting D_{ij} is the probability of merging cell (i, j) with cell $(i+1, j)$, and R_{ij} is the probability of merging cells (i, j) and $(i, j+1)$. For example, if three grid cells with index (i, j), $(i+1, j)$ and $(i+2, j)$ form into a spanning cell, it means those cells need to merge in the vertical direction and set $D_{ij} = D_{i+1,j} = 1$. Because there are only $(M - 1) \times N$ pairs in matrix D, elements in the last row should be all zero, similar to the last column of R. As shown in Fig. 2, the architecture of Formerge contains two main components: an encoder and a merged head.

Encoder. First, we adopt a RoIAlign operation [13] with 7×7 output size from P_2 at each table cell in grid $M \times N$. The output of this operation is a sequence of grid cell features $P_{grid} \in R^{(M \times N) \times C \times 7 \times 7}$. Next, P_{grid} is flattened and mapped to d-dimensions with one-layer MLP. We refer to the output of this transformation as the grid cell embeddings E_{cell}.

Position embeddings are added to the cell embeddings to retain positional information. We use 2-dimensional positional encoding according to occupied rows and columns in a grid of cells. In this case, two sets of embeddings, row-embedding and column-embedding, are learned with size 200, which is assumed to be the maximum number of rows/columns in a table. Then, based on the coordinate of the location of the cell in the grid, we concatenate the row and column embedding to get the final positional embedding E_{pos} for that cell.

We follow a standard design of the Transformer encoder from [20], which consists of alternating layers of a multi-head self-attention module and a feed-forward network (FFN). The embedding vector $E = E_{cell} + E_{pos}$, where $E_{cell}, E_{pos} \in R^{(M \times N) \times d}$, is considered as input vector to Transformer encoder.

Merge Head. The merge head includes two parallel branches for the row and column. Each branch is a classifier attached to enhanced context features from the Transformer encoder, implemented with two-hidden-layer MLP. There are 512 nodes at the hidden layer, and two layers are connected by a ReLU nonlinear. Finally, a sigmoid function predicts the probability of cell merging.

3.3   Loss Functions

This section describes the loss function for training a two-stage pipeline. There are four losses in total, both the Split and Merge models, each having two losses for row and column predictions.

Split Model. Let R_s, R_s^* be the predicted and ground truth of the row separator heat map, similar to C_s, C_s^* of the column heat map. Compared with split loss in [8], we take additional dice loss along with binary cross entropy to form into following function with two variables x, y:

$$L(x, y) = \alpha \times BCE(x, y) + (1 - \alpha) \times DICE(x, y) \quad (1)$$

where α is a hyper-parameter for balancing loss, we set $\alpha = 0.4$.
   The loss for the split model is then:

$$L_{split} = L(R_s, R_s^*) + L(C_s, C_s^*) \quad (2)$$

Merge Model. Let R_m, R_m^* be the predicted and ground truth on the merge row matrix, similar to C_m, C_m^* on the merge column matrix. We use only binary cross-entropy loss for this classification task:

$$L_{merge} = BCE(R_m, R_m^*) + BCE(C_m, C_m^*) \quad (3)$$

Overall. With the above definitions, the training loss for the TSR pipeline in our work can be defined as follow:

$$L_{TSR} = L_{split} + L_{merge} \quad (4)$$

4   Experiments

4.1   Dataset and Metric

This paper uses two well-known public datasets for the experiment and evaluation analysis.

1. **SciTSR** [14] is a large-scale table structure recognition dataset from scientific papers. It contains 15,000 tables in PDF format and their corresponding high-quality structure labels obtained from LaTeX source files. SciTSR splits 12,000 for training and 3,000 for testing, with 2,885 and 716 complicated

tables in the training and test set. The author extracts all the 716 complicated tables from the test set to create a new test subset called SciTSR-COMP. Because most tables in the SciTSR dataset are still simple grid-like tables that do not take advantage of our proposed method, we only use SciTSR-COMP in the evaluation scheme. Follow [14]; the cell adjacent relationship score is used as the evaluation metric of this dataset[1].

2. **ICDAR19-cTDaR modern** [17]. This dataset focuses on current documents from scientific journals, forms, and financial statements. We collect 300 pages (track A) for training and 100 pages (Track B1) for evaluation. Because of the lack of table structure annotation in the train set, we manually labeled ourselves. Besides, the convex hull of the content describes a cell region, so we further prepared the coordinate of text boxes when re-constructing table cells to ensure that evaluation is more meaningful. The official competition metric[2] is the weighted average F-Score (WAvg-F1) which, based on cell adjacency relation, is used in this dataset.

4.2   Implementation Details

All experiments are implemented using PyTorch 1.11.0 and conducted on GPU NVIDIA V100 32GB. The weights of the ResNet-18 are initialized from the pre-trained model of ImageNet. We employ the AdamW algorithm [26] for optimization, with the following hyper-parameters: $\beta_1 = 0.9$, $\beta_2 = 0.999$, $\epsilon = 1e-08$, $\lambda = 5e - 4$. In the Transformer encoder, we set the input dimension, encoder layers, head number, and the dimension of feedforward networks to 512, 3, 8, and 512, respectively.

In the training phase, we rescale the longer side of the table image to 640 while keeping the aspect ratio. We first train the split module in 50 epochs and then freeze split to train the merge module for another 50 epochs. Finally, the whole pipeline is further fine-tuned for 15 epochs. In the testing phase, the scale method is the same as in training. The separation maps in the split module are binarized dynamically with the Otsu algorithm [31]. The grid cells are merged based on the two merge matrixes with a threshold of 0.9.

Table 1. Table structure recognition performance on SciTSR-COMP

Series System	Series Precision	Series Recall	Series F1
Split only	94.93	85.43	89.72
Split + merge [5]	95.47	92.81	94.02
Split + SEM [7]	96.52	93.82	95.15
Split + relation [8]	96.51	94.41	95.37
Split + Formerge	96.31	95.05	95.62

---

[1] https://github.com/Academic-Hammer/SciTSR.
[2] https://github.com/cndplab-founder/ctdar_measurement_tool.

Table 2. Table structure recognition performance on ICDAR19-cTDaR modern

Method	IoU@0.6			IoU@0.7			IoU@0.8			IoU@0.9			WAvgF1
	P	R	F1	P	R	F1	P	R	F1	P	R	F1	
Split + merge [5]	56.9	44.0	49.0	52.6	39.8	45.3	40.9	30.9	35.2	16.4	12.4	14.1	33.9
Split + relation [8]	57.4	45.2	50.6	52.9	41.7	46.6	40.9	32.2	36.1	16.3	12.9	14.4	34.9
Split + Formerge	58.8	46.9	52.2	54.1	43.1	48.0	41.9	33.4	37.2	17.1	13.2	15.2	36.1

4.3 Experiment Results

We compare our method with other state-of-the-art spanning cell recovery methods on SciTSR-COMP and ICDAR19-cTDaR modern datasets. We leverage the same trained model weighted of the split module in the first step of the training phase and implement other spanning cell recovery methods. We experimented with the split module and got the result $F1 = 89.72\%$, which is a comparable result in [7] with $F1 = 89.77\%$ on SciTSR-COMP. Starting with the same grid cell structure, we followed the training strategy in [5,8] to get a fair comparison with our approach. Besides, we modify slightly merge module in SPLERGE [5] to enable an end-to-end manner. We refer to the system T3 for SEM results, which uses only visual features reported in [7]. The quantitative results are shown in Table 1 and Table 2. On SciTSR-COMP and ICDAR19-cTDaR modern, our cell merging approach has archived the best score of 95.62% and 36.1%, which outperformed others in the same two-stage TSR system.

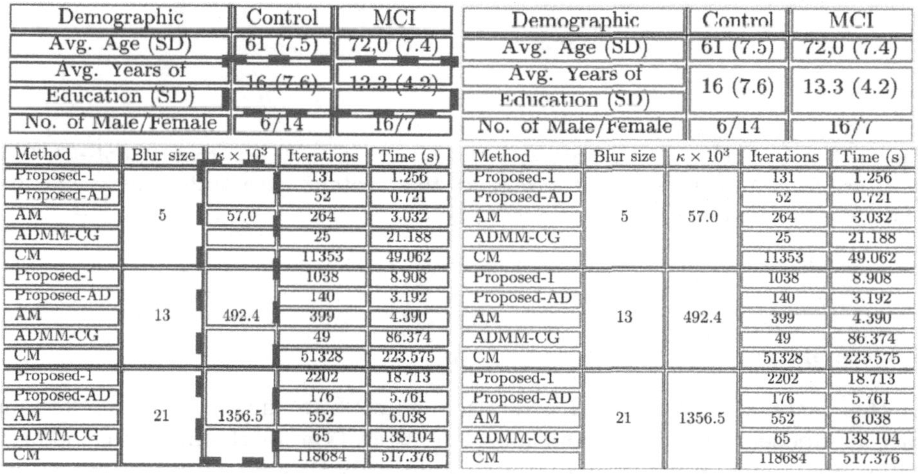

Demographic	Control	MCI		Demographic	Control	MCI
Avg. Age (SD)	61 (7.5)	72,0 (7.4)		Avg. Age (SD)	61 (7.5)	72,0 (7.4)
Avg. Years of Education (SD)	16 (7.6)	13.3 (4.2)		Avg. Years of Education (SD)	16 (7.6)	13.3 (4.2)
No. of Male/Female	6/14	16/7		No. of Male/Female	6/14	16/7

Method	Blur size	$\kappa \times 10^3$	Iterations	Time (s)	Method	Blur size	$\kappa \times 10^3$	Iterations	Time (s)
Proposed-1			131	1.256	Proposed-1			131	1.256
Proposed-AD			52	0.721	Proposed-AD			52	0.721
AM	5	57.0	264	3.032	AM	5	57.0	264	3.032
ADMM-CG			25	21.188	ADMM-CG			25	21.188
CM			11353	49.062	CM			11353	49.062
Proposed-1			1038	8.908	Proposed-1			1038	8.908
Proposed-AD			140	3.192	Proposed-AD			140	3.192
AM	13	492.4	399	4.390	AM	13	492.4	399	4.390
ADMM-CG			49	86.374	ADMM-CG			49	86.374
CM			51328	223.575	CM			51328	223.575
Proposed-1			2202	18.713	Proposed-1			2202	18.713
Proposed-AD			176	5.761	Proposed-AD			176	5.761
AM	21	1356.5	552	6.038	AM	21	1356.5	552	6.038
ADMM-CG			65	138.104	ADMM-CG			65	138.104
CM			118684	517.376	CM			118684	517.376

Fig. 3. Quality results of relation method (left column) and Formerge (right column) on SciTSR-COMP. The green dash boxes denote the inaccurate prediction results. (Color figure online)

4.4   Result Analysis

Some examples in Fig. 3 compare our method with the relation method in the SciTSR-COMP set. According to the result, Formerge works robustly in cases that span cells located in non-header positions. These locations rarely appear in the table structure; spanning cells are usually the table's header because it is a critical factor in determining the table's structure. This result demonstrates the generalization of our method.

Some qualitative results of our approach on two benchmark datasets are presented in Fig. 4. Regardless of any lingual document or spanning cell position, Formerge can perform well and has an accurate prediction. However, our method still has limitations under some situations, such as over-segmentation error of split model (Fig. 4-d) and cells containing multiple-line text (Fig. 4-g).

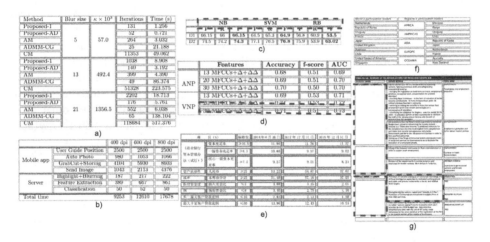

Fig. 4. Experiment results of TSR with the Formerge. (a-d) are from SciTSR-COMP, (e-g) are from ICDAR19-cTDaR. Examples that contain green dash boxes are failure cases. (Color figure online)

4.5   Ablation Study

In this section, we conducted several experiments to investigate the effect of each component and find an optimized configuration for the proposed module. All experiment result is evaluated on SciTSR-COMP set.

Number of Transformer Encoder Layers. We hypothesize that the encoder is vital to enhance feature sequence in the long range, and we processed changing the number of encoder layers. The result in Fig. 5 reflects that increasing layers lead to a slight accuracy drop. The size of the grid structure is not too large, not succeed 200 in each dimension, so this probably caused inefficiency in scaling.

Table 3. Compare positional encoding methods in Transformer encoder.

Series Positional Encoding	Series Precision	Series Recall	Series F1
None	96.19	94.45	95.23
2D Fixed Sin-Cos	96.29	94.64	95.39
2D Learned Embedding	96.31	95.05	95.62

Positional Encoding. To study the impact of 2D positional encoding, we compare the result with two variants of 2D positional encoding. Table 3 shows that using learned embedding encoding is better than fixed sine-cosine encoding about 0.23% of F1. Perhaps learned positional encoding helps to overcome the problem of size variation of the table structure. We find that not passing spatial encodings in the encoder only leads to a minor F1 drop of 0.16%.

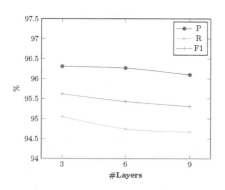

Fig. 5. Effect of encoder layer

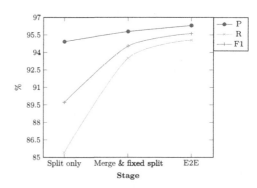

Fig. 6. Performance in training stages

Effectiveness of End-to-End Training Manner. In the design of the training scheme, there are three stages to train the whole TSR model: split module and merge module alternately update weights; when one updates, the other is frozen. Then, merge module trained jointly with split to take advantage of end-to-end. The result in Fig. 6 shows that in the last stage, the F1-score increase significantly from 94.54% to 95.62%. This experiment demonstrates that the end-to-end manner is more robust than the separated design, which is why recent research has prioritized the design of a homogeneous model rather than isolated parts.

5   Conclusion

This paper presents Formerge, a new approach for spanning cell recovery in complex table structures. We adopt a Transformer network to take advantage of global grid-level self-attention from the Transformer encoder for cell merging. Our Transformer-based approach can achieve higher performance when integrated into a two-stage TSR system than previous merge module approaches. Additionally, Formerge is appropriate and robust with an end-to-end manner method. Experimental results demonstrate that our method has outperformed current spanning cell recovery methods in two complicated table structure public benchmarks.

Acknowledgments. We acknowledge Ho Chi Minh City University of Technology (HCMUT), VNU-HCM for supporting this study.

References

1. Tran, T.A., Tran, H.T., Na, I.S., Lee, G.S., Yang, H.J., Kim, S.H.: A mixture model using Random Rotation Bounding Box to detect table region in document image. J. Vis. Commun. Image Represent. **39**, 196–208 (2016)
2. Chi, Z., Huang, H., Xu, H.D., Yu, H., Yin, W., Mao, X.: Complicated table structure recognition. arXiv preprint arXiv:1908.04729 (2019)
3. Khan, S.A., Khalid, S.M.D., Shahzad, M.A., Shafait, F.: Table structure extraction with bi-directional gated recurrent unit networks. In: 2019 International Conference on Document Analysis and Recognition, ICDAR 2019, pp. 1366–1371 (2019)
4. Kieninger, T., Dengel, A.: Table recognition and labeling using intrinsic layout features. In: International Conference on Advances in Pattern Recognition, pp. 307–316 (1999)
5. Tensmeyer, C., Morariu, V.I., Price, B.L., Cohen, S., Martinez, T.R.: Deep splitting and merging for table structure decomposition. In: 2019 International Conference on Document Analysis and Recognition, ICDAR 2019, pp. 114–121 (2019)
6. Li, Y., et al.: Rethinking table structure recognition using sequence labeling methods. In: Document Analysis and Recognition-ICDAR 2021: 16th International Conference, Lausanne, Switzerland, 5–10 September 2021, Proceedings, Part II 16, pp. 541–553 (2021)
7. Zhang, Z., Zhang, J., Du, J., Wang, F.: Split, embed and merge: an accurate table structure recognizer. Pattern Recogn. **126**, 108565 (2022)
8. Ma, C., Lin, W., Sun, L., Huo, Q.: Robust table detection and structure recognition from heterogeneous document images. Pattern Recogn. **133**, 109006 (2023)
9. Lin, W., et al.: TSRFormer: Table Structure Recognition with Transformers. In: Proceedings of the 30th ACM International Conference on Multimedia, pp. 6473–6482 (2022)
10. He, K., Zhang, X., Ren, S., Sun, J.: Deep residual learning for image recognition. In: Proceedings of the IEEE Conference on Computer Vision and Pattern Recognition 2016, pp. 770–778 (2016)
11. Devlin, J., Chang, M.W., Lee, K., Toutanova, K.: BERT: Pre-training of deep bidirectional transformers for language understanding. In: Proceedings of the 2019 Conference of the North American Chapter of the Association for Computational Linguistics: Human Language Technologies, vol. 1 (Long and Short Papers), pp. 4171–4186 (2019). https://doi.org/10.18653/v1/N19-1423
12. Zhang, J., Elhoseiny, M., Cohen, S., Chang, W., Elgammal, A.: Relationship proposal networks. In: Proceedings of the IEEE Conference on Computer Vision and Pattern Recognition, pp. 5678–5686 (2017)

13. He, K., Gkioxari, G., Dollár, P., Girshick, R.: Mask r-cnn. In: Proceedings of the IEEE international Conference on Computer Vision, pp. 2961–2969 (2017)
14. Chi, Z., Huang, H., Xu, H.D., Yu, H., Yin, W., Mao, X.L.: Complicated table structure recognition. arXiv preprint arXiv:1908.04729 (2019)
15. Zhong, X., ShafieiBavani, E., Jimeno Yepes, A.: Image-based table recognition: data, model, and evaluation. In: Vedaldi, A., Bischof, H., Brox, T., Frahm, J.-M. (eds.) ECCV 2020. LNCS, vol. 12366, pp. 564–580. Springer, Cham (2020). https://doi.org/10.1007/978-3-030-58589-1_34
16. Chen, J., Lopresti, D.: Model-based tabular structure detection and recognition in noisy handwritten documents. In: 2012 International Conference on Frontiers in Handwriting Recognition, pp. 75–80 (2012)
17. Gao, L., et al.: ICDAR 2019 competition on table detection and recognition (cTDaR). In: 2019 International Conference on Document Analysis and Recognition (ICDAR), pp. 1510–1515 (2019)
18. Xue, W., Yu, B., Wang, W., Tao, D., Li, Q.: Tgrnet: A table graph reconstruction network for table structure recognition. In: Proceedings of the IEEE/CVF International Conference on Computer Vision, pp. 1295–1304 (2021)
19. Lin, T. Y., Dollár, P., Girshick, R., He, K., Hariharan, B., Belongie, S.: Feature pyramid networks for object detection. In: Proceedings of the IEEE Conference on Computer Vision and Pattern Recognition, pp. 2117–2125 (2017)
20. Vaswani, A., et al.: Attention is all you need. In: Advances In Neural Information Processing Systems, vol. 30, pp. 5998–6008 (2017)
21. Itonori, K.: Table structure recognition based on textblock arrangement and ruled line position. In: Proceedings of 2nd International Conference on Document Analysis and Recognition (ICDAR 1993), pp. 765–768 (1993)
22. Wang, Y., Phillips, I.T., Haralick, R.M.: Table structure understanding and its performance evaluation. Pattern Recogn. **37**(7), 1479–1497 (2004)
23. Tran, T.A., Tran, H.T., Na, Kim, S.H.: A hybrid method for table detection from document image. In: 2015 3rd IAPR Asian Conference on Pattern Recognition (ACPR), pp. 131–135 (2015)
24. Pan, X., Shi, J., Luo, P., Wang, X., Tang, X.: Spatial as deep: Spatial cnn for traffic scene understanding. In: Proceedings of the AAAI Conference on Artificial Intelligence, vol. 32 (2018)
25. Prasad, D., Gadpal, A., Kapadni, K., Visave, M., Sultanpure, K.: CascadeTabNet: An approach for end to end table detection and structure recognition from image-based documents. In: Proceedings of the IEEE/CVF Conference on Computer Vision and Pattern Recognition Workshops, pp. 572–573 (2020)
26. Loshchilov, I., Hutter, F.: Decoupled weight decay regularization. arXiv preprint arXiv:1711.05101 (2017)
27. Li, M., Cui, L., Huang, S., Wei, F., Zhou, M., Li, Z.: Tablebank: Table benchmark for image-based table detection and recognition. In: Proceedings of the Twelfth Language Resources and Evaluation Conference, pp. 1918–1925 (2020)
28. Siddiqui, S.A., Fateh, I.A., Rizvi, S.T.R., Dengel, A., Ahmed, S.: Deeptabstr: Deep learning based table structure recognition. In: 2019 International Conference on Document Analysis and Recognition (ICDAR), pp. 1403–1409 (2019)
29. Schreiber, S., Agne, S., Wolf, I., Dengel, A., Ahmed, S.: Deepdesrt: Deep learning for detection and structure recognition of tables in document images. In: 2017 14th IAPR International Conference on Document Analysis and Recognition (ICDAR), pp. 1162–1167 (2017)

30. Smock, B., Pesala, R., Abraham, R.: PubTables-1M: towards comprehensive table extraction from unstructured documents. In: Proceedings of the IEEE/CVF Conference on Computer Vision and Pattern Recognition, pp. 4634–4642 (2022)
31. Otsu, N.: A threshold selection method from gray-level histograms. IEEE Trans. Syst. Man Cybern. **9**, 62–66 (1979)
32. Siddiqui, S.A., Khan, P.I., Dengel, A., Ahmed, S.: Rethinking semantic segmentation for table structure recognition in document. In: 2019 International Conference on Document Analysis and Recognition (ICDAR), pp. 1397–1402 (2019)
33. Qiao, L., et al.: Lgpma: complicated table structure recognition with local and global pyramid mask alignment. In: Document Analysis and Recognition-ICDAR 2021: 16th International Conference, pp. 99–114 (2021)
34. Deng, Y., Rosenberg, D., Mann, G.: Challenges in end-to-end neural scientific table recognition. In: 2019 International Conference on Document Analysis and Recognition (ICDAR), pp. 894–901 (2019)
35. Carion, N., Massa, F., Synnaeve, G., Usunier, N., Kirillov, A., Zagoruyko, S.: End-to-end object detection with transformers. In: Vedaldi, A., Bischof, H., Brox, T., Frahm, J.-M. (eds.) ECCV 2020. LNCS, vol. 12346, pp. 213–229. Springer, Cham (2020). https://doi.org/10.1007/978-3-030-58452-8_13
36. Zhang, J., et al.: Watch, attend and parse: an end-to-end neural network based approach to handwritten mathematical expression recognition. Pattern Recogn. **71**, 196–206 (2017)

GriTS: Grid Table Similarity Metric for Table Structure Recognition

Brandon Smock[ID], Rohith Pesala[✉][ID], and Robin Abraham[ID]

Microsoft, Redmond, WA, USA
{brsmock,ropesala,robin.abraham}@microsoft.com

Abstract. In this paper, we propose a new class of metric for table structure recognition (TSR) evaluation, called grid table similarity (GriTS). Unlike prior metrics, GriTS evaluates the correctness of a predicted table directly in its natural form as a matrix. To create a similarity measure between matrices, we generalize the two-dimensional largest common substructure (2D-LCS) problem, which is NP-hard, to the 2D most similar substructures (2D-MSS) problem and propose a polynomial-time heuristic for solving it. This algorithm produces both an upper and a lower bound on the true similarity between matrices. We show using evaluation on a large real-world dataset that in practice there is almost no difference between these bounds. We compare GriTS to other metrics and empirically validate that matrix similarity exhibits more desirable behavior than alternatives for TSR performance evaluation. Finally, GriTS unifies all three subtasks of cell topology recognition, cell location recognition, and cell content recognition within the same framework, which simplifies the evaluation and enables more meaningful comparisons across different types of TSR approaches. Code will be released at https://github.com/microsoft/table-transformer.

1 Introduction

Table extraction (TE) [3,14,17] is the problem of inferring the presence, structure, and—to some extent—meaning of tables in documents or other unstructured presentations. In its presented form, a table is typically expressed as a collection of cells organized over a two-dimensional grid [5,13,16]. Table structure recognition (TSR) [6,7] is the subtask of TE concerned with inferring this two-dimensional cellular structure from a table's unstructured presentation.

While straightforward to describe, formalizing the TSR task in a way that enables effective performance evaluation has proven challenging [9]. Perhaps the most straightforward way to measure performance is to compare the sets of predicted and ground truth cells for each table and measure the percentage of tables for which these sets match exactly—meaning, for each predicted cell there is a matching ground truth cell, and vice versa, that has the same rows, columns, and text content. However, historically this metric has been eschewed in favor of measures of *partial* correctness that score each table's correctness on a range of $[0, 1]$ rather than as binary correct or incorrect. Measures of partial correctness

G. A. Fink et al. (Eds.): ICDAR 2023, LNCS 14191, pp. 535–549, 2023.
https://doi.org/10.1007/978-3-031-41734-4_33

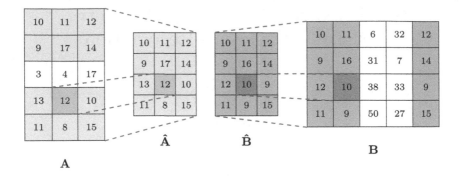

Fig. 1. Grid table similarity (GriTS) is based on computing the two-dimensional most similar substructures (2D-MSS) between two matrices (see Sect. 3 for details). In this example, the 2D-MSS of matrices **A** and **B** are substructures **Â** and **B̂**. **Â** and **B̂** create a direct correspondence (or alignment) between entries in the original matrices, depicted here as darker shading for one pair of corresponding entries.

are useful not only because they are more granular, but also because they are less impacted by errors and ambiguities in the ground truth. This is important, as creating unambiguous ground truth for TSR is a challenging problem, which can introduce noise not only into the learning task but also performance evaluation [10, 15].

Designing a metric for partial correctness of tables has also proven challenging. The naive approach of comparing predicted cells with ground truth cells by their absolute positions suffers from the problem that a single mistake in cell segmentation can offset all subsequent cells by one position, which may result in a disproportionate penalty. Several metrics have been proposed that instead consider the *relative* positions of predicted cells [4, 6, 12, 18]. However, these metrics capture relative position in different ways that do not fully account for a table's global two-dimensional (2D) structure. Metrics also offer differing perspectives on what constitutes the task to be measured, what property of a predicted cell is evaluated, and whether predicted cells can be partially correct.

To address these issues, in this paper we develop a new class of metric for TSR called grid table similarity (GriTS). GriTS attempts to unify the different perspectives on TSR evaluation and address these in a modular way, both to simplify the evaluation and to make comparisons between approaches more meaningful. Among our contributions:

- GriTS is the first metric to evaluate tables directly in their matrix form, maintaining the global 2D relationships between cells when comparing predictions to ground truth.
- To create a similarity between matrices, we extend the 2D largest common substructure (2D-LCS) problem, which is NP-hard, to 2D most similar

substructures (2D-MSS) and propose a polynomial-time heuristic to solve it. This algorithm produces both an upper and lower bound on its approximation, which we show have little difference in practice.

– We outline the properties of an ideal TSR metric and validate empirically on a large real-world dataset that GriTS exhibits more ideal behavior than alternatives for TSR evaluation.

– GriTS is the first metric that addresses cell topology, cell content, and cell location recognition in a unified manner. This makes it easier to interpret and compare results across different modeling approaches and datasets.

2 Related Work

Table 1. Comparison of metrics proposed for table structure recognition.

Name	Task/Cell Property	Data Structure	Cell Partial Correctness	Form
DAR_{Con} [6]	Content	Set of adjacency relations	Exact match	F-score
DAR_{Loc} [4]	Location	Set of adjacency relations	Avg. at multiple IoU thresholds	F-score
BLEU-4 [12]	Topology & function	Sequence of HTML tokens	Exact match	BLEU-4
TEDS [18]	Content & function	Tree of HTML tags	Normalized Levenshtein similarity	TEDS
$GriTS_{Top}$	Topology	Matrix of cells	IoU	F-score
$GriTS_{Con}$	Content	Matrix of cells	Normalized LCS	F-score
$GriTS_{Loc}$	Location	Matrix of cells	IoU	F-score

A number of metrics exist for evaluating table structure recognition methods. These include the cell adjacency-based metric used in the ICDAR 2013 Table Competition, which we refer to as directed adjacency relations (DAR) [4,6], 4-gram BLEU score (BLEU-4) [12], and tree edit distance similarity (TEDS) [18]. In Table 1 we categorize these metrics across four dimensions: subtask (cell property), data structure, cell partial correctness, and overall score formulation.

2.1 Subtask/Property

Each metric typically poses the table structure recognition task more specifically as one of the following:

1. *Cell topology recognition* considers the layout of the cells, specifically the rows and columns each cell occupies over a two-dimensional grid.

2. *Cell content recognition* considers the layout of cells and the text content of each cell.
3. *Cell location recognition* considers the layout of cells and the absolute coordinates of each cell within a document.

One way to characterize these subtasks is in terms of the *property* of the cell that is considered most central to the recognition task. For cell topology, this can be considered the colspan and rowspan of each cell. Each perspective is useful. Cell content recognition is most aligned with the end goal of TE, but for table image input it can depend on the quality of optical character recognition (OCR). Cell location recognition does not depend on OCR, but not every TSR method reports cell locations. Cell topology recognition is independent of OCR and is applicable to all TSR methods, but is not anchored to the actual content of the cells by either text or location. Thus, accurate cell topology recognition is necessary but not sufficient for successful TSR.

Functional analysis is the subtask of table extraction concerned with determining whether each cell is a key (header) or value. While it is usually considered separate from structure recognition, metrics sometimes evaluate aspects of TSR and functional analysis jointly, such as TEDS and BLEU-4.

2.2   Data Structure

A table is presented in two dimensions, giving it a natural representation as a grid or matrix of cells. The objective of TSR is usually considered to be inferring this grid structure. However, for comparing predictions with ground truth, prior metrics have used alternative abstract representations for tables. These possibly decompose a table into sub-units other than cells and represent them in an alternate structure with relationships between elements that differ from those of a matrix. The metrics proposed by Göbel et al. [6] and Gao et al. [4] deconstruct the grid structure of a table into a *set* of directed adjacency relations, corresponding to pairs of neighboring cells. We refer to the first metric, which evaluates cell text content, as DAR_{Con}, and the second, which evaluates cell location, as DAR_{Loc}. Li et al. [12] represent a table as a token *sequence*, using a simplified HTML encoding. Zhong et al. [18] also represent a table using HTML tokens but use the nesting of the tags to consider a table's cells to be *tree*-structured, which more closely represents a table's two-dimensional structure than does a sequence.

Metrics based on these different representations each have their own sensitivities and invariances to changes to a table's structure. Zhong et al. [18] investigate a few of these sensitivities and demonstrate that DAR mostly ignores the insertion of contiguous blank cells into a table, while TEDS does not. However, largely the sensitivities of these metrics that result from their different representations have not been studied. This makes it more challenging to interpret and compare them.

2.3   Cell Partial Correctness

Each metric produces a score between 0 and 1 for each table. For some metrics this takes into account simply the fraction of matching sub-units between a prediction and ground truth. Some metrics also define a measure of correctness for each sub-unit, which is between 0 and 1. For instance, TEDS incorporates the normalized Levenshtein similarity to allow the text content of an HTML tag to partially match the ground truth. Defining partial correctness at the cell level is useful because it is less sensitive to minor discrepancies between a prediction and ground truth that may have little or no impact on table extraction quality.

2.4   Form

The form of a metric is the way in which the match between prediction and ground truth is aggregated over matching and non-matching sub-units. DAR uses both precision and recall, which can be taken together to produce the standard F-score. BLEU-4 treats the output of a table structure recognition model as a sequence and uses the 4-gram BLEU score to measure the degree of match between a predicted and ground truth sequence. TEDS computes a modified tree edit distance, which is the cost of transforming partially-matching and non-matching sub-units between the tree representations of a predicted and ground truth table.

2.5   Spreadsheet Diffing

A related problem to the one explored in this paper is the identification of differences or changes between two versions of a spreadsheet [2, 8]. In this case the goal is to classify cells between the two versions as either modified or unmodified, and possibly to generate a sequence of edit transformations that would convert one version of a spreadsheet into another.

3   Grid Table Similarity

To motivate the metrics proposed in this paper, we first introduce the following attributes that we believe an ideal metric for table structure recognition should exhibit:

1. *Task isolation*: the table structure recognition task is measured in isolation from other table extraction tasks (detection and functional analysis).
2. *Cell isolation*: a true positive according to the metric corresponds to exactly one predicted cell and one ground truth cell.
3. *Two-dimensional order preservation* [1]: For any two true positive cells, tp_1 and tp_2, the relative order in which they appear is the same in both dimensions in the predicted and ground truth tables. More specifically:
   (a) The maximum true row of tp_1 < minimum true row of tp_2 $\longleftrightarrow$ the maximum predicted row of tp_1 < minimum predicted row of tp_2.

(b) The maximum true column of tp_1 < minimum true column of tp_2 $\iff$ the maximum predicted column of tp_1 < minimum predicted column of tp_2.

(c) The maximum true row of tp_1 = minimum true row of tp_2 $\iff$ the maximum predicted row of tp_1 = minimum predicted row of tp_2.

(d) The maximum true column of tp_1 = minimum true column of tp_2 $\iff$ the maximum predicted column of tp_1 = minimum predicted column of tp_2.

4. *Row and column equivalence*: the metric is invariant to transposing the rows and columns of both a prediction and ground truth (i.e. rows and columns are of equal importance).

5. *Cell position invariance*: the credit given to a correctly predicted cell is the same regardless of its absolute row-column position.

The first two attributes are considered in Table 1. In Sect. 4, we test the last two properties for different proposed metrics. While not essential, we note again that in practice we believe it is also useful for a TSR metric to define partial correctness for cells and to have the same general form for both cell content recognition and cell location recognition. In the remainder of this section we describe a new class of evaluation metric that meets all of the above criteria.

3.1   2D-LCS

We first note that Property 3 is difficult to enforce for cells that can span multiple rows and columns. To account for this, we instead consider the matrix of *grid cells* of a table. Exactly one grid cell occupies the intersection of each row and each column of a table. Note that as a spanning cell occupies multiple grid cells, its text content logically repeats at every grid cell location that the cell spans.

To enforce Property 3 for grid cells, we consider the generalization of the longest common subsequence (LCS) problem to two dimensions, which is called the two-dimensional largest (or longest) common substructure (2D-LCS) problem [1]. Let $\mathbf{M}[R, C]$ be a matrix with $R = [r_1, \ldots, r_m]$ representing its rows and $C = [c_1, \ldots, c_n]$ representing its columns. Let $R' \mid R$ be a subsequence of rows of R, and $C' \mid C$ be a subsequence of columns of C. Then a substructure $\mathbf{M}' \mid \mathbf{M}$ is such that,

$$\mathbf{M}' \mid \mathbf{M} = \mathbf{M}[R', C'].$$

2D-LCS operates on two matrices, $\mathbf{A}$ and $\mathbf{B}$, and determines the largest two-dimensional substructures, $\hat{\mathbf{A}} \mid \mathbf{A} = \hat{\mathbf{B}} \mid \mathbf{B}$, the two have in common. In other words,

$$\text{2D-LCS}(\mathbf{A}, \mathbf{B}) = \underset{\mathbf{A}'\mid\mathbf{A}, \mathbf{B}'\mid\mathbf{B}}{\arg\max} \sum_{i,j} f(\mathbf{A}'_{i,j}, \mathbf{B}'_{i,j}) \tag{1}$$

$$= \hat{\mathbf{A}}, \hat{\mathbf{B}}, \tag{2}$$

where,

$$f(e_1, e_2) = \begin{cases} 1, & \text{if } e_1 = e_2 \\ 0, & \text{otherwise} \end{cases}.$$

3.2   2D-MSS

While a solution to the 2D-LCS problem satisfies Property 3 for grid cells, it assumes an exact match between matrix elements. To let cells partially match, an extension to 2D-LCS is to relax the exact match constraint and instead determine the two most *similar* two-dimensional substructures, $\tilde{\mathbf{A}}$ and $\tilde{\mathbf{B}}$. We define this by replacing equality between two entries $\mathbf{A}_{i,j}$ and $\mathbf{B}_{k,l}$ with a more general choice of similarity function between them. In other words,

$$\text{2D-MSS}_f(\mathbf{A}, \mathbf{B}) = \underset{\mathbf{A}'|\mathbf{A}, \mathbf{B}'|\mathbf{B}}{\arg\max} \sum_{i,j} f(\mathbf{A}'_{i,j}, \mathbf{B}'_{i,j}) \tag{3}$$

$$= \tilde{\mathbf{A}}, \tilde{\mathbf{B}}, \tag{4}$$

where,

$$0 \le f(e_1, e_2) \le 1 \qquad \forall e_1, e_2.$$

Taking inspiration from the standard F-score, we define a general similarity measure between two matrices based on this as,

$$\tilde{S}_f(\mathbf{A}, \mathbf{B}) = \frac{2\sum_{i,j} f(\tilde{\mathbf{A}}_{i,j}, \tilde{\mathbf{B}}_{i,j})}{|\mathbf{A}| + |\mathbf{B}|}, \tag{5}$$

where $|\mathbf{M}_{m \times n}| = m \cdot n$.

3.3   Grid Table Similarity (GriTS)

Finally, to define a similarity between tables, we use Eq. (5) with a particular choice of similarity function and a particular matrix of entries to compare. This has the general form,

$$\text{GriTS}_f(\mathbf{A}, \mathbf{B}) = \frac{2\sum_{i,j} f(\tilde{\mathbf{A}}_{i,j}, \tilde{\mathbf{B}}_{i,j})}{|\mathbf{A}| + |\mathbf{B}|}, \tag{6}$$

where $\mathbf{A}$ and $\mathbf{B}$ now represent tables—matrices of grid cells—and f is a similarity function between the grid cells' properties. Interpreting Eq. (6) as an F-score, then letting $\mathbf{A}$ be the ground truth matrix and $\mathbf{B}$ be the predicted matrix, we can also define the following quantities, which we interpret as recall and precision: $\text{GriTS-Rec}_f(\mathbf{A}, \mathbf{B}) = \frac{\sum_{i,j} f(\tilde{\mathbf{A}}_{i,j}, \tilde{\mathbf{B}}_{i,j})}{|\mathbf{A}|}$ and $\text{GriTS-Prec}_f(\mathbf{A}, \mathbf{B}) = \frac{\sum_{i,j} f(\tilde{\mathbf{A}}_{i,j}, \tilde{\mathbf{B}}_{i,j})}{|\mathbf{B}|}$.

A specific choice of grid cell property and the similarity function between them defines a particular GriTS metric. We define three of these: $\text{GriTS}_{\text{Top}}$ for cell topology recognition, $\text{GriTS}_{\text{Con}}$ for cell text content recognition, and $\text{GriTS}_{\text{Loc}}$ for cell location recognition. Each evaluates table structure recognition from a different perspective.

The matrices used for each metric are visualized in Fig. 2. For cell location, $\mathbf{A}_{i,j}$ contains the bounding box of the cell at row i, column j, and we use IoU to compute similarity between bounding boxes. For cell text content, $\mathbf{A}_{i,j}$ contains

Group	Sequence of Administration		
	Phase I	Phase II	Phase III
I	C	A	B
II	B	C	A
III	A	B	C

(a) An example presentation table from the PubTables-1M dataset.

Group	Sequence of Administration	Sequence of Administration	Sequence of Administration
Group	Phase I	Phase II	Phase III
I	C	A	B
II	B	C	A
III	A	B	C

(b) GriTS_Con

[0, 0, 1, 2]	[0, 0, 3, 1]	[-1, 0, 2, 1]	[-2, 0, 1, 1]
[0, -1, 1, 1]	[0, 0, 1, 1]	[0, 0, 1, 1]	[0, 0, 1, 1]
[0, 0, 1, 1]	[0, 0, 1, 1]	[0, 0, 1, 1]	[0, 0, 1, 1]
[0, 0, 1, 1]	[0, 0, 1, 1]	[0, 0, 1, 1]	[0, 0, 1, 1]
[0, 0, 1, 1]	[0, 0, 1, 1]	[0, 0, 1, 1]	[0, 0, 1, 1]

(c) GriTS_Top

[136.42, 477.25, 160.62, 501.45]	[185, 477.25, 470.89, 487.22]	[185, 477.25, 470.89, 487.22]	[185, 477.25, 470.89, 487.22]
[136.42, 477.25, 160.62, 501.45]	[185, 491.48, 271.9, 501.45]	[284.5, 491.48, 371.39, 501.45]	[384, 491.48, 470.89, 501.45]
[136.42, 505.82, 160.62, 515.72]	[185, 505.82, 271.9, 515.72]	[284.5, 505.82, 371.39, 515.72]	[384, 505.82, 470.89, 515.72]
[136.42, 515.73, 160.62, 525.63]	[185, 515.73, 271.9, 525.63]	[284.5, 515.73, 371.39, 525.63]	[384, 515.73, 470.89, 525.63]
[136.42, 525.64, 160.62, 535.53]	[185, 525.64, 271.9, 535.53]	[284.5, 525.64, 371.39, 535.53]	[384, 525.64, 470.89, 535.53]

(d) GriTS_Loc

Fig. 2. An example presentation table from the PubTables-1M dataset, along with corresponding ground truth grid cell matrices for different GriTS metrics. Each matrix entry corresponds to one grid cell. Entries that correspond to spanning cells are shaded darker for illustrative purposes.

the text content of the cell at row i, column j, and we use normalized longest common subsequence (LCS) to compute similarity between text sequences.

For cell topology, we use the same similarity function as cell location but on bounding boxes with size and relative position given in the grid cell coordinate system. For the cell at row i, column j, let $\alpha_{i,j}$ be its rowspan, let $\beta_{i,j}$ be its colspan, let $\rho_{i,j}$ be the minimum row it occupies, and let $\theta_{i,j}$ be the minimum column it occupies. Then for cell topology recognition, $\mathbf{A}_{i,j}$ contains the bounding box $[\theta_{i,j} - j, \rho_{i,j} - i, \theta_{i,j} - j + \beta_{i,j}, \rho_{i,j} - i + \alpha_{i,j}]$. Note that for any cell with rowspan of 1 and colspan of 1, this box is $[0, 0, 1, 1]$.

3.4 Factored 2D-MSS Algorithm

Computing the 2D-LCS of two matrices is NP-hard [1]. This suggests that all metrics for TSR may necessarily be an approximation to what could be considered the ideal metric. We propose a heuristic approach to determine the 2D-MSS by factoring the problem. Instead of determining the optimal subsequences of rows and columns jointly for each matrix, we determine the optimal subsequences of rows and the optimal subsequences of columns independently. This uses dynamic programming (DP) in a nested manner, which is run twice: once to determine the optimal subsequences of rows and a second time to determine the optimal subsequences of columns. For the case of rows, an inner DP operates on sequences of grid cells in a row, computing the best possible sequence alignment of cells between any two rows. The inner DP is executed over all pairs of predicted and ground truth rows to score how well each predicted row can be aligned with each ground truth row. An outer DP operates on the sequences of rows from each matrix, using the pairwise scores computed by the inner DP to determine the best alignment of subsequences of rows between the predicted and ground truth matrices. For the case of columns, the procedure is identical, merely substituting columns for rows. The nested DP procedure is $O(|\mathbf{A}| \cdot |\mathbf{B}|)$. Our implementation uses extensive memoization [11] to maximize the efficiency of the procedure.

This factored procedure is similar to the RowColAlign algorithm [8] proposed for spreadsheet diffing. Both procedures decouple the optimization of rows and columns and use DP in a nested manner. However, RowColAlign attempts to optimize the number of pair-wise exact matches between substructures, whereas Factored 2D-MSS attempts to optimize the pair-wise similarity between substructures. These differing objectives result in the two procedures having differing outcomes given the same two input matrices.

3.5 Approximation Bounds

The outcome of the procedure is a valid 2D substructure of each matrix—these just may not be the *most similar* substructures possible, given that the rows and columns are optimized separately. However, given that these are valid substructures, it follows that the similarity between matrices $\mathbf{A}$ and $\mathbf{B}$ computed by this procedure is a *lower bound* on their true similarity. It similarly follows that because constraints are relaxed during the optimization procedure, the lowest similarity determined when computing the optimal subsequences of rows and the optimal subsequences of columns serves as an *upper bound* on the true similarity between matrices. We define GriTS as the value of the lower bound, as it always corresponds to a valid substructure. However, the upper bound score can also be reported to indicate if there is any uncertainty in the true value. As we show in Sect. 4, little difference is observed between these bounds in practice.

4 Experiments

In this section, we report several experiments that assess and compare GriTS and other metrics for TSR. Given that due to computational intractability no algorithm perfectly implements the ideal metric for TSR as outlined in Sect. 3, the main goal of these experiments is to assess how well each proposed metric matches the behavior that we would expect in the theoretically optimal metric.

Table 2. In this experiment we evaluate models of varying strengths on the PubTables-1M test set and compare the upper and lower bounds for GriTS$_{Con}$ that are produced. This shows how closely GriTS approximates the true similarity between two tables using the Factored 2D-MSS algorithm.

Epoch	$GriTS_{Con}$	Upper Bound	Difference	Equal Instances (%)
1	0.8795	0.8801	0.0005	81.2%
2	0.9347	0.9348	0.0002	87.6%
3	0.9531	0.9532	0.0001	91.1%
4	0.9640	0.9641	< 0.0001	93.0%
5	0.9683	0.9683	< 0.0001	93.2%
10	0.9794	0.9795	< 0.0001	96.1%
15	0.9829	0.9829	< 0.0001	96.9%
20	0.9850	0.9850	< 0.0001	97.4%

GriTS computes a similarity between two tables using the Factored 2D-MSS algorithm, which produces both an upper and a lower bound on the true similarity. The difference between the two values represents the uncertainty, or maximum possible error, there is with respect to the true similarity. In the first experiment, we measure how well GriTS approximates the true similarity between predicted and ground truth tables in practice. To do this, we train the Table Transformer (TATR) model [15] on the PubTables-1M dataset, which contains nearly one million tables, for 20 epochs and save the model produced after each epoch. This effectively creates 20 different TSR models of varying strengths with which to evaluate predictions. We evaluate each model on the entire PubTables-1M test set and measure the difference between GriTS$_{Con}$ and GriTS$_{Con}$ upper bound, as well as the percentage of individual instances for which these bounds are equal.

We present the results of this experiment in Table 2. As can be seen, there is very little difference between the upper and lower bounds across all models. The uncertainty in the true table similarity peaks at the worst-performing model, which is trained for only one epoch. For this model, GriTS$_{Con}$ is 0.8795 and the measured uncertainty in the true similarity is 0.0005. Above a certain level of model performance, the average difference between the bounds and the percentage of instances for which the bounds differ both decrease quickly as performance improves, with the difference approaching 0 as GriTS approaches a score of 1. By epoch 4, the difference between the bounds is already less than 0.0001, and for at least 93% of instances tested the GriTS score is in fact the true similarity. These results strongly suggest that in practice there is almost no difference between GriTS and the true similarity between tables.

In the next set of experiments, we compare GriTS and other metrics for TSR with respect to the properties outlined in Sect. 3. The goal of these experiments is to assess how well each metric matches the behavior that we would expect in the theoretically optimal metric. We evaluate each metric on the original ground truth (GT) and versions of the ground truth that are modified or corrupted

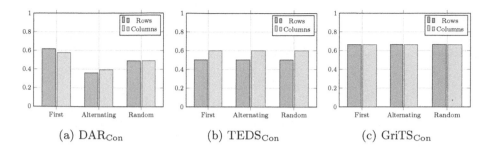

Fig. 3. In this experiment, we compare the response of each metric to a predicted table with half of its rows missing (blue) or half of its columns missing (red) under different schemes for creating missing rows and columns. Results are averaged over 44,381 tables from the test set of PubTables-1M. (Color figure online)

in controlled ways, which shows the sensitivity or insensitivity of each metric to different underlying properties. To create corrupted versions of the GT, we select either a subset of the GT table's rows or a subset of its columns, where each row or each column from the GT is selected with probability $p(x)$, preserving their original order and discarding the rest.

For the experiments we use tables from the test set of the PubTables-1M dataset, which provides text content and location information for every cell, including blank cells. To make sure each remaining grid cell has well-defined content and location after removing a subset of rows and columns from a table, we use the 44,381 tables that do not have any spanning cells. In order to make the metrics more comparable, we define a version of TEDS that removes functional analysis from the evaluation, called TEDS$_{Con}$, by removing all header information in the ground truth.

In the first experiment, the goal is to test for each metric if rows and columns are given equal importance and if every cell is credited equally regardless of its absolute position. To test these, we create missing columns or missing rows in a predicted table according to three different selection schemes. In the first part of the experiment we select each row with probability 0.5 using the following three different selection schemes:

- First: select the first 50% of rows.
- Alternating: select either every odd-numbered row or every even-numbered row.
- Random: select 50% of rows within a table at random.

In the second part of the experiment, we select columns using the same three schemes as for rows. In all six cases, exactly half of the cells are missing in a predicted table whenever the table has an even number of rows and columns.

We compare the impact of each selection scheme on the metrics in Fig. 3. For a metric to give equal credit to rows and columns, it should be insensitive to (produce the same value) whether half the rows are missing or half the columns are missing. For a metric to give equal credit to each cell regardless of absolute

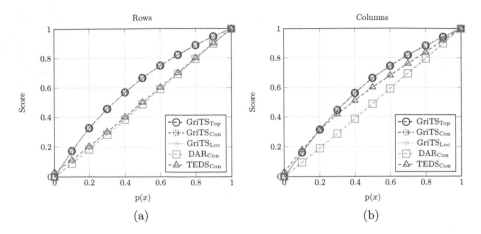

Fig. 4. In this experiment, we compare the response of each metric to a predicted table with random rows missing (Fig. 4a) or random columns missing (Fig. 4b) as we vary the probability that a row or column is missing. Results are averaged over 44,381 tables from the test set of PubTables-1M.

position, it should be insensitive to which row or column the missing cell occurs in. The results show that DAR_{Con} is sensitive both to whether rows or columns are selected and to which rows or columns are selected. $TEDS_{Con}$ is sensitive to whether rows are selected or columns are selected, but is not sensitive to which rows or columns are selected. On the other hand, $GriTS_{Con}$ produces a nearly identical value no matter which scheme is used to select half of the rows or half of the columns.

In the second experiment, we select rows and columns randomly, but vary the probability $p(x)$ from 0 to 1. We also expand the results to include all three GriTS metrics. We show the results of this experiment in Fig. 4. Like in the first experiment, DAR_{Con} produces a similar value when randomly selecting rows or randomly selecting columns, and we see that this holds for all values of $p(x)$. Likewise, this is true not just for $GriTS_{Con}$ but all GriTS metrics. On the other hand, for $TEDS_{Con}$, we see that the metric has a different sensitivity to randomly missing columns than to randomly missing rows, and that the relative magnitude of this sensitivity varies as we vary $p(x)$.

In Fig. 5, we further split the results for DAR and GriTS by their precision and recall values. Here we see that for GriTS, not only are rows and columns equivalent, but recall and precision match the probability of rows and columns being in the prediction and ground truth, respectively. On the other hand, DAR has a less clear interpretation in terms of precision and recall. Further, for DAR we notice that there is a slight sensitivity that shows up to the choice of rows versus columns for precision, which was not noticeable when considering F-score. Overall these results show that GriTS closely resembles the ideal metric for TSR and exhibits more desirable behavior than prior metrics for this task.

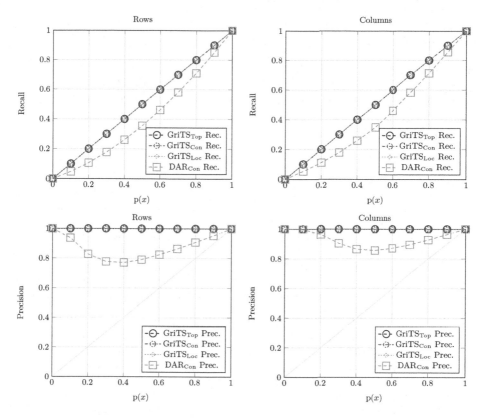

Fig. 5. In this experiment, we compare the response of each metric split into precision and recall to a predicted table with random rows missing or random columns missing as we vary the probability that a row or column is missing. Results are averaged over 44,381 tables from the test set of PubTables-1M.

5   Conclusion

In this paper we introduced *grid table similarity* (GriTS), a new class of evaluation metric for table structure recognition (TSR). GriTS unifies all three perspectives of the TSR task within a single class of metric and evaluates model predictions in a table's natural matrix form. As the foundation for GriTS, we derived a similarity measure between matrices by generalizing the two-dimensional largest common substructure problem, which is NP-hard, to 2D most-similar substructures (2D-MSS). We proposed a polynomial-time heuristic that produces both an upper and a lower bound on the true similarity and we showed that in practice these bounds are tight, nearly always guaranteeing the optimal solution. We compared GriTS to other metrics and demonstrated using a large dataset that GriTS exhibits more desirable behavior for table structure recognition evaluation. Overall, we believe these contributions improve the interpretability and stability of evaluation for table structure recognition and make it easier to compare results across different types of modeling approaches.

Acknowledgements. We would like to thank Pramod Sharma, Natalia Larios Delgado, Joseph N. Wilson, Mandar Dixit, John Corring, Ching Pui WAN, and the anonymous reviewers for helpful discussions and feedback while preparing this manuscript.

References

1. Amir, A., Hartman, T., Kapah, O., Shalom, B.R., Tsur, D.: Generalized LCS. Theor. Comput. Sci. **409**(3), 438–449 (2008)
2. Chambers, C., Erwig, M., Luckey, M.: SheetDiff: a tool for identifying changes in spreadsheets. In: 2010 IEEE Symposium on Visual Languages and Human-Centric Computing, pp. 85–92. IEEE (2010)
3. Corrêa, A.S., Zander, P.O.: Unleashing tabular content to open data: a survey on pdf table extraction methods and tools. In: Proceedings of the 18th Annual International Conference on Digital Government Research, pp. 54–63 (2017)
4. Gao, L., et al.: ICDAR 2019 competition on table detection and recognition (cTDaR). In: 2019 International Conference on Document Analysis and Recognition (ICDAR), pp. 1510–1515. IEEE (2019)
5. Gatterbauer, W., Bohunsky, P., Herzog, M., Krüpl, B., Pollak, B.: Towards domain-independent information extraction from web tables. In: Proceedings of the 16th International Conference on World Wide Web, pp. 71–80 (2007)
6. Göbel, M., Hassan, T., Oro, E., Orsi, G.: A methodology for evaluating algorithms for table understanding in PDF documents. In: Proceedings of the 2012 ACM Symposium on Document Engineering, pp. 45–48 (2012)
7. Göbel, M., Hassan, T., Oro, E., Orsi, G.: ICDAR 2013 table competition. In: 2013 12th International Conference on Document Analysis and Recognition, pp. 1449–1453. IEEE (2013)
8. Harutyunyan, A., Borradaile, G., Chambers, C., Scaffidi, C.: Planted-model evaluation of algorithms for identifying differences between spreadsheets. In: 2012 IEEE Symposium on Visual Languages and Human-Centric Computing (VL/HCC), pp. 7–14. IEEE (2012)
9. Hassan, T.: Towards a common evaluation strategy for table structure recognition algorithms. In: Proceedings of the 10th ACM Symposium on Document Engineering, pp. 255–258 (2010)
10. Hu, J., Kashi, R., Lopresti, D., Nagy, G., Wilfong, G.: Why table ground-truthing is hard. In: Proceedings of Sixth International Conference on Document Analysis and Recognition, pp. 129–133. IEEE (2001)
11. Jaffar, J., Santosa, A.E., Voicu, R.: Efficient memoization for dynamic programming with ad-hoc constraints. In: AAAI, vol. 8, pp. 297–303 (2008)
12. Li, M., Cui, L., Huang, S., Wei, F., Zhou, M., Li, Z.: Tablebank: table benchmark for image-based table detection and recognition. In: Proceedings of The 12th Language Resources and Evaluation Conference, pp. 1918–1925 (2020)
13. Oro, E., Ruffolo, M.: TREX: an approach for recognizing and extracting tables from PDF documents. In: 2009 10th International Conference on Document Analysis and Recognition, pp. 906–910. IEEE (2009)
14. Pinto, D., McCallum, A., Wei, X., Croft, W.B.: Table extraction using conditional random fields. In: Proceedings of the 26th Annual International ACM SIGIR Conference on Research and Development in Information Retrieval, pp. 235–242 (2003)

15. Smock, B., Pesala, R., Abraham, R.: PubTables-1M: towards comprehensive table extraction from unstructured documents. In: Proceedings of the IEEE/CVF Conference on Computer Vision and Pattern Recognition (CVPR), pp. 4634–4642 (2022)
16. Wang, X.: Tabular abstraction, editing, and formatting (1996)
17. Yildiz, B., Kaiser, K., Miksch, S.: pdf2table: a method to extract table information from pdf files. In: IICAI, pp. 1773–1785. Citeseer (2005)
18. Zhong, X., ShafieiBavani, E., Yepes, A.J.: Image-based table recognition: data, model, and evaluation. arXiv preprint arXiv:1911.10683 (2019)

Author Index

G. A. Fink et al. (Eds.): ICDAR 2023, LNCS 14191, pp. 551–552, 2023.
https://doi.org/10.1007/978-3-031-41734-4

Printed in the United States
by Baker & Taylor Publisher Services